T0175699

Phytosterols as Functional Food Components and Nutraceuticals

NUTRACEUTICAL SCIENCE AND TECHNOLOGY

Series Editor

FEREIDOON SHAHIDI, PH.D., FACS, FCIC, FCIFST, FRSC
University Research Professor
Department of Biochemistry
Memorial University of Newfoundland
St. John's, Newfoundland, Canada

1. Phytosterols as Functional Food Components and Nutraceuticals, *edited by Paresh C. Dutta*

ADDITIONAL VOLUMES IN PREPARATION

Biotechnology for Performance Foods, Functional Foods, and Nutraceuticals, *edited by Jean-Richard Neeser and Bruce J. German*

Phytosterols as Functional Food Components and Nutraceuticals

edited by
Paresh C. Dutta
Swedish University of Agricultural Sciences, SLU
Uppsala, Sweden

CRC Press
Taylor & Francis Group
Boca Raton London New York

CRC Press is an imprint of the
Taylor & Francis Group, an **informa** business

First published 2004 by Marcel Dekker, Inc.

Published 2019 by CRC Press
Taylor & Francis Group
6000 Broken Sound Parkway NW, Suite 300
Boca Raton, FL 33487-2742

© 2004 by Taylor & Francis Group, LLC
CRC Press is an imprint of Taylor & Francis Group, an Informa business

First issued in paperback 2019

No claim to original U.S. Government works

ISBN 13: 978-0-367-44662-8 (pbk)
ISBN 13: 978-0-8247-4750-3 (hbk)

**Visit the Taylor & Francis Web site at
http://www.taylorandfrancis.com**

**and the CRC Press Web site at
http://www.crcpress.com**

Library of Congress Cataloging-in-Publication Data
A catalog record for this book is available from the Library of Congress.

Series Introduction

The Nutraceutical Science and Technology series provides a comprehensive and authoritative source of the most recent information for those interested in the field of nutraceuticals and functional foods. There is a growing body of knowledge, sometimes arising from epidemiological studies and often substantiated by preclinical and clinical studies, demonstrating the relationship between diet and health status. Many of the bioactives present in foods, both from plant and animal sources, have shown to be effective in disease prevention and health promotion. The emerging findings in the nutrigenomics and proteomics areas further reflect the importance of diet in a deeper sense, and this, together with the increasing burden of prescription drugs in treatment of chronic diseases such as cardiovascular ailments, certain types of cancer, diabetes, and a variety of inflammatory diseases, have all pushed interest in functional foods and nutraceuticals to a new high. The interest is quite widespread from producers to consumers, regulatory agencies, and health professionals.

In this series, particular attention is paid to provide the most recent and emerging information on a range of topics covering the chemistry, biochemistry, epidemiology, nutrigenomics and proteomics, engineering, formulation, and processing technologies related to nutraceuticals, functional foods, and dietary supplements. Quality management, safety, and toxicology, as well as disease prevention and health promotion aspects of products of interest, are addressed. The series also covers relevant aspects related to preclinical and clinical trials as well as regulatory and labeling issues.

This series provides much needed information on a variety of topics. It addresses the needs of professionals, students, and practitioners in the fields of food science, nutrition, pharmacy, and health, as well as leads conscious consumers to the scientific origin of health-promoting substances in foods, nutraceuticals, and dietary supplements. Each volume covers a specific topic of related foods or prevention of certain types of diseases, including the process of aging.

Fereidoon Shahidi

Preface

Phytosterols (plant sterols) have been known to lower chloesterol levels in humans since the 1950s, a finding that has resulted in the development of various food products enriched with these compounds. Widespread consumption of such food products, popularly known as *functional foods*, would decrease blood cholesterol levels and consequently decrease the occurrence of coronary heart diesase in certain populations in Western societies. Volumes of data have been accumulated in the area of foods enriched with phytosterols.

This comprehensive volume covers a wide range of issues related to plant sterols, including occurrence, analysis, biological effects, currently available functional foods containing phytosterols, and the prospects of increasing phytosterol levels in plants. This book presents extensive literature reviews on the cholesterol-lowering properties and safety aspects of phytosterols as functional food ingredients. There are other possible biological effects of phytosterols. Cancer is another major cause of death in Western societies, and increasing evidence suggests that phytosterols may have effects on the development of this disease. Research on the role of phytosterols in cancer prevention is at quite an early stage, and increasing research in this area is expected. Chapter 5 discusses recent developments in the area of phytosterols and cancer.

Research on the formation and biological effects of cholesterol oxidation products in foods and in biological samples is plentiful. However, similar research on phytosterol oxidation products has been downplayed until now for various reasons, e.g., the argument of low levels of phytosterol absorption in humans. Recent developments in the areas of occurrence, analysis, and the biological effects of phytosterol oxidation products are also highlighted in this book. Based on the present knowledge of various aspects of phytosterols, it can be confidently anticipated that these interesting natural compounds are going to play a positive role in improving public health around the world. Research in the areas of other potential biological funtions of phytosterols is expected to increase in the future.

As the editor of this book, I would like to express my profound gratitude to all the contributors for taking some of their valuable time to complete this volume. Unlimited and very helpful support from the staff members of Marcel Dekker, Inc., is also gratefully acknowledged.

Paresh C. Dutta

Contents

Contributors

Susan W. Andersson   Department of Clinical Nutrition, Sahlgrenska Academy at Göteborg University, Göteborg, Sweden

Leon C. Boyd   Department of Food Science, North Carolina State University, Raleigh, North Carolina, U.S.A.

Guus S. M. J. E. Duchateau   Unilever Health Institute, Unilever Research and Development, Vlaardingen, The Netherlands

Paresh C. Dutta   Department of Food Science, Swedish University of Agricultural Sciences, SLU, Uppsala, Sweden

Jiri Frohlich   Healthy Heart Program, University of British Columbia, Vancouver, British Columbia, Canada

Helena Gylling   University of Kuopio, and Kuopio University Hospital, Kuopio, Finland

Hans-Gerd M. Janssen   Unilever Research and Development, Vlaardingen, The Netherlands

David Kritchevsky   The Wistar Institute, Philadelphia, Pennsylvania, U.S.A.

Arnis Kuksis   University of Toronto, Toronto, Ontario, Canada

Anna-Maija Lampi Department of Applied Chemistry and Microbiology, University of Helsinki, Helsinki, Finland

Arjan J. H. Louter Unilever Research and Development, Vlaardingen, The Netherlands

Tatu A. Miettinen Biomedicum Helsinki, University of Helsinki, Helsinki, Finland

Robert A. Moreau Eastern Regional Research Center, Agricultural Research Service, U.S. Department of Agriculture, Wyndmoor, Pennsylvania, U.S.A.

Lena Normén Healthy Heart Program, University of British Columbia, Vancouver, British Columbia, Canada

Lisa Oehrl Dean Department of Food Science, North Carolina State University, Raleigh, North Carolina, U.S.A.

Vieno Piironen Department of Applied Chemistry and Microbiology, University of Helsinki, Helsinki, Finland

W. M. Nimal Ratnayake Food Directorate, Nutrition Research Division, Health Canada, Ottawa, Ontario, Canada

Jari Toivo National Technology Agency of Finland, Helsinki, Finland

Elke Trautwein Unilever Health Institute, Unilever Research and Development, Vlaardingen, The Netherlands

Elizabeth J. Vavasour Food Directorate, Chemical Health Hazard Assessment Division, Health Canada, Ottawa, Ontario, Canada

1

Occurrence and Levels of Phytosterols in Foods

Vieno Piironen and Anna-Maija Lampi
University of Helsinki, Helsinki, Finland

I. INTRODUCTION

Phytosterols are present in all plants and in foods containing plant-based raw materials. In normal diets vegetable oils and products based on them are generally acknowledged to be the richest sources of phytosterols (1,2). However, the significance of other foods, especially cereal products and vegetables, depends on dietary patterns. Some foods generally consumed only in low quantities but containing considerable amounts of sterols, such as nuts, may contribute significantly to the dietary phytosterol intakes of some individuals or population groups. On the other hand, food items with rather low levels of phytosterols but consumed as major food items may become significant sources.

When various foods are evaluated as phytosterol sources, the main interest is generally in the levels of individual sterols, particularly different desmethyl sterols such as sitosterol, campesterol, stigmasterol, avenasterols, and stanols, which comprise the majority of phytosterols in normal foods (2,3). More rarely, monomethyl and dimethyl sterols are also determined. In addition to the parent sterol composition, the distribution of the various steryl conjugates is also of interest. These conjugates, i.e., esters with fatty acids (SEs), esters with phenolic acids (SPHEs), glycosides (SGs), and acylated glycosides (ASGs), may have different chemical, technological, and nutritional properties. Cholesterol often accounts for 1–2% of the total sterols in plants and may comprise 5% or more in certain plant families,

1

species, organs, or tissues (3). However, cholesterol levels of plants are not discussed further in this chapter.

The serum cholesterol–lowering effect of phytosterols, when consumed at levels of 1.5–2 g/day, has led to great interest in phytosterol-enriched foods and their development. However, little is known about the effects of phyto-sterol intake from nonenriched foods. Phytosterols were suggested to be at least partially responsible for the differences in plasma cholesterol levels and synthesis observed when 16 normolipidemic subjects were given experimental diets with corn or olive oil (4). Ågren et al. (5) showed that both serum total and low-density lipoprotein (LDL) cholesterol levels of patients with rheu-matoid arthritis were significantly decreased by a vegan diet containing on average 732 mg/day of total phytosterols. Furthermore, Ellegard et al. (6) concluded that the effect of the current dietary recommendation to reduce saturated fat and increase dietary fiber may partly be explained by the phytosterol content of the diet. Later, Otslund (7) compared corn oil and corn oil purified free from sterols, and reported that cholesterol absorption was significantly increased when the oil was purified from phytosterols. In a case-control study of De Stefani et al. (8), there was a strong inverse relationship between the total phytosterol intake and stomach cancer; the relationship remained after control for antioxidants, such as vitamin C. On the other hand, a higher dietary intake of phytosterols was not associated with a lower risk of colon or rectal cancer in a prospective epidemiological study (9). More research is clearly needed to clarify the potential role of phytosterols in nonenriched diets. For this task, reliable food composition data based on phytosterol levels in foods are needed (9–12).

Recent estimates of phytosterol intake from nonenriched foods range from 138 to 358 mg/day (9,12–16). However, lower estimates have also been published; analysis of 3-day composite diets gave intakes of 78 mg (general U.S. population), 89 mg (Seventh Day Adventists, pure vegetarians), 344 mg (Seventh Day Adventists, lacto-ovo-vegetarians), and 230 mg (Seventh Day Adventists, nonvegetarians) (17). The significance of different food groups as phytosterol sources varies in different populations. Cereal products were the main contributors (38–43%), followed by margarines and oils both in the Netherlands and in Finland (9,12). The contribution of vegetables, fruits, and berries was also significant, i.e., 20–25%. In Uruguay fruits were calculated to contribute as much as 36.4%, followed by vegetables (15.9%) and tubers (11.3%) (8). On the other hand, in the United Kingdom oils and fats were calculated to contribute 87 mg/day cereals 62 mg/day and vegetables 15 mg/day (13). Sitosterol is the main dietary phytosterol, with a reported proportion of 56–79% of the total dietary phytosterol intake (9,13,14). Campesterol and stigmasterol contributed 18% and 9% and stanols 9% to the total phytosterols (9).

II. PHYTOSTEROLS IN VEGETABLE OILS AND FATS

Vegetable oils are in general rich in free phytosterols and their fatty acid esters, although some differences between oils as dietary phytosterol sources are evident. Furthermore, effects of various processes applied in vegetable oil refining and in producing oil-based products must also be taken into account.

A. Phytosterol Contents and Compositions of Vegetable Oils

Most crude vegetable oils contain 1–5 g kg^{-1} of total phytosterols (1,2,18). For example, crude soybean oil contains 3.0–4.4 g kg^{-1} of phytosterols (Table 1). Among the most commonly used oils, corn and rapeseed oils are exceptions. In recent studies, crude corn oil was reported to contain 7.8–11.1 g kg^{-1} and rapeseed oil 6.8–8.8 g kg^{-1} of total sterols (20–22) (Table 1). Earlier, corn oil was reported to contain as much as 13.9 g kg^{-1} of total phytosterols (1). In addition, some special oils used in lower quantities are still richer in sterols; wheat germ and corn germ oils were reported to contain 17–26 g kg^{-1} (1,20) and 10.7 g kg^{-1} of phytosterols, respectively (20). On the other hand, lower amounts of sterols are found in palm oil (0.7–0.8 g kg^{-1}) (20,22) and coconut oil (0.7 g kg^{-1}) (22).

Refining of oils leads to somewhat lower phytosterol levels (Table 1). As an example, the recently reported total sterol contents in refined rapeseed and corn oils, available commonly for consumers, are 6.4–7.7 and 6.9–7.7 g kg^{-1}, whereas the corresponding values for crude oils are 6.8–8.8 and 7.8–11.1 g kg^{-1}, as described above. Effects of different refining steps are discussed in more detail below in section C.

The most important desmethyl sterol in vegetable oils is sitosterol, which in two recent comprehensive studies accounted for 38% (borage oil) to 91% (avocado oil) (24) and 51% (rapeseed oil) to 95% (walnut oil) of phytosterols (22). The range for rapeseed and soybean oils is rather similar, 51–60% (22–24,27,28) and 52–61% (22,24,28), respectively, whereas higher proportions have been reported for various olive oils, 66–89% (22–25,28–30). Other desmethyl sterols occurring in significant amounts include stigmasterol, campesterol, and Δ5-avenasterol. Borage, sesame, and evening primrose oils contained substantially more Δ5-avenasterol than the other analyzed oils (24). The same authors also measured low concentrations of stanols (<50 mg kg^{-1}) in all oils, but a relatively higher concentration occurred in corn and crude evening primrose oils. Brassicasterol is typical for rapeseed oil, occurring in amounts of 7–13% of total phytosterols (22, 24,25). In plants belonging to the family Cucurbitaceae, Δ7-sterols predominate (see Sec. IV). Therefore, pumpkin seed oil is rich in Δ7-sterols and its sterol compostion can be used to detect adulteration (31).

Table 1 Phytosterols in Crude and Refined Vegetable Oils (g kg^{-1})

Sample	Brassicasterol	Campesterol	Sitosterol	Stigmasterol	Avenasterols	Stanols	Total	Ref.[a]
Corn, crude	—	1.69–2.01	5.41–6.46	0.58–0.68	0.10–0.11	—	7.80–11.14	20–22
	tr	2.59	9.89	0.98	0.36	—	13.90	1
Corn, refined	nd	1.23–1.64	4.54–5.43	0.46–0.59	0.10–0.41	0.23–0.33	6.86–7.73	22–24
	nd	1.58	6.90	0.76	0.22	—	9.52	1
Olive extravirgin	–/nd	0.045–0.050	1.18–1.33	0.009–0.013	0.17–0.22	0.003.5–0.007	1.44–1.62	23–25
Olive, cold pressed	nd	0.02–0.05	1.22–1.30	nd–0.03	0.16–0.60	0.03–0.04	1.56–1.93	22, 24
Palm, crude	—	0.14–0.20	0.43–0.52	0.07–0.10	nd–0.03	—	0.69–0.79	20, 22
Palm, refined	nd	0.14–0.18	0.35–0.41	0.07–0.10	nd–0.03	nd–tr	0.60–0.68	22, 24, 26
Peanut, refined	0.01	0.24–38	1.15–1.69	0.12–0.22	nd–0.13	0.03	1.67–2.29	22, 24
	nd	0.31	1.31	0.19	0.21	—	2.06	1
Rapeseed, crude	1.11	2.93	4.20	nd	—	—	6.82–8.78	20–22
	0.55	1.56	2.84	0.02	0.13	—	5.13	1
Rapeseed, refined	0.51–0.92	1.64–3.00	3.58–3.95	nd–0.16	0.14–0.36	0.02–0.12	6.39–7.67	22, 24, 27
	0.27	0.76	1.38	0.01	0.06	—	2.50	1
Soybean, crude	—	0.57–0.71	1.73–1.84	0.58–0.61	0.11–0.14	—	3.02–4.44	20–22
	nd	0.68	1.83	0.64	0.07	—	3.27	1
Soybean, refined	nd–0.007	0.34–0.82	1.24–1.73	0.37–0.64	0.04–0.14	nd–0.07	2.03–3.28	22, 24, 26
	nd	0.47	1.23	0.47	0.01	—	2.21	1
Sunflower, refined	0.02	0.27–0.55	1.94–2.57	0.18–0.32	0.19–0.56	0.04	2.63–3.76	22, 24

[a] From Refs. 20 and 21 only total phytosterol contents. In Ref. 22 stanols not reported; brassicasterol reported only for rapeseed oil. In Refs. 1 and 27 stanols not reported.

—, Not reported; nd, not detected; tr, traces.

Monomethyl and dimethyl sterols are found in lower amounts in oils (18,32,33). However, they have been of interest in rice bran oil (see Sec. III below). Among the more commonly used oils, the proportion of dimethyl sterols was substantial in olive and linseed oils (18). Cycloartenol and 24-methylene cycloartanol were the main components (18,32). Among the monomethyl sterols of virgin olive oils, obtusifoliol, gramisterol, cyclo-eucalenol, and citrostadienol were identified (32). They were also the main monomethyl sterols in linseed and sesame oils (18,33). Desmethyl, mono-methyl, and dimethyl sterols contributed 75–93%, 3–20%, and 2–5% of total phytosterols in the oil of four *Sesamum* species (33).

Both heredity and growing conditions affect phytosterol contents and compositions. The total sterols ranged from 1.76 to 3.48 g kg^{-1} and sitosterol from 0.93 to 1.71 g kg^{-1} of oil in genetically modified soybeans differing in their fatty acid compositions (34). Among soybean, sunflower, and canola cultivars total sterols varied twofold, although the composition was consist-ent within a crop (35). Furthermore, the total phytosterol levels of canola were markedly affected by genetic modification (36). In oils derived from genetically modified varieties of one canola line, brassicasterol, campesterol and sitosterol levels were consistently decreased but no systematic trend was found in five modified varieties of the other studied line. Brassicasterol in the analyzed varieties ranged from 0.85 to 3.60 g kg^{-1} of oil, campesterol from 2.05 to 4.79 g kg^{-1} and sitosterol from 4.57 to 8.79 g kg^{-1}. In another study, the total phytosterol contents of 9 canola lines varied between 4.59 and 8.07 g kg^{-1} of oil, and those in 12 sunflower and 11 soybean lines between 2.10 and 4.54 g kg^{-1} and 2.35 and 4.05 g kg^{-1}, respectively (37). The sterol composition also varied between individual lines. Furthermore, both planting location and temperature affected phytosterol accumulation in soybean. The total phyto-sterol content increased consistently with increasing temperature. Phytosterol composition also changed; a greater percentage of campesterol and a lower percentage of stigmasterol and sitosterol occurred at higher temperatures.

B. Phytosteryl Conjugates in Vegetable Oils

The distribution of phytosterols to free sterols (FSs) and steryl fatty acid esters (SEs) has been shown to differ greatly in different oils (18,21,22,24, 38,39). FSs accounted for 54–85% of total phytosterols in soybean, sesame, olive, cottonseed, coconut, palm, borage, safflower, expeller-pressed sun-flower, and cold-pressed peanut oils, whereas their proportion was only 32–44% in canola, rapeseed, corn, refined peanut, avocado, evening primrose, and refined sunflower oils (24). Verleyn et al. (22) also showed that in corn and rapeseed oils phytosterols mainly occurred as SEs (56–60%), whereas the majority of other vegetable oils (soybean, sunflower, palm, etc.) contained much lower SE proportions (25–40%) (Fig. 1). They also found some differences in the phytosterol compositions of FSs and SEs. In corn and soybean oils Δ5-avenasterol occurred only in SEs, whereas in refined sunflower oil a low amount of Δ5-avenasterol was also present in FSs. On the other hand, Phillips et al. (24) found Δ5-avenasterol in both fractions of corn and soybean oils. More brassicasterol occurred in FSs than in SEs of rapeseed (22,24).

C. Effects of Processing

1. Effects of Oil Refining on Phytosterol Contents

Phytosterols are partly removed with other components of crude oils in vegetable oil refining. They may also react by oxidation, isomerization, and

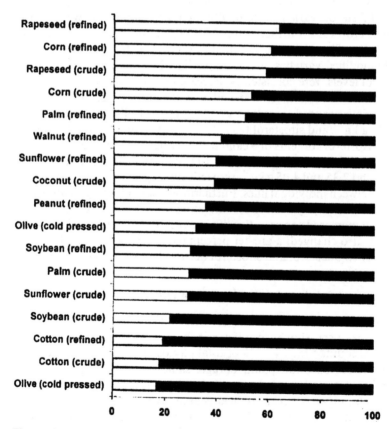

Figure 1 Proportions (%) of esterified (open bar) and free (solid bar) phytosterols in vegetable oils. (From Ref 22.)

other intermolecular transformation reactions; dehydroxylation; hydrolysis; and dehydrogenation (2). The magnitude of sterol losses depends on the process conditions. Therefore, reported losses vary between different studies. On the other hand, refining processes are constantly being developed and modified, and evaluation of the different process steps, based on earlier studies, should be made with caution.

In the extensive review of the earlier literature, Kochhar (19) concluded that, depending on the type of oil and conditions employed for processing, a loss of sterols between 10% and 70% has generally occurred. Jawad et al. (40) studied the effects of physical refining on soybean oil phytosterols and reported that sterols decreased progressively with increasing processing temperature or time. The total phytosterol content decreased from 3.90 g kg^{-1} in crude oil to 2.60 g kg^{-1} in refined oil, when phosphoric acid–degummed and bleached oil was physically refined at 240°C for 2 h, and to 0.40 g kg^{-1} after refining at 300°C for 2 h. The total sterol content was reduced by 15% and a further 1% on degumming and bleaching of canola oil (41), and by 36%, 18% and 24%, respectively, in fully refined corn, soybean and rapeseed oil compared with the crude oils (21). Phillips et al. (24) compared sterols of crude and refined oils for six oil types. The mean reduction in total and free sterols was 9% and 31%, respectively. However, only for two oil types did the samples represent the same batch of raw materials, thus reflecting changes due to the refining process only.

Generally, no significant differences in the relative proportions of the individual phytosterols in crude and refined vegetable oils have been observed (21,22). However, Jawad et al. (40) reported that some sterols are more labile than others. Campesterol and stigmasterol both decreased but the sitosterol level remained almost constant. They also noticed that the level of Δ7-stigmastenol increased and concluded that this was due to isomerization of sitosterol. There were some indications that severe physical refining also brings about isomerization of Δ5-avenasterol to Δ7-avenasterol.

Significant changes in the proportions of FS and SE occur during refining (21,22,24). Refining reduced the total phytosterol and FS contents of evening primrose and borage oils, whereas the concentration of SE increased (24). A similar trend was observed in crude and refined oil samples from different origins. The results of Verleyn et al. (22) for several oil types confirm these results. Furthermore, steryl esters of corn, soybean, and rapeseed oils were reduced by only 3–16% when the total sterols were reduced by 18–36% (21). These results show that refining causes preferential removal of FS and that esterification of sterols also occurs (22,24). It was proposed that sterol esterification is promoted by increasing temperatures during the deodorization process (22).

SGs (19) and SPHEs (42,43) are largely removed during refining. Oryzanols (a sterol mixture composed mainly of dimethyl and desmethyl sterols as ferulic acid esters; see Sec. III.D) were unchanged in crude and degummed rice bran oils but decreased in alkali-refined, bleached, and deodorized oils (42). The total loss of oryzanols was 51% and that caused by alkali-refining alone was 47%. During storage of deodorized oil for 7 weeks an additional 19% of oryzanols were lost. Later, sterol losses were shown to be highly dependent on the processes applied (44). Oil subjected to physical refining retained the original amount of oryzanols after refining, whereas chemically refined oils had considerable lower amounts. Oryzanols were carried into the soapstock, leading to losses of 83–95% during alkali refining.

Phytosterols removed from the oil to the deodorization distillates are utilized in increasing amounts for phytosterol enrichment of foods. Therefore, methods to separate phytosterols from the distillate have been developed (45,46).

2. Other Effects of Refining on Phytosterols

Analysis of steradienes, which are formed during refining, may be useful as proof that fats or oils have undergone a refining process and for detecting of the presence of processed fat or oil in a product claimed to be unrefined (21,47–51). The term steradiene is commonly used to include dehydration products of sterols. All Δ5-unsaturated sterols having an OH group in position 3 can form 3,5-steradienes as dehydration products. In the case of sitosterol, Δ3,5-stigmastadiene is formed. Stigmasterol, which has an additional double bond in the side chain, yields a diene. Steryl esters can only be converted to steradienes after their hydrolysis to free sterols (51).

Sitosterol reacted mainly during bleaching earth treatment, and temperature and bleaching earth activity were the main factors affecting the extent of the reaction (47). During deodorization some steradienes were removed by distillation and, depending on the conditions, new steradienes were formed. On the other hand, in the study of Ferrari et al. (21), deodorization was the principal step increasing the level of these compounds. The contents of steradienes were from not detected (corn oil, soybean oil) to 0.6 (rapeseed oil) mg kg^{-1} in the crude oils, 1.2 (soybean), 10.3 (corn), and 225 (rapeseed) mg kg^{-1} after bleaching, and increased after deodorization by about 15- to 20-fold in corn and soybean oils and by about 2-fold in rapeseed oil, resulting in steradiene concentrations of 28–393 mg kg^{-1}.

Later, factors determining steradiene formation in bleaching and deodorization were studied in more detail with the use of mathematical model equations (51). The reaction was strongly affected by temperature, bleaching time, and bleaching earth activity; the highest measured steradiene content after bleaching was 41.9 mg kg^{-1}. As the deodorization temperature of corn oil rose from 200°C to 260°C, a gradual increase in the steradiene content from 71.9 to 309.3 mg kg^{-1} was observed. A great proportion of steradienes was distilled and recovered in the deodorization distillate.

3. Other Processes

Vegetable oils and fats are modified to be better suitable for various purposes. Some of these processes, used generally in industry, may affect phytosterol compositions. On the other hand, phytosterols may be oxidized during food processing and storage.

The proportion of SEs may change due to interesterification procedures (52). Sodium methoxide–catalyzed interesterification of corn, soybean, and rapeseed oils led to some increase in SE contents, showing that moderate esterification of sterols occurred during chemical interesterification. A nonspecific lipase preparation from *Candida cylindracea* as a catalyst for enzymatic interesterification of rapeseed oil also resulted in formation of SEs. On the other hand, during interesterification using *sn*-1,3-specific lipozyme (lipase from *Rhizomucor miehei*), some cleavage of SEs occurred.

Hydrogenation of vegetable oils and fats may decrease phytosterol contents. For cottonseed and soybean oil, total sterols were reduced in hydrogenated oils by 9% and 12%, respectively, and FSs were decreased by 12% and 14% (24). Hydrogenated palm and soybean oils contained 0.28 and 1.99 g kg^{-1} of phytosterols, whereas the corresponding figures for oils without hydrogenation were 0.61 and 3.28 g kg^{-1} (26). Structural modification of dimethyl sterols during hydrogenation of oils has also been reported (53). Positional isomerization of the double bonds in the side chain was the predominant reaction. The main compounds formed were cyclobranol and cyclosadol from 24-methylene cycloartanol and 9β,19-cyclo-5α-lanost-25-en-3β-ol from cycloartenol.

Being lipids, phytosterols may be oxidized during food processing and storage. Heating of oils at high temperatures, such as in deep-frying conditions, accelerates oxidation of lipids and leads to sterol losses. Oxidation of phytosterols to various oxidation products is discussed in Chapter 11. In this chapter a short summary from the point of view of sterol losses is given.

The reactivity of sterols and the extent of their degradation were shown to depend on the sterol structure, mainly unsaturation of the ring structure, temperature, and matrix composition (54–58). Oehrl et al. (55) heated canola, coconut, peanut, and soybean oils under simulated frying conditions at 100, 150, and 180°C for 20 h. They observed that the heating system affected the results; metal contact increased losses. Heating of canola oil in metal pans in a forced-draft oven for 4 h decreased sitosterol from 3.23

to 2.16 g kg^{-1} at 100°C and to 0.18 g kg^{-1} and 0.12 g kg^{-1} at 150°C and 180°C, respectively. In another study, corn and soybean oils, heated in an oxidograph apparatus at 120°C for 4 h, lost 15% of phytosterols, whereas heating at 180°C led to losses of 24% and 21% (59). For rapeseed oil the loss was 8% and for sunflower 12% at both temperatures. When stigmasterol was heated in an oven at 180°C for 2 h, the total amount of oxidation products was 18% when heated in bulk and 14% and 18% when heated in tripalmitate and purified rapeseed oil, respectively (58). Considerable losses were observed when soybean oil was heated at 180°C for 8 h/day with cooling to room temperature overnight and taking samples at regular intervals up to 96 h (60). Microwave heating or conventional heating of oils did not decrease their phytosterol contents (61).

D. Margarines

The phytosterol contents of margarines depend on their fat content, the oils and fats used as raw materials, and processes applied in the production (1,62–65). In the early review of Weihrauch and Gardner (1), the total phytosterol contents of margarines with 80% fat ranged between 1.36 and 5.86 g kg^{-1}. Slover et al. (62) reported that margarines, including imitation and diet margarines, contain 0.42–4.13 g kg^{-1} of sitosterol, 0.11–1.06 g kg^{-1} of campesterol, and 0.13–0.71 g kg^{-1} of stigmasterol. In a later study, total phytosterol contents of soft margarines with 40–80% fat ranged from 1.30 to 5.40 g kg^{-1} and those of blended margarines, containing both vegetable oils and butter, and hard margarines for baking from 1.80 to 3.55 g kg^{-1} and from 2.30 to 2.60 g kg^{-1}, respectively (63). In the German food composition table margarines with 80% fat are reported to contain 3.10–3.80 g kg^{-1} of total phytosterols (64). In the U.S. food composition database the range for hard regular margarines is from 1.36 to 5.71 g kg^{-1} and that for soft margarines from 1.44 to 4.83 g kg^{-1} (65).

III. CEREALS AND CEREAL PRODUCTS

Cereals are generally regarded as good sources of phytosterols. However, some differences exist in the phytosterol compositions and levels of different cereals. Furthermore, phytosterols are highly localized within the kernel, which leads to significant differences in the phytosterol levels of various commercial products available for consumers. Thus, the role of cereal products in the total sterol intake depends on the dietary pattern and on the way in which cereals are consumed, i.e., as whole-meal products or products made from highly refined flours. Of special interest is also the fact that not only various sterols but also their conjugates are localized the kernel.

A. Total Phytosterol Contents in Cereals

The total phytosterol contents of various cereals range mainly from about 350 to 1200 mg kg^{-1} fresh weight (fw), although especially in the older literature higher phytosterol contents have also been reported (Table 2). The studies have often focused on only a few samples of one cereal and only percentage sterol compositions have been given (73). Therefore, they do not support the comparison of different cereals. Both progress in the analytical methods and differences in genetic factors, growing conditions, and post-harvest handling may affect the reported results. Only recently have analytical methods capable of liberating sterols from glycosides and thus including them in the total sterol contents been applied (66,69,72).

In Finland, rye was shown to contain on average 955 mg kg^{-1} of phytosterols, whereas the corresponding figures for barley, wheat, and oats were 761, 690, and 447 mg kg^{-1} fw, respectively, when two cultivars of each cereal were sampled from the same location in the same year (66). Similarly, the total phytosterol contents given as mg g^{-1} lipids showed that oats contains less phytosterols than the other three cereals; the total sterol contents for rye, barley, wheat, and oats were 71.2, 35.6, 42.0, and 12.1 mg g^{-1}, respectively (26). On the other hand, two oat cultivars with high fat contents (14–16% fat) contained more sterols than two regular oat cultivars

Table 2 Phytosterol in Cereals (mg kg^{-1} fw)

Sample	Campesterol	Sitosterol	Stigmasterol	Avenasterols	Stanols	Total	Ref.[a]
Barley	150–192	437–484	24–36	56–69	17–19	720–801	66
Buckwheat	93	775	tr	40	23	963	66
	200	1640	80	—	—	1980	1
Corn	—	—	—	—	—	662–1205	67, 68
	320	1200	210	—	—	1780	1
Millet	112	371	18	87	nd	770	66
Oats	32–46	237–321	11–21	15–56	8–9	350–491	66, 69
	36–51	258–323	23–38	158–212	—	480–611	70
Rice	146	375	104	20	32	723	66
Rye	128–210	358–607	22–37	5–42	122–220	707–1134	66, 71
Wheat	108–150	288–486	15–24	nd–22	151–171	447—830	66, 71, 72
	270	400	—	—	—	690	1

[a] In refs. 67, 68, and 71, only total sterols reported. Stanols and avenasterols not reported in Refs. 1 and 72. Stanols not reported in Ref. 69.
—, Not reported; nd, not detected; tr, traces.

(8.2–8.6% fat), 610 *vs.* 480 mg kg^{-1} (70). For corn, total sterol contents of 662–1205 mg kg^{-1} were recently reported (67,68).

Both genetic factors and growing location may affect phytosterol contents of cereals. Määttä et al. (69) found statistically significant differences in the total sterol contents of seven oat cultivars, which were grown at three locations in Sweden, whereas no differences were observed between locations. Only minor differences were found in the sterol contents of two cultivars of barley, oats, rye, or wheat (7–11%), or later when 10 rye cultivars were compared (total range 707–856 mg kg^{-1} fw) (66). All the compared samples were grown at the same location and in the same year. However, the results obtained in the latter study indicated that the total variation in the cereal raw materials, caused by growing conditions, affected by growing location and year, and by wider genetic variation, may be more significant. The total range in various rye samples was from 707–1134 mg kg^{-1} fw. In two yellow dent corn hybrids the total sterol contents differed markedly; the values were 752 and 1205 mg kg^{-1} (67). Two other corn hybrids contained 662 and 703 mg kg^{-1} of sterols (68). Moreau et al. (74) compared seeds of 49 accessions of corn; the total sterols varied substantially, from 1.8% to 4.4% of oil with a mean value of 2.77%.

B. Phytosterol Composition of Cereals

Sitosterol dominates in the phytosterols of cereals. Its proportion in wheat, rye, barley, and oats ranged from 49% to 64% (66). Similarly, in a review of earlier studies, Chung and Ohm (73) concluded that the proportion of sitosterol in corn, oats, rice, rye, sorghum, and wheat is generally over 55%. Other desmethyl sterols occurring in significant amounts are campesterol, stigmasterol, sitostanol, campestanol, and the two avenasterols, Δ5- and Δ7-avenasterols. Stanols are found especially in rye, wheat, and corn (26,66,75). In rye and wheat, sitostanol accounted for 9–14% and 12–16% and campestanol for 8–9% and 8–11% of sterols, respectively (66). Similarly, Dutta and Applequist (26) reported that the proportion of stanols was 23% in rye and 23–26% in wheat. In oats and barley stanols have either been found in trace amounts or have not been detected (26,66). The proportion of avenasterols is substantial especially in oats (26,69,73,76). Δ5-Avenasterol accounted for 21–26% and Δ7-avenasterol for 5–8% of total phytosterols (26,66).

Monomethyl and dimethyl sterols occur in lower amounts. The ratio of 4-desmethyl, 4-monomethyl, and triterpene alcohols/dimethyl sterols was reported to be 85:3:12 in wheat grain (77). Sorghum grains contained 394 mg kg^{-1} dry weight (dw) of desmethyl sterols when the content of monomethyls sterols was 12.54 mg kg^{-1} (78). Gramisterol, citrostadienol, and obtusifoliol were identified as the main monomethyl sterols in wheat (77) and in rice bran (79). Cycloartenol and 24-methylene cycloartanol are dimethyl sterols generally identified in cereals (66,73,78,79).

C. Phytosterols in Milling Products

Considerable differences are found in the phytosterol contents of different parts of the kernel and thus in various milling products. Germ and bran fractions are known to be the best sterol sources among milling products. Generally, the phytosterol contents of rye and wheat milling products were shown to correlate with their ash content (Fig. 2). The total sterol content of 726–839 mg kg^{-1} fw in wheat grains led to total sterol contents of 398–430 mg kg^{-1} in the most refined flour (0.6% ash) and 1680–1770 mg kg^{-1} in the bran (4% ash). In rye, the total sterol contents of the most refined flours (ash 0.7%) ranged from 474 to 621 mg kg^{-1} and those of the bran (ash 4%) from 1760 to 1880 mg kg^{-1} when the grains contained 1090–1130 mg kg^{-1} of sterols. Much higher levels, 4114 mg kg^{-1}, were found in wheat germ (66). For wheat bran the range in recent studies was 1479–1951 mg kg^{-1} and for rye bran 1478–1881 mg kg^{-1} (66–71). On the other hand, sterols are not concentrated in the bran fraction of oats (26,66,71). The content measured in oat bran was only 446 mg kg^{-1} (66) and 557–620 mg kg^{-1} (71). Dutta and Appelqvist (26) determined a concentration of 9.4 mg g^{-1} of oat bran lipids, whereas that of wheat bran was 44.9 mg g^{-1}.

It is interesting to note that in addition of the total phytosterol levels, sterol compositions also differ in different parts of the kernel (occurrence of bound conjugates, see next section). In wheat, rye, and corn, stanols are concentrated in the outer layers of the kernel, whereas they are almost absent from the germ. In rye their proportion was 15% in flours with 0.7% ash, 19% in whole-meal flour, and 29% in bran. In wheat the corresponding figures for flours with 0.6% and 1.2% of ash were 14–15% and that for bran was 32% (66). In commercial wheat germ stanols were detected only in low amounts, i.e., less than 1–3% (26,66). No stanols were found in hand-dissected corn germ in which sitosterol accounted for 71% of sterols (75). In aleurone and fiber fractions sitostanol was the main sterol (51% and 43%), and sitosterol accounted for 21% and 34%, respectively. The proportion of campestanol was also significant, 19% and 15%.

D. Phytosteryl Conjugates in Cereal Grains and Their Fractions

Phytosterols are found in cereals as free sterols (FSs), esters with fatty acids (SEs), and phenolic acids (SPHEs; mainly steryl ferulates, SFEs), glycosides

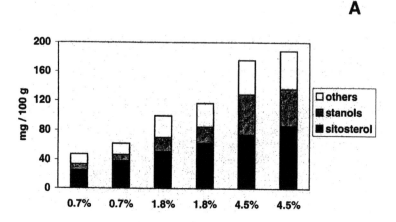

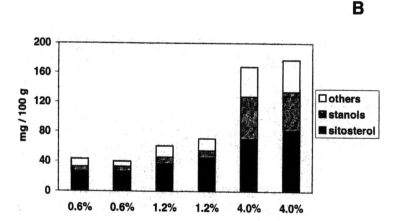

Figure 2 Fractionation of phytosterols in milling process of rye (A) and wheat (B). Samples with different ash contents were taken twice. (From Ref 66.)

(SGs), and acylated glycosides (ASGs). The occurrence of these classes varies between different cereals and in the various parts of the kernel.

Chung and Ohm (73) reviewed earlier studies and calculated that in millet the proportion of FSs was the highest, 5.4% of total lipids, whereas SEs, SGs, and ASGs accounted for 3%, 0.4%, and 1.0%, respectively. In two later studies on corn, SEs were shown to be the main class of steryl lipids. Two corn hybrids with total phytosterol contents of 1.78% and 2.23% of oil were shown to contain 1.06–1.28% of SEs, 0.56–0.71% of FSs, and 0.16–0.24% of SFEs in the oil (67). Two other hybrids contained 300–318 mg kg^{-1} of SEs, 268–272 mg kg^{-1} of FSs, and 94–112 mg kg^{-1} of SFEs (68). In seeds of 49 accessions of corn both SEs and SFEs varied markedly; SEs from 0.76% to 3.09% of oil (mean 1.5%) and SFEs from 0.087% to 0.839% (mean 0.27%). The proportion of FS was 1% (74). In addition, a comparison of rye, barley, wheat, and oats showed that the distribution of steryl conjugates differs between these cereals; in rye and barley SEs were the main sterol class whereas in wheat FSs dominated (80). On the other hand, in oats the proportion of sterols liberated only after acid hydrolysis was much higher than in the other cereals analyzed.

Localization of the conjugates differs greatly in different parts of corn kernels. Their distribution in hand-dissected aleurone, fiber, germ, and pericarp fractions of corn was studied by Moreau et al. (75). In aleurone (oil content 5.9%) the concentrations of SEs, FSs, and SFEs were 3.37%, 0.51%, and 6.31% of oil; in fiber (oil content 1.9%) 5.88%, 2.14%, and 5.37%; in germ (oil content 40%) 0.47%, 0.48%, and 0.02%; and in pericarp (oil content 0.6%) 5.83%, 12.09%, and 1.72%. Thus, in aleurone and fiber fractions SEs and SFEs dominated, whereas in germ SEs and FSs dominated and occurred in equal amounts. In pericarp FS dominated. Furthermore, recovery of steryl classes to milling products of corn has been shown to depend markedly on the processes applied (67,68,81–83). Comparison of nonstarch and starch lipids of wheat flour showed that the SE + triacylglycerol fraction accounted for 11.5% of nonstarch lipids and 4.7% of starch lipids, whereas FSs (proportion 4.1% of lipids), ASGs (2.5%), and SGs (1.8%) were only detected in nonstarch lipids (84). Furthermore, Farrington et al. (85) reported that some changes may occur in the sterol class composition during storage. During prolonged storage (5 years) of wheat flours (three types), the total sterol contents remained unchanged but free sterols decreased and esters increased; FS contents were 28–218 mg kg^{-1} at the beginning and 8–90 mg kg^{-1} after 5 years, whereas SE contents increased from 86–306 mg kg^{-1} to 214–321 mg kg^{-1}.

Similarly to the total sterol contents, steryl conjugates also vary within the same cereal. In fiber oil of 16 corn hybrids the SE content ranged from 6.5% to 9.5%, FSs from 1.9% to 4.3%, and SFEs from 2.9% to 9.2% (86). A significant relationship existed between the amounts of the three phytosterol classes; an increase in the amount of any of the three classes indicates an increase in the amount of the other two classes. In addition, significant differences were observed in the amounts of the three phytosterol classes for the same hybrid grown at different locations. These differences could be attributed to a number of factor, including weather, soil fertility, agronomic practices, or other factors. The proportions of all the phytosterol classes increased with decrease in the moisture content at harvest (87).

Sterol compositions may differ to some extent in FSs and the conjugates. In sorghum, desmethyl sterols occurred as FS, SE, and SG but monomethyl sterols were entirely free. More stigmasterol occurred free than in ester or glycoside form (78). For wheat the proportion of sitosterol was 65.6% in FS and 71.8% in SE, and the corresponding figures for campesterol were 26.0% and 22.8% (77). These authors also observed some differences in monomethyl and dimethyl sterols. As reviewed later, sterol compositions of SPHEs differ greatly from those of the other steryl classes.

SPHEs are a special class of phytosterols that have been of interest particularly in corn and rice. In the case of rice, this interest has led to a commercial product γ-oryzanol, which is a mixture containing mainly dimethyl and desmethyl sterols as ferulic acid esters (43,88–90). Effects of γ-oryzanol on health have been widely studied (91–94). SPHEs are also studied because of their antioxidant activity (95–98).

In corn, the main SPHEs were shown to be sitostanyl and campestanyl ferulates; lower amounts of sitosteryl and campesteryl ferulates as well as minor amounts of sitostanyl and campestanyl p-coumarates were also identified (99). Later, a total of 16 compounds were identified in corn bran (100). Esters of sitostanyl and campestanyl accounted for 80.8% of sterols in SPHEs, whereas only 10.7% was accounted for by sitosterol, campesterol and stigmasterol, which are the most important sterols in other steryl classes. Norton (43) showed that sitostanyl ferulate accounted for 64–85% of SPHEs in corn bran and related fractions.

SPHEs are highly localized within the corn kernel. Seitz (99,101) reported that they are localized in the inner pericarp layer. Therefore, SPHEs were regarded as candidates for kernel pathogen resistance compounds for both fungi and insects (100). Recent studies by Moreau et al. (75) and Singh et al. (82) have shown that SFEs of corn are concentrated in the single-cell aleurone layer. Hand dissection studies revealed that 96% of SFEs, 64% of FSs, and 90% of SEs of coarse fiber were located in the aleurone layer.

The reported total SPHE contents of corn grains vary to some extent. Seitz (99) found concentrations from 31 to 70 mg kg^{-1}, whereas in two later studies the contents ranged in three hybrids from 98 to 113 mg kg^{-1} (102,103) and in two other hybrids from 94 to 112 mg kg^{-1} (68). In corn grains with different moisture levels at harvest the contents ranged between 76 and 84 mg kg^{-1} (87).

Considerable differences are found between corn bran (produced by dry milling) and fiber (produced by wet milling). In bran the SPHE content was 93.3 mg kg^{-1} (100) and in bran and related products 70–540 mg kg^{-1} (43). Later, the effects of corn milling processes on the SFE yield have been studied extensively (67,68,81–83,103,104). Oil extracted from corn fiber may contain up to 6.75% of SFEs (104). Different processes led to levels of 200–260 mg kg^{-1} in bran (1.3–1.8% in oil) and 440–1530 mg kg^{-1} in fiber (1.7–5.7% of oil) (103). Other studies on milling procedures have led to corn fiber with SFE levels of 1.06–5.89% in oil (67,83). Using supercritical fluid extraction coupled with supercritical fluid fractionation, SFEs of corn bran (105) and corn fiber (106) were further enriched.

In rice bran SPHEs the sterol composition is exceptional compared with the total sterol composition of rice grains. Dimethyl sterols dominate in the SPHE fraction (43,88–90). The three major compounds are cycloartenyl, 24-methylene cycloartanyl, and campesteryl ferulates (43,89,90). In addition, Xu and Godber (90) identified seven other ferulates in γ-oryzanol. The total SPHE concentration in brown rice (without hull) was 456 mg kg^{-1} seeds (102) and in rice bran 3.4 g kg^{-1} of bran or 15.7 g kg^{-1} of oil (43). Much lower levels were measured in commercial oils (43). In the by-product of rice milling, pearling dust of rice, the total SPHE level was 1.19 g kg^{-1} (71), in extrusion stabilized rice bran 3.0–3.1 g kg^{-1} (107), and in refined rice bran oil 115–787 mg kg^{-1} (89). Oryzanols were relatively stable to extrusion temperature, but increase in extrusion temperature reduced the retention of oryzanols during storage of rice bran (107). Raw rice bran lost 16.4% of oryzanols during storage for 35 days and 62.7% after 1 year. Extraction methods have been developed to obtain a good yield of γ–oryzanol (108–111). Using supercritical fluid extraction the yield of γ-oryzanol was as high as 5.39 g kg^{-1} of rice bran (111).

SPHEs are also found in considerable amounts in rye, wheat, and triticale (71,99,102). In wheat and rye campestanyl ferulate was the main component, followed by sitostanyl ferulate (71,99). In addition, campesteryl and sitosteryl ferulates are also found. The same compounds, campestanyl and sitostanyl ferulate, were the major components in triticale (99). For wheat grain, total SFE concentrations of 62–123 mg kg^{-1} (Seitz), 53 mg kg^{-1} (wheat without hull) (102), and 62–63 mg kg^{-1} (71) were determined. The corresponding values for rye were 29 mg kg^{-1} (99) and 55–64 mg kg^{-1} (71) and for triticale 52 mg kg^{-1} (99). In wheat and rye, steryl ferulates were localized in the bran fraction; in wheat bran the content was 30–39 mg kg^{-1} whereas only traces were detected in the flour with 0.6% ash (71). Similarly, 15–25 mg kg^{-1} of SFEs were found in rye bran and only traces in flour with 0.7% ash. Interestingly, no steryl ferulates were found in oat bran (71) or oat or millet grains (102). In barley without hulls low amounts were detected (102).

E. Phytosterols in Bakery Products

Phytosterol contents in bakery products depend particularly on the type of flour and the amount and type of added fat. The contribution of cereal raw materials to the total phytosterol contents of bakery products depends on the grain fraction utilized. The total phytosterol contents of bread ranged from 410 to 824 mg kg^{-1} (112) and from 392 to 902 mg kg^{-1} fw, being highest in bread baked mainly with whole-meal flour (66). In other bakery products concentrations from 500 to 848 mg kg^{-1} were measured. In an earlier investigation, different types of bread contained 447–640 mg kg^{-1} of phytosterols (72).

IV. VEGETABLES, FRUITS, AND BERRIES

The total phytosterol concentrations given on a fresh weight basis are markedly lower in vegetables, fruits, and berries than in vegetable oils and cereals. However, their significance as dietary phytosterol sources has been shown to be considerable (9,10,12).

A. Phytosterols in Vegetables

1. Phytosterol Contents and Compositions

In vegetables the phytosterol contents given on a fresh weight basis vary from low to moderate (Table 3). However, surprisingly few studies on vegetables as phytosterol sources have been carried out recently with the aim of providing data suitable for food composition databases and tables (113,114). In this type of studies, representative sampling, detailed sample description, and validity of the analytical method are all crucial. The focus has mainly been on the role of sterols in plants or on postharvesting changes, which has required different sampling methods.

In two recent studies, the best sources of phytosterols among the analyzed vegetables were broccoli, Brussels sprout, cauliflower and dill; their total phytosterol contents were more than 300 mg kg^{-1} fw (113,114). The content was less than 100 mg kg^{-1} in potted lettuce, onion, potato, and tomato; within the range 100–200 mg kg^{-1} in carrot, Chinese cabbage, leek, red beet, swede, and white cabbage; and in the range 200–300 mg kg^{-1} in pea, sweet pepper, and parsley (114). The median content among 22 vegetables was 160 mg kg^{-1} (113). The lowest contents among analyzed vegetables were found in potato (38–51 mg kg^{-1}) and the highest in Brussels sprouts (367–430 mg kg^{-1}; 113,114). However, high consumption may raise an item with a low sterol content to the status of a significant sterol source.

The range in the total phytosterol contents was surprisingly high, i.e., 246–4100 mg kg^{-1} dw, when the total sterol contents were given on dry weight basis. The analyzed contents were more than 4000 mg kg^{-1} dw in

Table 3 Phytosterols in Vegetables (mg kg^{-1} fw)

Sample	Campesterol	Sitosterol	Stigmasterol	Avenasterols	Stanols	Total	Ref.[a]
Broccoli	67–69	285–310	8–11	2	nd–18	367–390	113, 114
Brussels sprout	71–80	277–340	3–3.8	nd	nd/8	370–430	113, 114
	60	170	nd	—	—	240	1, 64, 65
Carrot	20–22	104–110	27–28	nd	tr–0.8	153–160	113, 114
	10	70	30	—	—	120	1, 64, 65
Cauliflower	72–95	216–260	16–37	tr	tr–0.6	310–400	72, 113, 114
	30	120	20	—	—	180	1, 64, 65
Lettuce	6–11	37–106	24–45	2	5	85–174	72, 114
	10	50	40	—	—	100	1, 64, 65
Onion	6–8.2	70	5.7–12	nd	nd	84–93	113, 114
	10	120	nd	—	—	150	1, 64, 65
Potato	tr–2.3	27–45	3–5	2	nd–5.6	38–73	72, 113, 114
	tr	30	10	—	—	50	1, 64, 65
Sweet pepper	42	164	2	nd	5	220	114
Tomato	2.8–5	24–33	16–17	4	2.8–8	47–74	113, 114
	10	30	30–35	—	—	70	1, 64, 65
White cabbage	28–31	94–114	nd–2	nd	2–3.5	130–148	113, 114
	20	70	nd	—	—	110	1

[a] Avenasterols not reported in Ref. 113. Avenasterols and stanols not reported in Refs. 1, 64, and 72. Only total sterols from Ref. 65.
—, Not reported; nd, not detected; tr, traces.

cauliflower and 3000–4000 mg kg^{-1} in broccoli, whereas they were less than 1000 mg kg^{-1} in leek, onion, and potato (114). These differences may partly be explained by the anatomical structures of the tissues, e.g., the proportion of membrane-rich tissues. It was interesting to note that the highest contents were at the level of many vegetable oils and higher than in cereal grains.

In addition to the general factors affecting phytosterol contents in plants, such as genetic factors, growing conditions, and postharvest changes, sample description and separation of the sample taken for analysis, i.e., separation of the edible portion, may also cause variation in the published phytosterol contents of vegetables. Phytosterol contents and compositions in different parts of a given plant may be different. As an example, flower heads of cauliflower and broccoli were significantly better sources of sterols than other parts of these vegetables; their total phytosterol contents were 4180 and 4210 mg kg^{-1} dw, whereas the corresponding contents in other parts were 1210–1730 and 2960 mg kg^{-1}, respectively (115). In addition, the peel of many vegetables contained about twofold higher levels than the edible portion.

Sitosterol is generally the main phytosterol in vegetables (1,72,113,114). Its contribution to the total phytosterols of about 20 analyzed vegetables was 43–86% (114). Campesterol was the second most important sterol in most vegetables; however, in some vegetables the proportion of stigmasterol was the second highest (up to a maximum of about 40%). Stanols generally accounted for less than 2%. Species belonging to the family Cucurbitacae, such as cucumber and pumpkin, and spinach are exceptional in their sterol composition; in these plants Δ7-sterols dominate (31,116–118). The phytosterol composition of Cucurbitacae species is regarded as one of the most complex in the plant kingdom. Spinasterol and dihydrospinasterol were determined as the main sterols in cucumber and spinach (114).

2. Phytosterol Conjugates

FSs, SEs, SGs, and ASGs are found in vegetables and their various tissues and organs in varying proportions. Furthermore, changes during maturation of the tissue and postharvest treatments are also found (119–128). These aspects were excellently reviewed by Moreau et al. (3).

Steryl class compositions in plasma membranes of spinach leaves and cauliflower inflorensces were shown to differ (127). Steryl lipids were dominated in cauliflower by FSs (31.7% of total lipids). Smaller amounts (0.7–7.3%) of SGs, ASGs, and SEs were also found. On the other hand, spinach plasma membranes contained less FSs (7% of lipids) and lower ratio of FSs to phospholipids. Typical for spinach was a relatively high content of ASGs (13%), which were found in much lower amounts in cauliflower. In addition, no SEs or SGs were detected. Drumm et al. (121) studied dry edible beans of market classes; in neutral lipids the proportion of FSs was 6.0–10.5% and that of SEs 2.0–4.8%, whereas glycolipids contained 21.0–25.2% of SGs and 57.3–64.4% of ASGs. The contents of ASGs, SGs and FSs in carrot were 880, 300, and 830 mg kg^{-1} dw, respectively (129).

Tomato and bell pepper also differ greatly in the distribution of steryl conjugates. Furthermore, effects of ripening and postharvest treatments on steryl conjugates of tomato and bell pepper have been studied extensively (119,120,122–126,128). Major differences have been reported in their sterol metabolism during ripening (120). In bell pepper, the content and composition of steryl lipids in pericarp tissue changed only slightly with ripening (120). FSs were predominant at all stages of ripening (mature-green, turning, red-ripe) of three bell pepper cultivars constituting from 66–75% of the total sterols. ASGs and SGs together composed 20–25%, and the ASG/SG ratio clearly decreased during ripening. On the other hand, ASGs accounted for more than half, and ASGs plus SGs composed 85–90%, of the total sterols in mature-green tomato fruit (119). With ripening a major increase in the content of steryl lipids as well as a dramatic change in their composition occurred; the content of FSs and SGs and particularly of SEs increased in the outer pericarp. The level of SEs was 10-fold higher in red-ripe than in mature-green fruit. In addition, the stigmasterol/sitosterol ratio increased considerably, especially in the FS fraction.

Both in tomato and bell pepper an adjustment in the level of steryl conjugates occurs during chilling, with a subsequent readjustment upon rewarming (122–125). During chilling of tomato, FSs increased much more than SGs, whereas ASGs decreased. On the other hand, in bell pepper SGs increased whereas FSs decreased more than ASGs. Furthermore, changes in the stigmasterol/sitosterol ratio were dependent on the steryl class, temperature, and plant.

3. Effects of Processing and Storage

Phytosterols of vegetables have been shown to be stable in industrial and home processing. No sterol losses were found in industrial blanching (130,131) or cooking (113). After storage of the bleached vegetables at −18°C for longer periods (up to 12 months), somewhat lower contents were however measured (130,131).

B. Phytosterols in Fruits and Berries

The total phytosterol contents of fruits, analyzed recently by Normen et al. (113), ranged from 13 (watermelon) to 440 mg kg^{-1} fw (passion fruit), with a median content of 160 mg kg^{-1} fw for 14 analyzed fruits. When the phytosterol contents of seven fruits were compared, they ranged between 116 (banana) and 228 mg kg^{-1} fw (orange) in all other fruits except avocado, which contained significantly more sterols, i.e., 752 mg kg^{-1} fw (114). The phytosterol contents in fresh berries ranged from 60 (red currant) to 279 mg kg^{-1} fw (lingonberry), being significantly lower in cultivated berries, i.e., currants and strawberries, than in wild berries, i.e., lingonberry and blueberry. Cultivated raspberry was richer in sterols than the other cultivated berries. These differences between berries could not be explained by the seed size, the proportion of seeds in berries, or the oil content of the berries (114). Phytosterol content of some fruits and berries are compared in Table 4. As in the case of vegetables, the range in the contents on a dry weight basis was substantial both for fruits and for berries: for fruits 471–2929 mg kg^{-1} and for berries 372–2160 mg kg^{-1} (114).

Table 4 Phytosterols in Fruits and Berries (mg kg^{-1} fw)

Sample	Campesterol	Sitosterol	Stigmasterol	Avenasterols	Stanols	Total	Ref.[a]
Apple	3.6–9	130–157	tr–1	7	nd–8	130–183	72, 113, 114
	10	110	nd	—	—	120	1, 64, 65
Avocado	41	618	3	39	5	752	114
Banana	13–15	84–120	13–18	nd	nd	116–161	72, 113, 114
	20	110	30	—	—	160	1, 64, 65
Grapes	14	143	2	tr	15	200	114
	tr	30	tr	—	—	40	1, 64, 65
Orange	30–34	170–200	9–10	4	nd	228–240	113, 114
	40	170	20	—	—	240	64, 65
Black currant	5	81	tr	nd	nd	88	114
Raspberry	9	233	tr	10	2	274	114
Strawberry	2	73	nd	3	nd	100	114
	tr	100	tr	—	—	120	1, 64

[a] Avenasterols not reported in Ref. 113. Avenasterols and stanols not reported in Refs. 1, 64, and 72. Only total sterols from Ref. 65.
—, Not reported; nd, not detected; tr, traces.

Sitosterol is the main sterol both in fruits and in berries (1,72,113,114). Its proportion ranged in fruits between 72% and 86% and in berries between 61% and 93% (114). Campesterol and stigmasterol were the two other major sterols, and stanols and avenasterols were also detected in most fruits and berries. However, considerable differences were observed in the sterol compositions of various fruits and berries. The proportion of stanols was in some berries as high as 6.5% (blueberry) and 12% (lingonberry). Furthemore, the proportions of other minor desmethyl sterols, monomethyl sterols and dimethyl sterols, were considerable especially in raspberry and strawberry, in which they accounted for more than 20% of the total sterols.

Berry seeds, which are separated in substantial amounts as by-products of the food industry, have also been of interest as potential sources of phytosterols (115). The phytosterol contents of some Bulgarian berry seed oils differed considerably, being 1.2 g kg^{-1} in chokeberry seed oil, 1.4 g kg^{-1} in black currant seed oil, and only 0.4 g kg^{-1} in rose hip seed oil (132). FSs accounted for 78–95% of total phytosterols. Recently, seeds of sea buck-thorn berry were shown to contain 1.20–1.80 g kg^{-1} of phytosterols, whereas the corresponding value for berries was 340–520 mg kg^{-1} and that for fresh pulp/peel 240–400 mg kg^{-1}, respectively (133). In blueberry and lingonberry seeds the total sterol contents ranged from 9.3 to 11.3 g kg^{-1} (134).

Both genetic differences, ripening, growing location, and postharvest storage may change sterol contents and compositions in fruits and berries.

In apples, regulation of ripening by reduction of oxygen concentration and temperature led to accumulation or decrease of sitosterol depending on the conditions (135). Fur et al. (136) studied the development of sterols during the last stages of ripening of Chardonnay grape skin. The maturation induced a loss of phytosterols in grape skin; an increase occurred at peak maturity and could be related to overmaturation phenomena. Development and ripening of peach fruits were monitored until day 113 after full bloom (137). FSs of mesocarp declined gradually. At the same time, the stigma-sterol/sitosterol ratio increased. In four varieties of avocado fruit the sterol content of oil was always higher in immature fruits (1.1–6.2%) than in mature fruits (0.8– 2.0%) (138). In the case of sea buckthorn, little variation was observed between subspecies and growing location. Harvesting date affected some sterols (133).

As in the case of vegetables, postharvest changes in steryl conjugates may occur during storage of fruits and berries (139,140). The molar ratio FS/SG/ASG changed from 59:40:1 at harvest to 65:33:2 after 15 weeks at 0°C plus 1 week at 20°C (140). Longer storage at 0°C increased the level of ASGs but no significant changes were observed in FSs or SGs (139).

V. OTHER FOODS

Various peanuts and almonds are rich in phytosterols. The total phytosterol contents in raw peanuts with shells were between 600 and 1608 mg kg^{-1} fw and those of shelled peanuts between 551 and 1269 mg kg^{-1}. Sitosterol contributed about 80% of sterols. In dry-roasted peanuts the range was 607–1135 mg kg^{-1} and in oil-roasted peanuts 614–1038 mg kg^{-1} (141). The samples represented four varieties of peanuts. Other published values for peanuts range from 1176 mg kg^{-1} to 2200 mg kg^{-1} (64,65,72,114). For almonds the total phytosterol contents range from 1384 to 1430 mg kg^{-1} (64,65,114) and for various nuts from 220 to 1580 mg kg^{-1} (1,65,64).

VI. FUTURE RESEARCH

Appreciation of the beneficial effects of phytosterols on human health has increased interest in their natural dietary levels. A number of studies on phytosterol levels and compositions in different plant-based materials and products have been carried out. However, the number of studies carried out to produce data specifically for food composition databases is still small. Reliable food composition data are essential for more detailed investigation of the role of phytosterols in the human diet. Therefore, more studies are needed on phytosterol levels and compositions as well as on the distribution of various steryl conjugates in foods, using well-designed sampling systems and validated analytical methods.

REFERENCES

1. JL Weihrauch, JM Gardner. Sterol content of foods of plant origin. J Am Dietet Assoc 73:39–47, 1978.
2. V Piironen, D Lindsay, TA Miettinen, J Toivo, A-M Lampi. Plant sterols: biosynthesis, biological function and their importance to human nutrition. J Sci Food Agric 80:939–966, 2000.
3. RA Moreau, BD Whitaker, KB Hicks. Phytosterols, phytostanols, and their conjugates in foods: structural diversity, quantitative analysis, and health-promoting uses. Progr Lipid Res 41:457–500, 2002.
4. TJ Howell, DE MacDougall, PJH Jones. Phytosterols partially explain differences in cholesterol metabolism caused by corn or olive oil feeding. J Lipid Res 39:892–900, 1998.
5. JJ Ågren, E Tvrzicka, MT Nenonen, T Helve, O Hänninen. Divergent changes in serum sterols during a strict uncooked vegan diet in patients with rheumatoid arthritis. B J Nutr 85:137–139, 2001.
6. L Ellegård, I Bosaeus, H Andersson. Will recommended changes in fat and fibre intake affect cholesterol absorption and sterol excretion? An ileostomy study. Eur J Clin Nutr 54:306–313, 2000.
7. RE Ostlund, SB Racette, A Okeke, WF Stenson. Phytosterols that are naturally present in commercial corn oil significantly reduce cholesterol absorption in humans. Am J Clin Nutr 75:1000–1004, 2002.
8. E de Stefani, P Boffetta, AL Ronco, P Brennan, H Deneo-Pel Legrini, JC Carzoglio, M Mendilaharsu. Plant sterols and risk of stomach cancer: a case-control study in Uruguay. Nutr Cancer 37:140–144, 2000.
9. AL Normén, HAM Brants, LE Voorrips, HA Andersson, PA van den Brandt, RA Goldbohm. Plant sterol intakes and colorectal cancer risk in the Netherlands Cohort Study on Diet and Cancer. Am J Clin Nutr 74:141–148, 2001.
10. PC Pillow, CM Duphorne, S Chang, JH Contois, SS Strom, MR Spitz, SD Hursting. Development of a database for assessing dietary phytoestrogen intake. Nutr Cancer 33:3–19, 1999.
11. JAT Pennington. Food composition databases for bioactive food components. J Food Comp Anal 15:419–434, 2002.
12. L Valsta, A Lemström, M-L Ovaskainen, A-M Lampi, J Toivo, T Korhonen, V Piironen. Estimation of plant sterol and cholesterol intake in Finland: quality of new values and their effect on intake. Br J Nutr, submitted.
13. GM Morton, SM Lee, DH Buss, P Lawrance. Intakes and major dietary sources of cholesterol and phytosterols in the British diet. J Hum Nutr Dietet 8:429–440, 1995.

14. JHM de Vries, A Jansen, D Kromhout, P van de Bovenkamp, WA van Staveren, RP Mensink, MB Katan. The fatty acid and sterol content of food composites of middle-aged men in seven countries. J Food Comp Anal 10:115–141, 1997.
15. KM Phillips, MT Tarragó-Tani, KK Stewart. Phytosterol content of experimental diets differing in fatty acid composition. Food Chem 64:415–422, 1999.
16. RC Schothorst, AA Jekel. Oral sterol intake in the Netherlands: evaluation of the results obtained by GC analysis of duplicate 24-h diet samples collected in 1994. Food Chem 64:561–566, 1999.
17. PP Nair, N Turjman, G Kessie, B Calkins, GT Goodman, H Davidovitz, G Nimmagadda. Diet, nutrition intake, and metabolism in populations at high and low risk for colon cancer. Dietary cholesterol, β-sitosterol, and stigmasterol. Am J Clin Nutr 40:927–930, 1984.
18. A Johansson-Kornfeldt. Sterols in vegetable oils: aspects of biological variation, influence of technological treatment, and the use of sterols to identify vegetable oils. Doctoral thesis. Uppsala: Swedish University of Agricultural Sciences, 1979.
19. SP Kochhar. Influence of processing on sterols of edible vegetable oils. Prog Lipid Res 22:161–188, 1983.
20. E Homberg, B Bielefeld. Sterinzusammensetzung und Steringehalt in 41 verschiedenen pflanzlichen und tierischen Fetten. Fat Sci Technol 91:23–27, 1989.
21. RAP Ferrari, E Schulte, W Esteves, L Brühl, KD Mukherjee. Minor constituents of vegetable oils during industrial processing. J Am Oil Chem Soc 73:587–592, 1996.
22. T Verleyen, M Forcades, R Verhe, K Dewettinck, A Huyghebaert, W De Greyt. Analysis of free and esterified sterols in vegetable oils. J Am Oil Chem Soc 79:117–122, 2002.
23. RJ Reina, KD White, D Firestone. Sterol and triterpene diol contents of vegetable oils by high-resolution capillary gas chromatography. JAOAC Int 82:929–935, 1999.
24. KM Phillips, DM Ruggio, JI Toivo, MA Swank, AH Simpkins. Free and esterified sterol composition of edible oils and fats. J Food Comp Anal 15: 123–142, 2002.
25. RJ Reina, KD White, EGE Jahngen. Validated method for quantitation and identification of 4,4-desmethylsterols and triterpene diols in plant oils by thin-layer chromatography–high resolution gas chromatography–mass spectrometry. JAOAC Int 80:1272–1280, 1997.
26. PC Dutta, L-Å Appelqvist. Saturated sterols (stanols) in unhydrogenated and hydrogenated edible vegetable oils and in cereal lipids. J Sci Food Agric 71: 383–391, 1996.
27. J Toivo, V Piironen, P Kalo, P Varo. Gas chromatographic determination of major sterols in edible oils and fats using solid-phase extraction in sample preparation. Chromatographia 48:745–750, 1998.
28. AC Bello. Rapid isolation of the sterol fraction in edible oils using a silica cartridge. JAOAC Int 75:1120–1123, 1992.

29. M Amelio, R Rizzo, F Varazini. Determination of sterols, erythrodiol, uvaol and alkanols in olive oils using combined solid-phase extraction, high-performance liquid chromatographic and high-resolution gas chromatographic techniques. J Chromatogr 606:179–185, 1992.
30. L Alonso, J Fontecha, L Lozada, M Juarez. Determination of mixtures in vegetable oils and milk fat by analysis of sterol fraction by gas chromatography. J Am Oil Chem Soc 74:131–135, 1997.
31. A Mandl, G Reich, W Lindner. Detection of adulteration of pumpkin seed oil by analysis of content and composition of spesific Δ7-phytosterols. Eur Food Res Technol 209:400–406, 1999.
32. D Chryssafidis, P Maggos, V Kiosseoglou, D Boskou. Composition of total and esterified 4α-monomethylsterols and triterpene alcohols in virgin olive oil. J Sci Food Agric 58:581–583, 1992.
33. A Kamal-Eldin, L-Å Appelqvist. Variations in the composition of sterols, tocopherols and lignans in seed oils from four Sesamum species. J Am Oil Chem Soc 71:149–156, 1994.
34. TL Mounts, SL Abidi, KA Rennick. Effect of genetic modification on the content and composition of bioactive constituents in soybean oil. J Am Oil Chem Soc 73:581–586, 1996.
35. G Cole, S Coughlan, N Frey, J Hazebroek, C Jennings. New sunflower and soybean cultivars for novel vegetable oil types. Fett/Lipid 100:177–181, 1998.
36. SL Abidi, GR List, KA Rennick. Effect of genetic modification on the distribution of minor constituents in canola oil. J Am Oil Chem Soc 76:463–467, 1999.
37. C Vlahakis, J Hazebroek. Phytosterol accumulation in canola, sunflower, and soybean oils: effects of genetics, planting location, and temperature. J Am Oil Chem Soc 77:49–53, 2000.
38. RE Worthington, HL Hitchcock. A method for the separation of seed oil steryl esters and free sterols: application to peanut and corn oils. J Am Oil Chem Soc 61:1085–1088, 1984.
39. MH Gordon, RE Griffith. A comparison of the steryl esters of coconut and palm kernel oils. Fat Sci Technol 94:218–221, 1992.
40. IM Jawad, SP Kochhar, BJF Hudson. The physical refining of edible oils. 2. Effect on unsaponifiable components. Lebensm Wiss Technol 17:155–159, 1984.
41. EM Prior, VS Vadke, FW Sosulski. Effect of heat treatments on canola press oils. 1. Non-triglyceride components. J Am Oil Chem Soc 68:401–406, 1991.
42. HS Yoon, SK Kim. Oxidative stability of high-fatty acid rice bran oil at different stages of refining. J Am Oil Chem Soc 71:227–229, 1994.
43. RA Norton. Quantitation of steryl ferulate and p-coumarate esters from corn and rice. Lipids 30:269–274, 1995.
44. AG Gopala Krishna, S Khatoon, PM Shiela, CV Sarmandal, TN Indira, A Mishra. Effect of refining of crude rice bran oil on the retention of oryzanol in the refined oil. J Am Oil Chem Soc 78:127–131, 2001.
45. S Ramamurthi, A McCurdy. Enzymatic pretreatment of deodorizer distillate

for concentration of sterols and tocopherols. J Am Oil Chem Soc 70:287–295, 1993.

46. JW King, NT Dunford. Phytosterol-enriched triglyceride fractions from vegetable oil deodorizer distillates utilizing supercritical fluid fractionation technology. Sep Sci Technol 37:451–462, 2002.

47. A Cert, A Lanzón, AA Carelli, T Albi. Formation of stigmasta-3,5-diene in vegetable oils. Food Chem 49:287–293, 1994.

48. K Grob, A Artho, C Mariani. Determination of raffination of edible oils and fats by olefinic degradation products of sterols and squalene, using coupled LC-GC. Fat Sci Technol 94:394–400, 1992.

49. E Schulte. Determination of edible fat refining by HPLC of Δ 3,5-steradienes. Fat Sci Technol 96:124–128, 1994.

50. T Verleyen, A Szulczewska, R Verhe, K Dewettinck, A Hyghebaert, W De Greyt. Comparison of steradiene analysis between GC and HPLC. Food Chem 78:267–272, 2002.

51. T Verleyen, E Cortes, R Verhe, K Dewettinck, A Hyghebaert, W De Greyt. Factors determining the steradiene formation in bleaching and deodorization. Eur J Lipid Sci Technol 104:331–339, 2002.

52. RAp Ferrari, W Esteves, KD Mukherjee. Alteration of steryl ester content and positional distribution of fatty acids in triacylglycerols by chemical and enzymatic interesterification of plant oils. J Am Oil Chem Soc 74:93–96, 1997.

53. A Strocchi, G Marascio. Structural modifications of 4,4'-dimethyl sterols during the hydrogenation of edible vegetable oils. Fat Sci Technol 95:293–299, 1993.

54. AM Lampi, J Toivo, RL Hovi, V Piironen. Stability of plant sterols and formation of oxidation products in oils during heating. Chemical Reactions in Foods IV, Prague, September 20–22, 2000. Czech J Food Sci 18:208–209, 2000.

55. LL Oehrl, AP Hansen, CA Rohrer, GP Fenner, LC Boyd. Oxidation of phytosterols in a test food system. J Am Oil Chem Soc 78:1073–1078, 2001.

56. PC Dutta, GP Savage. Formation and content of phytosterol oxidation products in foods. In: F Guardiola, PC Dutta, R Codony, GP Savage, eds. Cholesterol and Phytosterol Oxidation Products: Analysis, Occurrence, and Biological Effects. Champaign, IL: AOAC Press, 2002, pp. 319–334.

57. A-M Lampi, L Juntunen, J Toivo, V Piironen. Determination of thermo-oxidation products of plant sterols. J Chromatogr 777:83–92, 2002.

58. L Soupas, L Juntunen, A-M Lampi, V Piironen. Distribution of oxidation products of stigmasterol heated in saturated and unsaturated lipid matrices. Pol J Food Nutr Sci 11/52:98–99, 2002.

59. M Rudzińska, H Jeleń, E Wasowicz. The content of phytosterols and their oxidized derivatives in heated plant oils. Pol J Food Nutr Sci 11/52:129–134, 2002.

60. M Ghavami, ID Morton. Effect of heating at deep-fat frying temperature on the sterol content of soya bean oil. J Sci Food Agric 35:569–572, 1984.

61. T Albi, A Lanzón, A Guinda, MC Pérez-Camino, M Léon. Microwave and

conventional heating effects on some physical and chemical parameters of edible fats. J Agric Food Chem 45:3000–3003, 1997.

62. HT Slover, RH Thompson, CS Davis, GV Merola. Lipids in margarines and margarine-like foods. J Am Oil Chem Soc 62:775–786, 1985.

63. V Piironen, J Toivo, A-M Lampi. Natural sources of dietary plant sterols. J Food Comp Anal 13:619–624, 2000.

64. H Scherz, F Senser, Souci-Fachmann-Kraut. Food Composition and Nutrition Tables. 6th revised and compiled edition. Stuttgart: Medpharm Scientific Publishers, 2000.

65. U.S. Department of Agriculture. USDA Nutrient Data Base, Release 13. Washington, DC, 2000

66. V Piironen, J Toivo, A-M Lampi. Plant sterols in cereals and cereal products. Cereal Chem 79:148–154, 2002.

67. V Singh, RA Moreau, LW Doner, SR Eckhoff, KB Hicks. Recovery of fiber in the corn dry-grind ethanol process: a feedstock for valuable coproducts. Cereal Chem 76:868–872, 1999.

68. V Singh, RA Moreau, AE Haken, KB Hicks, SR Eckhoff. Effect of various acids and sulfites in steep solution on yields and composition of corn fiber and corn fiber oil. Cereal Chem 77:665–668, 2000.

69. K Määttä, A-M Lampi, J Petterson, BM Fogelfors, V Piironen, A Kamal-Eldin. Phytosterol content in seven oat cultivars grown at three locations in Sweden. J Sci Food Agric 79:1021–1027, 1999.

70. RC Zambiazi, R Przybylski. Changes in lipid composition in oats with elevated fat content. Adv Plant Lipid Res, 22–25, 1998.

71. P Hakala, A-M Lampi, V Ollilainen, U Werner, M Murkovic, K Wähälä, S Karkola, V Piironen. Steryl phenolic acid esters in cereals and their milling fractions. J Agric Food Chem 50:5300–5307, 2002.

72. D Jonker, GD van der Hoek, JFC Glatz, C Homan, MA Posthumus, MB Katan. Combined determination of free, esterified and glycosilated plant sterols in foods. Nutr Rep Int 32:943–951, 1985.

73. OK Chung, J-B Ohm. Cereal lipids. In: K Kulp, JG Ponte Jr, eds. Handbook of Cereal Science and Technology. New York: Marcel Dekker 2000, 417–477.

74. RA Moreau, V Singh, KB Hicks. Comparison of oil and phytosterol levels in germplasm accessions of corn, teosinte, and Job's tears. J Agric Food Chem 49:3793–3795, 2001.

75. RA Moreau, V Singh, A Nuñez, KB Hicks. Phytosterols in the aleurone layer of corn kernels. Biochem Soc Trans 28:803–806, 2000.

76. J Kesselmeier, W Eichenberger, B Urban. High performance liquid chromatography of molecular species from free sterols and sterylglycosides isolated from oat leaves and seeds. Plant Cell Physiol 26:463–471, 1985.

77. M Ohnishi, S Obata, S Ito, Y Fujino. Composition and molecular species of waxy lipids in wheat grain. Cereal Chem 63:193–196, 1986.

78. MA Palmer, BN Bowden. The pentacyclic triterpene esters and the free, esterified and glycosylated sterols of Sorghum vulgare grain. Phytochem 14:1813–1815, 1975.

79. N Kuroda, M Ohnishi, Y Fujino. Sterol lipids in rice bran. Cereal Chem 54:997–1006, 1977.
80. J Toivo, K Määttä, A-M Lampi, V Piironen. Free, esterified, and glycosylated sterols in Finnish cereals. Proceedings of Euro Food Chem X: Functional Foods. A New Challenge for the Food Chemists, FECS-Event No. 234. Hungary: Budapest, 22–24 September 1999, pp 509–512.
81. RA Moreau, KB Hicks, MJ Powell. Effect of pretreatment on the yield and composition of oil extracted from corn fiber. J Agric Food Chem 47:2869–2871, 1999.
82. V Singh, RA Moreau, PH Cooke. Effect of corn milling practices on aleurone layer cells and their unique phytosterols. Cereal Chem 78:436–441, 2001.
83. V Singh, RA Moreau, KB Hicks, SR Eckhoff. Effect of alternative milling techniques on the yield and composition of corn germ oil and corn fiber oil. Cereal Chem 78:46–49, 2001.
84. FD Conforti, CH Harris, JT Rinehart. High-performance liquid chromatographic analysis of wheat flour lipids using an evaporative light scattering detector. J Chromatogr 645:83–88, 1993.
85. WHH Farrington, MJ Warwick, G Shearer. Changes in the carotenoids and sterol fractions during the prolonged storage of wheat flour. J Sci Food Agric 32:948–950, 1981.
86. V Singh, RA Moreau, AE Haken, SR Eckhoff, KB Hicks. Hybrid variability and effect of growth location on corn fiber yields and corn fiber oil composition. Cereal Chem 77:692–695, 2000.
87. V Singh, P Yang, RA Moreau, KB Hicks, SR Eckhoff. Effect of harvest moisture content and ambient air drying on maize fiber oil yield and its phytosterol composition. Starch/Stärke 53:635–638, 2001.
88. RP Evershed, N Spooner, MC Prescott, LJ Goad. Isolation and characterisation of intact steryl ferulates from seeds. J Chromatogr 440:23–35, 1988.
89. EJ Rogers, SM Rice, RJ Nicolosi, DR Carpenter, CA McClelland, LJ Romanczyk. Identification and quantitation of γ-oryzanol components and simultaneous assessment of tocols in rice bran oil. J Am Oil Chem Soc 70:301–307, 1993.
90. Z Xu, S Godber. Purification and identification of components of γ-oryzanol in rice bran oil. J Agric Food Chem 47:2724–2728, 1999.
91. GS Seetharamaiah, N Chandrasekhara. Studies in hypocholesterolemic activity of rice bran oil. Atherosclerosis 78:219–223, 1989.
92. N Rong, LM Ausman, RJ Nicolosi. Oryzanol decreases cholesterol absorption and aortic fatty streaks in hamsters. Lipids 32:303–309, 1997.
93. K Yasukawa, T Akihisa, Y Kimura, T Tamura, M Takido. Inhibitory effect of cycloartenol ferulate, a component of rice bran, on tumor promotion in two-stage carcinogenesis in mouse skin. Biol Pharm Bull 21:1072–1076, 1998.
94. JA Weststrate, GW Meijer. Plant sterol-enriched margarines and reduction of plasma total-and LDL-cholesterol concentrations in normocholesterolaemic and mildly hypecholesterolaemic subject. Eur J Clin Nutr 52:334–343, 1998.
95. A Koski, K Wähälä, A Hopia, V Piironen, A-M Lampi, J Toivo, M Heinonen.

Antioxidant properties of phenolic acids esterified with β-sitosterol. XXth International Conference on Polyphenols, Freising-Weihenstephan, Technische Universität München: Germany, September 10–15, 2000, pp. 387–388.

96. Z Xu, N Hua, JS Godber. Antioxidant activity of tocopherols, tocotrienols, and γ-oryzanol components from rice bran against cholesterol oxidation accelerated by 2,2′-azobis(2-methylpropinamidine). J Agric Food Chem 49:2077–2081, 2001.

97. Z Xu, JS Godber. Antioxidant activities of major components of γ-oryzanol from rice bran using a linoleic acid model. J Am Oil Chem Soc 78:645–649, 2001.

98. J-S Kim, JS Godber, JM King, W Prinyawiwatkul. Inhibition of cholesterol autoxidation by the nonsaponifiable fraction in rice bran in an aqueous model system. J Am Oil Chem Soc 78:685–689, 2001.

99. LM Seitz. Stanol and sterol esters of ferulic and p-coumaric acids in wheat, corn, rye and triticale. J Agric Food Chem 37:662–667, 1989.

100. RA Norton. Isolation and identification of steryl cinnamic acid derivates from corn bran. Cereal Chem 71:111–117, 1994.

101. LM Seitz. Sitostanyl ferulate as an indicator of mechanical damage to corn kernels. Cereal Chem 67:305–307, 1990.

102. RA Moreau, MJ Powell, KB Hicks, RA Norton. A comparison of the levels of ferulate phytosterol esters in corn and other seeds. Adv Plant Lipid Res 472–474, 1998.

103. RA Moreau, V Singh, SR Eckhoff, MJ Powell, KB Hicks, RA Norton. Comparison of yield and composition of oil extracted from corn fiber and corn bran. Cereal Chem 76:449–451, 1999.

104. RA Moreau, MJ Powell, KB Hicks. Extraction and quantitative analysis of oil from commercial corn fiber. J Agric Food Chem 44:2149–2154, 1996.

105. SL Taylor, JW King. Optimization of the extraction and fractionation of corn bran oil using analytical supercritical fluid instrumentation. J Chromatogr 38:91–94, 2000.

106. SL Taylor, JW King. Enrichment of ferulate phytosterol esters from corn fiber oil using supercritical fluid extraction and chromatography. J Am Oil Chem Soc 77:687–688, 2002.

107. TS Shin, JS Godber, DE Martin, JH Wells. Hydrolytic stability and changes in E vitamers and oryzanol of extruded rice bran during storage. J Food Sci 62:704–708, 1997.

108. Z Shen, MV Palmer, SST Ting, RJ Fairclough. Pilot scale extraction of rice bran oil with dense carbon dioxide. J Agric Food Chem 44:3033–3039, 1996.

109. M Saska, GJ Rossiter. Recovery of γ-oryzanol from rice bran oil with silica-based continuos chromatography. J Am Oil Chem Soc 75:1421–1427, 1998.

110. NT Dunford, JW King. Phytosterol enrichment of rice bran oil by a supercritical carbon dioxide fractionation technique. J Food Sci 65:1395–1399, 2000.

111. Z Xu, JS Godber. Comparison of supercritical fluid and solvent extraction methods in extracting γ-oryzanol from rice bran. J Am Oil Chem Soc 77:547–551, 2000.

112. S Bryngelsson. Plant sterols in cereal products in Sweden. Diploma thesis in food chemistry, Department of Food Science, Chalmers University of Technology, 1997.

113. L Normén, M Johnsson, H Andersson, Y van Gameren, P Dutta. Plant sterols in vegetables and fruits commonly consumed in Sweden. Eur J Nutr 38:84–89, 1999.

114. V Piironen, J Toivo, R Puupponen-Pimiä, A-M Lampi. Plant sterols in vegetables, fruits and berries. J Sci Food Agric 83:330–337, 2003.

115. AM Lampi, R Puupponen-Pimiä, J Toivo, V Piironen. By-products of food industry: possible sources of plant sterols. Stability of plant sterols in foods and model systems. Bioactive compounds in plant foods. COST Action 916: bioactive compounds in plant foods, Tenerife, Canary Island, April 26–28, 2001. Proceedings 171–172, 2002

116. T Akihisa, P Ghosh, S Thakur, FU Rosenstein, T Matsumoto. Sterol composition of seeds and mature plants of family Cucurbitaceae. J Am Oil Chem Soc 63:653–658, 1986.

117. VK Garg, WR Nes. Occurrence of Δ5-sterols in plants producing predominantly Δ7-sterols: studies on the sterol compositions of six Cucurbitaceae seeds. Phytochemistry 25:2591–2597, 1986.

118. T Akihisa, N Shimizu, P Ghosh, S Thakur, FU Rosenstein, T Tamura, T Matsumoto. Sterols of the Cucurbitaceae. Phytochemistry 26:1693–1700, 1987.

119. BD Whitaker. Changes in the steryl lipid content and composition of tomato fruit during ripening. Phytochemistry 27:3411–3416, 1988.

120. BD Whitaker, WR Lusby. Steryl lipid content and composition in bell pepper fruit at three stages of ripening. J Am Soc Hort Sci 114:648–651, 1989.

121. TD Drumm, JI Gray, GL Hosfield. Variability in the major lipid components of four market classes of dry edible beans. J Sci Food Agric 50:485–497, 1990.

122. BD Whitaker. Changes in lipids of tomato fruit stored at chilling and non-chilling temperatures. Phytochemistry 30:757–761, 1991.

123. BD Whitaker. Lipid changes in mature-green tomatoes during ripening, during chilling, and after rewarming subsequent to chilling. J Am Soc Hort Sci 119: 994–999, 1994.

124. BD Whitaker. A reassessment of heat treatment as a means of reducing chilling injury in tomato fruit. Postharvest Biol Technol 4:75–83, 1994.

125. BD Whitaker. Lipid changes in mature-green bell pepper fruit during chilling at 2°C and after transfer to 20°C subsequent to chilling. Physiol Plant 93:683–688, 1995.

126. RE McDonald, TG McCollum, EA Baldwin. Prestorage heat treatments influence free sterols and flavor volatiles of tomatoes stored at chilling temperature. J Am Soc Hort Sci 121:531–536, 1996.

127. CP Rochester, P Kjellbom, C Larsson. Lipid composition of PLASMA membranes from barley leaves and roots, spinach leaves and cauliflower inflorescences. Physiol Plant 71:257–263, 1987.

128. RE McDonald, TG McCollum, EA Baldwin. Heat treatment of mature-green

tomatoes: differential effects of ethylene and partial ripening. J Am Soc Hort Sci 123:457–462, 1998.
129. GA Picchioni, AE Watada, BD Whitaker, A Reyes. Calcium delays senescence-related mambrane lipid changes and increases net synthesis of membrane lipid components in shredded carrots. Postharvest Biol Technol 9:235–245, 1996.
130. V Piironen, R Puupponen-Pimiä, J Toivo, A-M Lampi. Stability of plant sterols in foods and model systems. Bioactive compounds in plant foods. COST Action 916: Bioactive compounds in plant foods, Tenerife, Canary Island, April 26–28, 2001. Proceedings 197–198, 2002.
131. R Puupponen-Pimiä, S Salonvaara, M Aarni, AM Nuutila, T Suortti, AM Lampi, V Piironen, M Eurola, KM Oksman-Caldentey. Blanching and long-term freezing affect differently on various bioactive compounds of vegetables. J Sci Food Agric. In press.
132. MD Zlatanov. Lipid composition of Bulgarian chokeberry, black currant and rose hip seed oils. J Sci Food Agric 79:1620–1624, 1999.
133. B Yang, RM Karlsson, PH Oksman, HP Kallio. Phytosterols in sea buckthorn (Hippophaë rhamnoides L.) berries: identification and effects of different origins and harvesting times. J Agric Food Chem 49:5620–5629, 2001.
134. J Koponen, H Kallio, R Tahvonen. Plant sterols in Finnish blueberry (Vaccinium myrtillus L.) and lingonberry (Vaccinium vitis-idea L.) seed oils. In: W Pfanhauser, GR Fenwick, S Kochar, eds. Biologically-Active Phytochemicals in Food. Cambridge: Royal Society of Chemistry, 2001, pp. 233–236.
135. IM Bartley. Changes in sterol and phospholipid composition of apples during storage at low temperature and low oxygen concentration. J Sci Food Agric 37:31–36, 1986.
136. Y Le Fur, C Hory, M-H Bard, A Olsson. Evolution of phytosterols in Chardonnay grape berry skins during last stages of ripening. Vitis 33:127–131, 1994.
137. R Izzo, A Scartazza, A Masia, L Galleschi, MF Quartacci, F Navari-Izzo. Lipid evolution during development and ripening of peach fruits. Phytochemistry 39:1329–1334, 1995.
138. YF Lozano, CD Mayer, C Bannon, EM Gaydou. Unsaponifiable matter, total sterol and tocopherol contents of avocado oil varieties. J Am Oil Chem Soc 70:561–565, 1993.
139. GA Picchioni, AE Watada, WS Conway, BD Whitaker, CE Sams. Phospholipid, galactolipid, and steryl lipid composition of apple fruit cortical tissue following postharvest CaCl$_2$ infiltration. Phytochemistry 39:763–769, 1995.
140. BD Whitaker, JD Klein, WS Conway, CE Sams. Influence of prestorage heat and calcium treatments on lipid metabolism in "golden delicious" apples. Phytochemistry 45:465–472, 1997.
141. AB Awad, KC Chan, AC Downie, CS Fink. Peanuts as source of β-sitosterol, a sterol with anticancer properties. Nutr Cancer 36:238–241, 2000.

2
Analysis of Phytosterols in Foods

Anna-Maija Lampi and **Vieno Piironen**
University of Helsinki, Helsinki, Finland

Jari Toivo
National Technology Agency of Finland, Helsinki, Finland

I. INTRODUCTION

Analysis of phytosterols in foods is a complex task because these compounds occur both as free alcohols and as various conjugates, and therefore may be easily extractable or tightly bound to the food matrix. Due to the positive health effects associated with consumption of phytosterol-enriched foods, interest in natural phytosterols present in foods has also increased (1–3). Considerable effort has been made to develop reliable, powerful, simple, and economical methods to analyze phytosterols. Traditional methods used hitherto were originally developed for cholesterol analysis and contained several purification and enrichment steps. Improvements in sample preparation and chromatographic techniques have led to significant progress in recent years (e.g., 1,4–6). In this chapter, analysis methods for studying phytosterol profiles and contents of plant materials and foods will be discussed, as well as various phytosterol classes including free sterols and their conjugates.

II. SAMPLE PREPARATION METHODS

A. Extraction of Phytosterols from Food Materials

Efficient extraction of phytosterols from food matrix is the most important step in the sample preparation procedure. Losses caused by the use of an inadequate extraction solvent and technique cannot be compensated later in

33

the analysis. When comparing extraction methods, attention should be paid to the various conjugates in which phytosterols may exist, i.e., as free alcohols, esters, glycosides, or acylated glycosides. The solubilities of various phytosterol compounds vary considerably. Nonpolar lipid solvents are efficient in extracting free and fatty acid–esterified phytosterols. The more polar phytosteryl glycosides are poorly soluble in nonpolar solvents and, thus, more polar solvents should be used to extract them.

The food matrix also has a major effect on how easily phytosterols are liberated. Phytosterols in crude and refined oils can simply be dissolved together with the sample in a nonpolar solvent. It is also relatively easy to liberate most phytosterols from oil grains, nuts, and other high-fat tissues by solvent extraction, whereas it is much more difficult to extract phytosterols through hard cell walls or from the complex polysaccharide–protein matrices found in wet vegetable tissues and cereals. Moreover, enzyme activities should be avoided to minimize phytosterol losses, conversions, and artifact formation.

One approach to improving phytosterol extractability from complex food materials is by degrading the matrix by alkaline, acid, or enzymatic hydrolysis. Alkaline solvents effectively dissolve lipid and protein matrices, whereas acid hydrolysis is needed to decompose some polysaccharides such as starch. Since alkaline hydrolysis is commonly used as a means to partition sterols from saponifiable lipids, extraction of lipids prior to hydrolysis may be omitted. It should be remembered that these treatments also hydrolyze phytosteryl conjugates, which makes them unsuitable when intact molecular species of phytosterol conjugates are to be studied. Hydrolytic procedures may also have adverse effects on some phytosterols, such as $\Delta 7$-phytosterols. Although phytosterols are relatively stable compounds against oxidation compared with polyunsaturated lipids, some precautions should be taken during sample preparation. Exposure to heat, light, and air should be avoided, and some investigators add antioxidants to their extraction mixtures (e.g., 7,8).

A variety of methods are available to extract phytosterols from food matrix, differing in efficiency, specificity, robustness, and cost efficiency. Thus, validation (e.g., recovery studies) of phytosterols should be included to optimize each sample extraction procedure for each food matrix type.

1. Extraction of Lipids

When phytosterols are directly extracted from a food matrix, sample particle size should be homogeneous and small to improve extraction efficiency. Seeds and dried plant materials are ground with pulverized glass (9), in a coffee grinder (10), mortar (11), mill (8,12), or blade grinder (13). In some studies, particle sizes are carefully characterized by passing the samples through sieves (13,14). Wet tissues may be homogenized in solvent.

Phytosterols may be extracted from air-dried plant material or from wet tissues. With dried material, water-immiscible solvents, such as hexane, can be used, but with wet tissues water-soluble solvents and solvent mixtures are required for efficient lipid extraction (15). When nonpolar solvents are used for extraction, drying of the sample is very important because residual moisture might isolate phytosterols from the solvent.

Both nonpolar and polar solvents and their mixtures have been used to liberate lipids from foods. Nonpolar solvents extract phytosterols as a part of lipids and are commonly used to extract oil grains. Polar solvents are more selective because they release polar phytosterol compounds better than nonpolar solvents. Since most of the phytosterols in oil grains exist as free alcohols or fatty acid esters, they are readily extracted with nonpolar solvents, whereas other plant materials such as cereals also contain glycosylated phytosterols, which are soluble in polar solvents.

Extraction with hexane has been used to liberate phytosterols from poppy and sunflower seeds (16), grape must (17), corn fiber (8) and oil seeds (18), and petroleum ether from pumpkin seed (7,19). Hexane and methylene chloride were equally efficient in extracting major phytosterols from dried gingsen seeds because the yields were comparable (12). After extraction, crude oil is obtained after centrifugation or filtration, e.g., through cellulose acetate or glass fiber filters (12,18).

Depending on its efficiency, the extraction procedure should be repeated to ensure total recovery of lipids. Generally, extractions are performed in triplicate. Solvent extraction becomes more efficient when the sample is heated and extraction is continuous. In a Soxhlet apparatus, dried and homogenized samples are weighed in thimbles that are subjected to steady extraction with heated solvent for several hours. Continuous extraction of lipids with heated hexane has been used for sesame seeds (11), avocado mesocarps (20), roasted coffee (21) and berry seeds (22), and with more polar acetone for cocoa butter (23). Lipid extraction may also be improved by using an ultrasonic bath (7,13), and by grinding a wet sample with sodium sulfate prior to extraction (24).

The most common lipid extraction solvent mixture, consisting of chloroform and methanol, has been applied to extract phytosterols from various plant materials, although this method was originally developed to extract lipids from animal tissues (25,26). Extraction with chloroform and methanol followed by purification with partitioning into chloroform and aqueous phases has been applied to extract lipids from seeds and leaves of oats (27), seeds and mature plants of Curcubitaceae (19), plant tissues (28), leafs of cucumber seedling (29), bell pepper fruits (30,31), edible beans (14), lyophilized celery cells (32), leaves of wood (33), snapbeans (34), tomato fruits (35), potato (36), blueberry and lingonberry seeds (37), and experimental diets (38). The extract is washed repeatedly with salt solutions or water to remove nonlipid contaminants. When Fishwick and Wright (39) compared three extraction procedures or extraction of total lipids and lipid classes from various plant materials, they found that chloroform and methanol extractions were very efficient in extracting lipids of tissues with high water content, e.g., tomato, spinach, potato, and pea.

Since hydrolytic and oxidative enzymes present in plant tissues may maintain their activity during chloroform-methanol extraction, denaturation should be considred prior to extraction. Most commonly this is performed by boiling in isopropanol, which is regarded as the safest and most efficient solvent (30,34,39–41), or alternatively in ethanol (42), methanol (28) or water (9,43). Inactivation of hydrolytic enzymes by heating in a microwave oven has also been done (44).

In some applications chloroform has been replaced with the less toxic solvent dichloromethane, which has extraction properties comparable to those of chloroform (24,45). As an alternative to chloroform-methanol extraction, Hara and Radin (46) introduced a lipid extraction method with a hexane-isopropanol mixture, which has the advantage of being less toxic. The procedure is less laborious because it is not necessary to wash the lipid extract and emulsion formation is avoided. However, when Dutta and Appelqvist (47) used this solvent mixture to extract phytosterols with other lipids from cereal samples, they removed water-soluble contaminants from the extract using the biphasic chloroform-methanol-water system of Folch and coworkers (25). Washing procedures have also been used after other extraction methods (10,43). Other solvent mixtures to release phytosterols from various plant materials include hexane with ethanol (48) or ethyl acetate (49). In order to extract specifically the more polar phytosterol conjugates (i.e., steryl glycosides and esters with phenolic acids), more polar solvents are used. Methanol was used to liberate steryl glycosides from rice bran (50) and *Cycos circinalis* seeds (51), and acetone to extract steryl ferulates from cereals (52,53).

The solubility properties of cereal lipids have been used to classify them as free or bound lipids. Classification is based on their sequential extractability (54,55). Free lipids are readily extracted with nonpolar solvents such as hexane. Bound lipids may further be divided into two categories: those weakly bound on the surface of polysaccharides and extracted with polar solvents, and those strongly bound within the polysaccharide matrix and liberated only with such solvent mixtures that penetrate the polysaccharide matrix, e.g., water-saturated *n*-butanol (WSB) (54). In another study, nonstarch lipids were extracted from wheat flour with WSB at room temperature and starch lipids with the same solvent mixture at 90–100°C (56). Generally, refluxing in WSB is considered to be the most efficient extraction method for cereal lipids because it efficiently penetrates the starchy matrix (10). It was better than chloroform-methanol solvent extraction in extracting lipids from dried peas and wheat flour (39). In one study, both solvents were used to extract total lipids from wheat grains (43). An alternative method to include bound lipids from cereals in phytosterol analysis is to acid hydrolyze the residue obtained after removal of free lipids with a hexane-isopropanol mixture. Using this method phytosterols bound to polysaccharides are also liberated (47).

One means to decrease the use of organic solvents in sample preparation to apply supercritical fluid extraction (SFE) to extract and concentrate phytosterols from oil- and fat-containing matrices. Supercritical CO_2 is the most commonly used solvent that may be modified by cosolvents, e.g., methanol and methyl *tert*-butyl ether (57). When using SFE, it is important to maintain adequate flow and pressure, and optimal experimental parameters to obtain complete and rapid extraction of phytosterols (6). Nevertheless, adequate enrichment of phytosterols needs repetitive extractions (6). When combining SFE using carbon dioxide and methyl *tert*-butyl ether as cosolvent with amino adsorbent, it was possible to enrich phytosterols from soybean oil (57). Since lipids possess unique extractabilities in SFE, it is also possible to use this technique to fractionate the extract. For example, total extract yields of rice bran in SFE reached their maximum in 10 min extraction, whereas γ-oryzanol concentrations were the highest after 15 min (58).

2. Hydrolysis of Phytosteryl Conjugates in Total Lipids, Fats, and Oils

In high-fat foods, e.g., oils, fats, and oil seeds, phytosterols exist mainly as free sterols and esters. To enable analysis of free sterols in these samples only cleavage of ester bonds is needed, which is easily performed by alkaline hydrolysis. Phytosterol glycosides and acetylated glycosides are stable during alkaline hydrolysis. When free phytosterols are to be liberated from these derivatives, the acetal bonds should be hydrolyzed by acid or enzymatic hydrolysis.

Numerous procedures have been used to saponify extracted lipids or refined oils and fats. Most commonly, lipids are hydrolyzed in potassium hydroxide methanol or ethanol for up to 1 h, after which the mixture is cooled and diluted, and the saponifiable lipids are extracted with a solvent. Depending on the sample size, incubations are carried out either under reflux or in test tubes. Phytosterols may be protected against oxidation during saponification by adding 3% pyrogallol or ascorbic acid as an antioxidant (e.g., 5,58). In a comprehensive study on methodology, an 8-min incubation at 80°C was found sufficient to hydrolyze all cholesterol esters in an oil sample (5). Later, saponification times from 10 to 60 min gave equal phytosterol recoveries (59). The authors pointed out that the mixtures should be vigorously shaken several times during saponification. As a gentle alternative to saponification to release free sterols from esters, enzymatic hydrolysis with triacylglycerol esterase may be used (60).

After saponification the mixture should be diluted with water prior to liquid–liquid partitioning of unsaponifiable lipids into a solvent. The proportions of alkaline alcohol, water, and solvent should be carefully controlled to avoid losses of phytosterols in the aqueous phase. Solvents used to extract phytosterols include hexane (13,36,47,61), heptane (62), cyclohexane (24,38), chloroform (63), isopropanol (19), ethyl ether (18,64–66), and tert-butyl methyl ether (12). When recovery of added radiolabeled cholesterol and cholesteryl oleate from edible oils after saponification and diethyl ether extractions was studied, it was found that recoveries of about 60% and 90% were obtained after one and two extractions, respectively (67). Only after five diethyl ether extractions was the recovery 100.5 ± 1.4%. When Toivo and coworkers (59) compared hexane, cyclohexane, and a hexane-diethyl ether (1:1) mixture for the extraction of nonsaponified lipids, they found no statistical differences between the solvents but preferred cyclohexane in their method. One advantage of using cyclohexane instead of diethyl ether to extract nonsaponified lipids is that it dissolves much less water than diethyl ether, i.e., 1 vs. 100 μL in 11 mL, respectively, which makes further steps in the analysis simpler (5). During washing of extracts, water-immiscible solvents do not easily form emulsions and afterward they do not need drying with sodium sulfate prior to evaporation of the solvent (4). In our laboratory, we have found that phytosterols are better extracted from the saponification mixture of high-fat samples with a diethyl ether-hexane (1:1) mixture than with cyclohexane. We have found that high amounts of soaps may trap phytosterols in the aqueous phase unless polar solvents are used to disrupt these interactions. Alternatively, the extraction should be repeated several times (4).

The unsaponifiable lipids may also be isolated from the mixture using adsorption chromatography by aluminum oxide solid phase extraction (68–70). By this means that emulsion formation is avoided. Fatty acid soaps are adsorbed on the surface and form insoluble soaps. In one method, the eluted solution is further purified in a quaternary amine column (69).

A few studies have been reported in which extracted lipids or oils are subjected to acid hydrolysis, which is performed when free phytosterols should be released from glycosides and acylated glycosides. Steryl glycoside fractions and total lipid extracts from plant materials were hydrolyzed with either hot hydrochloric or sulfuric acids (23,36,42,45). Kesselmeier and coworkers preferred enzymatic digestion with β-glucosidase to acid hydrolysis (27).

Jonker and coworkers (71) showed that the most common phytosterols, i.e., campesterol, sitosterol, and stigmasterol, were stable during the acid hydrolysis step and no acid-induced dehydration at C-3 occurred. This is in contrast to one report that acid decomposes up to 80% of phytosterols (27). The risks and benefits of acid hydrolysis should be carefully compared because some groups of phytosterols are sensitive to acidic conditions. Phytosterols in a few plant families, e.g., Curcubitaceae and Theaceae, consist mainly of Δ7-sterols that are easily decomposed or isomerized during acid hydrolysis. Therefore, acid hydrolysis should be omitted when analyzing these materials and replaced by enzymatic digestion (72,73). Some decomposition of Δ5- and Δ7-avenasteryl glycosides and isomerization of Δ5-avenasterol has also been shown to occur (27,74). Δ5-Avenasterol contents of whole wheat flour were 1.9 and 2.7 mg/100 g with and without acid hydrolysis, whereas the contents of total phytosterols increased by 25% with acid hydrolysis (59). Thus, a compromise between some minor changes in phytosterol profiles and improvements in total yields should be made. In our laboratory, we have found that the food matrix shields phytosterols during acid hydrolysis, i.e., isolated phytosteryl glycosides are prone to decomposition.

It has been suggested that food samples should be subjected to acid hydrolysis prior to alkaline hydrolysis because phytosteryl glycosides, being relatively polar compounds, may remain in the aqueous phase and be lost after saponification. Thus, they would not be extracted with the unsaponified material (15).

3. Extraction of Free Phytosterols After Hydrolysis of Food Materials

Phytosterols can be analyzed from food samples without extracting lipids first. Although the essential objective of various hydrolysis methods is to liberate free sterols from conjugates, such methods are also valuable in enhancing extraction efficiency by degrading the food matrix. There are several advantages in the use of direct saponification. In comparison with the method including lipid extraction it is more economical, faster, and safer because toxic lipid solvents can be replaced by less toxic ones (4).

Direct saponification methods to determine free and esterified phytosterols derive from those of phytosterol analysis of oils and fats, and of cholesterol analysis of various food items (75,76). Indyk (77) validated a simple direct saponification method followed by extraction of the unsaponifiable lipids to analyze several types of food products for cholesterol, phytosterols, and tocopherols. Using this method, phytosterol contents of vegetable oil and infant formula samples and two certified reference materials of the European Commission Community Bureau of Reference (BCR) were comparable to those found in the literature and to the indicative values given by BCR. Their finding that initial isolation of lipids prior to saponification is unnecessary in phytosterol and cholesterol analysis was later confirmed by others (59,78). Moreover, it has been stated that direct saponification is superior to indirect methods to analyze cholesterol in dairy products because during lipid extraction some sterols may be partitioned into the aqueous phase and thus lost (77,79).

Direct saponification may be applied to dry and liquid samples (77). First, samples are carefully dispersed in ethanol or methanol, after which concentrated potassium hydroxide solution is added. In one study, an alcohol volume of 4 mL was found to be sufficient for 1 g of sample (78). Saponification temperatures vary from 30°C to 85°C and time from 8 min to 60 min, and the final concentration of potassium hydroxide is 0.33 to 0.5 M in ethanol (4,77,78,80).

When a direct saponification method to analyze ergosterol from cereals and other feedstuffs was compared to one including a methanol extraction step prior to saponification, the direct saponification method gave 12–377% higher ergosterol yields (81). Similar phytosterol and cholesterol contents were obtained when a whole wheat flour and diet composite sample were subjected to direct saponification and total lipid extraction by chloroform-methanol followed by saponification (59).

Lyophilized and frozen-diet samples, mushrooms, and feedstuffs have been analyzed for phytosterols using direct saponification. The phytosterols were recovered from the saponified mixture either by diethyl ether (82), cyclohexane (80,83), or hexane extraction (81).

To liberate glycosidic phytosterols that are important conjugates in, say, cereals, acid hydrolysis is needed prior to saponification. Already in 1985, Jonker and coworkers introduced a method to analyze phytosterols directly from plant materials. They hydrolyzed food samples in 6 M hydrochloric acid in a boiling water bath for 30 min. Thereafter, the mixture was cooled and neutralized with potassium hydroxide pellets, lipids were saponified in ethanolic potassium hydroxide for another 30 min, and finally the unsaponifiable lipids were extracted with toluene and purified by a few washing steps. Later, the method was modified by adding a lipid extraction step after acid hydrolysis and optimizing the hydrolysis steps and lipid extraction solvents (59).

Inclusion of an acid hydrolysis step in phytosterol analysis increased the measured contents in various food items and diet composites by 1–156% (Table 1). As expected, acid hydrolysis had only a minor affect on phytosterol contents of oils and oil grains because they contain mainly free and esterified sterols. Vegetable and cereal products had higher amounts of phytosteryl glycosides and, thus, acid hydrolysis increased their phytosterol amounts by at least 21%. With acid hydrolysis, the phytosterol contents of three Dutch experimental diet composites increased by 20%, 26%, and 38 % (71), and of one U.S. diet composite by 15% (59). It should be noted that some of the increased phytosterol amounts could be due to their better release after hydrolysis. In the Dutch study, cholesterol contents also increased slightly (<4%), although cholesterol is not thought to form glycosides (71). If the sample was rich in acid-resistant cellulose, as are, for example, bran fractions, it would be advantageous to enhance phytosterol extraction by predigesting the sample with cellulase enzymes (9). However, phytosterol contents of whole wheat flour and diet composite samples were equal when they were analyzed with or without total lipid extraction prior to acid and alkaline hydrolyses (59), which means that both methods are suitable for the analysis of common food items.

Table 1 Comparison of Two Sample Preparation Methods to Analyze Phytosterol Contents in Foods (A: Direct Acid Hydrolysis Followed by Saponification; B: Direct Saponification)

Food sample	Sample preparation method	mg / 100 g fw				Ref.
		Campesterol	Stigmasterol	Sitosterol	Total phytosterols	
Cauliflower	A	7.4	1.7	21.6	31.4	71
	B	6.4	1.4	173	26.0	
Lettuce	A	1.1	4.5	10.6	17.4	71
	B	1.0	4.3	7.9	14.0	
Light rye bread	A	21.2	2.2	37.7	64.0	71
	B	13.0	1.5	23.7	41.2	
Whole wheat flour	A	10.1.	1.6	38.9	65.5	59
	B	8.5	1.4	29.2	52.1	
Rapeseed oil	A	242	2.8	375	704	59
	B	238	2.5	370	694	
Sunflower kernel	A	19.0	20.7	128	176	59
	B	17.8	20.5	124	170	
Dried onion	A	5.6	3.4	45.8	55.9	59
	B	2.1	1.3	18.4	21.8	

Direct acid and alkaline hydrolysis methods have been used to study phytosterol contents of vegetables, fruits, berries, and cereals consumed in Sweden and the Netherlands (84,85) and in Finland (86,87).

B. Enrichment and Purification of Free Phytosterols

The enrichment and purification methods needed after the lipid extraction and liberation of free sterols depend on the final analytical method used and the scope of the study. Modern chromatographic techniques enable us to use various specific enrichment and purification procedures. Chromatographic methods are commonly used during sample preparation to purify and concentrate phytosterols from other lipids. Normal-phase (NP), reversed-phase (RP), and argentation chromatography are used in sample clean-up procedures. Phytosterols are most commonly purified from the unsaponifiable matter or total lipid extracts with silicic acid and alumina. RP materials, e.g., Lipidex 5000 and octadecyl-bonded silica, may also be used. As a general rule, when the amount of lipid is at least 200 mg, column chromatography techniques are used, whereas with smaller sample amounts thin-layer chromatography (TLC) or solid-phase extraction (SPE) may be applied (6). Thus, in analytical work, TLC and SPE are the two most commonly used techniques.

Saponification may also be considered as a purification step because it removes triacylglycerols, phospholipids, and other esters from the lipid extract (15). Prior to gas chromatography (GC) analysis, saponifiable lipids are generally removed, whereas this is not necessary prior to NP-HPLC because the oil matrix does not interfere with the analysis. To improve separation of phytosterols in RP-HPLC and to avoid contamination of the column, saponification of oil sample is recommended (88).

To confirm that all lipids have been saponified, TLC with silica may be performed using, for example, hexane-diethyl ether-acetic acid with volume ratios of 400:45:4 or 85:15:1 (5,89). Clean-up of phytosterols by preparative TLC is still a common step prior to GC analysis (65). The unsaponifiable lipids are applied on a silica TLC plate, eluted with, say, hexane-diethyl ether (1:1) (70) or one of the above-mentioned mixtures. After development of the plate, phytosterols can be visualized by both destructive and non-destructive methods (6,90). Phytosterol standards can be charred after spraying with sulfuric or phosphomolybdic acid, or immersed in iodine vapor to determine relative retentions (R_f) to allow collection of sample phytosterols. Nondestructive means to visualize the compounds include spraying with 2,7-dichlorofluorescein (91), berberin hydrochloride (45), or methanol (70). Methanol makes the plate translucent and the phytosterols are detected as white zones. After nondestructive spraying, phytosterols can be scraped off the plate, extracted from silica with, for example, diethyl ether or chloroform-diethyl ether mixture, and used for further analysis.

Phytosterols can be purified from the unsaponifiable lipids with silica SPE, which is much faster than TLC clean-up. For example, the unsaponifiable lipid sample can be applied on a silica cartridge in a low-polarity solvent. Phytosterols are adsorbed on the silica surface and interfering compounds are washed with solvent mixtures of increasing polarities. Finally, free phytosterols are released with a polar solvent such as 50–70% diethyl ether in hexane (86,92). Quaternary ammonium–bonded phase cartridges were used to enrich free sterols from a lipid extract (60), and with aluminum oxide cartridges to enrich free sterols after saponification (69). Cholesterol and, later, phytosterols were isolated from acidified saponification mixture by octadecyl silyl cartridges instead of by extraction of the unsaponifiable lipids (93,94).

Saponification of lipid extracts can also be avoided. Phytosterols can be separated from the washed neutral lipid extract with digitonin precipitation (29). In a more modern approach, free sterols and tocopherols of vegetable oil samples were directly silylated. This resulted in reduced polarity, and silylated phytosterols and tocopherols could be isolated together with nonpolar phytosteryl esters from the slightly more polar triacylglycerol matrix by silica SPE (95).

Recovery of phytosterols after purification and enrichment steps is an important factor in quantitative analyses. There are several means to quantitatively collect phytosterols after saponification and solvent extraction (e.g., 67), but the yields from precipitation and chromatographic methods vary considerably. For example, phytosterol from concentrates gave recoveries of 85% and 95% during silica gel TLC and florisil column chromatography, respectively (96). Higher and more reproducible recoveries have been measured for SPE techniques. Recoveries of added sito- and stigmasterols were 100% and 99% when they were applied on a silica cartridge and eluted with hexane-ethyl acetate (60:40) (97).

III. CHROMATOGRAPHIC ANALYSIS OF FREE PHYTOSTEROLS

Individual free phytosterols are most commonly analyzed by capillary GC. Separation based on numbers of carbon atoms is easily obtained. Nowadays Δ5-sterols and their respective stanols, Δ7-sterols, and some isomerization products can also be separated in efficient capillary GC columns. HPLC techniques are less often used to separate free phytosterols than GC, but some applications exist (2,6). One of the reasons for limited use of HPLC may be that there is not such a general and sensitive HPLC detector for phytosterols as the flame ionization detector in GC. However, with new separation techniques and detection possibilities, attempts have been made to apply HPLC to the analysis of free phytosterols in foods. In this section the principles of chromatographic separation will be discussed. During the past few years several extensive reviews of chromatographic analysis of phytosterols and related compounds have been published (4,6,90,98,99), and they are warmly recommended to those readers specifically interested in this area.

A. Preparative Chromatography of Free Sterols

To apply preparative chromatographic methods is recommended before analytical separation when identifying phytosterols of samples with an unknown or complicated phytosterol composition. For example, when a sample is known to contain a range of various free sterols, e.g., des-, mono-, and dimethyl sterols, $\Delta 5$- and $\Delta 7$-sterols, some of them are likely to overlap with each other in the analytical separation. TLC with silica as the adsorbent can also be applied to separate various free sterols from each other on the basis of their polarity (6). The use of TLC with chloroform-diethyl ether (9:1, v/v) as the mobile phase resulted in separation of desmethyl sterols (R_f 0.25–0.34), monomethyl sterols (R_f 0.34–0.40), and dimethyl sterols (R_f 0.40–0.46) (11). When TLC was eluted with benzene-chloroform (3:1, v/v), relative mobilities of des-, mono-, and dimethyl sterols were 0.16, 0.22, and 0.27, respectively (50). A more detailed separation was achieved with n-hexane-diethyl ether (1:1, v/v) elution. The following R_f values were obtained: $\Delta 7$-sterols 0.33, $\Delta 5$-sterols 0.35, monomethyl sterols 0.45; and dimethyl sterols 0.53 (70,100).

Argentation TLC has been applied to separate free sterols according to their number and configuration of double bonds. Desmethyl and monomethyl 24-ethylidene sterols extracted from Solanaceae seed oils and derivatized to acetates were fractionated on silver nitrate impregnated silica plates with methylene chloride-carbon tetracholoride (1:5) elution (101). Akihisa and colleagues used this technique to separate acetylated desmethyl, monomethyl, and dimethyl sterols extracted from various plant materials prior to mass spectroscopy (MS) and nuclear magnetic resonance (NMR) identification (50,102,103). Sterols with different unsaturation in the ring structure can also be separated using argentation TLC. 7-Dehydrocholesterol, desmosterol, and cholesterol had relative mobilities of 0.29, 0.69, and 0.72 when silver nitrate silica plates were eluted with chloroform-acetone (85:15, v/v) (104). After visualization of phytosterol groups with nondestructive stains

(See Sec. II C), phytosterol fractions are transferred from silica and used for further analysis.

B.   Chromatographic Separation and Analysis of Phytosterols

1.   Analytical Gas Chromatography

Several GC methods are available for the analysis of phytosterols from food materials. Some of them are capable of separating only the few most common phytosterols, whereas others can also separate several dozens of minor compounds. Analysis of multicomponent foods containing substantial amounts of other nonsaponifiable lipids, especially tocopherols, needs effective GC separation (4). Earlier, coelution of α-tocopherol with sterols caused misleading results. The choice of method should be based on the research problem.

GC protocols of cholesterols can be adapted to the analysis of phytosterols, but a better resolution is needed due to the variety of phytosterols present in plant materials. Separations of phytosterols, started on apolar packed (3–8 mm i.d.) columns, are nowadays performed on capillary columns (0.1–0.32 mm i.d.) of apolar or intermediate polarity (4,6,90). Phytosterol samples are injected by various techniques including split, split/splitless, and on-column. The most cautious researchers use cold on-column injections that ensure greater stability of phytosterols, and that are more reliable for quantitative analysis and less discriminating than split injections (4). Runs are performed either isothermally or with temperature programming. Phytosterols are most commonly detected with a flame ionization detector or by MS. Flame ionization detection (FID) is a universal detector that has good sensitivity, response linearity, and generality, and MS is suitable for structural identification and quantitation (6).

Underivatized phytosterols are poorly volatile and relatively labile at high temperatures, which makes their analysis at high temperatures unreliable. However, advances in modern capillary column technologies have made it possible for some investigators to analyze phytosterols as free alcohols (e.g., 62,80,105). To improve volatilization, stability, and resolution on GC, phytosterols are commonly converted to their trimethylsilyl ether (TMS) or acetate derivatives. The derivates show less tailing on chromatograms than the free sterols. TMS derivatives are more suitable to quantitative analysis, and simpler and faster to produce than acetate derivatives (6).

N,O-Bis(trimethylsilyl)trifluoroacetamide (BSTFA) with trimethylchlorosilan or *tert*-butylmethyl ether are the most commonly used derivatization mixtures. Since sterols are poorly soluble in these mixtures, pyridine is added to the mixtures (4). *N*-Methyl-*n*-trimethylsilyl heptafluorobutyramide

(MSHBFA) with 1-methylimidazole and trimethylsilylimidazole are other silylating reagents used. In general, incubation times vary from 15 min to overnight and temperatures from room temperature to 80°C. In a German method, MSHBFA is recommended because it is suitable for routine work and completely derivatizes both phytosterols and the betulin that is used as internal standard (70). However, BSTFA is also suitable to derivatize betulin (65). The high fluorine content of MSHBFA also protects the detector and the use of pyridine as solvent can be omitted. BSTFA reagent gave better precision and recovery of phytosterols than Sylon HTP reagent (24). After derivatization with BSTFA, the mixture contains excess reagent and by-products that are volatile and may be removed by evaporating under a nitrogen stream. The residue is dissolved in, say, hexane prior to GC analysis (5,13,86,106). Provided that TMS derivatives are protected from moisture, they can be stored in derivatization reagent under refrigeration for several weeks (5) and in hexane at −20°C for 3 days (89,106). Acetate derivatives are produced by incubating in, for example, freshly prepared acetic anhydride-pyridine solution at 75–80°C for 1 h (91).

Separation of the major phytosterols in, say, edible oils and diet composites can be obtained using a nonpolar or low-polarity column. As an example, a 5% phenylmethylsilicone column (e.g., SPB-5, HP-5) with dimensions of 25–30 m × 0.25 mm i.d. and 0.25 μm film thickness separates phytosterol TMS in less than 30 min (24,107,108). The analysis time can be even shorter, since only 9 min was needed to separate the major free sterols from edible oils on a low-polarity column (5% cross-linked phenylmethyl-silicone; CP-sil-5CB, 10 m × 0.32 mm, 0.12 μm) with a 1-m deactivated pre-column (62). In this some minor compounds coeluted with others. Selectivity of columns can be improved by increasing their polarity. Sitostanyl and campestanyl TMS could be separated rather well from the respective phytosterols using an apolar methylsilicone column (OV-1; 25 m × 0.32 mm, 0.52 μm) (47), whereas a baseline separation was achieved using a medium-polarity 14% cyanopropylphenylmethylpolysilozane column (DB-1701; 30 m × 0.25 mm, 0.25 μm) (106). Separation of stanyl derivatives from steryl derivatives can also be obtained with longer low-polarity columns such as 5% cross-linked phenylmethylsilicone columns of 60 m (5,59).

GC analysis of pumpkin seed and other oils containing Δ7-sterols is more challenging than analysis of those containing mainly Δ5-sterols and cannot be performed using an apolar column. With a medium-polarity cross-linked 65% dimethyl–35% diphenylpolysiloxane column (HP-35; 30 m × 0.25 mm, 0.25 μm), four Δ7-steryl TMS derivatives were resolved from sitosteryl and 24-methylcholesteryl TMS (7,73). Another medium-polarity column (33% phenyldimethylpolysiloxane; OV-61; 25 m × 0.25 mm, 0.1 μm) was used to separate isomerization products of Δ7-sterols, e.g., Δ8(14)-sterol and Δ14-sterols, from sunflower oil sterols derivatized to their TMS ethers (72). Separation could also be obtained by a column of high polarity (100% cyanopropylsilicone; CP-sil-88, 100 m × 0.25 mm, 0.1 μm).

When a sample contains monomethyl and dimethyl sterols in addition to Δ5- and Δ7-sterols, they are preparatively fractionated using, for example, TLC prior to GC analysis. Thereafter, analysis of desmethyl, monomethyl, and dimethyl sterols is obtained on a regular apolar (11) or low-polarity column (61). Up to 25 TMS derivatives of various phytosterols from sea buckthorn berries could be separated on a 14% cyanopropylphenylmethyl-polysilozane column (DB-1701; 30 m × 0.25 mm, 0.25 μm) in 60 min without prior fractionation into des-, mono-, and dimethyl sterols (66).

Simultaneous analysis of phytosterols and tocopherols can be performed using capillary GC without the risk of coelution of cholesterol and tocopherols. Silylated phytosterols and tocopherols were well separated on a low-polarity column (DB-17HT, 13 m × 0.32 mm; HP-5MS; 30 m × 0.25 mm) (95,109).

2. Analytical High-Performance Liquid Chromatography

RP-HPLC is more selective for separating free sterol homologues and unsaturated analogues than NP-HPLC. In RP-HPLC separation is based on the hydrophobic interactions between the analytes and the stationary phase, which is influenced by molecular size and the number of double bonds, whereas in NP-HPLC separation is mainly based on the adsorptive interactions and polarity differences (6).

One of the earliest separations of free sterols was conducted using C6 column with gradient elution from 50% to 100% acetonitrile in water and UV detection at 200 nm (27). The method was used to separate free sterols extracted from oat leaves and seeds. Phytosterol samples were subjected to TLC purification prior to HPLC analysis, but no derivatization step was needed. Since some phytosterols coeluted in this system, e.g., Δ7-stigmastenol and stigmasterol, the results were confirmed by GC analysis. However, the authors concluded that this RP-HPLC method could be used as a rapid method to analyze phytosterols in foods. When the applicabilities of two RP-HPLC systems with UV detection at 206 nm were compared in the analysis of eight common phytosterols, a C18 column and 80–85% methanol in water as eluent was found to be better than a C8 column and 99% methanol in water (110). With the C18 column method, baseline separation of major phytosterols from saponified oil extracts and algae extracts could be obtained, whereas with the C8 column, no phytosterols coeluted. The selectivities of these two column systems for phytosterols were different, which was shown by the dissimilar elution orders of compounds. Since the chromatograms contained several unidentified peaks, there was a need to confirm phytosterol identification by GC analysis. Later, stigmasterol and sitosterol could be well separated using a narrow-bore C8 column, isocratic acetonitrile-water (86:14, v/v) elution, and UV absorption at 208 nm for detection (97). The RP-HPLC method was found to be sufficiently sensitive to detect these phytosterols at the concentrations present in soybean oil. Although intraday precision and interday reproducibility of standard compounds were good, and recoveries of added stigmasterol and sitosterol from oil matrix were also good at 1.08% and 0.97%, respectively, the overall chromatogram contained considerable amounts of other compounds that might interfere with the analysis.

Although simpler sample preparation is considered to be an important advantage of HPLC over GC analysis, it is recommended that samples should be saponified to remove glycerol esters that would otherwise contaminate the column and impair the separation when using RP-HPLC. This is more important with high-fat than with low-fat samples. Warner and Mounts found facile saponification of vegetable oil samples prior to RP-HPLC analysis to be necessary prior to phytosterol analysis but not before tocopherol analysis (88). They used a C18 column, methanol-water (98:2, v/v) elution, and evaporative light scattering detection (ELSD). In another study, RP-HPLC was used to simultaneously analyze phytosterols, cholesterol, and tocopherols in dairy and nondairy foods and oils. The samples were saponified before analysis. Here three major phytosterols and cholesterol could be determined separately, whereas two of the tocopherols coeluted with each other. The determination limit of the analytes was about 10 mg/kg sample (77).

Ergosterol present in, for examples, mushrooms and foods contaminated by molds has two double bonds ($\Delta 5,7$) in its sterol ring. Thus, is has UV absorption at 282 nm, which enables it to be specifically and sensitively detected by a UV detector. Schwadorf and Müller (81) developed a sensitive and reliable analysis method to study ergosterol in cereals and mixed feeds using NP-HPLC. Other phytosterols lacking the sterol ring structure do not interfere with the analysis. They used direct saponification without any further purification methods, which made the method very rapid to detect mold contamination.

3. Other Separation Techniques

Supercritical fluid chromatography (SFC) is a technique that is suitable for separating nonvolatile and thermolabile compounds using supercritical carbon dioxide as the mobile phase. Separation is achieved rapidly, costs are relatively low, and either UV detection at low wavelength (e.g., 190 nm) or FID can be used (6,90). Despite these benefits, SFC has only been used in free phytosterol analysis in a few studies. This may be because of its relatively poor separation power, which is lower than that of HPLC (6). However, Snyder and coworkers (57) successfully used SFC technique with a capillary column (Dionex SB-Octyl-50, 10 m x 0.1 mm, 0.5 μm) at 100°C and with a carbon dioxide pressure gradient to separate four major phytosterols of plant lipid mixtures extracted with SFE. Adequate separation of the phytosterols from each other and from other lipids present was obtained. Moreover, FID responses for phytosterols were linear over wide range.

Capillary electrochromatography (CEC) is another relatively new technique that has been applied to lipid analysis recently. CEC is driven by electroosmotic force, as is eletrophoresis, and is easily modified by the conditions used as in HPLC. It has better separation power than SFC and HPLC (6). Three major phytosterols of rapeseed oil were separated as 1-methylpyridinium 2-alkoxy *p*-toluenesulfonates. The derivative bore a positive charge and possessed UV absorption at 284 nm. Analysis was performed in a 77-cm silica capillary tube with methanol-acetonitrile-acetic acid buffer with methylated β-cyclodextrin (111). Abidi et al. (112) succeeded in baseline separating four sterols, i.e., lanosterol, ergosterol, stigmasterol, and sitosterol, using a 40-cm ODS column. They concluded that the side chain structure was the main determinant of the separation because the sterols eluted according to their side chain hydrophobicity.

4. Identification of Phytosterols Using Chromatography and Spectroscopy

GC combined with MS detection is a commonly used method to identify phytosterols. Electronic impact ionization (EI) produces positively charged molecular ions ($[M^+]$) that decompose to typical fragmentation ions. Ionization is most often carried out at an ionization voltage of 70 eV. Both TMS and acetate derivatives of phytosterols have typical fragmentation patterns that are valuable in determining their molecular weights and structural features. High ionization temperatures may modify fragmentation because hydroxyl groups at C-3 are labile and thus easily dehydrated. Excellent reviews written by Rahier and Benveniste (113) and Goad (15) are referred to when a detailed description of mass spectral identification of phytosterols is needed. In 1967, Knights (114) published GC-MS data of 15 sterols analyzed both as free alcohols and derivatized to TMS, acetates, and trifluoroacetates. The data included base peaks, most abundant peaks and fragment peaks above m/z 210. He also described formation of each ion by fragmentation. This section provides a short introduction to this area and an overview of the studies in which GC-MS has been used to confirm identification of phytosterols in food analysis.

Table 2 Example of Studies Where Food Phytosterols Have Been Identified by GC Retention and MS Fragmentation Patterns

Analyzed phytosterols: free or derivatized	Source of phytosterols	GC column	Ionization conditions	GC retention	Mass spectral data presented	Number of identified sterols	Ref.
Free							
	Sunflower oil	OV-240 (40 m × 0.25 mm i.d.)			MS spectra	6[a]	72
	Pineapple, passionfruit, orange and grapefruit juices	DB5-MS (12.5 m × 0.20 i.d., 0.33 µm)	70 eV	Retention time	Ions	7	48
	Jojoba oil	HP-1 (30 m × 0.25 mm i.d., 0.25 µm)		RRT to cholesterol	Ions	12	105
	Borage oil	CP-sil 8CP (30 m × 0.25 mm i.d., 0.25 µm)	70 eV, 200°C	RRT to α-cholestane	Ions	17	61
Trimethylsilyl ether							
	Sesame seed	BP5 (19 m × 0.33 mm i.d.)	70 eV, 250°C	RRT to α-cholestane	Ions	17	11
	Sunflower oil	OV-240 (40 m × 0.25 mm i.d.)			MS spectra	7[a]	72

					2[b]	47
Vegetable oils and cereal lipids	BP5 (25 m × 0.33 mm i.d., 0.5 μm)	70 eV, 250°C	RRT to α-cholestane	MS spectra and ions	2[b]	47
Oat and sunflower oil	DB-1701 (30 m × 0.25 mm i.d., 0.25 μm)	70 eV, 200°C	RRT to α-cholestane	MS spectra	8	106
Rice bran oil	SPB-5 (30 m × 0.25 mm i.d., 0.1 μm)	70 eV, 280°C		Ions	10	49
Sea buckthorn berries	DB-1701 (30 m × 0.25 mm i.d., 0.25 μm)	70 eV, 180°C	RRT to cholesterol	Ions	25	66
Blueberry and lingonberry	DB-1701 (30 m × 0.25 mm i.d., 0.25 μm)	70 eV, 180°C	RRT to cholesterol	Ions	10	37
Ginseng seed oil	DB-5 (60 m × 0.32 mm i.d., 0.25 μm)	70 eV	RRT to α-cholestane	Ions	12	12
Sterculiaceae seed	MEGA SE54 (25 m × 0.32 mm i.d., 0.15 μm)	70 eV, 300°C		Ions	5	13

RRT, relative retention time.
a Δ7-sterols and their isomerization products.
b stanols.

In general, free Δ5-sterols produce intensive M^+ that may be base peaks. Molecular ions of TMS and acetate derivatives are most commonly desilylated or deacetylated (ROH) and show a strong $[M^+\text{-ROH}]$ ion (114). In mass spectra of phytosteryl TMS, the $[M^+]$ exist, whereas they are absent in those of acetates (113). There are some characteristic ions that can be used to differentiate Δ5-sterols, Δ7-sterols, and stanols from each other. Δ5-Sterols produce typical ions at $[M^+\text{-}129]$, when a part of the A ring is lost; Δ7-sterols always have an intensive peak at m/z 255 that is a fragment of the four-ring structure; and stanols have a base peak at m/z 215 that is formed after removal of the side chain and part of the D ring. The structure of side chains can be concluded from ions when the whole side chain (SC) or the SC and the ROH are cleaved from M^+. Δ5-Sterols lose water more easily than methyl groups, whereas loss of water seldom occurs with Δ7-sterols (72). Acetate derivatives are suitable to detect unsaturation in the SC because fragmentation occurs more often in the SC if it contained an unsaturation, whereas the ring structure was disrupted with a saturated SC (91).

Several studies have produced characteristic mass fragmentation patterns and GC retention data of phytosterols present in foods. Some of the most recent ones are presented in Table 2. More mass spectra of phytosterols are also available in various MS libraries.

Although GC-MS has been an effective tool to explain phytosterol structures, other techniques are needed to determine, for example, the location of double bonds in the ring, the pattern of side chain substitution, and configurations of epimers at C-24 (15). NMR spectroscopy has resolved many such problems. Both ^{1}H NMR and ^{13}C NMR spectra have been used for structural analysis of phytosterols (115). Improvements in RP-HPLC techniques have made it a powerful tool for the isolation of individual sterols prior to NMR. Thus NMR spectral data in combination with retention time data from various GC separations and mass spectral data from GC-MS enables identification and structure elucidation of food phytosterols (115). For example, tables of NMR chemical shifts of phytosterols in general (15) and in Curcurbitaeac (102), and of dimethyl sterols (103), have been published.

C. Quantitative Analysis of Phytosterols

To perform quantitative phytosterol analysis, all steps in the procedure including sample preparation and final chromatographic analysis must be carefully planned and validated. All extraction steps should be repeated many times for high recovery, and losses must be avoided. The most common means to control the procedure is to add an internal standard to compensate for losses.

Recovery studies of a large range of phytosterols cannot be performed because only a few phytosterols are commercially available with reasonable purity and amounts. Therefore, many method performance studies have been conducted using cholesterol and some of the most easily available phytosterols, namely, stigmasterol and sitostanol.

In general, phytosterols are quantified during the chromatographic analysis. In some studies, phytosterols were quantitated spectrophotometrically although qualitative separation by GC was performed (22).

1. Internal Standard

A European study comparing methods for cholesterol analysis pointed out that the internal standards were not added sufficiently early to control losses during analysis of edible oils (65). It is even more important to add an internal standard early to a food sample that is subjected to several hydrolysis, extraction, and purification treatments and not only in the final chromatographic analysis.

At present, only a few internal standards are commonly used (Table 3). They are added to sample mixtures at various steps during the analytical procedure. Two obvious reasons for differences in addition steps are the internal standards' chemical dissimilarity with the phytosterols and their varying stabilities during the procedure. Cholesterol and its corresponding stanol cholestanol are chemically most similar to phytosterols, containing 3β-hydroxyl groups in the sterol structure, which makes their "fate" during various sample preparation steps very similar to that of phytosterols. It should be remembered that cholesterol may also be present in small amounts in plant-based materials and thus should be compensated for when used as an internal standard. Cholestanol would be a better choice for an internal standard because its content in foods may be neglected. It could also be used when samples include animal-based materials and contain cholesterol. However, high-resolution GC is needed to separate cholesterol from cholestanol. The C3 epimer of cholesterol, namely, epicholesterol, has also been used as an internal standard for phytosterol analysis (116). Unfortunately, epicholesterol is labile toward acid hydrolysis.

Cholestane, having the same sterol ring structure as cholesterol but lacking the β-hydroxyl group, is also a commonly used internal standard. It is well separated from cholesterol and phytosterols under any GC conditions, but it also behaves differently in the sample preparation steps. Being less polar than phytosterols it does not coelute with sterols in preparative chromatographic methods. Furthermore, it is not derivatized as phytosterols are because it does not have a hydroxyl group. Betulin (Lup-20[29]-

Table 3 Internal Standards (ISTDs) used for Quantitative GC Analysis

Compound	Sample	Addition of ISTD	Major steps after ISTD addition	Derivatization	Ref.
Betulin	Total diet	After saponification	TLC	Silylation	82
	Oils and fats	Prior to saponification	Saponification, Al_2O_3 and TLC purification	Silylation with MSHFBA	70
	Oils	Prior to saponification	Saponification	Silylation with BSTFA/TMCS	108
Cholestane	Plant tissue	Prior to homogenization	Acid and alkaline hydrolysis	Silylation[a]	42
	Mixed diet	Prior to acid hydrolysis	Acid and alkaline hydrolysis	Silylation with dimethylformamide/bis-silyl-trifluoroacetamide[a]	71
	Cooked dishes	Prior to saponification	Saponification	Silylation with BSTFA/TMCS[a]	21
	Oil and meat	Prior to saponification	Saponification	Silylation with BSTFA/TMCS[a]	18
	Oil	Prior to saponification	Saponification	Silylation with Tri-Sil reagent[a]	12
	Crude oil of seeds	Prior to saponification	Saponification	Silylation with BSTFA/TMCS[a]	109
Dihydrocholesterol	Ground pumpkin seed	Prior to extraction	Lipid extraction, silica chromatography, alkaline hydrolysis	Silylation with BSTFA in TBME	73

ISTD, internal standard; MSHFBA, N-methyl-N-(trimethylsilyl)heptafluorobutyramide; BSTFA/TMCS, N,O-bis(trimethylsilyl)trifluoroacetamide with 1% trimethylchlorosilane; TBME, tert-butylmethyl ether.
[a] ISTD is not derivatized.

ene-3β,28-diol), the fourth common internal standard, contains two hydroxyl groups and is thus more polar than phytosterols. Both its chromatographic and solubilization characteristics differ from those of phytosterols. Betulin is not soluble in nonpolar hydrocarbons such as hexane, but needs a more polar solvent such as di-isopropyl ether or chloroform. Differences in TLC behavior of sterols and the internal standard were demonstrated and overcome in a German phytosterol analysis method. Since betulin had a smaller relative retention factor (0.3) than that of sterols (0.33–0.53), all the bands were collected for further analysis (68,70,100). Moreover, we have found that betulin is degraded more under acid hydrolysis than phytosterols, which makes it an unsuitable internal standard when acid hydrolysis is used.

2. Chromatographic Responses of Phytosterols

GC combined with FID is considered to provide a general and linear response for similar compounds provided that chromatographic parameters are optimized. Experimentally derived response factors should be determined to compensate for differences in responses in FID and losses during the GC procedure, and have frequently been determined for cholesterol analysis (4). However, it is not possible to measure response factors for all phytosterols because only a limited number of compounds are commercially available at relatively high purity. To calculate the actual purity of a sterol standard, the normalization method, i.e., assuming that all sterols present give an equal response on the chromatogram, has been recommended (117). However, the sterol standards may contain solvents, water, or other compounds that cannot be detected by GC. For example, sitosterol crystals may exist as anhydrate, semihydrate, and monohydrate (118), which causes variation in the actual sterol concentration of the standard.

Responses of phytosterols have been determined in relation to the internal standards. Although α-cholestane has no and betulin two hydroxyl groups that may be derivatized, the TMS derivate does not contribute to the FID response (70). Thus the choice of internal standard should not have a profound effect on the relative responses. In one study comparing the responses of cholesterol and betulin, the responses were found to be linear, repeatable, and almost equal (1.028 ± 0.056, $N = 21$) (65), whereas in another the responses changed from day to day (79). Thus, in a European method to detect sitosterol or stigmasterol in butter, a standard mixture of the phytosterol and betulin (the internal standard) is analyzed before and after each sampling (117). However, response factors of betulin and sterols were found to be identical within the limits of experimental error if the overall procedure was carefully followed and betulin was completely derivatized to TMS (70). Response factors relative to the internal standard used in quantitative analysis of all phytosterols may be based on, for example, cholesterol (108), stigmasterol (95), or sitosterol (66,107). Sitosterol and stigmasterol calibrations were performed daily to analyze phytosterols in vegetable oils (18). Standard response curves of four phytosterols on six levels have also been determined and the responses for other phytosterols were taken from the standard compound with the closest number of total carbon atoms (38,86). Very good linear detection was obtained for cholesterol and stigmasterol between 0.05 and 0.5 mg/sample (24), for campesterol and sitosterol between 5 and 300 µg/mL (18), and for four phytosterols between 1 and 200 µg (86). Many investigators use equal responses of phytosterols to α-cholestane (12, 22,47,119), which is relevant based on the good performance data of GC-FID described above. It is recommended to check from time to time that the response factors are close to 1.00. In our laboratory, on a daily basis we analyze a sample including equal amounts of cholesterol, cholestanol, and stigmasterol to verify the separation and detection efficacy of our chromatographic system (86).

In one study in which sterols were analyzed as acetate derivatives, the responses of six sterols were linear over a range of 2–15 ng, and absolute responses were so similar that no correction factors were used (91).

There are several possible HPLC detection methods for phytosterols. UV and ESLD are the ones used most often (6,90). Since phytosterols do not have any specific chromophores in their structures, absorption at 190–210 nm is measured by UV. At these wavelengths, mainly double bonds between carbon atoms absorb light and, thus, molar absorptivities of different phytosterols vary notably.

UV detection at λ = 210–214 nm was used to analyze major phytosterols and cholesterol from food samples with an external standard method. The calibration curve for cholesterol was linear over two orders of magnitude and was also used for quantitation of phytosterols (77). Later, stigmasterol and sitosterol were detected by UV absorption at 208 nm (97). HPLC-UV linearity of stigmasterol and sitosterol were excellent ($\geq$0.999 and 1.000) at concentration levels of 0.7–70 µg/mL, and slopes in calibration fittings were 0.102 and 0.089, respectively. The limit of detection, 0.5 µg/mL, was low enough to enable determination of phytosterols at the levels present in edible oils.

ELSD gives a very stable baseline and is insensitive to changes in solvent composition during runs once the operating conditions have been optimized. Calibration of each compound is needed for quantification because responses are highly dependent on molecular mass (6,56,90). Phytosterols in vegetable oils were analyzed by HPLC-ELSD (88). Calibration curves for four phytosterols were measured and found to be linear over the range (10–100 µg) studied, but detector responses varied remarkably. Calibration curve slopes for brassicasterol and stigmasterol were at about the same level and were about 10% that of sitosterol, whereas that for campesterol was in-between. Quantitative analysis of sitosterol and stigmasterol from oil samples has also been performed using atomic pressure chemical ionization (APCI) mass spectrometry, which in addition to quantitative data also provides identification data of the compounds studied (97).

These few exsamples show that quantitative analysis of phytosterols is more reliably performed by GC-FID than by HPLC, especially when a pure reference standard is not available for accurate calibration.

IV. REFERENCE METHODS AND MATERIALS

Before a phytosterol analysis method is applied, it must be validated. The sample preparation procedure should be efficient enough to isolate all sterols, to liberate them from conjugates and purify the extracts when necessary without losses, and, finally, to analyze them chromatographically (1). One approach is to compare the method to a reference method that has been validated by international or national authorities. Another is to use reference samples of which the phytosterol contents have been confirmed. Unfortunately, neither of these is directly applicable for the analysis of complex food materials because reference methods and materials are available only for oil and fat samples. Traditionally there has been no major interest in validating phytosterol analysis because phytosterols are neither nutrients nor harmful compounds. Reference methods for analysis of cholesterol in foods have been applied to phytosterol analysis. It should be remembered that plant materials contain numerous compounds that might interfere in chromatographic quantitation.

The International Union of Pure and Applied Chemistry (75) published a standard method to identify and determine sterols in animal and vegetable oils and fats. This method did not include addition of internal standard, and quantitive analysis was poorly explained. In a European method for analysis of sterols of fatty substances (64), cholestanol as the internal standard was added to sample before saponification, which enabled quantitative analysis. The method included preparative separation of sterols on basic silica TLC plates, derivatization to TMS ethers, and analysis by capillary GC. Responses of phytosterols were assumed to be equal to that of cholestanol. The German sterol method (70,100) is more complex than the above-mentioned methods, including several purification steps. Betulin, the internal standard, is added to a fat or oil sample prior to saponification. The unsaponifiable lipids are isolated by aluminum oxide column chromatography and silica gel TLC. Sterols are derivatized to TMS and analyzed by GC using an internal standard method without correction factors for detector responses. All the steps in the method are thoroughly explained and discussed. Another European reference method was published in 2001 to analyze sitosterol or stigmasterol in butter (120). In this method, after betulin has been added as the internal standard to the saponification mixture, sterols are extracted and derivatized to TMS prior to GC. Standard solutions are analyzed before and after each sampling and the response factors derived from them are used in calculating the sterol contents of butter samples. The Association of Official Analytical Chemists has published two official methods to analyze cholesterol in multi-component foods. The first one included a separate lipid extraction step (76), whereas the second was based on direct saponification (121).

At present, only two reference materials for phytosterols are available. They are both supplied by the BCR. One of them is a mixture of soybean and maize oils (BCR 162), and the other is butter oil enriched with phytosterols (BCR 164).

Validation of a phytosterol analysis method should be performed using recovery studies of added phytosterols and conjugates, as well as studies on repeatability and reproducibility. Moreover, the performance of the method should be followed using appropriate in-house reference samples, test mixtures, and blanks; in addition, especially when applied to new types of foods the whole method should be evaluated.

V. CHROMATOGRAPHIC ANALYSIS OF INTACT PHYTOSTEROL CONJUGATES

Oils, fats, and other plant-based foods contain various phytosterol conjugate classes, namely, free alcohols (FSs), esterified fatty acids (SEs) or phenolic acids (SPHEs), glycosides (SGs), and acylated steryl glycosides (ASGs). Esters with fatty acids are commonly found in various food materials, whereas esters with phenolic acids (mainly ferulic acid), occur, for example, in some bran factions of cereals. Chemical properties and physiological effects of the phytosterols classes differ substantially from each other. Thus, there is a need to characterize in which conjugates phytosterols in food exist. There are numerous means to separate different phytosterol classes, as reviewed by Moreau and coworkers (2). Moreover, several studies have been published in which phytosterol conjugates from food samples have been separated (2,6). Most of them have been performed using NP-HPLC.

A. Preparative Separation of Phytosterol Conjugates

Preparative separation is needed when phytosterol contents and compositions of each class are to be studied separately so as to isolate them from other compounds in the lipid extract to be studied as intact molecules. In vegetable oils and fats, phytosterols occur mainly as FSs and SEs, whereas in other materials SGs, ASGs and SPHEs are also present.

It is relatively easy to separate FSs and SEs from each other by adsorption chromatography because they have different polarities. Earlier, phytosterol conjugates were separated using open column chromatography. When silica was used as adsorbent and vegetable oil in hexane was applied to the column, SE were eluted with 2% ethyl acetate in hexane followed by triacylglycerols eluted with 10% ethyl acetate in hexane and finally FS that were eluted with 100% ethyl acetate (122). To compensate for losses occurring during chromatography, cholesterol and cholesteryl heneicosanate were added as internal standards. Later, two-step open column chromatography was used to separate quantitatively FSs and SEs (108). Vegetable oil was dissolved in hexane and applied on a silica column from which SEs were eluted with hexane-ethyl acetate (90:10, v/v) and FSs with hexane-diethyl ether-ethanol (25:25:50, by vol). Model sterol compounds were used to confirm separation selectivity.

Similarly, preparative separation of phytosterol classes can be obtained by SPE. For example, silica cartridges were used to purify steryl esters from 1- to 20-mg total lipids. Steryl esters were eluted from silica with light petroleum-diethyl ether (98.5:1.5, v/v) (28). A small-scale separation of FSs and SEs from vegetable oils can also be performed by SPEs using neutral alumina cartridges (116). Aliquots of 100–200 mg of sample oils were dissolved in hexane together with cholesterol and cholesteryl ester as internal standards. SEs were collected by eluting the cartridge with hexane and hexane-diethyl ether (80:20, v/v). FSs were eluted with ethanol-hexane-diethyl ether (50:25:25, by vol). The SPE separation procedure was efficient and caused no losses because recoveries of added markers for free and esterified sterols were 100% and 105%, respectively, and phytosterol contents of vegetable oil samples were closer when analyzed by SPE separation than by direct analysis.

Separation of lipids in gram scale should be performed in open columns. Lipids extracted from red bell pepper were separated into neutral lipids, glycolipids, and phospholipids using silica gel column chromatography. These three lipid classes were eluted from the column with chloroform, acetone, and methanol, respectively (31). Column chromatography on alumina was used for isolation from 0.1–2 g of lipid extracts or oils (28).

Purity of the phytosterol classes may easily be confirmed on silica gel TLC plates. Silica gel TLC plates (20 cm × 20 cm) were developed first with isopropyl ether-acetic acid (96:4, v/v) for 10 cm and then with diethyl ether-hexane-acetic acid (90:10:1, v/v/v). In this system, the relative mobilities of steryl esters were 50% higher than those of free sterols (123). Four phytosterol classes from plant material could be separated on silica gel TLC with dichloromethane-methanol-water (85:15:0.5, v/v/v) as developing solvent, yielding relative retentions of 0.35 for SG, 0.55 for ASG, 0.7 for FS, and 0.9 for SE (45). TLC has also been used as a preparative means to isolate steryl esters (28).

A multistep sample preparation method was developed and validated to analyze intact Δ5- and Δ7-steryl glycosides in plant materials (73). Steryl glycosides were isolated from lipid extracts using silica gel chromatography, and purified by dissolving the residue in 1,2-dimethoxyethane containing 10% ethylamine and applied on a phenylboronic acid SPE cartridge. After washing the column, SGs were eluted with methanol containing sorbitol (1 mmol/3mL). The eluate was dried and dissolved in an appropriate solvent prior to HPLC analysis. Moreover, the authors used a two-step chromatographic procedure to isolate SGs from a saponification mixture of pumpkin seed lipids to be used as reference compounds.

Phytosterols esterified with phenolic acids, namely ferulates, were fractionated from rice bran extracts in silica column using hexane and diethyl ether or ethyl acetate as eluent mixtures. Phytosteryl ferulates were isolated using 100% diethyl ether (124) and 50% ethyl acetate in hexane (50) after sequential elution of other compounds with increasing polarity.

B. Analysis of Lipid Profiles

Numerous preparative TLC procedures, some of which were described above, have traditionally been used to qualitatively separate various lipid classes. Nowadays, total lipid profiles from extracts are commonly analyzed by NP HPLC procedures. HPLC separation has become more popular since ELSD was introduced in the mid-1980s. With qualitative analysis the shortcomings of unequal UV detector responses are avoided. Various adsorbents and elution solvents are used (2,6).

Lipids of soft plant tissues were separated into 23 classes using a silica column (LiChrosorb Si-60, 100 × 3 mm, 7 μm) and a ternary gradient system consisting of iso-octane-tetrahydrofuran mixture, isopropanol, and water (125). FSs, ASGs, and SGs were obtained as separate peaks, whereas SEs coeluted with carotenoids. When nonstarch and starch lipids from wheat flour were analyzed using a related method, 17 and 7 lipid classes were separated, respectively. SEs coeluted with triacylglycerols (56). In both studies, a good separation of lipid classes with a wide range of polarities could be obtained, but the separation and equilibrium of the HPLC system took 80 min. A somewhat faster separation of lipid classes extracted from corn fiber was performed on a diol column (LiChrosorb Diol, 100 x 3 mm, 5 μm) with a binary gradient elution (hexane-acetic acid and isopropanol) (8). In this system, SEs were resolved from triacylglycerols, free fatty acids, FSs, and

SPHEs. At least 10 peaks were detected from a tomato lipid extract after separation on a silica column (LiChrospher Si-60, 125 × 4 mm, 5 μm) with a gradient elution (chloroform and methanol-water (95:5, v/v) (44). Alumina (LiChroCART, 125 × 4 mm, 5 μm) with a linear binary gradient elution (0.5% tetrahydrofuran in hexane and tetrahydrofuran-isopropanol-hexane (20:20:60, by vol)) separated the nonpolar lipids rapidly (about 10 min) and efficiently, resulting in excellent resolution of SEs from wax esters and triacylglycerols (126). Separation of nonpolar lipid classes was also achieved with an alumina column and a binary gradient system at 75°C (127). In all these HPLC systems, ELSD was used as a general detector. However, detector responses between lipid classes varied. Steryl glycosides gave a threefold response compared to acylated steryl glycosides (44).

SFC has also been used to separate lipid classes from each other. One of the advantages of SFC over HPLC is that FID can be used to detect compounds after SFC. An SPC unit with a capillary column (Dionex SB-Octyl-50, 10 m × 0.1 mm, 0.5 μm) and carbon dioxide as a carrier gas was used to separate FSs and SEs from free fatty acids and triacylglycerols of vegetable oil deodorizer distillates (128).

C. Analysis of Specific Phytosterol Conjugate Classes

1. Phytosterols Esterified with Fatty Acids

Phytosterols esterified with fatty acids (SEs) are thought to function as storage compounds. They are studied to understand the biochemistry of phytosterols in plants and to identify the authenticity of edible oils, among other reasons. Thus, analytical methods to study these compounds are needed. Attempts have been made to resolve these complex mixtures by HPLC, GC, and their combinations and by preparative chromatographic techniques.

To identify natural SE in vegetable oils and to make assumptions of SE structures, an RP-HPLC method with UV detection at λ = 205 nm was developed (129). Relative retention volumes of synthetized SEs together with an internal standard cholesteryl oleate were determined before the contributions of unsaturation and carbon atom numbers were calculated. Chemical structure of the fatty acid moiety had a stronger effect on retention behavior than the sterol moiety, which means that the esters of a specific fatty acid were resolved to a set of partly overlapping peaks containing various phytosterol moieties.

To improve separation, SEs were first fractionated into esters bearing short-chain and long-chain fatty acids by adsorption HPLC (28). Unsaturation of the fatty acyl moieties was further characterized by preparative and analytical argentation chromatography. Esters with saturated fatty acyls with at least eight carbon atoms had the highest mobilities ($R_f = 0.75$–0.8), followed by monounsaturated ($R_f = 0.65$), diunsaturated ($R_f = 0.55$), triunsaturated ($R_f = 0.35$), and tetraunsaturated esters ($R_f = 0.20$), on silver nitrate–impregnated silica plates eluted with hexane-toluene (7:3, v/v). Thereafter, the fractions were subjected to RP-HPLC or GC-MS used in negative ion chemical ionization mode with ammonia as the reagent gas. The MS system used produced specific fragment ions for sterol and fatty acid moieties.

Later, the separation efficiency of both GC and HPLC has been improved. Up to 25 intact SE molecules were at least partly separated by high-temperature GC on a medium-polarity column consisting of 50% phenyl- and 50% methylpolysiloxane (25 m × 0.25 mm, 0.1 μm). Injection was performed by cold on-column technique to avoid degradation and sample partitioning (130). The same samples were subjected to RP-HPLC–mass detection systems consisting of an octadecyl-bonded silica column (Spherisorb ODS-2, 100 x 3 mm, 0.5 μm) with gradient elution of acetonitrile, water, and ethyl acetate. With HPLC, about 16 peaks were detected. When combining data from these two analyses, SE profiles of vegetable oils could be resolved because in GC the retention times increased with degree of unsaturation whereas in HPLC the effect was opposite.

Improvements in techniques to combine HPLC and GC instruments made it possible to analyze SE using on-line HPLC-GC without tedious preparative chromatographic procedures. NP-HPLC with, for example, cyanopropyl silica or silica column is used to remove triacyl glycerols and other interfering compounds prior to high-temperature GC. Oil samples have usually been derivatized to piconyl esters (131) or TMS ethers (e.g., 132,133) so as to be able to include free sterols in the analysis. With on-line HPLC-GC, reproducible results for free sterols are obtained, whereas esterified sterols may partly be degraded during GC due to high temperatures (131). When transfer temperatures and efficiencies were carefully optimized, the method was shown to be applicable to qualitative and quantitative analysis of FSs and SEs in vegetable oil methyl esters (133). Thermal degradation of SEs during GC requires that correction factors for esters should be determined separately from those for free sterols. A calibration function for cholesteryl stearate in reference to betulin was measured as 1.65 ± 0.03 (133). When cholesteryl laurate was used as an internal standard, a relative response factor of 1.0 was used for all SEs (134).

Vegetable oil SEs from steryl ester fractions were analyzed by direct electrospray tandem mass spectrometry with a triple quadrupole LS-MS-MS instrument (135). The sample was dissolved in chloroform-methanol containing ammonium acetate, and precursor ion scans in positive ion mode were measured. SEs were identified and quantified based on $[M + NH_4]^+$

ions. The system was carefully calibrated with 11 cholesteryl esters from which molar correction factors for phytosteryl esters were calculated. Molar proportions of SEs were confirmed by separate phytosterol analysis by GC after saponification. After SEs were separated on a high-temperature GC column (DB-5ht; 15 m × 0.25 mm, 0.1 μm) using cold on-column injection, MS was used to confirm the identities of synthetized esters of phytosterols (134). The authors used MS in positive chemical ionization mode with ammonia as the reagent, because they found this to be the most effective means to generate intensive $[M + NH_4]^+$ adduct ions and two fragments corresponding to losses of fatty acid with and without water.

2. Phytosterols Esterified with Phenolic Acids

The two most common phytosterols esterified with phenolic acids (SPHEs) are *trans*-ferulic acid and *trans*-coumaric acid. Ferulates and coumarates are present in the outer layers of many cereals and are important phytosterol constitutes in bran fractions and, especially, in oils from rice and corn bran (1,2). A mixture of phytosteryl ferulates called γ-oryzanol is commercially extracted from rice bran. Their analysis is most commonly performed by RP-HPLC, in which separation is based on differences in molecular mass and hydrophobicity. The phenolic acid moieties of phytosterol ferulates and coumarates absorb light at 290–340 nm, which makes it possible to monitor them with a UV detector. The ferulates and coumarates can be distinguished by their slightly different UV spectra (52). There are several RP-HPLC systems to analyze ferulates and coumarates derived from cereals (e.g. 52,124). Up to 10 conjugates were separated from rice and corn extracts using a C18 RP column (Deltabond ODS, 250 x 4.6 mm, 5 μm) and acetonitrile-*n*-butanol-acetic acid-water (82:3:2:13, by vol) as the solvent system (136). Phytosteryl ferulates of γ-oryzanol have been separated and their phytosteryl moieties identified by GC-MS after hydrolysis (49). Later, several structures have been identified by UV, MS, and ^{1}H NMR spectroscopic data, NP- and RP-HPLC retention data, and by comparison with synthesized compounds (50,137). Steryl ferulates from cereals have also been identified by HPLC-MS (53).

3. Phytosteryl Glycosides

Phytosteryl glycosides (SGs), also referred to as sterolins, are widely distributed in edible plants and are important constituents in our diet. For example, the sum of SGs and ASGs of wheat flour and pea were 31.0 and 42.6 mg/100 g dry product, respectively (44). SGs of toxic cycad seeds may also be associated with neurodegenerative disorders (51). Analysis of intact

SG is performed by RP-HPLC after preparative fractionation to glycoside class by silica chromatography (e.g., (27). UV and ELSD have been used to monitor the eluting compounds. Fatty acid moieties may be liberated from ASGs by mild saponification to improve separation on RP-HPLC (27,73), or ASG may also be analyzed intact (31). Hitherto glucose has been the only carbohydrate moiety in phytosteryl glycosides. When SGs of oat leaves and seeds were separated on a C6 column (Sperisorb hexyl 125 × 4.6 mm, 5 µm) with water and acetonitrile gradient elution, five peaks were identified, two of which contained two compounds (27). Later, 10 SGs were separated and identified from synthesized soya and pumpkin seeds (73). Molecular species of SGs and ASGs extracted from red bell pepper were analyzed by RP HPLC combined with APCI-MS and confirmed by GC analysis of the hydrolyzed phytosterol and fatty acid moieties (31). Campesteryl and sitosteryl glucosides were the two SGs identified. In AGSs, campesteryl and sitosteryl glucosides were esterified with palmitic, stearic, linoleic, and linolenic acids. In another study, glucosides of campesterol or dihydrobrassicasterol, sitosterol, and stigmasterol were identified by MS and NMR spectroscopy (51).

VI. CONCLUDING REMARKS

When analyzing phytosterol contents in foods both the sample preparation and the analytical method should be carefully evaluated. There are a number of routine methods available to quantitatively isolate phytosterols from various food matrices, to liberate free sterols from their conjugates, to purify the extract, and, finally, to determine free phtyosterols. When intact phytosterol conjugates are studied, preparative separation of conjugate classes should be performed before intact molecules or free phytosterols can be analyzed. Instead of general procudures to analyze intact phytosterol classes in foods, there are only a few specific applications reported. In the future, more efforts should be focused on developing validated and automatized sample preparation and analytical methods for routine work as well as novel methods for specific needs.

REFERENCES

1. V Piironen, D Lindsay, TA Miettinen, J Toivo, A-M Lampi. Plant sterols: biosynthesis, biological function and their importance to human nutrition. J Sci Food Agric 80:939–966, 2000.
2. RA Moreau, BD Whitaker, KB Hicks. Phytosterols, phytostanols, and their

conjugates in foods: structural diversity, quantitative analysis, and health-promoting uses. Progr Lipid Res 41:457–500, 2002.
3. RE Ostlund Jr. Phytosterols in human nutrition. Annu Rev Nutr 22:533–549, 2002.
4. M Fenton. Chromatographic separation of cholesterol in foods. J Chromatogr 624:369–388, 1992.
5. RH Thompson, GV Merola. A simplified alternative to the AOAC official method for cholesterol in multicomponent food. JAOAC Int 76:1057–1068, 1993.
6. SL Abidi. Chromatographic analysis of plant sterols in foods and vegetable oils. J Chromatogr 935:173–201, 2001.
7. A Mandl, G Reich, W Lindner. Detection of adulteration of pumpkin seed oil by analysis of content and composition of specific delta-7-phytosterols. Eur Food ResTechnol 209:400–406, 1999.
8. RA Moreau, MJ Powell, KB Hicks. Extraction and quantitative analysis of oil from commercial corn fiber. J Agric Food Chem 44:2149–2154, 1996.
9. WR Morrison, SL Tan, KD Hargin. Methods for the quantitative analysis of lipids in cereal grains and similar tissues. J Sci Food Agric 31:329–340, 1980.
10. AU Osagie, M Kates. Lipid composition of millet (*Pennisetum americanum*) seeds. Lipids 19:958–965, 1984.
11. A Kamal-Eldin, L-Å Appelqvist, G Yousif, GM Iskander. Seed lipids of *Sesamum indicum* and related wild species in Sudan. The sterols. J Sci Food Agric 59:327–334, 1992.
12. THJ Beveridge, TSC Li, JCG Drover. Phytosterol content in American ginseng seed oil. J Agric Food Chem 50:744–750, 2002.
13. B Bruni, A Medici, A Guerrini, S Scalia, F Poli, C Romangnoli, M Muzzoli, G Sacchetti. Tocopherol, fatty acids and sterol distributions in wild Ecuadorian *Theobroma subincanum* (Sterculiaceae) seeds. Food Chem 77:337–341, 2002.
14. TD Drumm, JI Gray, GL Hosfield. Variability in the major lipid components of four market classes of dry edible beans. J Sci Food Agric 50:485–497, 1990.
15. LJ Goad. Phytosterols. In: BV Charlwood, DV Banthorpe, eds. Methods in Plant Biochemistry, Vol 7, Terpenoids. San Diego: Academic Press, 1991, pp. 369–434.
16. A Johansson. The content and composition of sterols and sterol esters in sunflower and poppy seed oils. Lipids 14:285–291, 1979.
17. C Cocito, C Delfini. Simultaneous determination by GC of free and combined fatty acids and sterols in grape musts and yeasts as silanized compounds. Food Chem 50:297–305, 1994.
18. C Vlahakis, J Hazebroek. Phytosterol accumulation in canola, sunflower, and soybean oils: effects of genetics, planting location, and temperature. J Am Oil Chem Soc 77:49–53, 2000.
19. T Akihisa, P Ghosh, S Thakur, FU Rosenstein, T Matsumoto. Sterol compositions of seeds and mature plants of family Cucurbitaceae. J Am Oil Chem Soc 63:653–658, 1986.

20. YF Lozano, CD Mayer, C Bannon, EM Gaydou. Unsaponifiable matter, total sterol and tocopherol contents of avocado oil varieties. J Am Oil Chem Soc 70:561–565, 1993.

21. MS Valdenebro, M León-Camacho, F Pablos, AG González, MJ Martin. Determination of the arabica/robusta composition of roasted coffee according to their sterolic content. Analyst 124:999–1022, 1999.

22. MD Zlatanov. Lipid composition of Bulgarian chokeberry, black currant and rose hip seed oils. J Sci Food Agric 79:1620–1624, 1999.

23. K Staphylakis, D Gegiou. Free, esterified and glucosidic sterols in cocoa butter. Lipids 20:723–728, 1985.

24. M Rodriguez-Palmero, S dela Presa-Owens, AI Castellote-Bargallo, MC Lopez Sabater, M Rivero-Urgell, MC de la Torre-Boronat. Determination of sterol content in different food samples by capillary gas chromatography. J Chromatogr 672:267–272, 1994.

25. J Folch, M Lees, GH Sloane-Stanley. A simple method for the isolation and purification of total lipids from animal tissues. J Biol Chem 226:497–509, 1957.

26. EG Bligh, WJ Dyer. A rapid method of total lipid extraction and purification. Can J Biochem Physiol 37:911–917, 1959.

27. J Kesselmeier, W Eichenberger, B Urban. High performance liquid chromatography of molecular species from free sterols and sterylglycosides isolated from oat leaves and seeds. Plant Cell Physiol 26:463–471, 1985.

28. RP Evershed, VL Male, LJ Goad. Strategy for the analysis of steryl esters from plant and animal tissues. J Chromatogr 400:187–205, 1987.

29. BD Whitaker, CY Wang. Effect of paclobutrazol and chilling on leaf membrane lipids in cucumber seedlings. Physiol Plant 70:404–411, 1987.

30. BD Whitaker, WR Lusby. Steryl lipid content and composition in bell pepper fruit at three stages of ripening. J Am Soc Hort Sci 114:648–651, 1989.

31. R Yamauchi, K Aizawa, T Inakuma, K Kato. Analysis of molecular species of glycolipids in fruit pastes of red bell pepper (*Capsicum annuum* L.) by high-performance liquid chromatography–mass spectrometry. J Agric Food Chem 49:622–627, 2001.

32. L Dyas, MC Prescott, RP Evershed, LJ Goad. Steryl esters in a cell suspension culture of celery (*Apium graveolens*). Lipids 26:536–541, 1991.

33. M Alberdi, J Fernandez, R Cristi, M Romero, D Rios. Lipid changes in cold hardened leaves of *Northofagus dombeyi*. Phytochem 29:2467–2471, 1990.

34. BD Whitaker, EH Lee, RA Rowland. EDU and ozone protection: foliar glycerolipids and steryl lipids in snapbean exposed to O3. Physiol Plant 80: 286–293, 1990.

35. BD Whitaker. A reassessment of heat treatment as a means of reducing chilling injury in tomato fruit. Postharvest Biol Technol 4:75–83, 1994.

36. A Bergenstråhle, P Borgå, LMV Jonsson. Sterol composition and synthesis in potato tuber discs in relation to glycoalkaloid synthesis. Phytochemistry 41: 155–161, 1996.

37. B Yang, J Koponen, R Tahvonen, H Kallio. Plant sterols in seeds of two

species of *Vaccinium* (*V. myrtillus* and *V. vitis-idaea*) naturally distributed in Finland. Eur Food Res Technol 216:34–38, 2003.
38. KM Phillips, MT Tarragó-Tani, KK Stewart. Phytosterol content of experimental diets differing in fatty acid composition. Food Chem 64:415–422, 1999.
39. MJ Fishwick, AJ Wright. Comparison of methods for the extraction of plant lipids. Phytochemistry 16:1507–1510, 1977.
40. E Homberg, B Bielefeld. Sterin- und Fettsäurezusammensetzung in Keimölen. Fat Sci Technol 92:118–121, 1990.
41. JP Palta, BD Whitaker, LS Weiss. Plasma membrane lipids associated with genetic variability in freezing tolerance and cold acclimation of *Solanum* species. Plant Physiol 103:793–803, 1993.
42. C Willemot. Sterols in hardening winter wheat. Phytochemistry 19:1071–1073, 1980.
43. M Ohnishi, S Obata, S Ito, Y Fujino. Composition and molecular species of waxy lipids in wheat grain. Cereal Chem 63:193–196, 1986.
44. T Sugawara, T Miyazawa. Separation and determination glycolipids from edible plant sources by high-performance liquid chromatography and evaporative light scattering detection. Lipids 34:1231–1237, 1999.
45. M-A Hartmann, P Benveniste. Plant membrane sterols: isolation, identification, and biosynthesis. Meth Enzymol 148:632–650, 1987.
46. A Hara, NS Radin. Lipid extraction of tissues with a low-toxicity solvent. Anal Biochem 90:420–426, 1978.
47. PC Dutta, L-Å Appelqvist. Saturated sterols (stanols) in unhydrogenated and hydrogenated edible vegetable oils and in cereal lipids. J Sci Food Agric 71:383–391, 1996.
48. L-K Ng, M Hupé. Analysis of sterols: a novel approach for detecting juices of pineapple, passionfruit, orange and grapefruit in compounded beverages. J Sci Food Agric 76:617–627, 1998.
49. Z Xu, JS Godber. Purification and identification of components of γ-oryzanol in rice bran oil. J Agric Food Chem 47:2724–2728, 1999.
50. T Akihisa, K Yasukawa, M Yamaura, M Ukiya, Y Kimura, N Shimizu, K Arai. Triterpene alcohol and sterol ferulates from rice bran and their antiinflammatory effects. J Agric Food Chem 48:2313–2319, 2000.
51. I Khabazian, JS Bains, DE Williams, J Cheung, JMB Wilson, BA Pasqualotto, SL Pelech, RJ Andersen, Y-T Wang, L Liu, A Nagai, SU Kim, U-K Craig, CA Shaw. Isolation of various forms of sterol β-D-glucoside from the seed of *Cycas circinalis*: neurotoxicity and implications for ALS–parkinsonism dementia complex. J Neurochem 82:516–528, 2002.
52. LM Seitz. Stanol and sterol esters of ferulic and p-coumaric acids in wheat, corn, rye, and triticale. J Agric Food Chem 37:662–667, 1989.
53. P Hakala, A-M Lampi, V Ollilainen, U Werner, M Murkovic, K Wähälä, S Karkola, V Piironen. Steryl phenolic acid esters in cereals and their milling fractions. J Agric Food Chem 50:5300–5307, 2002.
54. M Zhou, K Robards, M Glennie-Holmes, S Helliwell. Oat lipids. J Am Oil Chem Soc 76:159–169, 1999.

55. OK Chung, J-B Ohm. Cereal lipid. In: K Kulp, JG Ponte Jr, eds. Handbook of Cereal Science and Technology, 2nd ed. New York: Marcel Dekker, 2000, pp 417–477.

56. FD Conforti, CH Harris, JT Rinehart. High-performance liquid chromatographic analysis of wheat flour lipids using an evaporative light scattering detector. J Chromatogr 645:83–88, 1993.

57. JM Snyder, JW King, SL Taylor, AL Neese. Concentration of phytosterols for analysis by supercritical fluid extraction. J Am Oil Chem Soc 76:717–721, 1999.

58. Z. Xu, JS Godber. Comparison of supercritical fluid and solvent extraction methods in extracting γ-oryzanol from rice bran. J Am Oil Chem Soc 77:547–551, 2000.

59. J Toivo, K Phillips, A-M Lampi, V Piironen. Determination of sterols in foods: recovery of free, esterified and glycosidic sterols. J Food Comp Anal 14:631643, 2001.

60. J Nourooz-Zadeh. Determination of the autoxidation products from free or total cholesterol: a new multistep enrichment methodology including the enzymatic release of esterified cholesterol. J Agric Food Chem 38:1667–1673, 1990.

61. I Wretensjö, B Karlberg. Characterization of sterols in refined borage oil by GC-MS. J Am Oil Chem Soc 79:1069–1074, 2002.

62. GSMJE Duchateau, CG Bauer-Plank, AJH Louter, M van der Ham, JA Boerma, JJM van Rooijen, PA Zandbelt. Fast and accurate method for total 4-desmethyl sterol(s) content in spreads, fat-blends, and raw materials. J Am Oil Chem Soc 79:273–278, 2002.

63. J Toivo, A-M Lampi, S Aalto, V Piironen. Factors affecting sample preparation in the gas chromatographic determination of plant sterols in whole wheat flour. Food Chem 68:239–245, 2000.

64. Commission Regulation (EEC). No 2568 / 91, Annex V. Determination of the composition and content of sterols by capillary-column gas chromatography. Off J Eur Commun L248:15–22, 1991.

65. G Lognay, M Severin, A Boenke, PJ Wagstaffe. Edible fats and oils reference materials for sterols analysis particular attention to cholesterol. Part 1. Investigation of some analytical aspects by experienced laboratorios. Analyst 117:1095–1097, 1992.

66. B Yang, RM Karlsson, PH Oksman, HP Kallio. Phytosterols in sea buckthorn (Hippophaë rhamnoides L.) berries: identification and effects of different origins and harvesting times. J Agric Food Chem 49:5620–5629, 2001.

67. G Lognay, P Dreze, PJ Wagstaffe, M Marlier, M Severin. Validation of a quantitative procedure for the extraction of sterols from edible oils using radiolabelled compounds. Analyst 114:1287–1291, 1989.

68. M Arens, H-J Fiebig, E Homberg. Sterine (Isolierung und GC-Untersuchung)–Gemeinschaftsarbeiten der DGF, 114. Mitteilung: Deutsche Einheitsmethoden zur Untersuchung von Fetten, Fettprodukten, Tensiden und verwandten Stoffen, 86. Mitt.: Analyse von Fettbeleistoffen VII. Fat Sci Technol 92:189–192, 1990.

69. M Amelio, R Rizzo, F Varazini. Determination of sterols, erythrodiol, uvaol and alkanols in olive oils using combined solid-phase extraction, high-performance liquid chromatographic and high-resolution gas chromatographic techniques. J Chromatogr 606:179–185, 1992.

70. K Aitzetmüller, L Brühl, H-J Fiebig. Analysis of sterol content and composition in fats and oils by capillary-gas liquid chromatography using an internal standard. Comments on the German sterol method. Fett/Lipid 100: 429–435, 1998.

71. D Jonker, GD van der Hoek, JFC Glatz, C Homan, MA Posthumus, MB Katan. Combined determination of free, esterified and glycosilated plant sterols in foods. Nutr Rep Int 32:943–951, 1985.

72. M Biedermann, K Grob, C Mariani, JP Schmidt. Detection of desterolized sunflower oil in olive oil through isomerized Δ7-sterols. Z Lebensm Unters Forsch 202:199–204, 1996.

73. P Breinhölder, L Mosca, W Lindner. Concept of sequential analysis of free and conjugated phytosterols in different plant matrices. J Chromatogr B 777: 67–82, 2002.

74. A Kamal-Eldin, K Määttä, J Toivo, A-M Lampi, V Piironen. Acid-catalyzed isomerization of fucosterol and Δ5-avenasterol. Lipids 33:1073–1077, 1998.

75. IUPAC. 2.403 Identification and determination of sterols by gas-liquid chromatography. In: C Paquot, A Hautfenne, eds. International Union of Pure and Applied Chemistry. Standard Methods for the Analysis of Oils, Fats and Derivatives, 7th ed. Oxford: Blackwell Scientific, 1987.

76. AOAC official method 976.26 cholesterol in multicomponent foods. In: W Horwitz, ed. AOAC Official Methods of Analysis of AOAC International, 17th ed., revision 1. Gaithersburg, MD: AOAC International, 2002.

77. HE Indyk. Simultaneous liquid chromatographic determination of cholesterol, phytosterols and tocopherols in foods. Analyst 115:1525–1530, 1990.

78. SM Al-Hasani, J Hlavac, MW Carpenter. Rapid determination of cholesterol in single and multicomponent prepared foods. JAOAC Int 76:902–906, 1993.

79. G Contarini, M Povolo, E Bonfitto, S Berardi. Quantitative analysis of sterols in dairy products: esperiences and remarks. Int Dairy J 12:573–578, 2002.

80. AA Jekel, HAMG Vaessen, RC Schothorst. Capillary gas-chromatographic method for determining non-derivatized sterols—some results for duplicate 24 h diet samples collected in 1994. Fresenius J Anal Chem 360:595–600, 1998.

81. K Schwadorf, HM Müller. Determination of ergosterol in cereals, mixed feed components, and mixed feeds by liquid chromatography. JAOAC 72:457–462, 1989.

82. GM Morton, SM Lee, DH Buss, P Lawrance. Intakes and major dietary sources of cholesterol and phytosterols in the British diet. J Hum Nutr Dietet 8:429–440, 1995.

83. P Mattila, A-M Lampi, R Ronkainen, J Toivo, V Piironen. Sterol and vitamin D2. contents in some wild and cultivated mushrooms. Food Chem 76:293–298, 2002.

84. L Normén, M Johnsson, H Andersson, Y van Gameren, P Dutta. Plant sterols in vegetables and fruits commonly consumed in Sweden. Eur J Nutr 38:84–89, 1999.
85. L Normén, S Bryngelsson, M Johansson, P Evheden, L Ellegård, H Brants, H Andersson, P Dutta. The phytosterol content of some cereal foods commonly consumed in Sweden and in the Netherlands. J Food Comp Anal 15: 693–704, 2002.
86. V Piironen, J Toivo, A-M Lampi. Plant sterols in cereals and cereal products. Cereal Chem 79:148–154, 2002.
87. V Piironen, J Toivo, R Puupponen-Pimiä, A-M Lampi. Plant sterols in vegetables, fruits and berries. J Sci Food Agric 83:330–337, 2003.
88. K Warner, TL Mounts. Analysis of tocopherols and phytosterols in vegetable oils by HPLC with evaporative light-scattering detection. J Am Oil Chem Soc 67:827–831, 1990.
89. VKS Shukla, PC Dutta, WE Artz. Camelina oil and its unusual cholesterol content. J Am Oil Chem Soc 79:965–969, 2002.
90. EB Hoving. Chromatographic methods in the analysis of cholesterol and related lipids. J Chromatogr B 671:341–362, 1995.
91. RJ Reina, KD White, EGE Jahngen. Validated method for quantitation and identification of 4,4-desmethylsterols and triterpene diols in plant oils by thin-layer chromatography–high resolution gas chromatography–mass spectrometry. JAOAC Int 80:1272–1280, 1997.
92. AC Bello. Rapid isolation of the sterol fraction in edible oils using a silica cartridge. JAOAC Int 75:1120–1123, 1992.
93. IC Tsui. Rapid determination of total cholesterol in homogenized milk. JAOAC 72:421–424, 1989.
94. J Toivo, V Piironen, P Kalo, P Varo. Gas chromatographic determination of major sterols in edible oils and fats using solid-phase extraction in sample preparation. Chromatogrphia 48:745–750, 1998.
95. M Lechner, B Reiter, E Lorbeer. Determination of tocopherols and sterols in vegetable oils by solid-phase extraction and subsequent capillary gas chromatographic analysis. J Chromatogr A 857:231–238, 1999.
96. E Kovacheva, G Ganchev, A Neicheva, I Ivanova, M Konoushlieva, V Andreev. Gas chromatographic determination of sterols in fat-soluble concentrates obtained from plant materials. J Chromatogr 509:79–84, 1990.
97. M Careri, L Elviri, A Mangia. Liquid chromatography–UV determination and liquid chromatography–atmospheric pressure chemical ionization mass spectrometric characterization of sitosterol and stigmasterol in soybean oil. J Chromatogr A 935:249–257, 2001.
98. G Lercker, MT Rodriguez-Estrada. Chromatographic analysis of unsaponifiable compounds of olive oils and fat-containing foods. J Chromatogr A 881: 105–129, 2000.
99. P Volin. Analysis of steroidal lipids by gas and liquid chromatography. J Chromatogr A 935:125–140, 2001.
100. H-J Fiebig, L Brühl, K Aitzetmüller. Sterine: Isolierung und gaschromatographische Untersuchung. Fett 100:422–428, 1998.

101.   T Itoh, S Sakurai, T Tamura, T Matsumoto. Occurrence of 24(E)-ethylidene sterols in two solanaceae seed oils and rice bran oil. Lipids 15:22–25, 1980.
102.   T Akihisa, N Shimizu, P Ghosh, S Thakur, FU Rosenstein, T Tamura, T Matsumoto. Sterols of the Cucurbitaceae. Phytochemistry 26:1693–1700, 1987.
103.   T Akihisa, Y Kimura, K Roy, P Ghosh, S Thakur, T Tamura. Triperpene alcohols and 3-oxo steroids of nine Leguminosae seeds. Phytochemistry 35: 1309–1313, 1994.
104.   GS Tint, M Seller, R Hughes-Benzie, AK Batta, S Shefer, D Genest, M Irons, E Elias, G Salen. Markedly increased tissue concentrations of 7-dehydrocholesterol combined with low levels of cholesterol are characteristic of the Smith–Lemli–Opitz syndrome. J Lipid Res 36:89–95, 1995.
105.   M Van Boven, P Daenens, K Maes, M Cokelaere. Content and composition of free sterols and free fatty alcohols in jojoba oil. J Agric Food Chem 45: 1180–1184, 1997.
106.   PC Dutta, L Normén. Capillary column gas-liquid chromatographic separation of delta-5-unsaturated and saturated phytosterols. J Chromatogr 816: 177–184, 1998.
107.   J Parcerisa, I Casals, J Boatella, R Codony, M Rafecas. Analysis of olive and hazelnut oil mixtures by high-performance liquid chromatography—atmospheric pressure chemical ionization mass spectrometry of triacylglycerols and gas–liquid chromatography of non-saponifiable compounds (tocopherols and sterols). J Chromatogr A 881:149–158, 2000.
108.   T Verleyen, M Forcades, R Verhe, K Dewettinck, A Huyghebaert, W De Greyt. Analysis of free and esterified sterols in vegetable oils. J Am Oil Chem Soc 79:117–122, 2002.
109.   M Du, DU Ahn. Simultaneous analysis of tocopherols, cholesterol, and phytosterols using gas chromatography. J Food Sci 67:1696–1700, 2002.
110.   R Holen. Rapid separation of free sterols by reversed-phase high performance liquid chromatography. J Am Oil Chem Soc 62:1344–1346, 1985.
111.   P Morin, D Daguet, JP Coïc, M Dreux. Usefulness of methylated-β-cyclodextrin-based buffers for the separation of highly hydrophobic solutes in non-aqueous capillary electrophoresis. J Chromatogr A 837:281–287, 1999.
112.   SL Abidi, S Thiam, IM Warner. Elution behavior of unsaponifiable lipids with various capillary electrochromatographic stationary phases. J Chromatogr A 949:195–207, 2002.
113.   A Rahier, P Benveniste. Mass spectral identification of phytosterols. In: WD Nes, EJ Parish, eds. Analysis of Sterols and Other Biologically Significant Steroids. San Diego: Academic Press, 1989, pp 223–250.
114.   BA Knights. Identification of plant sterols using combined GLC/mass spectrometry. J Gas Chromatogr 5:273–282, 1967.
115.   T Akihisa, WCMC Kokke, T Tamura. Naturally occurring sterols and related compounds from plants. In: GW Patterson, WR Nes, eds. Physiology and Biochemistry of Sterols. Champaign, IL: AOCS, 1991, pp 172–228.
116.   KM Phillips, DM Ruggio, JI Toivo, MA Swank, AH Simpkins. Free and esterified sterol composition of edible oils and fats. J Food Comp Anal 15:123–142, 2002.

117. Commision Regulation (EC) No 175/1999. Reference methods for the determination of certain tracers in butter, butteroil and cream. Offi J Eur Commun L20:22–25, 1999.

118. LI Christiansen, JT Rantanen, AK von Bondsdorff, MA Karjalainen, JK Yliruusi. A novel method of producing a microcrystalline β-sitosterol suspension in oil. Eur J Pharm Sci 15:261–269, 2002.

119. L Pizzoferrato, S Nicoli, C Lintas. GC-MS characterization and quantification of sterols and cholesterol oxidation products. Chromatographia 35:269–274, 1993.

120. Commission Regulation (EC) No 213/2001, Annex XIV. Determining sitosterol or stigmasterol in butter or concentrated butter by capillary-column gas chromatography. Offi J Eur Comm No L37:42–50, 2001.

121. AOAC official method 994.10 cholesterol in foods. In: W, Horwitz ed. AOAC Official Methods of Analysis of AOAC International, 17th ed., revision 1. Gaithersburg, MD: AOAC International, 2002.

122. RE Worthington, HL Hitchcock. A method for the separation of seed oil steryl esters and free sterols: application to peanut and corn oils. J Am Oil Chem Soc 61:1085–1088, 1984.

123. RT Lorenz, G Fenner, LW Parks, K Haeckler. Analysis of steryl esters. In: WD Nes, EJ Parish, eds. Analysis of Sterols and Other Biologically Significant Steroids. New York: Academic Press, 1989, pp 33–47.

124. RA Norton. Isolation and identification of steryl cinnamic acid derivatives from corn bran. Cereal Chem 71:111–117, 1994.

125. RA Moreau, PT Asmann, HA Norman. Analysis of major classes of plant lipids by high-performance liquid chromatography with flame ionization detection. Phytochemistry 29:2461–2466, 1990.

126. J Nordbäck, E Lundberg. High resolution separation of non-polar lipid classes by HPLC-ELSD using alumina as stationary phase. J High Resol Chromatogr 22:483–486, 1999.

127. RA Moreau, K Kohout, V Singh. Temperature-enhanced alumina HPLC method for the analysis of wax esters, sterol esters, and methyl esters. Lipids 37:1201–1204, 2002.

128. JW King, NT Dunford. Phytosterol-enriched triglyceride fractions from vegetable oil deodorizer distillates utilizing supercritical fluid fractionation technology. Sep Sci Technol 37:451–462, 2002.

129. JT Billheimer, S Avart, B Milani. Separation of steryl esters by reversed-phase liquid chromatography. J Lipid Res 24:1646–1651, 1983.

130. MH Gordon, RE Griffith. Steryl ester analysis as an aid to the identification of oils and blends. Food Chem 43:71–78, 1992.

131. K Grob, M Lanfranchi. Reproducibility of results from LC-GC of sterols and wax esters in olive oils. J High Resol Chromatogr 12:624–626, 1989.

132. A Artho, K Grob, C Mariani. On-line LC-GC for the analysis of the minor components in edible oils and fats—the direct method involving silylation. Fat Sci Technol 95:176–180, 1993.

133. C Plank, E Lorbeer. On-line liquid chromatography-gas chromatography for

the analysis of free and esterified sterols in vegetable oil methyl esters used as diesel fuel substitutes. J Chromatogr 683:95–104, 1994.

134. W Kamm, F Dionisi, L-B Fay, C Hischenhuber, H-G Schmarr, K-H Engel. Analysis of steryl esters in cocoa butter by on-line liquid chromatography–gas chromatography. J Chromatogr 918:341–349, 2001.

135. P Kalo, T Kuuranne. Analysis of free and esterified sterols in fats and oils by flash chromatography, gas chromatography and electrospray tandem mass spectrometry. J Chromatogr A 935:237–248, 2001.

136. RA Norton. Quantitation of steryl ferulate and p-coumarate esters from corn and rice. Lipids 30:269–274, 1995.

137. AM Condo Jr, DC Baker, RA Moreau, KB Hicks. Improved method for the synthesis of trans-feruloyl-β-sitostanol. J Agric Food Chem 49:4961–4964, 2001.

3
Plant Sterol Analysis in Relation to Functional Foods

Guus S. M. J. E. Duchateau, Hans-Gerd M. Janssen, and Arjan J. H. Louter
Unilever Research and Development, Vlaardingen, The Netherlands

I. INTRODUCTION

A. Historical Perspective

Sterols are high-melting alcohols with a typical structure based on the cholestane structure and are ubiquitous in nature. They perform important physiological roles in all living organisms. Information on the composition of sterols in various sources, animal as well as vegetable, has been an area of analytical interest for a long time, from both the physiological perspective and for their role as marker substances in identification. Comprehensive reviews on the different instrumental techniques for sterol analysis are collated in the book by Goad and Akihisa and the recent paper by Abidi (1,2). Sterols are classified into three main groups ("sterol classes") based on the presence or absence of methyl moieties at the 4-position in the A ring. The most frequently occurring natural plant sterols are the desmethyl sterols β-sitosterol, campesterol, and stigmasterol (Fig. 1). Dietary intake ranges from 250 to 500 mg/day with about 65% of intake as β-sitosterol, 30% as campesterol, and 5% as stigmasterol and low amounts of other sterols (3–5). Recently, the efficacy of plant sterols, predominantly the 4-desmethyl sterols, in lowering blood cholesterol levels has drawn renewed interest. This also resulted in a renewed interest in their analysis. Reviews on the cholesterol-lowering properties of plant sterols and stanols, being the fully saturated sterol analogues, focus on this cholesterol lowering effect

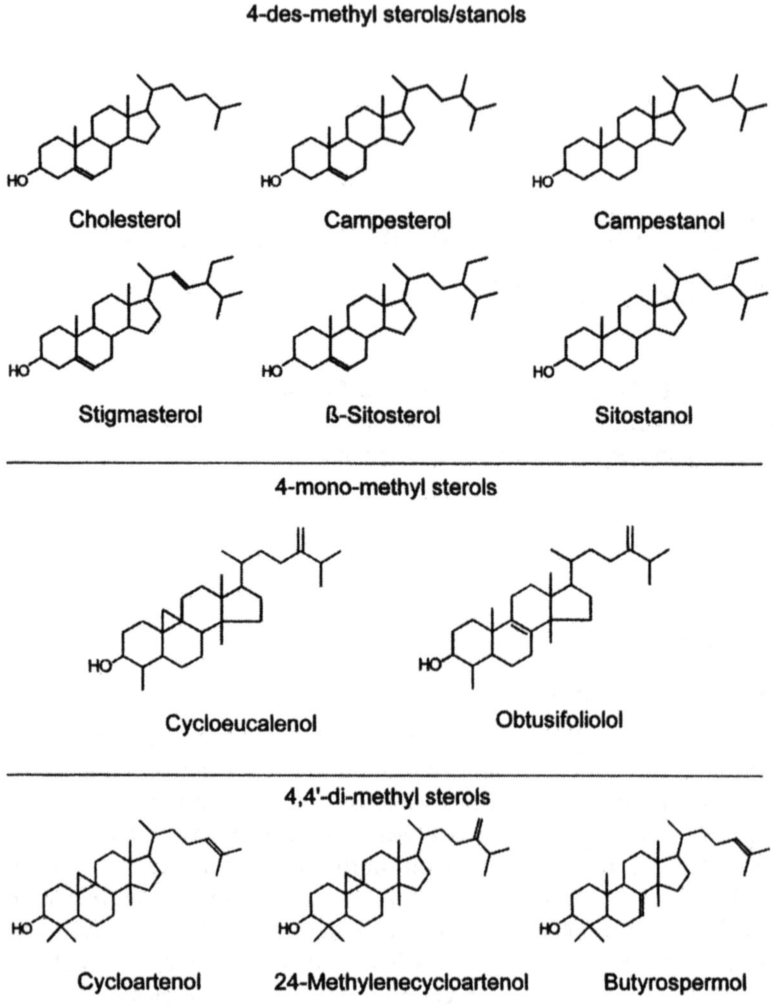

4-des-methyl sterols/stanols

Cholesterol   Campesterol   Campestanol

Stigmasterol   ß-Sitosterol   Sitostanol

4-mono-methyl sterols

Cycloeucalenol   Obtusifoliolol

4,4'-di-methyl sterols

Cycloartenol   24-Methylenecycloartenol   Butyrospermol

Figure 1 Main sterol/stanol structures.

(6–11). More information on the role of plant sterols in cholesterol lowering can be found in Chapter 6. The new interest in plant sterols in general is based on the far better physicochemical properties of the now used sterol/ stanol fatty acid esters as food ingredient compared to the free forms, while the cholesterol-lowering properties remain the same (Fig. 2) (12,13). Amounts used in clinical studies resulting in a significant blood cholesterol lowering effect (10–15% LDL cholesterol reduction) are in the range of 2–3

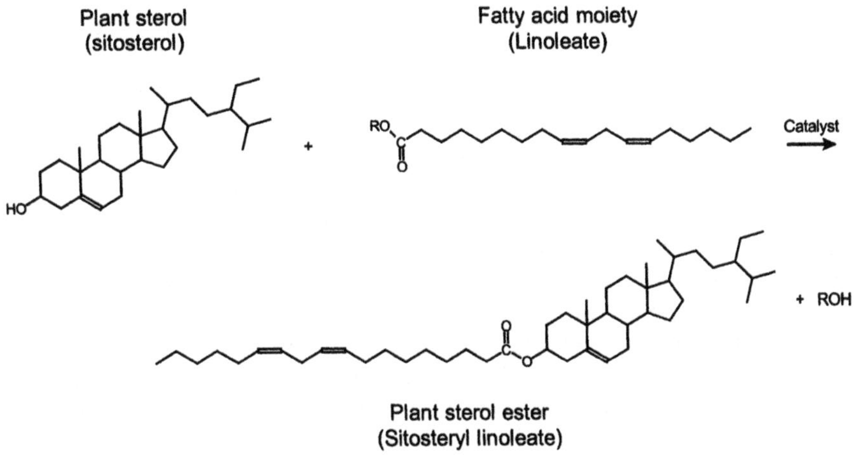

Plant sterol
(sitosterol)

Fatty acid moiety
(Linoleate)

Figure 2 Typical sterol structure (sitosterol) with the general reaction scheme for esterification with a fatty acid (linoleate) to the corresponding plant sterol ester.

g/day (14–16). Interestingly, plant sterols are hardly being absorbed them-selves, yet they are in molecular structure closely related to cholesterol. The reported levels of β-sitosterol absorption show an absorption of less than 5%, independent of the amount of total sterols fed (4,17). Plasma levels are about 1–5% relative to plasma cholesterol (10). Practically, consumer products formulated to ensure the required daily intake for the cholesterol lowering effect contain plant sterols as their vegetable oil–derived fatty acid esters, at the higher percentage level. As an example, 13% of plant sterol esters (equivalent to about 8% plant sterols) is used in spreadable, vegetable oil–based margarines, marketed in Europe, Australia, United States, and several other regions. Other food formats, such as milk, yogurt, chocolate, ground beef, and vegetable oils, have been suggested as carrier or are investigated to ensure the daily supply required to reduce the plasma cholesterol levels (18–20). The daily sterol amount required for this effect is thus obtained from products with a sterol level higher than the natural level in most vegetable oils or other food sources, which typically range from almost zero (refined oils) to about 1% (crude oils) and in some specific sources as high as 2%. Consequently, intake from these new products is higher than previously (21–23). With this new interest and at the higher product levels, plant sterol/stanols and their esters have shifted from a natural minor food component to an approved food ingredient. Analytical methods to describe the quality of plant sterol esters in the pure form, and to measure levels used in food products, had to change as well.

The analytical methodologies used for sterols/stanols in the past focused on the low levels as found in natural sources. These methods provided detailed information on the composition typical for the plant or animal source. It was used for identification purposes or detection of adulteration (24,25). Most methods as described in normative references are based on an extensive sample clean-up followed by a derivatization step, and finally a form of chromatography, e.g., the typical sequence of saponification, thin-layer chromatography (TLC), derivatization, and gas chromatography (GC) (26–28). These methods are time consuming and have a limited precision but provide detailed structural information. For low sample numbers this approach proved to be satisfactory. However, some authors looked for faster liquid chromatography (LC) approaches, but at the cost of a less optimal detection compared to GC flame ionization detection (FID) (29–31). Recently, a sample clean-up step, which has been used with success in environmental analysis, proved to be compatible with sterol analysis as well. Carstensen and Schwack used GPC to isolate the sterol-containing fraction from edible oils before a final GC run (32). Alternatively, LC methods are applied to answer very specific situations in the analysis of sterol/stanol epimers (33).

With sterols/stanols, and the corresponding fatty acid esters becoming more frequently used food ingredients, faster and more robust analytical methods are required to monitor the raw material quality, to follow processing steps, and to develop raw material and product specifications. All these questions are added to classical information required from sterol methods: information on oil origin (= source) and detection of adulteration. In addition, methods based on saponification do not differentiate between the free and esterified sterols. The new interest for the esters as being convenient to formulate cholesterol-lowering agents would call for such a method. Finally, consumer product levels should be confirmed to ensure proper label information, and measurements in biofluids are required to support bioavailability and mechanistic studies. Within our laboratory we have developed several methods with a focus on sterols and sterol esters present in a diverse set of matrices to meet these new requirements.

B. Analytical Requirements for Sterols Used as Functional Food Ingredient

In the subsequent development stages of a (functional) food product with a new ingredient added, inevitably a number of steps will be taken to ensure safety and quality of the new ingredient and hence the quality of the final product. Since specifications for plant sterol esters as food ingredients in the pure form are not published in common monograph collections such as listed by the FAO/WHO *Codex Alimentarius* or the *US Food Chemical Codex*, it was required to develop such a monograph and methods for each step in the process, keeping in mind the use of the data and the (expected) number of samples. The different steps that can be recognized in the development started with the gathering of basic information on the new

Table 1 Analytical Method Requirements and Characteristics for Sterols Intended as New Food Ingredients

Method/matrix	Level of information	Speed requirements	Expected sample number	Complexity allowed, robustness
PS raw material[a]	Detailed sterol composition within each class, sterol content	Moderate	Low	High, methods will only be used in initial stages until spec's are formalized, no restrictions on sample clean-up
PS esterification/ processing	Esters vs. free sterols, sterol content	High	Moderate	Low, should match with process control, robust
Preformulation studies/stability	Esters vs. free sterols, sterol content	Low	Low	High
Safety assessment	Detail, i.e., minor components[a]	Low	Low	High
Regulatory support	Composition	Low	Low	High
Efficacy studies	Main sterol, sterol content in products	Low	High	Low, robust methods with respect to analysis in products and biosamples
Production QA/QC	Low, sum level in products	High	High	Low, robust method, ring tested, at line analysis, batch release

[a] Standard methods as used for edible oils are also used to analyze for impurities, metals, different types of pesticides, PAHs via DACC, etc., in sterols (56–58).

ingredients in terms of the effects of the different processing steps, data on safety and quality, and finally data on the uptake in humans to support the functional claim. During the product development stages, the analyte(s) remain the same, but levels and matrices can very widely, e.g., raw material (nearly pure) vs. plasma samples (μmol/L level in a highly complex environment). Some methods might develop almost naturally into the next stage, while at other moments a complete redesign of the analytical strategy is required to accommodate the new situation. The requirements for each stage in the development are summarized in Table 1. The development stages of the sterol methods with their main requirements seen in our laboratory are presented in Fig. 3.

1. Raw Material Characterization and Sourcing Requirements

An extensive analytical review of raw material supplier batches is generally required to establish draft internal acceptance specifications because no external references describing sterols as raw material exist (34). By this stage the information on the sample should reveal the quality in terms of total level of sterols, the level of each sterol class and possibly each individual sterol, confirming the source, type, and level of impurities (if any). The latter

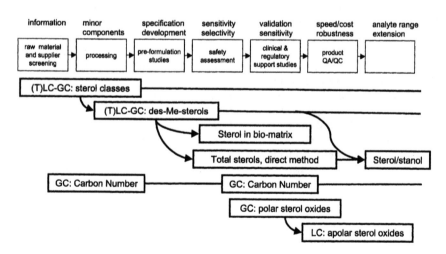

Figure 3 Schematic representation of the different sterol methods applied during each development step (top axis) of a sterol ester–containing food product. Time is from left to right and arrows indicate the subsequent evolution into a new method.

is not discussed in detail in this context. The classical methods based on TLC-GC provide sufficient detail on the sterol composition within each sterol class and would thus be fine for detailed raw material character- ization. However, for the vast number of samples expected, a dramatic improvement in method speed and a significant reduction of hands-on time would be required.

2. Esterification

In most common edible oils sterols exist both in the free and esterified form, with esters, such as the fatty acid, ferulic, and cinnamic acid esters, in specific plant sources (Fig. 4, Table 2) (23,35). Little information could be obtained from earlier work reported in the public domain on methods designed to differentiate between free and esterified sterols in raw materials.

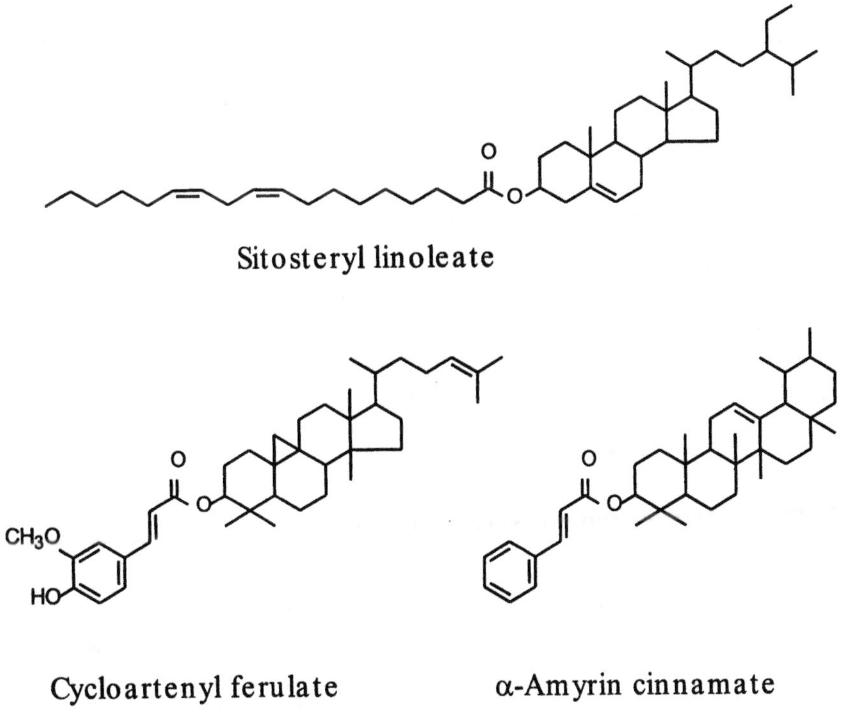

Sitosteryl linoleate

Cycloartenyl ferulate   α-Amyrin cinnamate

Figure 4 Different sterol esters.

Table 2   Relative Distribution of Different Free and Esterified Sterols in Some Vegetable Oil Sources[a]

Source	4-Desmethyl sterols	4-Monomethyl sterols	4,4-Dimethyl sterols	Ratio esters/free sterols
Palm	+ +	−	−	−1
Rape seed	+ + +	−	−	< 1
Soybean	+ +	+	−	> 1
Sunflower seed	+ +	+	−	> 1
Corn fiber	+ + +	−	−	< 1
Shea nut	−	+	+ +	−
Rice bran	+ + +	+	+ +	≪ 1

[a] Esters are fatty acid, ferulic or cinnamic acid type.

Based on the "new" use of sterol esters, the raw material has to be fully esterified with fatty acids in a food-grade process. Within our laboratory we have ample experience with a method generally used within the oils and fats industry, based on the boiling point separation of acylglycerides and free fatty acids by GC (36). Sterols and sterol esters fit perfectly within the established retention scheme.

3.   Preformulation and Product Development

Formulation of plant sterol esters into a food matrix would require access to relatively low amounts of the esters for processing runs and product property optimization. However, at this stage, sensory or efficacy data are generally not yet required. The prototypes are not used for taste panels or human consumption and the number of samples is low. The typical information requested is on degree of esterification, total sterol content, and stability as a function of applied unit operations or storage at various conditions. Analytical requirements are relatively low in terms of speed, sample capacity, or robustness of the method.

4.   Specification Setting and Safety

We used the newly developed sterol analysis methods together with methods typically applied to fully refined edible oils and fats for bulk parameters. The data were evaluated to obtain quality information and to support future specification settings for plant sterols and plant sterol esters as functional food ingredients.

5. Efficacy Studies in Humans and Animals

A critical transition in the method development is reached upon the change to biosample analysis. The relatively low amounts of the available biomatrix itself (milliliters per sample at maximum), the low expected concentration (μmol/L), and the presence of endogenously formed sterols (e.g., cholesterol and its precursors lathosterol, lanosterol, and desmosterol) or plant sterols from the background diet, require both sensitive and specific methods (see Sect. IIIB). A complicating factor in this is the wide concentration range between the high levels of plasma cholesterol (low millimolar range) on the one hand and the very low plant sterol levels in plasma (low micromolar range) on the other.

6. Consumer-Relevant Information on Product Levels

Consumer products require specific labeling information on the sterol level. The claimed content figure on the packaging has to be at the target and within a small, but still analytically, realistic range of "error," e.g., 8% $\pm$ 0.8% (target level $\pm$ 10%). To protect consumers from misleading information or from non-efficacious products, the level claimed could be verified by independent or government laboratories. This requirement brings with it the fact that methods used for label claim purposes should be generally available and transparent; moreover, they should be tested in different laboratories for its ruggedness. A formal round-robin test is then required.

Taking into account all of the above factors, we now recognize two axes of information, one representing the level of information to be supplied by the analysis, and the other axis the type of sample (matrix) to be analyzed. Going along the information axis, the information density increases from straightforward questions on the total sterol level to complex detailed sterol analyses giving the concentration levels of each of the individual sterols/stanols and their esters. At the same time on the other axis, matrix complexity will increase from the fairly simple pure sterol or sterol ester mixture to the extremely complex samples resulting from human or animal trials. This whole concept is explained in the overview of methods (Table 3) which serves as a starting point for each new method request. Table 3 is pivotal in our sterol analysis approach and should provide the analytical strategy for dealing with most problems. Both axes are explained in more detail, including technicalities in the following sections. Already at this point it is interesting to see that the chromatographic separation method is more or less only determined by the level of detail desired, whereas the sample pretreatment method required is almost exclusively determined by the sample matrix. The selection of the chromatographic conditions is hence

Table 3 Overview of Chromatographic Methods and Sample Pretreatment Steps for the Analysis of Sterols or Sterol/Stanol-Containing Samples[a]

	Sterol bulk product	Sterol esters bulk product	Food product (e.g. Spread)	Biomatrix
Total sterol content	None	Hot saponification	Hot saponification, or hot saponification and silylation	Hot saponification and silylation
	GC: ON-COLUMN OR SPLIT INJECTION, SHORT NONPOLAR COLUMN, FID DETECTION			
Sterol vs sterol esters	Silylation	Silylation	Silylation and LC fractionation	Extraction, silylation, and LC fractionation
	GC: ON-COLUMN INJECTION SHORT, NONPOLAR COLUMN PROGRAMMED TO HIGH FINAL TEMPERATURE, FID DETECTION			
Sterol classes (des-, mono, and dimethyl)	LC or TLC fractionation and silylation	Transesterification prior to LC fractionation: classes PS, DiMS, MMS, DeMS	Extraction and transesterification prior to LC fractionation: classes PS, TAGs, DiMS, MMS, DeMS	Extraction and transesterification prior to LC fractionation: classes PS, TAGs, DiMS, MMS, DeMS
	GC: ON-COLUMN OR SPLIT INJECTION, SHORT MEDIUM-POLARITY COLUMN, FID DETECTION			
Individual (desmethyl) sterols	Sample preparation: See Sterol classes (des-, mono and dimethyl)			
	GC: ON-COLUMN OR SPLIT INJECTION, MEDIUM TO LONG POLAR COLUMN, FID DETECTION			
Sterol oxidation products	LC isolation and silylation	Cold saponification, LC isolation and silylation	Cold saponification, LC isolation and silylation	Cold saponification, LC isolation and silylation
	GC: ON-COLUMN INJECTION, MEDIUM TO LONG POLAR COLUMN, FID DETECTION			
Nonpolar sterol oxidation products	None	Cold saponification	Cold saponification	Cold saponification
	HPLC: NORMAL PHASE (SILICA), ELSD DETECTION			

[a] Hot saponification = saponification in boiling ethanolic KOH; cold saponification = saponification in ethanolic KOH at room temperature overnight/dark.

to a large extent independent from the matrix in which the sterols are present and vice versa.

II. ACTUAL METHOD DESCRIPTIONS

A. Pretreatment and Preseparation Methods for Sterol Analysis

During the development stage and the eventual production of plant sterol containing functional foods, compositional information is crucial. As outlined in Fig. 3, the required level of detail will shift from "as high as possible" in the early knowledge–development stages, to a mere accurate quantification of the total sterol level in the final production stage. In each of these stages chromatography is called on extensively. The actual requirements imposed on the (total) chromatographic methods for each of these phases, however, will vary dramatically. Chromatography applied in this situation demonstrates its extremely versatile nature. It is equally suited for the detailed analysis of pure sterol mixtures as well as for the accurate quantification of sterols in complete food products or biomedical samples originating from human or animal trials.

Both LC and GC are extremely powerful methods for the characterization of lipids or, more specifically, sterol-containing samples (2). In many respects the two techniques are interchangeable. Determination of the exact sterol composition of a sterol raw material can be performed both by GC and by LC. A distinct drawback of HPLC is the lack of detectors, which are easy to use and sensitive enough. It is mainly for this reason, as well as because of its greater speed and the higher efficiency, that GC has now become the method of choice for most of the sterol analyses. In addition to that, GC is very sensitive, and response factors for the various sterols are more or less identical. Next to being interchangeable, GC and LC are also complementary as has convincingly been demonstrated by Grob (37). Similar to the classical TLC, normal-phase LC (NPLC) provides excellent possibilities for group-type separation of edible oils and related samples. On a silica material, a mixture of free sterols is separated into the main three sterol classes on the basis of the number of methyl groups at the 4-position and the accessibility of the 3-hydroxy group, with hardly any effect of the carbon number of the sterol or number and positions of double bonds in the sterol molecule. For more complex mixtures, such as spreads, NPLC can be used for group-type separation of nonpolar species as, for example, sterol esters from the significantly more polar free sterols. The ultimate analytical method evidently is a combination of LC class–type (pre-)separation followed by detailed GC analysis of the individual classes. The off-line combination of this is basically a more modern and faster alternative for off-line TLC-GC methods as employed in official methods such as the Deutsche Gesellschaft fur Fettwisschenschaft (DGF) method F-III-1 (38). The on-line combination of LC and GC is pioneered by Grob and is particularly attractive if large series of identical analyses have to be performed.

From the above the powerful nature of GC for sterol analysis is evident. The full potential of GC, however, is only realized in combination with the appropriate sample pretreatment techniques and/or preseparation methods. This is especially true for the more complex sterol–containing samples, such as vegetable oil–based spreadable margarines (spreads), or biosamples like blood and feces. When looking to the most frequently employed methods of pretreatment and preseparation, two groups of techniques can be distinguished: wet chemical methods and chromatographic preseparation. Representatives from the two groups either can be used as stand-alone sample pretreatment methods prior to GC analysis or can be combined to give maximal sensitivity or selectivity. The most important techniques of the two classes are summarized in Table 4.

The wet chemical sample pretreatment techniques are used to: (a) isolate sterols from complex samples (e.g., extraction or saponification), (b) modify the chemical nature of the sterols to give them a polarity similar to that of the sterol esters so that sterols and sterol esters can be collected in one LC fraction (silylation), or finally (c) reduce the polarity/adsorptivity of free sterols in a final GC analysis suppressing adsorption and peak tailing effects. The combination of one of the chromatographic preseparation methods with an efficient GC separation on a selected fraction results in a very powerful two-dimensional system, where the first dimension provides class-type selectivity, and the second dimension provides the separation within a class as well as sensitive detection. For large-scale routine use it is highly attactive if the multidimensional approaches outlined above can be performed on-line. This clearly is a reason to replace TLC by a column type of preseparation method. Irrespective of which combination is selected and whether it is performed in an off-line or on-line fashion, the typical difficulties commonly seen in multidimensional methods are also encountered here. Very often the fraction resulting from the first dimensional treatment is too dilute for direct injection into the second separation system. Moreover, solvent compatibility problems are occasionally encountered. The exact way in which sterol-containing samples have to be pretreated is determined by the actual question or the degrees of information required in combination with the composition of the matrix. As an example, if the total sterol content of a food product needs to be determined, saponification is a logical step. Upon saponification all sterol-containing molecules (e.g., sterol esters, sterol ferulates, etc.) are converted to free sterols. These can than be quantified using a "total sterol" GC method. If, on the other hand, the ratio of sterol

Table 4 Pretreatment Techniques Relevant for Sterol Analyses

Method	Principle
Wet chemical methods	
Extraction	Isolation of the lipid fraction containing the sterols using, e.g., liquid-solid extraction, Soxhlet, accelerated solvent extraction etc.
Saponification/extraction	Cleavage of ester bonds under the action of a strong alkaline solution. Effective method for removal of triacylglycerides if combined with an extraction using a nonpolar solvent. After cleavage sterols end up in the "nonsaponifiables" fraction.
Transesterification	Conversion of an ester into another ester, usually by reaction with an alcohol or alkyl-donating reagent. Also used to release sterols from sterol esters.
Silylation	Conversion of polar groups (hydroxy, amine, carboxylic acid, etc.) to trimethylsilyl groups. The resulting ether derivative has more favorable GC properties (e.g., lower elution temperature and reduced tendency to adsorb). In NPLC silylated sterols elute together with sterol esters.
Chromatographic methods for preseparation	
Thin-layer chromatography	Method for class-type separation of complex samples. The most widely used stationary phase material is silica. On this material separation is based on polarity (alkanes < sterol esters < TAGs < free sterols). Fractions can be recovered semiquantitatively from the TLC plate for further analysis by, e.g., GC.
Normal-phase HPLC	Comparable to TLC. In general the efficiency of the preseparation is better. The loadability and the ruggedness tend to be lower. Technique is readily automated.
Solid-phase extraction and column chromatography	Comparable to TLC and NPLC. Lower efficiency than HPLC, but disposable nature allows use of very "dirty" samples as, e.g., plasma extracts, etc.
Gel permeation chromatography	Preseparation method that can be used to separate molecules on the basis of size. Gained some popularity for pesticide analysis in edible oils and fats (32,57). Not widely used for sterol analysis.

Table 5 Requirements for Sample Pretreatment and Preseparation Prior to Gas Chromatography[a]

	Sterol bulk product	Sterol esters bulk product	Food product (e.g., spread)	Biomatrix
Total sterol content	*Aim*: Improve GC peak shape for free sterols (optional). *Technique*: Silylation.	*Aim*: Conversion of sterol esters into free sterols. *Technique*: Saponification or **transesterification**. *Aim*: Improve GC peak shape for free sterols (optional). *Technique*: Silylation.	*Aim*: Elimination of interfering analytes (optional). *Technique*: Extraction or **saponification**. *Aim*: Conversion of sterol esters into free sterols. *Technique*: **saponification** or transesterification. *Note*: Saponification accomplishes two aims simultaneously. *Aim*: Improve GC peak shape for free sterols (optional). *Technique*: **Silylation**.	*Aim*: Elimination of interfering analytes. *Technique*: Extraction or **saponification**. *Aim*: Conversion of sterol esters into free sterols. *Technique*: **saponification** or transesterification. *Note*: Saponification accomplishes two aims simultaneously. *Aim*: Improve GC peak shape for free sterols. *Technique*: **Silylation**.
Sterol vs. sterol esters	*Aim*: Improve GC peak shape for free sterols (optional). *Technique*: **Silylation**.	*Aim*: Improve GC peak shape for free sterols (optional). *Technique*: **Silylation**.	*Aim*: Elimination of interfering analytes sterols (optional). *Technique*: **Extraction**. *Aim*: Isolation of sterols and sterol esters from fat phase/extract (optional). *Technique*: LC/TLC/SPE preseparation after silylation.	*Aim*: Elimination of interfering analytes. *Technique*: **Extraction**. *Aim*: Isolation of sterols and sterol esters from fat phase/extract (optional).

Sterol classes (des-, mono- and dimethyl)

Aim: Group type separation of DeMS, MMS, and DiMS. *Technique:* **LC/TLC/SPE.** *Aim:* Improve GC peak shape for free sterols (optional). *Technique:* **Silylation.**

Aim: Conversion of sterol esters into free sterols. *Technique:* **Saponification** or transesterification. *Aim:* Improve GC peak shape for free sterols (optional). *Technique:* **Silylation.**

Aim: Elimination of interfering analytes (optional). *Technique:* Extraction or **saponification.** *Aim:* Conversion of sterol esters into free sterols. *Technique:* **saponification** or transesterification. *Note:* Saponification accomplishes two aims simultaneously. *Aim:* Improve GC peak shape for free sterols (optional). *Technique:* **Silylation.**

Aim: Improve GC peak shape for free sterols (optional). *Technique:* **Silylation.**

Technique: **LC/TLC/SPE** preseparation after silylation. *Aim:* Improve GC peak shape for free sterols (optional). *Technique:* **Silylation.** *Aim:* Elimination of interfering analytes. *Technique:* Extraction or **saponification.** *Aim:* Conversion of sterol esters into free sterols. *Technique:* **saponification** or transesterification. *Aim:* Improve GC peak shape for free sterols (optional). *Technique:* **Silylation.** *Note:* Saponification accomplishes two aims simultaneously.

Individual (desmethyl) sterols

Sample preparation: See sterol classes (des, mono and dimethyl)

[a] Procedures/techniques printed in bold face incorporated in the methods in use at Unilever Research and Development Vlaardingen.

esters to free sterols has to be determined in a sterol ester bulk product, saponification can evidently not be used. Table 5 gives an overview of the general requirements that have to be imposed on the sample pretreatment and preseparation methods as a function of the type of analysis that is desired and the matrix in which the sterols are present. The experimental details of the procedures are summarized in Sec. II.B and subsections thereof. Here it should be emphasized that the parameters given are general guidelines rather than fully optimized conditions. Some parameters tend to be rather uncritical, whereas others have to be carefully selected and controlled. A proper validation study should always be performed to confirm the validity of the approach selected and the experimental conditions chosen.

B. Experimental Details of Pretreatment and Preseparation Methods

1. Saponification

Saponification probably is the most frequently performed reaction in oil and fat research. Treatment of the sample with a strongly alkaline alcoholic solution converts esters into (dissociated) free acids and free alcohols. In complex samples, such as natural extracts and biomedical samples, part of the sterols might be present as complex conjugates. Upon saponification most of these will also be released, with possibly the exception of sterylglycosides which require acid hydrolysis (39). Because saponification effectively destroys solid matrices, it can also be used as a much faster alternative to liquid extraction because in the solid samples the sterols can be strongly bound to a solid matrix or even captured in matrix particles. To recover the sterols from the saponification mixture after reaction, an extraction with a nonpolar organic solvent has to be performed to obtain the "unsaponifiables" fraction. A wide range of conditions for saponification of fat and fatty food samples has been described in literature. In our laboratory basically three different saponification protocols are used in sterol analysis: (a) strong saponification with a sodium or potassium hydroxide solution in methanol or ethanol; (b) strong saponification with a sodium hydroxide in methanol solution followed by borontrifluoride/methanol esterification; or (c) mild saponification with potassium hydroxide in ethanol. Protocol b, saponification with subsequent methylation of the fatty acids, is only applied if both the sterol and the fatty acid composition of the sample must be characterized. The experimental protocols for these procedures are summarized in Table 6. The details given in this table are the conditions as used in our laboratory. Based on practical experience we are confident in

Table 6 Experimental Details for Three Saponification Methods (Including Extraction)

Procedure	Protocol for raw materials or spreads. For samples that contain the analytes of interest at lower levels higher sample weights can be processed.
Strong sodium hydroxide in ethanol saponification (alkaline saponification)	To approximately 0.2 g of sample (dry weight), 4 mL of a 2 mol/L solution of NaOH in ethanol is added. The sample is mixed and heated to 70°C for 90 min in a closed vial with occasional shaking. The vial is allowed to cool to room temperature. Extraction is performed in triplicate using hexane or dichloromethane (5 mL each). The choice of the extraction solvent is determined by practical parameters (coextracted interferents, upper vs. lower layer etc.).
Strong sodium hydroxide saponification with BF_3/methanol methylation	To approximately 0.2 g of sample (dry weight), 8 mL of a 0.5 mol/L methanolic sodium hydroxide solution is added. The sample is mixed and boiled under reflux for 30 min. Next, 10 mL of a 20% BF_3 in MeOH solution is added, the sample is boiled for another 4 min. Finally, 5 mL hexane is added. The resulting mixture is again boiled for 4 min. The vial is filled with sodium chloride–saturated water. The hexane layer is finally transferred into a test tube containing 1 g of anhydrous sodium sulfate to remove traces of water.
Mild saponification using potassium hydroxide in methanol at room temperature	To approximately 500 mg of sample add 5 mL of a 2 mol/L potassium hydroxide in ethanol solution. Seal vial and shake intensively for 15 s. Allow to react for 18 h (overnight) in the dark at room temperature. For extraction add 20 mL of dichloromethane and vortex for 15 s. Add 10 mL of a 0.1% w/w citric acid solution. Shake for 15 s. Discard the aqueous top layer. Repeat citric acid washing step until the organic layer is clear.

stating that the saponification process is very rugged. Slightly altered conditions are very likely to yield identical results.

For relatively simple sterol-containing samples, such as raw materials and spreadable margarines, alkaline saponification generally is without experimental difficulties. For more complex samples, especially those containing high levels of proteins, emulsifiers, and carbohydrates, emulsion formation or gelation of the organic solvent during the extraction can be an annoying experimental problem. Standard tricks, such as salt addition, addition of a low percentage of a lower alcohol, or repeated freezing/ thawing, usually suffice for breaking the emulsion or the gel. For samples that show a very strong gelation the use of a slightly more polar extraction solvent such as, for example, ethyl acetate might be necessary. For samples that show a tendency to form strong emulsions upon extraction, vortex mixing should be avoided as emulsions resulting from vortexing are particularly difficult to break.

2. Transesterification

The chemical process in which ester bonds are broken and the acid and/or the alcohol part of the ester are coupled to another alcohol or acid to yield new esters is called transesterification. Transesterification can be obtained either in a two-step approach of saponification with subsequent methylation (Table 6), or in a one-step approach. The one step method uses sodium alkanolates, e.g., sodium butanolate or sodium methanolate (40). The experimental details of the one-step sodium methanolate method as applied in our lab are given in Table 7. In sterol analysis there are two situations in which transesterification has to be applied. First, if detailed information on the fatty acid part of a sterol esters is desired. Transesterification of a sterol ester gives the free sterol and the methylated fatty acid. The various analytical methods for FAME analysis can now be used for detailed analysis of the fatty acid moiety of the sterol esters. Evidently, all other alkyl esters (TAGs!) have to be removed before transesterification. One-step transesterification can also be used as a rapid alternative for saponification with boiling alcoholic alkaline solutions. Transesterification of a sterol ester–containing spread, for example, would convert all sterols into free sterols, which would than allow isolation of the sterols by HPLC and subsequent analysis. The ultimate choice between alkaline saponification and one-step transesterification as two means for converting sterol esters into free sterols depends on a number of parameters. Alkaline saponification is generally a stronger method. It is hence more suited for the more complex samples as, for example, spreads, food products, plasma, or feces, in which the sterols might be present in other forms or bound to solid particles. To avoid

Table 7 Experimental Details for Transesterification of Sterol Esters and Sterol-Containing Fat Samples

Procedure	Protocol
Sodium methanolate transesterification	A solution of 20% sodium methylate is diluted in MTBE (ratio 4:6 v/v). 1 mL of this solution is added to 100 mg of sample. The sample is vortexed for 10 s. Reaction is performed at 70°C for 15 min with regular mixing. Allow mixture to cool to approx. 40°C. Add 1 mL of water, shake, and add 5 mL of hexane. Vortex for at least 1 min. Centrifuge at 2000 g. Transfer the upper hexane layer into a clean vial. Repeat the extraction twice and combine the extracts.

problems with incomplete release of the sterols, the recommendation is to use one-step transesterification only for sterol ester hydrolysis of simple samples, i.e., sterol esters, and preisolated fat phases from whole foods. For all other samples alkaline treatment is recommended. The drawbacks of alkaline saponification include, among other factors, the greater risk of emulsion formation upon extraction of the sterols and the need for an additional drying step if the sterol-containing extract has to be further fractionated using normal-phase HPLC. Alkaline saponification also has a slightly higher risk of cis/trans isomerization or migration of the double bond in the fatty acid chain (41).

3. Silylation

Silylation refers to the chemical reaction in which trimethylsilyl-donating reagents are used to convert polar hydrogen–containing groups, e.g., amine or hydroxy groups, into the corresponding trimethylsilyl ethers. See Table 8 for practical details. In GC there are two main reasons for the use of silylation: (a) silylation increases the volatility of the analytes and (b) it reduces interactions with active sites in the chromatographic columns resulting in an improved peak shape. Although on modern, well-deactivated GC stationary phases underivatized sterols can be eluted with excellent peak shapes, it can still be advisable to silylate free sterols prior to GC analysis. This is especially true if low levels of sterols have to be analyzed in complex samples, e.g., blood extracts, etc. Coinjected "dirt" from such samples tends to rapidly deteriorate column inertness, resulting in tailing peaks and increased errors in quantification if the sterols are analyzed without

Table 8 Experimental Details for Silylation of Dry Sterol–Containing Samples

Procedure	Protocol
N,O-Bis(trimethylsilyl) trifluoroacetamide (BSTFA) silylation	Weigh approximately 200 mg of a dry sample into a vial. Add 2 mL of a BSTFA/pyridine solution 1/4 v/v. The BSTFA contains 1% trimethylchlorosilane. Mild silylation can be performed at room temperature (1 h). If sterically hindered groups are to be silylated, or if the samples are not perfectly dry, it is recommended to work at elevated temperature (60–70°C) for 30 min.

derivatization. For relatively simple samples containing high levels of sterol(esters) such as raw materials or even spreads, the silylation step can also be safely omitted, as described in Sec. III.E.

In sterol analysis, silylation is not only performed to improve the GC behavior of the free sterols, it can also be used to modify the polarity of the free sterols in a way that free sterols and sterol esters obtain more or less identical polarities for liquid chromatography separations. Liquid-phase group-type separations on normal-phase silica-type materials can than be used to collect "free" sterols (but now silylated) and sterol esters in a single fraction. A simple GC separation than suffices to quantify the two groups of analytes: (silylated) free sterols and sterol esters. See Table 8 and Sec. III.C. This analysis is particularly useful if esterification degrees have to be determined for spread samples.

4. Liquid Chromatography

Besides being an excellent separation method for the analysis of sterols, liquid chromatography also is an extremely powerful sample pretreatment technique for subsequent sterol analysis by GC. Whereas the LC separation of sterols is almost exclusively performed using (nonaqueous) reversed-phase LC, LC sample pretreatment for GC in the area of sterol analysis is generally performed in the normal-phase mode. The main advantage of NPLC separations is its much greater potential for group-type selectivity, i.e., the possibility to isolate groups, or in common sterol terminology referred to as "classes," of compounds with similar chemical functionalities irrespective of their size. For example, in NPLC all free alcohols would elute as one band more or less independent of the alkyl chain length. For this reason NPLC is ideally suited to isolate a specific class of compounds from a complex sample. High-performance NPLC combines excellent group-type separation characteristics with a very high efficiency. This is a distinct advantage over TLC or SPE sample pretreatment. The original procedures for detailed sterol analysis often included a TLC step for isolation of the polarity group of interest (42). Because this TLC preseparation involves several manual sample handling steps it cannot be automated easily. With automation being a key requirement in our laboratory, TLC was replaced by HPLC on a similar stationary-phase material. Our fully automated LC system with autosampler and fraction collector allows preseparation of up to 30–40 samples per day. Other authors selected SPE as the replacement for TLC (43). Although SPE clearly has some advantages over HPLC prefractionation (*vide infra*), the main advantage of HPLC is the much higher separation efficiency. Sterol class–type separation is much better than in SPE and generally also better than in TLC. It is only in the last 10–15 years that HPLC has attracted considerable attention as a sample pretreatment technique for GC. A drawback of HPLC sample preparation is that the columns have a relatively low loadability in comparison to preparative TLC plates. This means that only small amounts can be injected and dilute fractions are obtained. Because of this, sample preconcentration prior to GC injection is generally required. With modern techniques for GC large-volume injection, this can be avoided resulting in a significantly reduced time for sample preparation (44). A second drawback of HPLC as a sample pretreatment method is the vulnerable nature of the columns. Unlike TLC and SPE where disposable separation media are used, HPLC columns have to be reused many times. For dirty samples TLC and SPE therefore remain the methods of choice. A final drawback of all NPLC preseparation systems is their sensitivity for water. Even only traces of water in the sample can significantly affect retention times of the analytes. Thus, care should be taken to avoid the introduction of water into the NPLC system.

Basically four different applications of NPLC sample pretreatment can be distinguished in sterol (class) analysis:

1. Isolation of sterol esters from a fat/oil or spread sample for quantification or further characterization of, for example, the fatty acid chains
2. Isolation of all sterols (i.e., free sterols *and* sterol esters) from a fat/oil or spread after silylation for (a) total sterol quantification, (b) determination of the sterol/sterolester ratio, or (c) further characterization of the sterols/sterol esters
3. Class-type separation of free sterols, i.e., separation of des-, mono-, and dimethyl sterols

    4.   Isolation of polar sterol oxidation products from the nonsaponifiables fraction.

The experimental details of the NPLC separations as used in our laboratory are given in Table 9.

    An important step in off-line LC isolation of specific sterol fractions is the determination of the elution window of the desired sterol fraction. This is generally done using a few readily available sterols, such as cholesterol, stigmasterol, and campesterol in case the desmethyl fraction is desired or cholesterol and stigmasterol esters if the elution window of the sterol ester fraction has to be established. In the determination of the elution windows it is recommended to include closely eluting compounds as well as to detect possible overlap as early as possible. As an example, wax esters elute only slightly before the sterol ester band. Injection of the pure compounds in LC allows accurate determination of the cut point between the fractions. In daily practice we use a real sample, e.g., a rape seed oil containing sterols and sterol esters for determining the HPLC windows. An example of a UV

Table 9   Experimental Details for Normal-Phase Liquid Chromatography Preseparation of Sterol-Containing (Fat) Samples

Isolation of free sterols or all sterols (free and esterified) from a spread. Last option requires prior silylation.	Column: Waters Spherisorb S5W (silica), 150 × 2.1 mm, particle size 5 μm (a column with a larger inner diameter is preferred if the sample contains low sterol levels or if more sample is needed for further experiments)	Gradient: 100% Hexane (5 min) to 50% MTBE at (30–40 min), to 100% Hx (41–60 min) Detection: UV at 210 nm Flow: 1 mL/min
Class type separation of free sterols into des-, mono-, and dimethyl sterols.	Column: Chrompack Spherisorb 5 Si, 150 × 4.6 mm, particle size 5 μm	Hexane/MTBE 90/10 v/v Detection: UV at 210 nm Flow: 1.5 m/min
Isolation of polar sterol oxidation products.	Column: Nucleosil 50, 100 × 4 mm, particle size 5 μm	Eluent: 1.5% isopropanol in hexane. Postrun conditioning: hexane/isopropanol 50:50 (v/v) for 7 min, start eluent for 20 min. Flow: 1 mL/min

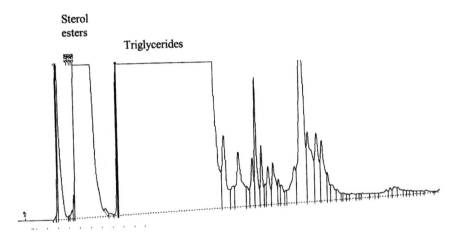

Figure 5 NP-LC-UV chromatogram of the group-type fractionation of the fat phase of a sterol ester–fortified spread.

trace (210 nm) obtained for such a sample is shown in Fig. 5. Although in theory retention in NPLC should only be a function of the number and nature of the polar groups in a molecule, in practice some effect of the alkyl chain length is seen as well. Within a compound group, larger molecules will generally elute first. Free sitosterol as an example will elute before cholesterol on LC. This means that elution windows should not be too narrow. This is even more necessary because retention times on normal-phase materials tend to be less stable than in reversed-phase separations.

C. Chromatographic Analysis of the Pre-Treated Sterol Samples

1. Separation Conditions

From the previous paragraphs it is clear that the extraction procedures for sample pretreatment and/or preseparation depend strongly on the type of question as well as on the sample matrix. In this respect sample pretreatment is more complex than the actual separation. The separation parameters depend almost exclusively on the degree of detail that is required, and are more or less independent of the original matrix of the sample. Detailed analysis of the desmethyl sterol composition of plasma, for example, can be performed using the same chromatographic separation conditions used for detailed sterol analysis of sterol concentrates (see also Chapter x). Evidently,

Table 10 Separation Conditions of the Various Chromatographic Methods for Sterol Analysis

Method	Column and precolumn	Injection and carrier gas (for GC)	GC temperature program or LC gradient	Detection
Total sterols (GC)	Column: CP-sil-5CB, 10 m × 0.32 mm × 0.12 μm. Precolumn: Nonpolar deactivated, 1.0 m × 0.53 mm Connected using press fit connector.	Injection: On-column, 0.5–4 μL. Carrier gas: Helium 1.5 ml/min.	60°C (1 min), 20°C/min to 300°C (3 min).	FID, 320°C
Sterol classes (GC)	Column: CP-sil-13CB, 50 m × 0.32 mm × 0.20 μm. Precolumn: Nonpolar deactivated, 1.0 m × 0.53 mm Connected using press fit connector.	Injection: On-column, 1 μL. Carrier gas: Hydrogen at 90 kPa.	170°C (2 min), 5°C/min to 250°C (0 min), 2°C/min to 300°C (5 min).	FID, 325°C
Free sterols vs. sterol esters (GC)	Column: J & W Scientific DB-5HT, 15 m × 0.32 mm × 0.10 μm. Precolumn: Nonpolar deactivated, 10 m × 0.53 mm Connected using press fit connector.	Large-volume injection: 20–100 μL. Carrier gas: Hydrogen at 3 ml/min.	100°C (12 min), 20°C/min to 240°C (0 min), 3°C/min to 260°C (0 min), 10°C/min to 370°C (5 min).	FID, 350°C

	Column	Injection/Carrier	Temperature program	Detector
Detailed desmethyl sterol (GC)	Column: CP-sil-13CB, 50 m × 0.32 mm × 0.20 μm. Precolumn: Nonpolar deactivated, 1.0 m × 0.53 mm Connected using press fit connector.	Injection: On-column, 1 μL. Carrier gas: Hydrogen at 90 kPa.	170°C (2 min), 5°C/min to 250°C (0 min), 2°C/min to 300°C (5 min).	FID, 325°C
Detailed sterols and stanols (GC)	Columns (in series): 1. CP-sil-13CB, 25 m × 0.32 mm × 0.25 μm, 2. CP-sil-8CB, 30 m × 0.25 mm × 0.20 μm. Precolumn: Nonpolar deactivated, 1.0 m × 0.53 mm Connected using press fit connectors.	Injection: On-column, 0.5 μL. Carrier gas: Hydrogen at 125 kPa.	60°C (1 min), 15°C/min to 250°C (0 min), 4°C/min to 310°C (12 min).	FID, 320°C
Sterol oxides, polar (GC)	Column: CP-sil-8CB, 25 m × 0.32 mm × 0.12 μm. Precolumn: Nonpolar deactivated, 1.2 m × 0.53 mm Connected using press fit connector.	Injection: On-column, 2–4 μL. Carrier gas: He at 75 kPa.	60°C (3 min), 20°C/min to 200°C (0 min), 4°C/min to 310°C (0 min).	FID, 320°C
Nonpolar sterol oxides (LC)	Column: Waters Spherisorb S5W (silica), 250 mm × 2.1 mm, particle size 5 μm.	Injection: 10 μL. Flow rate 0.2 mL/min.	A: 0.5% 2-propanol in n-hexane, B: 2-propanol in n-hexane (1:1, v/v). Gradient: 100% A (0–0.1 min) to 96% A (15 min) to 77%A (35 min), tot 40% A (36–44 min) to 100% A (45–70 min).	ELSD, tube temperature: 50°C, Nebulizer gas: N_2 at 1 L/min

the required sample preparation procedures are distinctly different. Moreover, in the plasma methods precautions should be taken to avoid rapid column contamination, such as by installing a retention gap or by using programmed temperature vaporiziation (PTV) injection.

Table 10 lists an overview of the analytical parameters of the various methods as they are used in our lab. Again it should be emphasized that similar stationary phases from other manufacturers or slightly different settings might be applicable as well.

2. Identification and Quantification

Identification and Peak Labeling. Special attention has to be paid on the proper peak labeling of sterols in GC methods. Within a relatively small carbon number inerval, most free sterols elute: the range of CN27–CN30, more specifically CN28 and CN29 [separation according to boiling point, closely correlating with the number of carbon atoms (CN) in the molecules]. Consequently, for the sterol esters the range is typically from CN46 to C48, if C18 is taken as the most common fatty acid. The sometimes small structural differences complicate a straightforward peak labeling with GC-FID methods. Unfortunately, direct GC-MS methods are not very helpful either, as mass spectra are very similar. Technically, GC-MS analysis of sterols and especially sterol-esters as standard technique in quantification/ identification was hampered by too low interface temperatures in the past. Identification based on retention times, on the other hand, is limited by the lack of (reliable) standard materials. Identification of the sterols in the GC chromatogram is thus by no means trivial. However, combining reported relative retention times (RRTs), a few model compounds and qualitative GC-MS data, a fairly detailed and validated peak labeling can be simply established. RRTs for a wide range of sterols are published in official documents and are remarkably consistent in elution order for the various GC stationary phases (Table 11)(27). Reference compounds can be obtained from a few suppliers specialised in sterols. The above described combination of RRTs, selected reference compounds and MS spectra, resulted in the peak-labelling for e.g. the des-methyl sterols as shown in Figure 6. An interesting phenomenon is observed for the elution properties of double-bond positional isomers. For several sterols the double bond in the B-ring is found at the 5 position ($\Delta 5$, most common) or on the 7-position ($\Delta 7$). The shift in double-bond position from 5 to 7 results for most free sterols in a more or less constant shift in retention time. For example, observe the retention time difference for the peak pairs $\Delta 5$-/$\Delta 7$-stigmasterol; $\Delta 5$-/$\Delta 7$-sitosterol; and $\Delta 5$-/$\Delta 7$-avenasterol. This phenomenon may also be used in peak

Table 11 Relative Retention Times (for the reference compound RRTs in bold face) of Sterols and Some Stanols for Various GC Stationary Phases

Sterol	Polymethyl(95)phenyl(5) siloxane (27)	Polymethyl(94)phenyl(5)vinyl(1) siloxane		OV-17 stat.phase packed column (26)	14% Phenyl 18% dimethyl polysiloxane, experimental (this chapter)	CP-sil-13 combined with CPSil-8
		(27)	(28)			
Cholesterol	0.63	0.67	**1.00**	**1.00**	0.98	0.99
Cholestanol			1.02	—	**1.00**	**1.00**
Brassicasterol	0.71	0.73	1.09	1.13–1.15	1.02	1.02
24-Methylene cholesterol	0.80	0.82	1.21	—	1.04	1.05
Campesterol	0.81	0.83	1.23	1.32–1.34	1.04	1.07
Campestanol	0.82	0.85	1.25		1.04	1.08
Stigmasterol	0.87	0.88	1.31	1.44–1.46	1.06	1.09
Δ7-Campesterol	0.92	0.93	1.38		1.08	1.12
Δ5,23-Stigmastadienol	0.95	0.95	1.40	—	—	—
Clerosterol	0.96	0.96	1.42	—	1.08	1.13
β-Sitosterol	**1.00**	**1.00**	1.47	1.66–1.68	1.08	1.14
Sitostanol	1.02	1.02	1.50	—	1.08	1.15
Δ5-Avenasterol	1.03	1.03	1.52	—	1.10	1.16
Δ5,24-Stigmastadienol	1.08	1.08	1.59	—	1.10	—
Δ7-Stigmastenol	1.12	1.12	1.65	1.85–1.95	1.10	1.19
Δ7-Avenasterol	1.16	1.16	1.70	—	1.13	1.23
Betulin	1.4	1.6	2.30	—	—	—

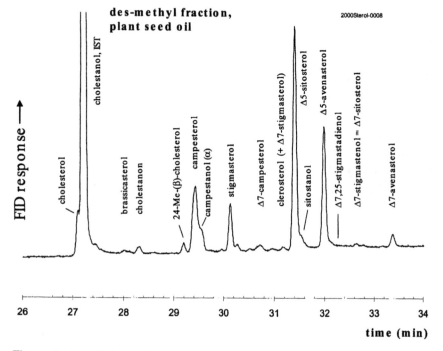

Figure 6 Detailed sterol composition of a vegetable oil sample analyzed after transesterification and LC sample clean-up on a CP-sil-13CB stationary phase capillary column. Please note that this sample was specifically selected to show a wide variety of different sterols. The composition is not representative for, e.g., soybean oil sterols (containing mainly campesterol, stigmasterol, sitosterol, and Δ7-stig-masterol (see Fig. 15)).

labeling and in interpreting the sterol chromatograms. Small differences in the retention time shifts for the selected peak pairs are caused by the isothermal elution.

Quantification. For quantification of sterols it is generally advisable to use an internal standard method. This is even more true if the sample requires elaborate sample pretreatment such as saponification or class-type fractionation by NPLC. The use of external standard quantification should be restricted to very simple analyses, such as rapid raw material screening. Standard addition methods for quantification would be ideal for situations where the sterols are present in a complex matrix as, e.g., a biofluid or a complex natural extract. Unfortunately, this route is hampered again by the lack of reliable standard materials at affordable prices. Moreover, standard addition methods tend to be more laborious than other methods for quantification and generally require the use of relatively large amounts of a sample. This could be difficult in, for example, sterol analysis in blood. For the latter type of samples where low levels of sterols are present in a very complex matrix, GC-MS using labeled analytes would be the ideal method for quantification. Again, however, the use of the method is precluded by the lack of suitable labeled reference materials. A notable exception here is D7-cholesterol that is readily available. Taking the above into consideration it becomes evident that the internal standard method should be the method of choice in most of the sterol analyses. When using this method one should always bear in mind that internal standard quantification will only give correct results if the internal standard is properly selected. The ideal internal standard should show a behavior identical to that of the sterols of interest during all steps of the analytical protocol. Moreover, it should never be present in the sample, should elute at a "free" position in the chromatogram, and should be available at a high purity and reasonable price.

In sterol analysis two internal standards are frequently used: β-cholestanol and 5α-cholestane. β-cholestanol is structurally very similar to the sterols and is hence a perfect internal standard for samples void of stanols that are analyzed using a high-resolution GC method (see, for example, Fig. 7). It cannot be used in combination with low resolution GC methods because on such systems it will coelute with cholesterol. Its use is also impossible if stanol mixtures are to be analyzed. Such mixtures are very likely to contain significant levels of the compound as a result of cholesterol hydrogenation. An alternative internal standard finally also is needed if low levels of sterols have to be quantified in samples that contain a very high level of cholesterol, such as blood samples. For such samples the separation of the β-cholestanol internal standard from the cholesterol is difficult at best. If β-cholestanol, the internal standard of first choice, cannot be used, 5α-

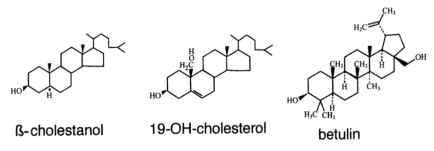

ß-cholestanol   19-OH-cholesterol   betulin

Figure 7   Internal standards commonly used in sterol chromatographic analysis.

Cholestane can be a reasonable alternative. Although this compound is structurally similar to the sterols, the lack of the hydroxy group on position 3 in the A-ring can make it chemically very different from the sterols or stanols. As a result of this it cannot, for example, be used in NPLC class type separations where it will elute significantly earlier than all sterols. In addition to the two internal standards mentioned above, alternative compounds such as betulin, are occasionally used (45). In contrast to the sterols and stanols this compound has two hydroxy groups, which means that it will behave slightly different from the normal sterols. Moreover, its FID response factor will be slightly different, both in the underivatized and in the silylated form. For sterol oxide analysis, other model compounds are evidently needed. A good candidate here is 19-OH cholesterol (46).

III. APPLICATIONS

A. Sterol Classes

At the initial stage of the research on plant sterols (PSs) detailed information was required on the types and concentration levels of all individual PSs in different raw materials next to the effects of processing. Since the most active compounds had not been identified at that stage, information on the entire "sterol spectrum," all three main sterol classes (Fig. 1), was needed. Because it is extremely difficult to separate all individual sterols from the three main classes in one GC analysis, an approach was developed to use their chemical differences for a preseparation into fractions based on the presence of none, one, or two methyl groups on the fourth position of the sterol backbone. Because only the total sterol equivalent of individual compounds had to be determined, samples were transesterified, which was a more rapid alternative to saponification for the samples of interest (mainly vegetables oils). The sample is transesterified using sodium methanolate (see Table 7 for details), which converts all sterol esters into free sterols and fatty acid methyl esters and the nonpolar components are extracted. Class separation of sterols was traditionally carried out by means of TLC. For the number of samples, which needed to be analyzed, the traditional separation was replaced by an NPLC approach, which circumvented a very time-consuming TLC step. By means of NPLC (see details in Table 9) PSs are well separated from TAGs and fractionated into the three classes: the 4,4-dimethyl PS fraction, the 4-monomethyl PS fraction, and the 4-desmethyl PS fraction (elution order). Each sterol class is collected from the column and analyzed by GC (see Table 10 for details). For quantification of the 4-desmethyl PS fraction an internal standard (β-cholestanol) is added prior to the HPLC separation. For quantification of the other two PS classes the internal standard is added after the HPLC fractionation. This method is especially useful if the overall sterol composition must be determined or the origin of the sterols must be identified. For example, rice bran oil containing all three PS classes (Fig. 8A–C), and for the sterol composition of some more "exotic" samples, e.g., oryzanol in rice bran oil.

The method was validated for the recovery and within-laboratory repeatability and reproducibility. The recovery was checked with a spiked sample of a sterol-free medium-chain triacylglyceride oil within the range of 0.05–0.5% and was 101.4% for the desmethyl sterols. The within-laboratory repeatability and reproducibility were determined with a regular PS. From these data the relative standard deviation (RSD) over a wide concentration range was determined. Both recoveries and RSD values are well within what would be appropriate for monitoring composition and level of PSE, keeping in mind that we focused on raw materials and products at about 10%. The RSD at that level is less than 2%.

B. Detailed Desmethyl Sterol Composition (Detailed Sterol Composition Method)

The des-methyl PS were identified as the most active cholesterol lowering ingredients, which prompted the need for a more dedicated approach. In this stage various raw material sources needed to be compared for concentration levels and identity of des-methyl PS. The sample prep and NPLC fractionation approach as described in Sect. IIB was applied, but only the desmethyl sterol fraction was analysed by GC (Table 10 for details). A typical example of the level of detail that can be obtained with this method, is shown in Fig. 6. Note that all peaks, which are labelled in this fraction, were identified by a combination of GC-MS and relative retention times (see Table 11). Please note that this sample was specifically selected to show a wide variety of different sterols. The composition is not representative for vegetable oil sterols used in cholesterol-lowering products (containing mainly campesterol, stigmasterol, sitosterol, and Δ7-stigmasterol. See Fig. 15 for a sterol profile of a vegetable oils sterol sample (47,49).

The relative retention order of the sterols is identical to that reported by several other authors, or as listed in official documents (26,27). The amount of each sterol is calculated against the internal standard (IST). Please note that for samples that contain high levels of cholesterol, a different IST such as 5α-cholestane may be required to avoid difficulties in accurately integrating the two potentially overlapping peaks. The method was validated over a wide concentration range (Fig. 9) giving a good repeatability and reproducibility. The method also yields the total sterol equivalent because all sterol derivatives are hydrolyzed prior to analysis.

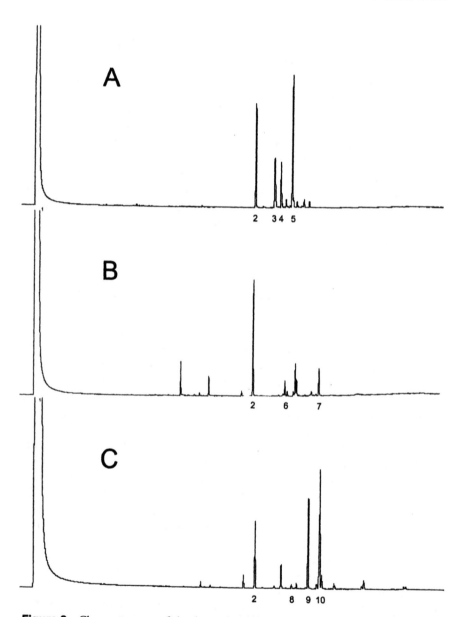

Figure 8 Chromatogram of the desmethyl (A), monomethyl (B), and dimethyl (C) fraction of rice bran oil. 1, solvent peak; 2, β-cholestanol (internal standard); 3, campesterol; 4, stigmasterol; 5, β-sitosterol; 6, obtusifoliol; 7, citrostadienol; 8, cycloartanol; 9, cycloartenol; 10, 24-Me-cycloartanol.

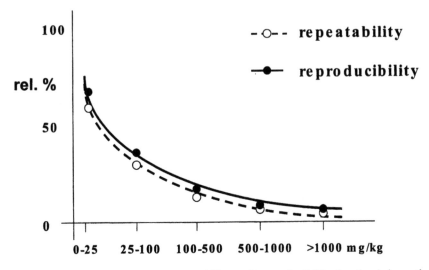

Figure 9 Within-laboratory repeatability and reproducibility for the 4-desmethyl detailed sterol composition method, over a wide concentration range.

Detailed sterol analysis is required at all stages of the functional foods development process and later production. Evidently the complexity of the matrix and the concentration levels differ significantly when comparing a raw material with a biofluid sample. However, in contrast to what one might expect at first glance, one single GC method suffices for all of these applications. The sample preparation of the various samples, on the other hand, will be very difficult. To a first approximation it could be stated that the sample preparation method is largely determined by the matrix, whereas the chromatographic method required is determined only by the degree of detail desired. In detailed sterol analysis saponification is widely applied. With only minor adjustments it can be used for the analysis of pure raw materials as one extreme of the analytical spectrum, to biofluids at the other extreme. The only analytically relevant difference between these samples when considering saponification is the concentration difference. Whereas pure materials contain the individual sterols at levels exceeding 50%, the concentrations in plasma samples can be as low as submicromolar concentration for an individual sterol. An example of the analysis of sterols in blood is shown in Fig. 10. Here approximately 0.5 g of sample, the maximum that is available in most animal or human trials, is saponified and the final extract is reconstituted in 50 μL solvent of which 1 μL is injected on-column. To compare this with raw material analysis: in raw

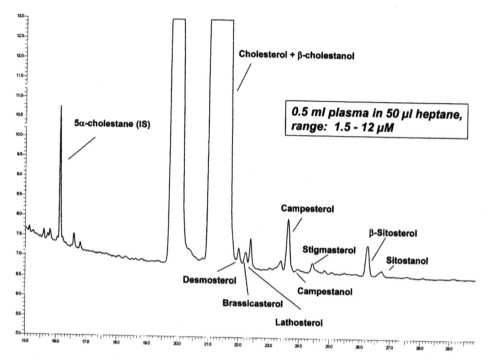

Figure 10 Detailed plasma sterol profile. The chromatogram is obtained with only 0.5 mL of plasma sample. Please note the difference in concentration for cholesterol vs. the different plant sterols and the cholesterol precursors desmosterol and lathosterol.

material analysis approximately 0.2 g of sample, the minimal mass that can be sampled without homogeneity problems, is processed. The nonsaponifiable matter is later extracted using approximately 10 mL of solvent. For raw materials the extract finally has to be diluted 10–50 times (for more details, see Table 6).

C. Free Sterols and Sterol Ester Levels (Free and Esterified Sterol Method)

In the food product under development, PSs were incorporated as fatty acid esters (PSEs). For optimization of the processing and preformulation studies, the need emerged for an accurate quantification of PSs and PSEs within one method, mainly to determine the conversion degree of the esterification reaction. For example, to analyze C18 esters of PSs, high-temperature GC using cold on-column injection (COC) is required because of their high boiling point. With this technique PSEs and PSs can be separated and quantified within in the same GC run, though for maximal sensitivity and selectivity they need to be isolated from the fat (mainly TAG) matrix before that step. The common approach of saponification with isolation and further analysis of the unsaponifiables could not be used here because information on PSE is lost when either saponification or transesterification methods are applied. Direct NPLC preseparation gives the problem of PSE eluting before the TAGs and PS eluting afterward. Moreover, for the latter fraction a strong sample dilution occurs. To circumvent the first problem samples were derivatised with TMS (see Table 8) in order to eliminate the polarity difference between PS and PSE. Subsequently NP-HPLC can be used to separate PSs and PSEs from TAGs and MAGs, which would otherwise interfere with the GC analysis (for details, see Table 9). The so-called HPLC polarity shift is illustrated in Fig. 11. All the PSs and PSEs are in the same elution fraction.

However, the HPLC fractionation causes sample dilution. In order to maintain the method sensitivity without further time-consuming concentration steps, large-volume injection (LVI) using so-called fully concurrent evaporation (37) was used, increasing the injection volume from 1 to 100 μL, thus increasing the sensitivity 100-fold. An automated on-line LC-GC setup was constructed, the device that is required for LVI in GC is shown in Fig. 12. The HPLC fraction that contains the PSs and PSEs is collected into a sample loop and injected into the GC system. For accurate GC quantification (see Table 10 for analysis details) of the PS-FA esters a C18 cholesteryl ester was applied as an internal standard.

An example of a sample containing free PSs and PSEs of FAs is shown in Fig. 11. The method is also capable of quantifying other than FA types of PSE, e.g., PS ferulates, acetates, and cinnamates. These sterol esters are present in rice bran oil and shea oil samples, respectively. The within-day repeatability of the method was evaluated; RSD values were below 3%. The trueness was better then 98% and the detection limit was 0.01%.

The advantages of this approach are obvious, but the main drawback is that very stable elution times on the NPLC system are required because the LVI interface has a limited elution window that can be transferred on-line to the GC. The on-line HPLC-GC approach was later converted to a more flexible off-line approach collecting a somewhat larger LC fraction but still using LVI-GC to maintain method sensitivity. In addition, when less accuracy is required a simpler profiling procedure could suffice.

An alternative method for the simultaneous determination of free and esterified sterols was developed in the framework of studies directed to understanding the fate of sterol esters in the gastrointestinal (GI) tract.

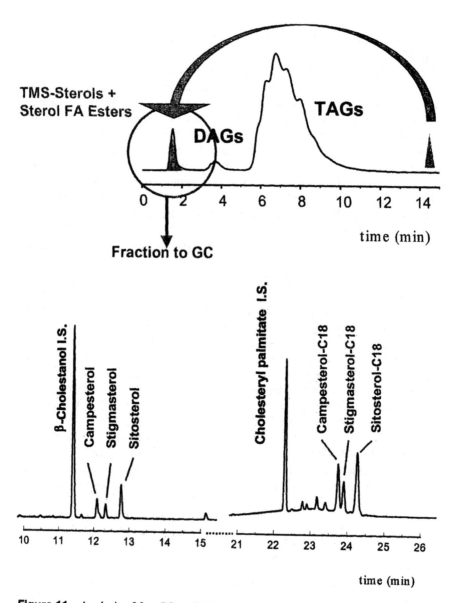

Figure 11 Analysis of free PS and PSE by HPLC isolation using the polarity shift of the sterol-TMS ethers (upper) and subsequent GC-FID analysis (lower).

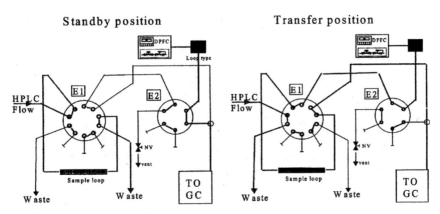

Figure 12 HPLC-GC instrument setup for on-line LC-GC analysis of sterols and sterol esters using LVI.

Although PSE should be handled by the various efficient esterases in the first part of the GI tract to result in the free sterols and fatty acids, a volunteer trial with stoma patients was conducted to confirm this. To understand the kinetics of the process, human trials have been conducted with otherwise healthy patients, ileostomized at different places in the GI tract (unpublished results). Determination of the levels of the free sterols and the sterols esters gives the desired information on the degree of hydrolysis as a function of GI tract transit time. Such studies require reliable methods for the determination of free vs. esterified sterols in the complex ileostomy excreta. From Table 5 it is clear that sample pretreatment should start with an extraction step. Saponification evidently cannot be used. The stoma fluid is a good example for the different and sometimes difficult matrices encountered in sterol analysis. Mild extraction conditions should be used, i.e., hexane extraction at ambient temperature under slow shaking overnight. Low temperatures have to be used to avoid reactions such as transesterification or oxidation of the compounds catalyzed by the stoma matrix as well as to minimize coextraction of other compounds that might interfere in subsequent quantification steps. The approach for sample clean-up as outlined in Table 5 resulted in the typical chromatograms as shown in Fig. 13. On average the study samples showed only low amounts of sterol esters still present at the end of the small intestine, indicating an efficient action of the GI enzymes (unpublished results). A similar approach can be used for the analysis of sterols in human or animal plasma. Plasma sterol analysis might be required to understand the absoption and uptake of sterols other than cholesterol. From the literature it is known that plant sterols, in contrast to

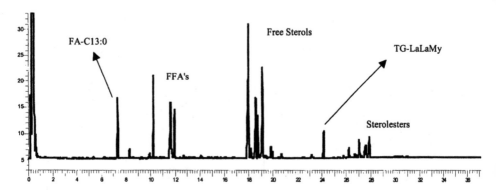

Figure 13 GC-FID chromatogram of a dichloromethane extract from the freeze-dried content of a stoma bag.

cholesterol, are hardly absorbed (3,10,17,48). The expected levels in plasma are in the micromolar range whereas cholesterol is in the millimolar range. A perfect example of this tremendous difference in concentration, is shown in the plasma sterol profile in Fig. 10.

D. Profiling of Main Fat Compounds

Foods fortified with phytosterols are complex samples containing a wide variety of compounds or compound classes present at widely differing concentration levels. In a typical PS containing vegetable oil–based spread for example, tri-acylglycerides can be present at levels as high as 60%. On the other hand, vitamins will generally not exceed the microgram to milligram/100 g product (ppb to ppm range). The applicability of gas chromatography for rapid profiling of the main compound groups in the fat fraction of a spread sample is illustrated in Fig. 14. This figure shows the analysis of a sample containing free sterols, sterol esters, free fatty acids (FFAs), monoacylglycerides (MAGs), diacylglycerides (DAGs), and tria-cylglycerides (TAGs). The derivatized compounds are separated on an apolar capillary column according to their boiling points, which correlate closely with the CN of the respective molecules. The chromatographic method used is the so-called carbon number method and is currently being reviewed for publication as an ISO standard method (36). Free fatty acids, free and esterified sterols, as well as acylglycerides can easily be detected and elute in clearly identifiable component groups. For PSs esterified with fatty acids covering a limited carbon number range, such as is the case for fatty acids derived from sunflower oil, no overlap in any of the component groups

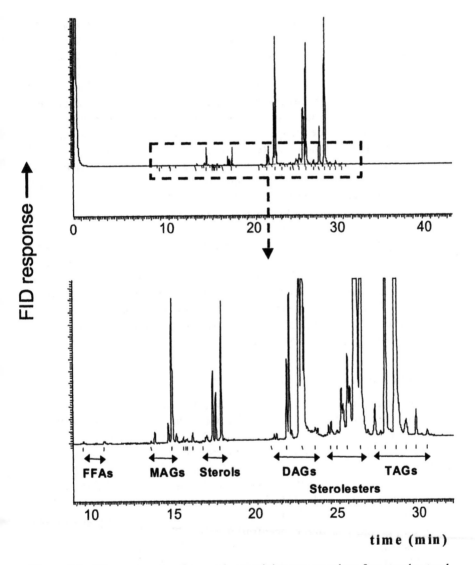

Figure 14 Chromatogram of a sample containing among others free sterols, sterol esters, free fatty acids, and fatty acid glycerol esters. The sample is fortified with certain component groups to show more detail. (Reprinted with permission from AJH Louter, CG Bauer-Plank, GSMJE Duchateau. Lipid Technol July 2002, 87–89.)

occurs. The free PSs are found within the CNs 26–28 and the PSEs in the present sample cover the CNs 42–48. The relative amounts of each component group can be derived from the relative area responses, which can be used to calculate the conversion degree. For the typical plant sterol–fortified spreads that contain sterol(ester) levels between 5% and 10%, this method can be used to get a first impression of the sterol and sterol ester content of a sample. For samples that contain lower levels or if a higher accuracy is needed, the method described above (Sec. III.C) should be used. The recovery and the within-laboratory reproducibility at the product level of 8% PS was tested with spiked samples and found to be satisfactory. Recovery for duplicates was 99.5% and 100.1%, and the relative standard deviation (for all $n = 5$) were 1.86% for a low-fat margarine (at 7.5% PS), 0.22% for a margarine (at 10.3% PS), 0.42% for a 12.6% fat blend, and 0.30% for a PSE concentrate.

For samples containing acylglycerides with a wider fatty acid distribution, severe overlap between the compound classes in the chromatogram occurs. In such samples, which in practice unfortunately means most real samples, a class separation is needed before the sterols or sterol esters can be quantified. This can be done using the HPLC-GC approach described above, where PSs and PSEs are separated from the main fat components (TAGs, MAGs) by means of NPLC.

E. Total Sterol Content, Direct Method

For analyzing the PS content of PSE food formulations used for clinical trials and final product quality control (QC), the only information that is generally required is the total desmethyl PS content. For this a fast and accurate method was required that covered levels ranging from approximately 5–10% (product formulation) to 60% (pure PSE). New methods had to be developed because the analytical methods for PS and PSE, though quite common and to be found in several normative method collections (e.g., ISO 12228, ref. 28) did not meet the requirements for ease of use, time involved per analysis, and concentration range. Typically, normative methods focus on the relatively low levels of PS and PSE as they occur naturally in edible oils and fats (typically below 1% level). Moreover, these methods are mostly time consuming and sometimes based on what is now considered outdated sample clean-up steps, such as TLC. However, for ingredient and product quality control, in many instances only the total level of desmethyl PS is required. The detailed desmethyl sterol method described above (Sec. III.B), involving transesterification and HPLC fractionation, is too time consuming for that task. As the total level of PSE is an important parameter in process monitoring as well as for label claim purposes, a fast alternative was needed. The speed of analysis could be increased dramatically by (a) simple liquid-liquid partitioning into hexane after alkaline hydrolysis, (b) direct injection of the sterols without derivatization, and (c) use of a fast temperature ramp in the GC program (see Table 10 for details). The resulting chromatogram is of course compressed in retention for the PSs with considerable overlap between the individual sterols. The relative retention times for PSs in this system are all close to each other (range 0.98–1.13 relative to the IST, Table 11) and the actual retention time is in the order of 7–8 min (Fig. 15).

This approach results in fast overall cycle times fitting well within QA/QC operations. An interesting remark can be made with respect to the limits to which analytical equipment can be pushed; in developing the fast GC oven program we encountered limitations in the heating rate of older GC types, limiting the method speed. The precision and accuracy was determined by repeated analyses of spiked samples in a collaborative test with 11 participants (49). The precision tests, collaborative test data, and

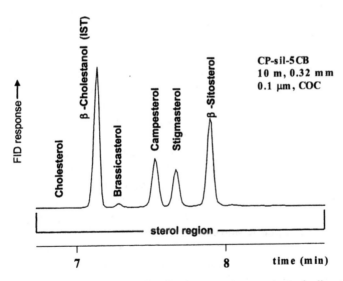

Figure 15 Detail of a chromatogram of a plant sterol ester mix, typically used in PS-enriched consumer products, after saponification. Shown is the elution window of the 4-desmethyl type of sterols. (Reprinted with permission from GSMJE Duchateau, CG Bauer-Plank, AJH Louter, M van der Ham, JA Boerma, JJM van Rooijen, PA Zandbelt. Fast and accurate method for total 4-desmethyl sterol(s) content in spreads, fat-blends, and raw materials. J Am Oil Chem Soc 79:273–278, 2002.)

control charts gave RSD values of 3% within laboratory and 5% between laboratories at a PSE level of ±10%. These results indicate that the product specification window can be monitored with the current method based on single-sample analysis. The method is currently in the process of ISO certification and has been accepted by the Dutch normalization authority.

F. Sterol Oxidation Products, Polar

PSs are very stable molecules but may undergo oxidation due to the presence of a double bond in the B ring (46,50). In addition, side chain oxidation can occur due to enzymatic reactions or excessive heating (51). In order to assess the extent to which this occurred, an analytical procedure was developed for the determination of concentration levels and identity of polar PS oxidation products (PSOPs). Sterol oxidation products appear at trace levels, which implies the use of sensitive methods. Due to the similarities between the structures of the different sterol oxidation products, the applied analytical tool requires high resolution. Furthermore, interfering compounds from the food matrix can complicate the trace analysis of sterol oxidation products. Moreover, the instability of some sterol oxidation products and the possibility of further oxidation of the sterols during sample handling in the analytical procedure requires mild enrichment procedures and careful sample handling.

The method is based on the isolation of PSOPs from the fat matrix by cold saponification of triglycerides and sterol esters. Cold saponification (see Table 6 for experimental details) was required to prevent formation of artifacts and loss of PSOPs, which has been observed previously in the analysis of cholesterol oxidation products (52).

Solvent extraction of the unsaponifiables is followed by isolation of sterol oxidation products. Because of PSOPs are more polar then the unoxidized PSs, NPLC proved to be an excellent tool for their separation (see Table 9 for details). With SPE less reproducible results were obtained, especially for the less polar PSOPs such as the 25-OH PSs, which are oxidized on the aliphatic side chain. The final analysis step is carried out by GC with FID. Because of their polarity the PSOPs need to be converted to their TMS ethers. Their thermal instability requires cold-on-column injection. Identification of the key PS oxidation products was performed by means of gas chromatography–mass spectrometry (GC-MS). The main products formed are 7-hydroxy-, 5,6-epoxy- and 7-keto-PS. 25-Hydroxy and 5,6-dihydroxy PS were only formed in trace amounts. The PSOP elution range in GC starts with 7-hydroxycampesterol and ends with 7-ketositosterol (see Fig. 16).

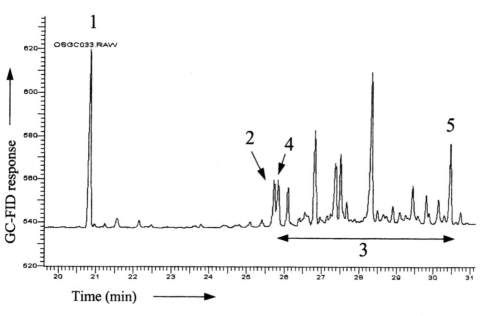

Figure 16 GC-FID chromatogram (PSOPs) of 19-OH cholesterol and phytosterol oxidation products obtained with the 25 m × 0.32 mm × 0.12 μm CP-sil-8CB column. Peak assignment: 1 = 5-α-Cholestane (I.S. II); 2 = 19-OH cholesterol (I.S. I); 3 = sterol oxidation products; 4 = 7-α-OH campesterol; 5 = 7 ketositosterol.

Isotope dilution MS was used to verify the absence of formation of potential artifacts by the method itself. To this end, different types of samples representing different stages in the whole analytical procedure were spiked with deuterated cholesterol. GC-MS was performed after the various stages of sample preparation. Figure 17 shows the GC-MS chromatogram obtained of a sample subjected to the entire protocol. Note that no new oxidation products have been formed; the 7-keto-Δ7-cholesterol was already present in the Δ7-cholesterol as purchased from Sigma. The most abundant oxidation product normally formed from cholesterol, 7-OH cholesterol, was not increased as a result of the sample preparation method.

The method is applicable to spreads (containing 20–65% water), oils, sterol esters, pure sterols, and fat extracts from food. The relative between-day reproducibility of the total content of sterol oxidation products in control samples is 8%, for the individual cholesterol oxidation products 6–15%. The recovery of sterol oxidation products is 91%. The limit of detection is 0.1 mg/kg.

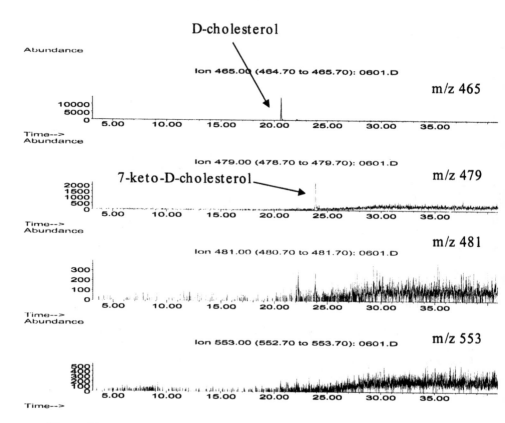

Figure 17 GC-MS chromatograms of a PS-containing spread sample spiked with Δ7-cholesterol. The traces represent the main ions of 7-keto- and 7-OH-Δ7-cholesterol. Note that no new oxidation products have been formed during sample treatment and analysis.

G. Sterol Oxidation Products, Apolar

For analysis of apolar PS oxidation products (ASOPs), the approach described above is not suitable. Due to the LC clean-up, apolar oxidation products such as ketoenes and ketodienes, which elute before the main sterol peak, are not collected and thus not detected by GC-FID. Oxidation products such as the hydroperoxides are not suitable for quantification by GC due to their thermal instability. To detect these apolar and polar PS oxidation products and to investigate fast options to quantify their levels a new method was developed based on NPLC with evaporative light scattering detection

(ELSD). This approach would circumvent the time-consuming LC sample clean-up and TMS derivatization. The NPLC approach provides a good separation between the different types of oxidation products owing to the differences in polarity. However, unlike the GC approach, there is no separation between the different PS oxidation products of one type, e.g., 7-hydroxy-PS. The detection limit that can be achieved is limited by the huge response for the unoxidized PS. Nevertheless, a good signal-to-noise ratio for both apolar and polar oxidation was observed (Fig. 18); the limit of detection for individual components was estimated at 5 mg/kg. This is still 50-fold higher than for the GC method, which was also reflected in the results when the same sample was analyzed for the concentration levels of the polar oxidation products by both the HPLC-GC-FID and the HPLC-ELSD approach. The latter technique gave systematically lower values because components below 5 mg/kg were not detected. These usually are compounds like 25-dihydroxy-PS and/or oxidation products of PS present at relatively low levels, e.g., brassicasterol and avenasterol. The main advantages of this approach remain speed of analysis and the ability to quantify apolar oxidation products.

H. Combined Sterol and Stanol Analysis

Initial detailed sterol analyses in our laboratory were performed using a slightly polar 14% phenyl–18% dimethylpolysiloxane stationary phase marketed by Chrompack/Varian under the name CP-sil-13. This column provides an excellent separation of the most relevant (desmethyl)sterols present in normal sterol samples, including the compounds present at low levels such as 24-methyl cholesterol, clerosterol, Δ5-avenasterol, and Δ7-stigmasterol. For the generally simpler stanol mixtures, this stationary phase also provides a good separation. For mixtures of sterols and stanols, however, the separation obtained is insufficient. In literature, GC stationary phases consisting of 5% phenyl–95% methylpolysiloxanes (known as SE-54, DB-5, Ultra-2, HP-5, CP-sil-8, etc.) have been shown to yield baseline separation for the sterol and the corresponding stanols (53,54). Unfortunately, on these 5% phenyl phases a number of critical pairs exists where a good separation is very difficult to obtain. As an example, sitostanol and Δ5-avenasterol can be nicely separated on a new 5% phenyl column at a carefully optimized temperature program. On a slightly aged column, however, the compounds frequently tend to coelute, making the system very critical to work with. Also in the steradiene/sterane/tocopherol region of the chromatogram the 5% phenyl phase gives a significantly poorer separation than a 14% phenyl-coated column. To combine the characteristics of the two columns we developed a method that uses two serially coupled columns. A combination

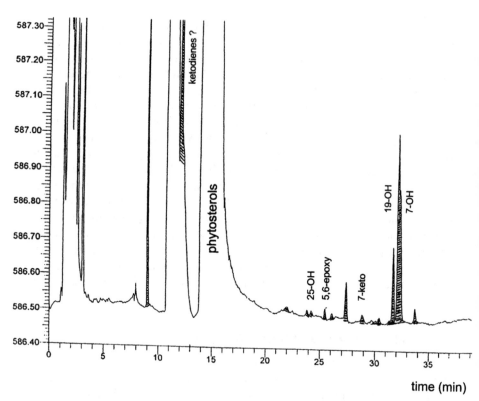

Figure 18 Apolar (ASOPs) and polar (PSOPs) sterol oxidation products detected with NPLC and ELSD. The ASOPs elute before and the PSOPs after the main PS peak. By lack of suitable standards, not all peaks are identified.

of a 25 m, 0.32 mm i.d. CP-sil-13CB (0.25 μm film thickness) column and a 30 m, 0.25 mm i.d. CP-sil-8CB column coated with a 0.25 μm film of stationary phase joined together using a press-fit connector gives an excellent combination of the selectivities of the individual columns. Note the order of the columns: CP-sil-13CB before CP-sil-8CB. In a temperature-programmed separation the exact elution order of the analytes depends on the order of the columns in the serial set. Examples of a separation obtained on the coupled column set are shown in Fig. 19.

The repeatability of the sterol/stanol method has been determined both for the sterols/stanols as well as for a number of other compounds often present in raw materials and related samples. The results are summarized in Table 12.

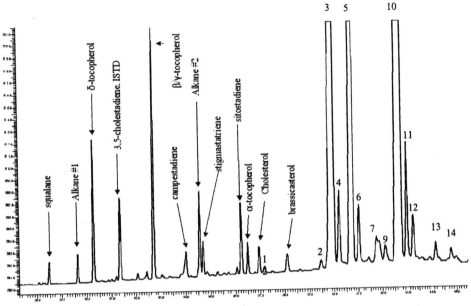

Figure 19 Chromatogram showing the separation of steradienes, tocopherols, alkanes, stanols, and sterols on the combined CP-sil-13 and CP-sil-8 column. Peak identification sterol region: 1 = cholestanol, 2 = 24-methyl cholesterol, 3 = campesterol, 4 = campestanol, 5 = stigmasterol, 6 = unknown, 7 = unknown, 8 = Δ7-campesterol, 9 = clerosterol (tentative), 10 = β-sitosterol, 11 = sitostanol, 12 = Δ5-avenasterol, 13 = Δ7 stigmastenol (tentative), 14 = Δ7 avenasterol (tentative).

I. Artefacts in Sterol Analysis

1. Matrix-Related Artifacts

Upon detailed analysis of the PS/PSE content and ratio in consumer products or fat-blends with the method described in Sect. III.C it might be observed that relatively high levels of analytes (though in absolute sense still extremely low), corresponding to the carbon number 40 area (CN40) are detected (Fig. 20). Initially CN40 analytes in a PSE-containing sample could be explained as PS esters with relatively short FA chains. That is, CN40 is the carbon number made up of a PSE with CN28 for the sterol part (campesterol, sitosterol) and a CN12 fatty acid (note: CN42 is the combination of a CN28 sterol with a C14 FA, and CN44 is the combination with C16, etc.). However, upon further inspection of the FA composition of the common fat blends used as FA source for PSEs this seems unlikely. The

Table 12 Repeatability of the Dual-Column Sterol/Stanol Method[a]

Component	Conc. (g/kg)	SD (%)
Part A: Early eluters		
Squalane	2.1	7.8
Alkane #1	2.9	1.3
δ-Tocopherol	15.6	7.2
β/γ-Tocopherol	29.2	6.7
Campestadiene	4.9	5.3
Alkane #2	10	2.8
Stigmastatriene	4.5	3.0
Sitostadiene	9.1	4.5
α-Tocopherol	3.8	7.8
Cholesterol	3.4	2.4
Cholestanol	0.9	8.9
Brassicasterol	3.2	1.8
Part B: Sterol/stanols		
Cholesterol	3.4	10.5
Brassicasterol	3.1	10.5
24-Methyl cholesterol	1.7	11.0
Campesterol	239	1.3
Campestanol	10.3	4.8
Stigmasterol	190	1.2
Δ7-Campesterol	4.1	9.1
Clerosterol	3.4	8.9
Sitosterol	344	1.1
Sitostanol	13.6	4.4
Δ5-Avenasterol	6.2	4.0

[a] The same method also provides data on a number of common impurities present in sterol/stanol raw materials.

level of CN12 FAs in most common vegetable oils is quite low. Therefore, the unexplained levels of CN40 in the analysis of sterol ester containing samples required another explanation. Further inspection of the oils used as FA source and the possible combinations that can result in a CN40 showed that oxidized FAs (hydroxy form) present in the PSE also may react with the BSTFA-derivatizing agents during the derivatization of the sample. Modern fat blends used in consumer products such as spreadable vegetable oil margarines are typically high in unsaturated fatty acids. The unsaturated FAs are vulnerable for some low level of oxidation, even when taking stringent production standards and specifications into account. A low level

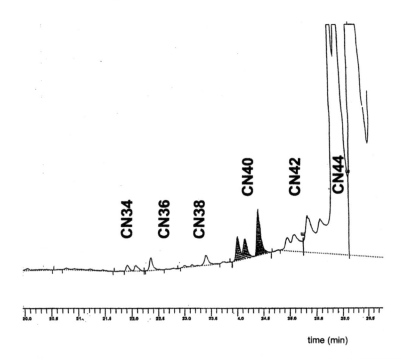

Figure 20 CN40 artifacts shown in carbon number analysis of a PS/PSE sample. Only the CN32 to CN46 part is shown. Please note the relatively high levels of the compounds eluting in the CN40 range.

of oxidized FAs (containing at some stage reactive hydroxyl functions) will always occur. Taking this into account, it becomes clear that there are quite some options to form the TMS derivative of the oxidized TAGs, which is equal to CN40 (Table 13).

Second, the vegetable oil in the matrix could also be a source of small amounts of partial glycerides, which may react to form analytes around CN40. For example, CN38 is not only the CN explained by any TAG with the FA combination $14+14+10$, which seems very unlikely in normal sources, but also the TMS derivative of the (nonoxidized) DAG $18+18$, or even the TMS derivative of the single oxidized DAG CN34 taking into account that each TMS group adds about two CNs to the original CN of the nonreacted molecule. Similarly, the number of combinations to explain CN40, 42, etc., is quite impressive.

This example shows that analysis of this type of samples requires careful inspection of the chromatograms, and detailed knowledge of the oil source and history is required for a full explanation.

Table 13 Carbon Numbers (CN) of Oxidized Partial Glycerides in Edible Oils

CN	FAs in DAG[a]	"true" CN	No. of TMS groups for nonoxidized form	CN including TMS groups upon FA oxidation			
				$1 \times OH$	$2 \times OH$	$3 \times OH$	$4 \times OH$
34	16 + 16	32	1	—	—	—	—
36	16 + 18	34	1	—	—	—	—
	16 + 18:1	34	1	38	—	—	—
	16 + 18:2	34	1	38	40	—	—
38	18 + 18	36	1	—	—	—	—
	18 + 18:1	36	1	40	—	—	—
	18:1 + 18:1	36	1	40	42	—	—
	18 + 18:2	36	1	40	42	—	—
	18:1 + 18:2	36	1	40	42	44	—
	18:2 + 18:2	36	1	40	42	44	46

[a] Ignoring the *sn*-specific position.

2. Silylation Artifacts

A second explanation for the CN40 artifact could be the derivatization reagent itself. It has been described that TMS radicals may form at high concentrations of the BSTFA reagent. The radicals could possibly react with double-bond systems in the plant sterols themselves. The then-formed double-derivatized sterol-TMS ethers would also elute in the higher CN range based on the estimated CN of 28 for the sterol and the 2xTMS effect (about CN32). This effect has been described, for example, for sterones (55); however, we have not observed this effect with our protocols.

3. Thermal Degradation of Sterol Esters

Despite their high molecular weight, sterol esters can be reliably separated and quantified using GC. Evidently, because of their high mass a sufficiently high GC (final) column temperature should be used. Typical final column temperatures as high as 375–400°C are applied. The thermal stability of the sterol esters is generally sufficient to allow the use of such temperatures. Under certain conditions, however, on-column degradation of the compounds can occur. This manifests itself by a slow increase of the baseline signal just before the sterol ester peak. This phenomenon is nicely illustrated in Fig. 21, which shows the GC-MS analysis of pure sitosteroloctadecanoate (C18 ester). The peak of the intact sterol ester (M_w 680.65) eluting at

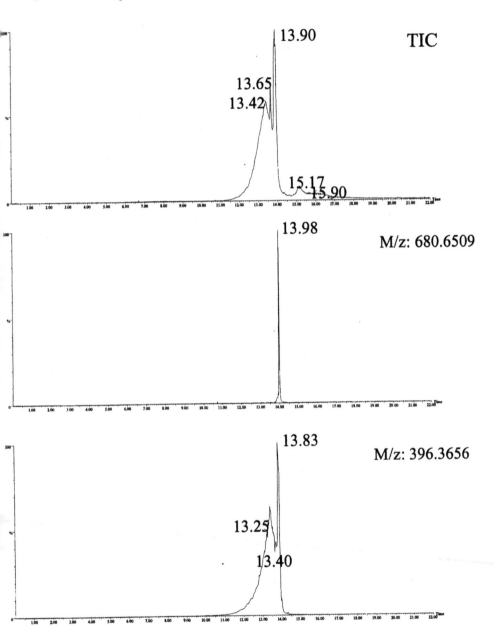

Figure 21 GC-MS analysis of sitosterol-octadecanoate showing severe degradation of the sterol ester.

13.98 min is preceded by a broad hump caused by lower molecular weight fragments generated by degradation of the sterol ester during elution. The mass spectra recorded over the hump show a very pronounced 396.37 fragment. This mass corresponds with the molecular mass of steradienes or sterols (M-H_2O fragment). If proper precautions are taken, on-column degradation of sterol esters can be avoided. Short columns operated at high linear velocities and constant flow conditions should preferably be employed. Column quality, or more specifically column inertness, and system contamination also seem to play an important role.

IV. GENERAL CONCLUSIONS AND FUTURE PERSPECTIVES

Upon entering the area of sterol analysis one might easily be overwhelmed with the high number of publications on this topic and the very detailed procedures described in official methods. Although each of these methods applies to a specific question, the practical approach might not be directly evident. For practical purposes within an industrial food research and development environment other criteria might be used in comparison with, say, clinical research. With the above-described methods we have defined within our laboratory a practical set of methods for sterol analysis which meet most of our requirements. New questions can be handled fairly quick by defining up front the desired level of detail and accuracy in the quantification. The answer required determines the separation strategy. The sample composition defines the clean-up strategy and the combination defines the solution to the question raised to the analyst.

With the examples described and the selection tools in the various tables, someone entering the area with basic knowledge on chromatography and sterols in general should be able to use the same tools and end up with a fit-for-purpose analytical method. Over the past several years we have seen and demonstrated ourselves that this approach results in less complicated analyses as would be expected from official methods and (older) sterol literature. Most likely, the approach taken by us will also work for many other matrices.

However, it is fair to remark that in the past years commercial instrumental developments allowed or facilitated the progress made in sterol analysis. GC-MS as essential for the identification is standard technique nowadays; sample-clean up strategies can be simplified without any loss in sensitivity (or even a gain!) with the introduction of large-volume injectors and on-line LC-GC. Sterol analysis has definitely moved from a highly specialized skill into a standard and mature analytical technique.

Taking this together, we are confident that GC techniques remain the methods of choice for sterol analysis. GCs are the common analytical tool in fats and oil analysis and present in most (industrial) laboratories in this field. Also, GC offers a high selectivity and thus detailed information in sterol analysis and good quantification options. Costs per sample are relatively low due to automation. It seems unlikely that these features of GC as analytical technique will be overruled by other techniques such as direct enzymatic or antibody-based methods ("kits") in the coming years.

ACKNOWLEDGMENTS

The authors thank Christina Bauer-Plank, Marnix van Amerongen, Arjan de Groot and Rob Diks for their discussions and critical view on the method requirements and methods. We thank the analytical staff at Unilever Research and Development Vlaardingen for their skill in the development and testing of the different methods discussed. Thanks also to Hubert Engels, Maria van der Ham, Ruud Poort, Herrald Steenbergen, and Toine van de Broek for contributing to the analytical development.

REFERENCES

1. LJ Goad, T Akihisa. Analysis of Sterols. 1st ed. London, 1997.
2. SL Abidi. Chromatographic analysis of plant sterols in foods and vegetable oils. J Chromatogr A 935:173–201, 2001.
3. G Salen, EH Ahrens Jr, SM Grundy. Metabolism of beta-sitosterol in man. J Clin Invest 49:952–967, 1970.
4. WH Ling, PJH Jones. Dietary phytosterols; a review of metabolism, benefits and side effects. Life Sci 57:195–206, 1995.
5. M Roth, P Favarger. La digestibilite des graisses en presence de certain sterols. Helv Physiol Pharmacol Acta 13:249–256, 1955.
6. PJH Jones, DE MacDougall, F Ntanios, CA Vanstone. Dietary phytosterols as cholesterol-lowering agents in humans. Can J Physiol Pharmacol 75:217–227, 1997.
7. MH Moghadasian, JJ Frohlich. Effects of dietary phytosterols on cholesterol metabolism and atherosclerosis: clinical and experimental evidence. Am J Med 107:588–594, 1999.
8. MH Moghadasian. Pharmacological properties of plant sterols—in vivo and in vitro observations. Life Sci 67:605–615, 2000.
9. M Law. Plant sterol and stanol margarines and health. BMJ 320:861–864, 2000.
10. MA Hallikainen, ES Sarkkinen, H Gylling, AT Erkkila, MIJ Uusitupa. Comparison of the effects of plant sterol ester and plant stanol ester-enriched mar-

garines in lowering serum cholesterol concentrations in hypercholesterolaemic subjects on a low-fat diet. Eur J Clin Nutr 54:715–725, 2000.

11. MA Hallikainen, ES Sarkkinen, MIJ Uusitupa. Plant stanol esters affect serum cholesterol concentrations of hypercholesterolemic men and women in a dose-dependent manner. J Nutr 130:767–776, 2000.

12. FH Mattson, SM Grundy, JR Crouse. Optimizing the effect of plant sterols on cholesterol absorption in man. Am J Clin Nutr 35:697–700, 1982.

13. A Sierksma, JA Weststrate, GW Meijer. Spreads enriched with plant sterols, either esterified 4,4-dimethylsterols or free 4-desmethylsterols, and plasma total- and LDL-cholesterol concentrations. Br J Nutr 82:273–282, 1999.

14. JA Weststrate, GW Meijer. Plant sterol-enriched margarines and reduction of plasma total- and LDL-cholesterol concentrations in normocholesterolaemic and mildly hypercholesterolaemic subjects. Eur J Clin Nutr 52:334–343, 1998.

15. HFJ Hendriks, JA Weststrate, T van Vliet, GW Meijer. Spreads enriched with three different levels of vegetable oil sterols and the degree of cholesterol lowering in normocholesterolaemic and mildly hypercholesterolaemic subjects. Eur J Clin Nutr 53:319–327, 1999.

16. RE Ostlund, CA Spilburg, WF Stenson. Sitostanol administered in lecithin micelles potently reduces cholesterol absorption in humans. Am J Clin Nutr 70:826–831, 1999.

17. DJ Sanders, HJ Minter, D Howes, PA Hepburn. The safety evaluation of phytosterol esters. Part 6. The comparative absorption and tissue distribution of phytosterols in the rat. Food Chem Toxicol 38:485–491, 2000.

18. OA Matvienko, DS Lewis, M Swanson, B Arndt, DL Rainwater, J Stewart, DL Alekel. A single daily dose of soybean phytosterols in ground beef decreases serum total cholesterol and LDL cholesterol in young, mildly hypercholesterolemic men. Am J Clin Nutr 76:57–64, 2002.

19. RP Mensink, S Ebbing, M Lindhout, J Plat, MMA van Heugten. Effects of plant stanol esters supplied in low-fat yoghurt on serum lipids and lipoproteins, non-cholesterol sterols and fat soluble antioxidant concentrations. Atherosclerosis 160:205–213, 2002.

20. J de Graaf, PRWD Nolting, M van Dam, EM Belsey, JJP Kastelein, PH Pritchard, AFH Stalenhoef. Consumption of tall oil-derived phytosterols in a chocolate matrix significantly decreases plasma total and low-density lipoprotein-cholesterol levels. Br J Nutr 88:479–488, 2002.

21. GM Morton, SM Lee, DH Buss, P Lawrance. Intakes and major dietary sources of cholesterol and phytosterols in the British diet. J Hum Nutr Diet 8:429–440, 1995.

22. L Normén, M Johnsson, H Andersson, Y van Gameren, P Dutta. Plant sterols in vegetables and fruits commonly consumed in Sweden. Eur J Nutr 38:84–89, 1999.

23. T Verleyen, M Forcades, R Verhe, K Dewettinck, A Huyghebaert, W De Greyt. Analysis of free and esterified sterols in vegetable oils. J Am Oil Chem Soc 79:117–122, 2002.

24. E Homberg. Sterol analysis as means to detection of mixed and adulterated fats. Fat Sci Technol 93:516–517, 1991.
25. M Biedermann, K Grob, C Mariani, JP Schmidt. Detection of desterolized sunflower oil in olive oil through isomerized delta 7-sterols. Z Lebensm Unters Forsch 202:199–204, 1996.
26. Anonymous. 2.403. Identification and determination of sterols by gas-liquid chromatography. In: A Dieffenbacher, WD Pocklington, eds. 1st Supplement to the 7th Edition of Standard Methods for the Analysis of Oils, Fats and Derivatives. London: Blackwell Scientific, 1992, pp 165–169.
27. Anonymous. 2.4.23. Sterols in fatty oils. European Pharmacopoeia. F-67029 Strasbourg, France: European Directorate for the Quality of Medicines, 2000.
28. Anonymous. Animal and vegetable fats and oils–determination of individual and total sterols contents: gas chromatographic method. ISO 12228. 1st, 1–16. 1991. Geneve, ISO.
29. HE Indyk. Simultaneous liquid chromatographic determination of cholesterol, phytosterols and tocopherols in foods. Analyst 115:1525–1530, 1990.
30. K Warner, TL Mounts. Analysis of tocopherols and phytosterol in vegetable oils by HPLC with evaporative light-scattering detection. J Am Oil Chem Soc 67:827–831, 1990.
31. M Careri, L Elviri, A Mangia. Liquid chromatography—UV determination and liquid chromatography—atmospheric pressure chemical ionization mass spectrometric characterization of sitosterol and stigmasterol in soybean oil. J Chromatogr A 935:249–257, 2001.
32. B Carstensen, W Schwack. GPC isolation and gas chromatogrphic determination of minor compounds in vegetable oils. Lipid Technol 135–138, 2002.
33. DJ Chitwood, GW Patterson. Separation of epimeric pairs of C24 alkylsterols by reversed-phase high-performance liquid-chromatography of the free sterols at subambient temperature. J Liq Chromatogr 14:151–163, 1991.
34. Anonymous. Codex Alimentarius. 2nd ed. Foods and Agricultural Organisation of the United Nations, 1999.
35. SP Kochhar. Influence of processing on sterols of edible vegetable-oils. Prog Lipid Res 22:161–188, 1983.
36. Anonymous. draft ISO/DIS22508: Animal and vegetable fats and oils—relative composition of oils/fats into free fatty acids, mono-, di-, and triacyl-glycerols by means of GLC (100% method) (carbon number analysis), 2002.
37. K Grob. On-line coupled LC-GC. Chromatographic Methods Series. Heidelberg: Huethig Verlag, 1991.
38. Anonymous. Sterols: isolation and gas chromatographic determination. Deutsche Gesellschaft fur Fettwissenschaft Standard Methods, Method F-III 1(98), 1998.
39. J Toivo, K Phillips, AM Lampi, V Piironen. Determination of sterols in foods: Recovery of free, esterified, and glycosidic sterols. J Food Comp Anal 14, 631–643, 2001.
40. F Ulberth. Determination of butanoic acid in milk fat and fat mixtures containing milk fat: a comparison of methods. Intl Dairy J 7:799–803, 1997.

41. M Yamasaki, K Kishihara, I Ikeda, M Sugano, K Yamada. A recommended esterification method for gas chromatographic measurement of conjugated linoleic acid. J Am Oil Chem Soc 76:933–938, 1999.
42. Anonymous. Determination of the composition of the sterol fraction of animal and vegetable oils and fats by TLC and capillary GC, official method Ch 6–91. Official Methods of the American Oil Chemists Society, 1991.
43. M Lechner, B Relter, E Lorbeer. Determination of tocopherols and sterols in vegetable oils by solid-phase extraction and subsequent capillary gas chromatographic analysis. J Chromatogr A 857:231–238, 1999.
44. HGJ Mol, M Althuizen, HG Janssen, CA Cramers, UAT Brinkman. Environmental applications of large volume injection in capillary GC using PTV injectors. J High Res Chromatogr 19:69–79, 1996.
45. K Aitzetmuller, L Bruhl, HJ Fiebig. Analysis of sterol content and composition in fats and oils by capillary-gas liquid chromatography using an internal standard. Comments on the German sterol method. Fett-Lipid 100:429–435, 1998.
46. AM Lampi, L Juntunen, J Toivo, V Piironen. Determination of thermooxidation products of plant sterols. J Chromatogr B 777:83–92, 2002.
47. AJH Louter, GSMJE Duchateau, C Bauer-Plank. Analysis of Plant Sterol Esters as Functional Food Ingredients. Lipid Technol 87–90, 2002.
48. MS Bosner, LG Lange, WF Stenson, RE Ostlund. Percent cholesterol absorption in normal women and men quantified with dual stable isotopic tracers and negative ion mass spectrometry. J Lipid Res 40:302–308, 1999.
49. GSMJE Duchateau, CG Bauer-Plank, AJH Louter, M van der Ham, JA Boerma, JJM van Rooijen, PA Zandbelt. Fast and accurate method for total 4-desmethyl sterol(s) content in spreads, fat-blends, and raw materials. J Am Oil Chem Soc 79:273–278, 2002.
50. PC Dutta, LA Appelqvist. Studies on phytosterol oxides. 1. Effect of storage on the content in potato chips prepared in different vegetable oils. J Am Oil Chem Soc 74:647–657, 1997.
51. F Guardiola, PC Dutta, R Codony, GP Savage. Cholesterol and Phytosterol Oxidation Products: Analysis, Occurence, and Biological Effects. 1st ed. Champaign IL: AOCS Press, 2002.
52. PW Park, F Guardiola, SH Park, PB Addis. Kinetic evaluation of 3 beta-hydroxycholest-5-en-7-one (7- ketocholesterol) stability during saponification. J Am Oil Chem Soc 73:623–629, 1996.
53. H Gylling, P Puska, E Vartiainen, TA Miettinen. Serum sterols during stanol ester feeding in a mildly hypercholesterolemic population. J Lipid Res 40:593–600, 1999.
54. LK Ng, M Hupe. Analysis of sterols: A novel approach for detecting juices of pineapple, passionfruit, orange and grapefruit in compounded beverages. J Sci Food Agr 76:617–627, 1998.
55. JL Little. Artifacts in trimethylsilyl derivatization reactions and ways to avoid them. J Chromatogr A 844:1–22, 1999.
56. F van Stijn, MAT Kerkhoff, BGM Vandeginste. Determination of polycyclic

aromatic hydrocarbons in edible oils and fats by on-line donor-acceptor complex chromatography and high-performance liquid chromatography with fluorescence detection. J Chromatogr A 750:263–273, 1996.

57. GA Jongenotter, MAT Kerkhoff, HCM van der Knaap, BGM Vandeginste. Automated on-line GPC-GC-FPD involving co-solvent trapping and the on-column interface for the determination of organophosphorus pesticides in olive oils. J High Res Chromatogr 22:17–23, 1999.

58. JJ Vreuls, RJJ Swen, VP Goudriaan, MAT Kerkhoff, GA Jongenotter, UAT Brinkman. Automated on-line gel permeation chromatography gas chromatography for the determination of organophosphorus pesticides in olive oil. J Chromatogr A 750:275–286, 1996.

4

Analysis of Phytosterols in Biological Samples

Arnis Kuksis
University of Toronto, Toronto, Ontario, Canada

I. INTRODUCTION

Phytosterols make up the greatest proportion of the unsaponifiable matter of plant lipids, but constitute only a minor proportion of the unsaponifiable matter of animal tissues, where cholesterol predominates. Both cholesterol and plant sterols naturally occur in the nonesterified and esterified form, along with conjugates with glycosides. Each plant species contains a characteristic phytosterol composition, which usually does not include cholesterol (1). The major plant sterol is sitosterol, whereas campesterol, stigmasterol, Δ^7-avenasterol, and brassicasterol are minor components. Depending on their structure and molecular weight, the phytosterols may be absorbed by the normal human intestine to a maximum of 5% of the absorption of free dietary cholesterol, but in patients with sitosterolemia the absorption of phytosterols may reach 30% of cholesterol absorption (2).

There exists a longstanding interest in the hypocholesterolemic effects of dietary phytosterols (3,4), which have been recognized for their hypocholesterolemic effects in human nutrition (5,6). Recent reports on the beneficial effects of incorporation of sitostanol esters into food fats has renewed interest in the plasma cholesterol–lowering activity of plant sterols (7,8), which had begun to vain because of poor patient compliance and large amounts of sterols required for measurable effect. Other benefits of plant sterol ingestion have been recognized in cancer prevention (9,10), and in modulation of inflammation and immunoresponse (11,12). Phytosterols have been claimed to be antiulcerative, antibacterial, antifungal (13), to heal injuries on skin and

133

mucosa (14,15), and to inhibit platelet aggregation (16). Due to this growing interest in the physiological properties of phytosterols, it is of great interest to obtain accurate quantitative data in the determination of these lipids both in food samples and in biological specimens.

The identification and quantification of plant sterols in biological materials is usually performed by capillary gas–liquid chromatography (GLC) and GLC with mass spectrometry (GC/MS) (2,17), which necessitate elaborate sample preparation and/or expensive equipment. However, recently published GLC methods with hydrogen flame ionization (GLC/FID) (18,19) provide for the precise measurement of sterols in diet samples and in human serum. High-performance liquid chromatography (HPLC) has also been proposed for the quantification of plant sterols in natural products, and the method has been extensively utilized. Careri et al. (20,21) have reviewed the field and have validated the methodology using HPLC/UV and HPLC with atmospheric pressure chemical ionization–mass spectrometry (LC/APCI/MS) methods, which they have developed for determination of sitosterol and stigmasterol in vegetable oils. Mezine et al. (22) have described the application of LC/APCI/MS to analysis of plant sterol and stanol esters in cholesterol-lowering food products.

This chapter reviews the progress in the analysis of plant sterols in biological samples with emphasis on eukaryotic systems. The principal topics

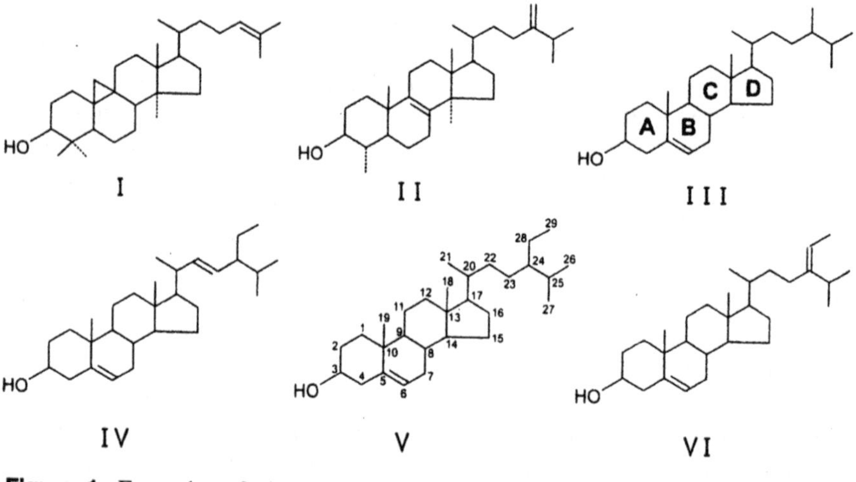

Figure 1 Examples of ubiquitous plant sterol structures: I, cycloartenol (4,4-dimethylsterol); II, obtusifoliol (4-methylsterol); III, campesterol (4-desmethylsterol); IV, stigmasterol; V, sitosterol; VI, Δ^5-avenasterol. (From Ref. 17.)

are the isolation of the plant sterols by chromatographic methods and their identification by a combination of chromatography and MS. The review emphasizes the crucial role of chromatography in the resolution and identification of isobaric components, that yield identical mass spectra. In contrast to most other reviews in the field, the current work includes analyses of intact plant steryl esters. The chapter concludes with a brief review of the metabolic transformations of plant sterols in animal tissues and cell cultures. In all instances, representative examples of applications of chromatographic and mass spectrometric techniques have been included to provide a practical context for the discussion.

Figure 1 gives the structures of major plant sterols isolated from animal tissues and cell cultures (17).

II. ISOLATION AND ENRICHMENT

Tissue and plasma sterols and steryl esters are usually isolated as part of the total lipid extract obtained by chloroform-methanol extraction, followed by hexane or petroleum ether refining, or solid-phase extraction. The plant sterols may be recovered in their native form (nonesterified plus esterified) or as part of the unsaponifiable matter (total sterol). The plasma and plasma lipoproteins may be prepared by conventional ultracentrifugation according to any of the established procedures, although this may cause sterol peroxidation (23). Subcellular (membrane) fractions of cells and tissues are obtained by homogenization and centrifugation (24,25).

The extraction of free plant sterols and steryl esters from yeasts and insects requires grinding of the material with glass beads before a total lipid extract can be satisfactorily obtained (26).

A. Solvent Extraction

There are two basic routines for the preparation of sterol extracts from tissues and cell cultures. Both nonesterified and esterified sterols are recovered along with other lipids during a total lipid extraction, which is usually performed with chloroform-methanol or a solvent of comparable polarity. The total lipid extract may be subjected to prefractionation into free sterol and steryl ester fractions by thin-layer or column chromatography as described for solid-phase extraction (see below). Frequently, however, a lipid extract is prepared along with saponification, which serves to enrich the plant sterols as a proportion of the total unsaponifiable material. The nonesterified sterols recovered by diethyl ether or hexane extraction under these conditions represent total rather than the naturally occurring free sterols.

1. Extraction Without Saponification

Plasma and tissue plant sterols and steryl esters are usually isolated as part of the total lipid extract obtained by chloroform-methanol partitioning according to Folch et al. (27) or Bligh and Dyer (28). Abidi (17) has reviewed other methods for plant sterol extraction from vegetable oils and fats and from foods and feeds, including supercritical fluid extraction, which, however, is not commonly employed for the extraction of the minor amounts of plant sterols from animal tissues or cells. To avoid peroxidation, the extraction, prefractionation, and saponification steps should be conducted under inert atmosphere and where feasible in the presence of chelating agents (29). The combined extracts are reduced to dryness and subjected to a suitable chromatographic resolution and mass spectrometric identification and quantification of the sterols.

2. Extraction with Saponification

For saponification, an aliquot of the lipid extract is stirred overnight at room temperature with 1 M ethanolic potassium hydroxide, which may be repeated one or more times.

Alkaline hydrolysis and extraction of cholesterol and plant sterols was used to quantify cholesterol, sitosterol, and sitostanol by GLC [as the trimethylsilyl (TMS) ethers in presence of 5α-cholestane] for the study of the mechanism of action of plant sterols on inhibition of cholesterol absorption (30). Complete separation of all the sterols was achieved using the method of Mattson et al. (31).

Beveridge et al. (32) have provided a recent detailed description of preparation of unsaponiable material from total lipid extracts of tissues or other sources. One milliliter of 0.1% (w/v) 5α-cholestane (internal standard) in tert-butyl methyl ether was added to 1 g of oil in a 100-ml stoppered flask. KOH (20 ml, 1 M) in methanol was added and stirred overnight (33). The solution was diluted with 40 ml of distilled water and extracted (three times) with 30 ml of tert-butyl methyl ether. The combined organic extract was washed with 15 ml of 0.5 M KOH in methanol, followed by repeated 30 ml distilled water. The saponification progress was monitored using TLC plates developed with hexane-ethyl ether-glacial acetic (80:20:1, by vol) (34). Plates were viewed under UV light (254 nm and 366 nm), then sprayed with 10% ethanolic phosphomolybdic acid, and charred (35).

The saponification may be carried out following an initial preparation of a total lipid extract, which may be first separated into nonesterified and esterified sterol containing fractions (see above).

The steryl ester–containing fraction isolated by thin-layer chromatography (TLC) or HPLC may be saponified using a miniaturized method for total sterol extraction with saponification as described for a preparation from a vegetable oil (20). A 200-mg amount of soybean oil was added to 250 μl of a solution of 6-ketocholestanol (1260 μg/ml) in hexane as internal standard and 2 ml of 2 M KOH in methanol in a 50-mL round-bottom flask. The mixture was heated under reflux at 90°C for 1 h. After cooling at room temperature, 4 ml of water and 4 ml of ethyl acetate was added to the mixture and vortex mixed. After phase separation, the aqueous phase was washed three times with diethyl ether. Finally, the diethyl ether solution was dried on anhydrous sodium sulfate, filtered, and then dried under a stream of dry nitrogen. Sample extract was stored at −18°C until analysis.

B. Solid-Phase Extraction and Fractionation

Solid-phase extraction utilizes samples solubilized in organic solvents and cannot be distinguished from clean-up or fractionation of total lipid extracts by normal or reversed-phase columns or cartridges. The total lipid extract may be resolved into nonesterified esterified sterols by TLC using a neutral lipid system, which retains the polar phospholipids at the origin. A single development with petroleum ether (b.p. 30–60°C)/diethyl ether 150:20 (v/v) yielded a single band for total free sterols and another single band for the steryl esters when analyzed on silica gel G (Merck) plates (20 × 20 cm, 250-μm-thick layer) (36). The free sterols and the steryl ester may then be analyzed separately. Total free sterols and steryl esters were isolated from plasma and the lipoprotein fractions by extraction with chloroform/methanol 2:1 (v/v) (37). The total lipid extract may be resolved into free sterols and steryl esters also by normal-phase HPLC. The free sterol and the steryl ester fractions can then be separately examined for the sterol composition by reversed-phase HPLC, TLC, or combinations of these chromatographic techniques with MS (37).

Thus, a crude lipid extract may be loaded onto a column of silicic acid and eluted with organic solvents of increasing polarity yielding various lipid fractions of increasing polarity. A 1:1 (v/v) mixture of hexane/diethyl ether elutes steryl esters, whereas diethyl ether yields 4,4-dimethylsterols, 4α-methyl sterols, and 4,4-desmethylsterols, but methanol recovers steryl glycosides (38,39). Alternatively, a gradient elution with 10–30% ether in hexane recovers steryl esters, 40–50% ether in hexane yields 4,4-dimethyl sterols and 4α-methyl sterols, while 60–70% ether in hexane yields 4,4-desmethylsterols, but ether/methanol recovers steryl glycosides (40,41). For quantification, each sterol subfraction may be further purified by reversed-phase column chromatography on Sephadex LH-20 or Lipidex 5000.

Careri et al. (20) described a solid-phase procedure for the extraction of phytosterols from a sample of soybean oil, which can be scaled down to work with animal cells and tissues. Silica gel (packing 500 mg/6 ml; Alltech, Milan, Italy) Solid-phase extraction (SPE) tubes were used after conditioning with 15 ml of hexane. Sample dissolved in a 5 ml hexane/ethyl acetate (95:5, v/v) mixture was added to the SPE cartridge at a flow rate of 1 ml/min. This solvent mixture (5 ml) was also used to perform the washing step, followed by elution with 6 ml of hexane/ethyl acetate (60:40, v/v). The eluate was dried under a stream of nitrogen and redissolved in 1 ml of hexane. The extracts were filtered on a polytetrafluoroethylene (PTFE) membrane filter (0.45 μm) and diluted with hexane (1:2) before injection into the HPLC system.

III. PURIFICATION TECHNIQUES

Regardless of the method of extraction (with or without saponification), the sterols and steryl esters require purification before further analysis and quantification. This is usually accomplished by TLC or flash chromatography, but other methods, including lipolysis (42) and preparation of digitonides (43,44), may also be employed.

A. Thin-Layer Chromatography

The unsaponifiable matter may also be subjected purification and subfractionation. Subclass fractionation of small amounts (< 200 mg) of lipid extracts or nonsaponifiable matter can be separated by silica gel TLC with suitable solvent systems. The TLC plate is normally sprayed with a nondestructive ethanolic solution of rhodamine or fluorescein, and the resolved bands are scraped off and extracted with diethyl ether for further identification and quantification by GLC or GC/MS. Separations are significantly improved by two-dimensional TLC or HPTLC. Specifically, Myher et al. (45) and Frega et al. (46) have shown that TLC of an unsaponifiable matter of a refined oil with hexane/diethyl ether (7:3, v/v) leads to distinctly separated spots of three sterol subclasses in an ascending order (4,4-desmethylsterol > 4-methyl sterol > 4,4-dimethyl sterol), which is consistent with decreasing solute polarity. Other mobile phases, such as benzene/diethyl ether (9:1, v/v), have been used for the separation of sterol subclasses and their esters from various sample sources (47,48). Crude TLC subfractions can be further resolved by reversed-phase HPLC.

In separate studies (49,50), the Δ^5-sterols, cholesterol, campesterol, and sitosterol were separated from their 5α-dihydro derivatives cholestanol, 5α-campestanol, and 5α-sitostanol by argentation TLC and then quantified as their TMS ether derivatives by GLC on a 6 ft × 4 mm glass columns packed with 3% QF-1 (Applied Science Laboratories). The retention times relative to

5α-cholestane are cholesterol 1.73, cholestanol 1.85, campesterol 2.52, 5α-campestanol 2.66, sitosterol 3.03, and 5α-sitostanol 3.17. It is necessary to separate the unsaturated sterols from their 5α-saturated derivatives by argentation TLC because only small differences exist between the unsaturated and 5α-saturated sterol peak retention times on QF-1. However, quantitative results by the two independent methods agreed within 10%.

B. Flash Chromatography

Total lipid extracts may be subjected to a rapid chromatographic step for removal of nonlipid impurities, glycolipids, and phospholipids. This method differs from both SPE and HPLC by being limited to one or two solvent steps permitting only instant fraction recovery. Miller and Gordon (42) used a 15 × 2 cm i.d. column containing C18 packing washed with methanol. An ether extract from triacylglycerol lipolysis was added to the top of the column and allowed to evaporate. The sample was eluted with methanol followed by hexane to yield two fractions, with the hexane fraction containing the steryl esters. Kalo and Kuuranne (51) used a 5.4 × 1.5 cm column packed with micropore filters (International Sorbent Technology, Hengoed, UK) and a mixture of dichloromethane/hexane (2:78, v/v) to recover sterol esters and a mixture of acetone/dichloromethane (6:94, v/v) to recover free sterols. Alternatively, a sample of total neutral lipids, including steryl esters, may be prepared by removing the phospholipids by rapid filtration of a chloroform solution through a zeolite or Florisil column (52).

C. Lipolysis

Kuksis et al. (53) have routinely removed phospholipids from total lipid extracts of plasma and plasma lipoproteins by hydrolysis with phospholipase C in order to improve the GLC profile of the neutral lipids, including steryl esters. The enzyme converts glycerophospholipids and sphingomyelins to diacylglycerols and ceramides, respectively, which are analyzed as the TMS ethers along with the other neutral lipids not affected by the enzyme digestion.

The separation of the steryl esters and triacylglycerols is critical for HPLC of steryl ester by UV or evaporative light scattering detection. Miller and Gordon (42) reviewed the previously utilized procedures and concluded that triacylglycerol contamination usually persists, which they have attempted to resolve by lipolysis of the triacylglycerols with porcine pancreatic lipase to which the steryl esters are resistant. The steryl esters can then be readily separated from the resulting free fatty acids, monoacylglycerols, and any remaining diacylglycerols by flash chromatography (see above).

D. Precipitation with Digitonin

A selective isolation of the 3β-hydroxysterols can be obtained by precipitation with digitonin. Werbin et al. (43) prepared digitonides to separate free and esterified sterols and to isolate 3β-hydroxysterols from tissue lipid extracts. Haust et al. (44) described a quantitative precipitation of nine 3β-hydroxy-sterols with digitonin. After removal of excess reagent, the sterol and digi-tonin moieties are dissolved in dry pyridine. The free sterols may be regener-ated for differential-photometric and GLC or TLC analyses. The recoveries of all 3β-hydroxysterols were nearly quantitative, except for coprostanol and 7-dehydrocholesterol, which were recovered in the 92–94% range.

IV. METHODS OF RESOLUTION

A. GLC and GC/MS

1. Resolution of Nonesterified Sterols

GLC is the most frequently used method for the analysis of sterols (2,17). Capillary columns with high-temperature liquid phases, such as phenyl-methylsilicone, provide a high degree of component resolution and high detection sensitivity. In a typical modern application, GLC is interfaced with FID to monitor analytes in the column effluents and to MS for structural identification and quantification via single-ion monitoring. The resolution of the analytes on the nonpolar GLC column is routinely improved by preceding the GLC analysis by a TLC or cartridge column separation including argentation chromatography. Alternatively, sterol mixture may be subjected to hydrogenation to convert the unsaturated sterols into stanols, or the unsaturated sterols may be oxidatively destroyed.

For the purpose of volatilization and stabilization of the analytes at the elevated temperatures, the hydroxyl functions of the sterols are silylated or acetylated, although free sterols have also been occasionally successfully analysed. The derivatization can be conveniently performed on the unsa-ponifiable material dissolved in dry pyridine in a screw-capped vial and treated with acetic anhydride or bis(trimethylsily)trifluoroacetamide in the presence of an internal standard.

Kuksis and Huang (54) reported GLC resolution of nonesterified cholesterol and plant sterols as the acetates on packed nonpolar columns. At the time campesterol was considered an isomer rather than a homologue of β-sitosterol and was named γ-sitosterol (44). The acetates were prepared by reacting the sterols with acetic anhydride-pyridine (1:1, v/v, 30 min at 80°C) and the steryl acetates isolated by TLC. Kuksis and Huang (55) also demonstrated the resolution of cholesterol, campsterol, sitosterol, and their

5β forms using similar GLC columns. The acetates of 5α and 5β isomers were resolved in an order that the 5β isomers preceded the 5α isomers without overlap, when analyzed as a mixture. Figure 2 shows near-baseline resolution of a mixture of standard 5α and 5β isomers of common sterols as the acetates (55). Octacosane was used as internal standard in the metabolic experiments conducted in parallel. Miettinen et al. (56) described the use packed GLC columns for the separation of cholesterol and plant sterols and their bacterial reduction products as the TMS ethers.

Ishikawa et al. (56) reported the resolution of cholestanol and 5α-saturated plant sterol derivatives in nonesterified form using 180 cm × 4 mm glass columns packed with 1% SP-1000 on Gas Chrom Q (Supelco) at a column temperature of 230°C and nitrogen as carrier gas. The method has been extensively employed in the resolution of stanols and stenols in sito-sterolemia (49,50). Figure 3 shows the resolution of the underivatized plasma sterols from a control subject, from a patient with cerebrotendinous xantho-matosis, and from a patient with sitosterolemia (49). An excellent resolution is realized for the corresponding sterols and stanols.

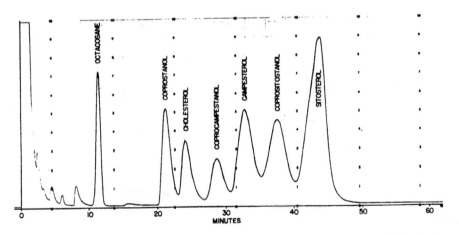

Figure 2 Gas chromatographic resolution of 5α- and 5β-sterols on QF-1. Peak identification: 1, coprostanol; 2, cholestanol; 3, coprocampestanol; 4, campestanol; 5, coprostigmastanol; 5, stigmastanol; octacosane, internal standard. Instrument: Aero-graph model A-100 with flame ionization kit. Column: stainless steel tube, 180 cm × 2 mm i.d., packed with 1% QF-1 on Gas Chrom P (100–120 mesh, Supelco). Carrier: N_2, 60 ml/min. On-column injector: 220°C; Detector: 240°C. Column temperature: 220°C isothermal. Sample: 1 μl of approximately 1% solution of sterol standards in chloro-form. (From Ref. 55.)

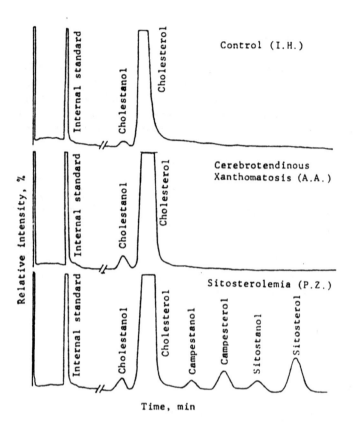

Figure 3 Gas chromatographic resolution of underivatized sterols present in plasma of control, cerebrotendinous xanthomatosis, and sitosterolemic patients that illustrate the separation of the Δ^5-unsaturated sterols from their 5α-saturated analogues on SP 1000 column. Peak identification is given in the figure. Instrument: Hewlett-Packard model 5830 gas chromatograph equipped with flame ionization detector. Column: glass tube, 180 cm × 4 mm i.d., packed with 1% SP-1000 on Gas Chrom Q (80–100 mesh, Supelco). Carrier: N₂, 30 ml/min. Flash heater, 250°C; Detector, 260°C; Column, 230°C isothermal. Sample: approximately 100 μg crude neutral sterols in presence of 140 μg of 5α-cholestane as an internal standard. (From Ref. 49.)

Figure 4 shows a polar capillary GLC resolution of the nonesterified blood plasma sterols from a patient with phytosterolemia before (A) and after (B) oxidative destruction of unsaturated sterols (45). A 15-m column length and an 8-min running time provided a good resolution with the saturated sterol emerging ahead of the unsaturated one (separation factor 1.035). Furthermore, the saturated 5β-sterol (coprostanol) emerged ahead of the

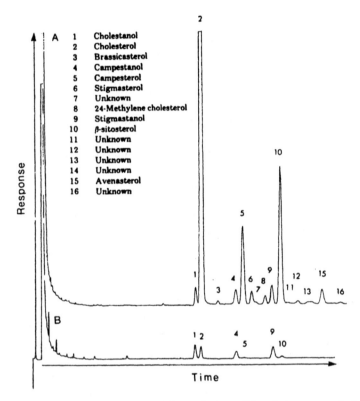

Figure 4 Gas chromatographic resolution of blood plasma sterols (as acetates) from a patient with phytosterolemia before (A) and after (B) oxidative destruction of unsaturated sterols. Peaks are identified as follows: 1, 3β-cholestanol; 2, cholesterol; 3, brassicasterol; 4, campestanol; 5, campesterol; 6, stigmasterol; 8, 24-methylenecholesterol; 9, stigmastanol; 10, sitosterol; 15, Δ^5-avenasterol; other peaks were not identified. Instrument: Hewlet-Packard model 5880 gas chromatograph equipped with flame ionization detector. Column: Supelcowax 10 flexible quartz capillary (15 m × 0.32 mm i.d.). Injector and detector, 270°C; column, 250°C isothermal. Carrier gas, H_2 at 2–5 psi head pressure. Sample: TMS ethers of sterols in hexane using a split ratio of 10:1. (From Ref. 57.)

saturated 5α-sterol, without overlapping any of the unsaturated sterols of lower molecular weight. This resolution is identical to that reported for the TMS ether of cholesterol and cholestanol on other polar liquid phases immobilized on capillary columns, including CP-wax-57CB (58). Similar separations are obtained for the saturated and unsaturated C_{28} and C_{29} plant sterols. When the sterols are chromatographed without derivatization, the sterol/stanol reproducibility of resolution tends to vary and sterol dehydration is an ever present danger. A simple but important confirmation of the presence and resolution of the stanols in the sterol mixture may be obtained by rechromatography of the acetates following oxidative destruction of the unsaturated sterols using the method of Serizawa et al. (59). Since GLC profiles are not as reproducible when underivatized sterols are chromatographed, Myher and Kuksis (52) recommended acetylation, especially since this approach is compatible with the destruction of the unsaturated sterols prior to rechromatography. In phytosterolemia the stanol peaks emerge as clearly resolved components despite a 50- to 100-fold excess of the Δ^5-sterols.

Recently, Salen et al. (60) reported the separation of plant sterols and stanols as the TMS ethers by GLC/FID using a capillary column (26 m × 0.32 mm) coated with 0.21 μm film of CP-wax-52CB (Chrompack, Bridgewater, NJ) at an isothermal temperature of 210°C and with helium as carrier gas. The plasma dietary sterol concentrations were measured following saponification of plasma or homogenized diet. 5α-Cholestane or coprostanol was added as internal standard prior to extraction of the neutral sterols and stanols with hexane. The solvent was evaporated, and TMS ether derivatives were prepared by the addition of Sil-Prep (Analtech, Deerfield, IL). After standing for 30 min, pyridine was evaporated and the residue was dissolved in hexane. The retention times relative to 5α-cholestane for TMS ether derivatives were as follows: cholesterol 1.86, campesterol 2.36, campestanol 2.24, sigmasterol 2.46, sitosterol 2.86, sitostanol 2.72, and avenasterol 3.46. The method was used to demonstrate hyperabsorption and retention of campstanol in a sitosterolemic homozygote (see below).

Phillips et al. (18) recently employed a 60-m capillary column with a moderately polar stationary phase (14% cyanopropylphenyl–86% dimethyl-polysiloxane) along with a one-step SPE to remove highly polar compounds from the saponified total lipid extract. Using epicholesterol as an internal standard, precise quantitative determination of phytosterols, stanols, and cholesterol metabolites in human serum was achieved. Figure 5 shows representative chromatograms (including epicholesterol internal standard) for reference standards (a and b), Accutrol control serum composite (c), and a sitostanol-containing serum sample (d) from a feeding trial (18). The chromatograms show unprecedented resolution of the sterols of interest, including that of the stanols and the corresponding stenols. Analytes were detected at levels of 120 ng/ml to 6 ng/ml with standard deviations of 0.02–0.12 μg/ml.

Gerst et al. (61) plotted the retention times relative to 5α-cholestane for TMS ethers of 26 unsaturated C_{27} sterols as their acetate and TMS ether derivatives on DB-5 (60 m) and CP-wax (30 m) columns. Gerst et al. (61) published a schematic representation of the GLC retention relative to 5α-cholestane for TMS ethers of unsaturated C_{27} sterols and related sterols on

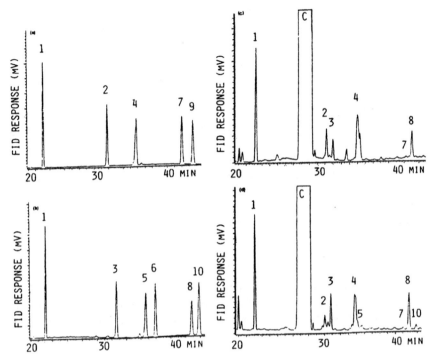

Figure 5 Representative chromatograms (including epicholesterol internal standard) for reference standards (a and b), Acutrol control serum composite (c), and sitostanol-containing serum sample from a feeding trial (d). Peaks are identified as follows: 1, epicholesterol; 2, desmosterol; 3, lathosterol; 4, campesterol; 5, campestanol; 6, stigmasterol; 7, lanosterol; 8, sitosterol; 9, fucosterol; 10, sitostanol; C, cholesterol. Column, RTx R-171 (14% cyanopropylphenyl-86% dimethylpolysiloxane; Restek, Bellefonte, PA), 60 m × 0.25 mm i.d., 0.25 μm film thickness; carrier gas: hydrogen; split ratio, 9:1; column head pressure, 21 psi; oven temperature, 265°C (45 min), then 10°C/min to 280°C hold 3.5 min. (From Ref. 18.)

DB-5 (60 m) and CP-wax columns. GLC retention times for authentic samples of unsaturated C_{27} sterol derivatives were determined using capillary columns coated with D-1 (60 m × 0.25 mm i.d.; 100% methylpolysiloxane; 0.1 μm film thickness, J&W Scientific, Folsom, CA), DB-55ms (30 or 60 m × 0.25 mm i.d.; 5% phenyl–95% methylsiloxane; 0.1 μm film thickness; J&W); Stabilwax (30 or 60 m × 0.25 mm i.d.; polyethylene glycol; 0.1 μm fm thickness; Restek; Bellefonte, PA); CP-wax-57CB (25 m × 0.32 mm i.d.; high-polarity polyethylene glycol; 0.2 μm film thickness; Chrompack, Raritan, NJ). The GLC analyses for DB-1, DB-5, Stabilwax, and CP-wax were done isothermally on a Shimadzu GC-9A instrument. Either nitrogen or helium was used as a carrier gas, with a split ratio of 50:1 or 20:1. Injector and flame ionization detector temperatures were held at 250°C and 290°C, respectively. GC/MS analyses were done isothermally at 250°C on a Hewlett-Packard 5890A chromatograph. The temperature of the injector and GC/MS interface was 270°C. Mass spectra (m/z 50–700) were measured on a ZAB-HF reversed-geometry double-focusing instrument at 0 eV and an electron impact ion source (200°C). The accelerating voltage was 8 kV and the resolution was 1000 (10% valley).

A simple, specific, and reproducible capillary column GLC method for the simultaneous determination of campesterol, sitosterol, and 7-ketocholesterol in small volumes of human lipoproteins has been proposed by Dyer et al. (62). The method involves extraction from lipoprotein samples using chloroform-methanol, saponification of sterol esters using cold KOH, purification, and derivatization to TMS ethers using bis(trimethylsilyl)trifluoroacetamide and 1% trimethylchlorosilane. Oxidation is prevented by drying under nitrogen and with the use of powerful antioxidants. Separation is achieved using a DB-1 capillary column and a two-stage temperature ramp from 180°C to 250°C and detection using FID. Table 1 gives the concentrations of campesterol, sitosterol, and 7-ketocholesterol in very-low-density lipoprotein

Table 1 Concentrations of Phytosterols and α-Ketosterol in Lipoprotein[a] Fractions of 10 Normolipemic Men (mmol/L)

Lipoprotein fraction	Campesterol	Sitosterol	7-Ketocholesterol
Conc. of sterol[b]			
VLDL	0.66 (0.25)	0.37 (0.10)	0.66 (0.49)
IDL	0.79 (0.28)	0.49 (0.16)	0.58 (0.38)
LDL	7.23 (3.20)	3.82 (2.05)	1.97 (1.28)
HDL	2.92 (1.18)	1.78 (0.64)	1.17 (0.55)
Conc. of sterol/conc. of cholesterol[c]			
VLDL	2.02 (1.03)	1.10 (0.51)	1.82 (1.90)
IDL	2.10 (0.93)	1.11 (1.04)	1.19 (0.81)
LDL	2.38 (0.77)	1.23 (0.38)	0.63 (0.34)
HDL	2.00 (0.75)	1.14 (0.42)	0.94 (0.78)

[a] Lipoproteins: VLDL, LDL, and HDL are very-low-density, low-density, and high-density lipoproteins, whereas IDL is a lipoprotein of a density intermediate between that of VLDL and LDL.
[b] Concentrations of plant sterols were determined by capillary GLC of TMS ethers of plant sterols and 7-ketocholesteerol using 5α-cholestane as internal standard.
[c] Concentration of cholesterol determined enzymatically. Mean (SD).
Source: From Ref. 62.

(VLDL), intermediate-density lipoprotein (IDL), low-density lipoprotein (LDL), and high-density lipoprotein (HDL) of 10 normal men. Concentrations of sterols are expressed as micromoles per liter of whole serum and also as a ratio to cholesterol concentration of each lipoprotein fraction (18). Nitrogen (oxygen free) was used as the carrier and auxiliary gas at flow rates of 1–2 ml/min. Peak height was used for quantification. Concentrations of cholesterol and other sterols were calculated from standard curves of the appropriate sterol using 5α-cholestane as internal standard.

The limit of quantification was determined by repeated analysis of sterols at low concentrations. The standard deviation of 10 measurements of a number of sterols was calculated. The concentration at which the mean exceeded zero by more than three standard deviations (SD) was taken as the limit of sensitivity.

Figure 6 shows the GLC resolution of the desmethyl, monomethyl, and dimethyl subfractions of rice bran sterols (45), which we have used as a secondary standard for the identification of plant sterols in plasma. The rice bran sterols were resolved into desmethyl, methyl, and dimethyl sterol fractions by TLC on silica gel H using hexane-diethyl ether-acetic acid (80:20:0.5, by vol) as the developing solvent. The plate was visualized by spraying with a solution of dichlorofluorescein and the steroids recovered by elution of the silica gel with chloroform-methanol (2:1, v/v).

Thompson et al. (63) examined the TMS derivatives of nine pairs of C_{24} epimeric sterols by polar capillary GLC and reported the physical separation of these epimeric compounds. Figure 7 shows the GLC separation of pairs of C_{24} epimeric sterol TMS ethers on SP-2340 liquid phase at 195°C (elution time from left to right). The analyses were performed on a 115 m × 0.25 mm glass capillary column coated with SP-2340 (Quadrex Corp., New Haven, CT), a cyanosilicone phase with chromatographic properties similar to those of polar polyester phases. With hydrogen as a carrier gas, the column provided about 230,000 theoretical plates. Substitution of helium for hydrogen doubled the retention times.

Child and Kuksis (64) used glass columns packed with 1–3% OV-225 (cyanopropylphenylsiloxane) on 100- to 120-mesh Gas Chrom Q for GLC separation of the benzoylated 7-dehydro derivatives of cholesterol, campesterol, and sitosterol. The separations were similar to those obtained for the TMS ethers and acetates, except that a considerably higher column temperature (255°C) was required for reasonable retention times. Child and Kuksis (65) also demonstrated a GLC resolution of the des-AB,8-one derivatives of cholesterol, campesterol, and sitosterol, which were comparable to those of the parent sterols. The GLC separations were obtained on a 10-m SP-2330 column with temperature programming from 140°C to 250°C. The des-AB,8-one derivatives of cholesterol and plant sterols were prepared to test the

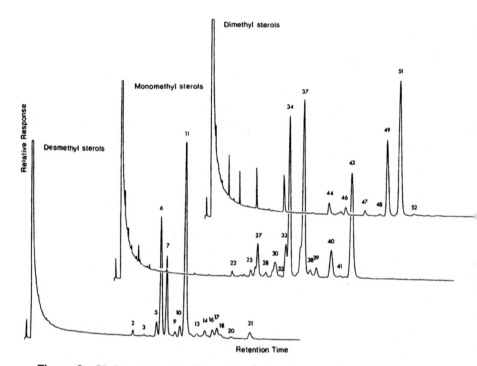

Figure 6   GLC resolution of desmethyl, monomethyl, and dimethyl subfractions of rice bran sterols (as acetates) used as reference compounds in identification of plant sterols in lymph and tissues of rats and in cell cultures. Peaks are identified as follows: (*desmethylsterols*): 2, cholesterol; 3, brassicasterol; 5, campestanol; 6, campesterol; 7, stigmasterol; 9, 24-methylenecholesterol; 10, stigmastanol; 11, sitosterol; 13, $\Delta^{7,22}$-stigmastanediol; 14, $\Delta^{7,22}$-methylene cholesterol; 16, Δ^5-avenasterol; 17, Δ^7-stigmastenol; 21, Δ^7-avenasterol; other peaks in this subfraction were not identified; (*monomethylsterols*): 25, campesterol; 27, obtusifoliol; 30, sitosterol; 33, methyl lophenol; 34, cycloeucalenol; 37, 24-methylene lophenol; 39, ethyl lophenol; 43, citrostadienol; other peaks in this subfraction were not identified; (*dimethylsterols*): 46, cycloartanol; 49, cycloartenol; 51, 24-methylene cycloartenol; other peaks in this subfraction were not identified. GLC conditions: Supelcowax 10 flexible quartz capillary column (15 m × 0.32 mm i.d.), installed in a Hewlett-Packard model 5880 gas chromatograph equipped with a flame ionization detector. The carrier gas was hydrogen with 2–5 psi head pressure. Column temperature was isothermal at 250°C, while the split injector temperature was 270°C. (From Ref. 57.)

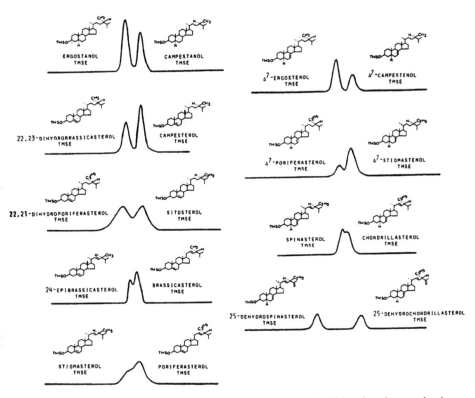

Figure 7 Gas chromatographic separation of pairs of C24 epimeric sterol trimethylsilyl ethers on SP-2340 at 195°C (elution time from left to right). Peaks are identified in the figure. Instrument: Hewlett-Packard model 5840 gas chromatogaph equipped with flame ionization detector. Column: glass capillary (115 m × 0.25 mm) coated with SP-2340, a cyanosilicone phase (Quadrex, New Haven, CT) providing 230,000 effective theoretical plates. Injector, 280°C; detector, 300°C; column, 185 or 195°C. Carrier gas, H_2 at 20–34 psi. Sample size: approximately 1.7 µl of a solution containing 2 µg/µl in isooctane. (From Ref. 63.)

critical role of the sterol ring structure in the differential absorption of cholesterol and plant sterols. The des-AB,8-one derivatives were prepared by permanganate peroxidation to facilitate study of the role of the integrity of the sterol ring in the differential absorption of cholesterol and plant sterols. The derivatives were obtained from the corresponding calciferol derivatives, which were obtained by irradiation of a solution of the 7-dehydrosterols.

The common plant sterols are tentatively identified based on the GLC retention times especially when the GC separations are preceded by a TLC or AgNO3-TLC separation. More complex mixtures require confirmation by

MS (66). Beveridge et al. (32) identified cholesterol, campesterol, stigmasterol, sitosterol, β-amyrin, sitostanol, lupeol, and erythrodiol by comparing their RRf values and their MS fragments with those of authentic samples Beveridge et al. (32) have prepared a summary of the fragmentation ions obtained for selected 4-desmethyl, 4-monomethyl, and 4,4-dimethyl sterols as obtained under GC/MS conditions (Table 2). The TMS ether derivatives of spinasterol, Δ^5-avenasterol, Δ^7-sitosterol, and Δ^7-avenasterol were identified by comparing the GC/MS and GC/FID values reported in the literature (66–68) and MS data (69–71). Clerosterol, $\Delta^{5,25(27)}$, was identified on the basis of its retention time relative to sitosterol acetate (71,72) and free sterol (73,74), and the characteristics of the allylic cleavage in $\Delta^{5,25(27)}$ sterols (75). The peak for sitosterol contained a minor peak identified on basis of retention time and an uncharacteristic mass spectra at m/z 343 as the Δ^7 isomer of sitosterol. The

Table 2 Molecular and Major Fragmentation Ions Obtained for Selected TMS Derivatives of American Ginseng (*P. quinquefolium*) Oil Unsaponifiables

		Main fragment ions, m/z (Relative intensity)[a]					
	RRT[b]	M+	M-15	M-90	M-105	M-129	Ot
Squalene	0.36	410(1)					
Oxidosqualene	0.45	426(1)					
4-Desmethylsterols							
Campesterol	0.84	472(18)	457(3)	382(42)	367(21)	343(53)	1
Stigmasterol	0.89	484(13)	469(3)	394(18)	379(7)	355(9)	
Clerosterol	0.96	484(18)	469(1)	394(15)	379(9)	355(15)	
Sitosterol	1.00	486(15)	471(5)	396(41)	381(20)	357(50)	1
Δ^5-Avenasterol	1.03	484(4)	469(3)	394(1)	379(4)	355(8)	
$\Delta^{5,24(25)}$-Stigmasterol	1.09	484(4)	469(4)	394(5)	379(9)	355(4)	
Δ^7-Sitosterol	1.12	486(57)	471(16)	396(3)	381(14)		2
Δ^7-Avenasterol	1.16	484(1)	469(5)	394(1)	379(4)		3
4-Monomethylsterols							
Citrostadienol	1.38	498(tr)	483(5)	408(1)	393(6)		3
4,4,-Dimethyl triterpene alcohols							
β-Amyrin	1.02	498(1)	483(tr)	408(tr)	393(1)		2
Lupeol	1.11	498(9)	483(2)	408(3)	393(6)		
24-Methylene cycloartenol	1.26	586(1)		496(3)	481(tr)		

[a] Intensity relative to base peak (percent); tr indicates trace.
[b] retention times relative to sitosterol TNS ether (35.9 min) using a DB-5 60-m column. The int standard 5α-cholestane elutes with a relative retention time of 0.40.
[c] Base peak (relative intensity = 100%). The peak for sitosterol also contained spinasterol; the peak f amyrin also contained sitostanol; the peak for Δ^5-avenasterol also contained butyrospermol; an peak for citrostadienol also contained erythrodiol, as minor components.
Source: Adapted from Ref. 32.

Δ^7 isomer of stigmasterol (spinasterol) was also present. The Δ^5-avenasterol overlapped with the minor component butyrospermol, which was identified based on relative retention of the acetate and TMS derivatives (76,77), the MS spectra of the acetate (78), and free sterol (78,79). Peak 9 was tentatively identified as $\Delta^{5,25(27)}$ stigmastadienol on the basis of the match of RRf and MS fragmentation with those reported by Kamal-Eldin et al. (71). The 24-methylene cycloartenol was identified on basis that RRf and MS data matched those of Kamal-Eldin et al. (71). Erythrodiol was identified by matching RRf and MS data with those of an authentic sample. The MS data of the main component of this peak, citrostadienol, matched the data provided by Kamal-Eldin et al. (71).

Yang et al. (80) have used GC/FID along with GC/MS for a most detailed investigation of the identity of the phytosterols in sea buckthorn berries of different origin and at differing harvesting times. Over 25 different sterols, including methyl sterols, dimethylsterols, and desmethylsterols were recognized. Sitosterol constituted 57–83% of the seed and pulp/peel sterols. The sterol content and composition showed little variation between sub-species and collection times.

Plank and Lorbeer (81,82) used high-temperature capillary GLC for the determination of the concentration of nonesterified sterols as well as their qualitative and quantitative composition and the concentration of the sterol esters in rapeseed oil methyl ester samples by GC/FID. Prior to analysis, the free sterols were silylated with N,O-bis(trimethylsilyl)trifluoroacetamide with 1% trichlorosilane; betulinol was used as an internal standard. The GLC separation was carried out on a 12 m × 0.32 mm i.d. fused silica capillary column coated with a 0.1-µm film of 5% phenylpolydimethylsiloxane (DB-5; J&W) connected in series with a 4 m × 0.53 mm i.d. uncoated, deactivated fused silica precolumn (Carlo Erba), a 3 m × 0.32 mm i.d. retaining precolumn coated with DB-5 of 0.1 µm film thickness, and an early vapor exit by means of a glass press-fit connection. Carrier gas inlet pressure behind the flow regulator was 250 kPa and the regulated flow rate was 1.2 ml/min at 40°C (hydrogen). Detector temperature was 370°C. Without going into further details of the exact operation of the system, it may be noted that after 8 min at an initial temperature of 130°C, the GLC oven was heated at 30°C/min to 260°C, then at 3°C/min to 270°C, and finally at 15°C/min to 345°C (held for 7 min). Total LC/GC run time was 33 min, with 5 min cooling. The free plant sterols, eluted in the order campesterol, sitosterol, and brassicasterol, were clearly dominant. In addition, cholesterol, stigmasterol, and Δ^5-avenasterol could be identified by analysis of samples spiked with reference substances and by comparison to literature data. The later emerging peaks represented the C_{18} fatty acid esters of campesterol and sitosterol.

Recently, Duchateau et al. (19) described a fast and accurate method for determining total 4-desmethyl sterol content in spreads, fat blends, and raw materials. It is based on whole-sample saponification, one-step liquid/liquid extraction, GLC determination without derivatization, and quantification using an internal standard. The calculated within-and between-laboratory reproducibility were 0.680 and 1.194 wt%, respectively, for sterol-containing spreads. The method has not been applied to sterol extracts of biological materials.

2. Resolution of Steryl Esters

Although subjecting the total lipid extract to saponification and isolating the sterols from the unsaponifiable fraction is common practice, it results in a loss of information about the fatty acid composition of the esters. While this approach may provide a short-cut to a limited objective, as will be shown below, it contains the potential of compromising the analyses of the minor sterols and, of course, destroying any information about the steryl and oxosteryl esters. Furthermore, methods based on free-sterol analysis in the unsaponifiable matter are time and labor consuming, and do not provide information regarding the fatty acid composition of the steryl esters. The total lipid extract may be advantageously subjected to chromatographic separation of the free sterols and the steryl esters using high-temperature GLC and HPLC with or without MS. The early work on the GLC separation of steryl esters was reviewed by Kuksis (83,84).

Plant steryl esters were first resolved by high-temperature GLC on short columns packed with nonpolar silicone liquid phases as part of total lipid profiling (83,85). Under these conditions, the campesteryl and sitosteryl esters of palmitic, oleic, and linoleic acids overlapped with the cholesteryl esters of comparable total acyl carbon number and were eluted with retention times between those of the TMS ethers of free ceramides and those of tripalmitoylglycerol in the total lipid profile. The palmitic acid esters of 5α and 5β derivatives of cholesterol and plant sterols were also fully resolved on the short columns over a 24-min period when run isothermally at 280°C (83). Normally, the plant steryl esters make up only a minor proportion of the total steryl esters of human subjects but can be recognized in the GC/MS profiles of total plasma lipids by searching for the M-fatty acid ions of the steryl esters of campesterol, stigmasterol, and sitosterol (86). The plant steryl esters of normolipemic subjects averaged two to three times the free plant sterol level, which was estimated at 0.4–0.6 mg/100 ml of plasma or red blood cells. Subsequently, capillary GC and GC/MS provided more extensive resolution and quantification of the plants steryl esters in patients with sitosterolemia. The short nonpolar capillary GLC columns provide essentially carbon number or molecular weight distribution (37). The steryl ester mixtures were also analyzed by GC/MS with electron impact and hydrogen chemical ionization (84). Since high-temperature GLC tends to degrade the polyunsaturated fatty acid esters of sterols, it is recommended that such samples be hydrogenated before GLC analysis of their carbon number distribution.

Intact cholesteryl and plant steryl esters may be resolved into individual molecular species by GLC on polar capillary columns. Kuksis et al. (37) used short (10 m) glass capillary columns coated with SP-2330 polysiloxane liquid phase (Supelco) and hydrogen as a carrier gas to resolved mixed cholesteryl and plant steryl esters at an isothermal temperature of 250°C. The peaks were identified by reference to standard steryl esters, AgNO$_3$-TLC fractions of steryl esters, and LC/MS. Figure 8 shows a polar capillary GLC resolution of the mixed cholesteryl and plant steryl esters recovered from plasma lipo-

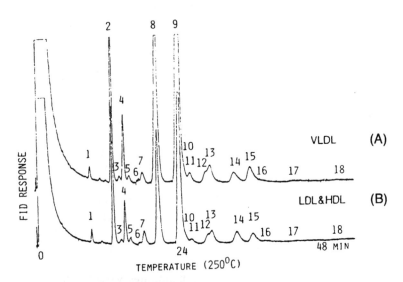

Figure 8 GLC elution profile of steryl esters of VLDL (A) and LDL + HDL (B) fractions of plasma from a patient with phytosterolemia. Peak identification: 1, cholesteryl myristate; 2, cholesteryl palmitate; 4, cholesteryl palmitoleate; 5, campesteryl palmitate; 7, sitosteryl palmitate + cholesteryl stearate; 8, cholesteryl oleate; 9, cholesteryl linoleate; 10, campesteryl oleate; 12, sitosteryl oleate; 13, campesteryl linoleate; 14, sitosteryl oleate; 15, cholesteryl arachidonate. GLC conditions: Hewlett-Packard gas chromatograph equipped with an SP-2330 glass capillary column (10 m × 0.25 mm i.d.) and a hydrogen flame ionization detector. The split injector and detector temperatures were 270°C. Column temperature, 150°C isothermal; carrier gas, hydrogen at 8 psi inlet pressure. Sample: intact steryl esters isolated by TLC. (From Ref. 37.)

proteins of a patient with severe sitosterolemia (37). There is complete separation of the corresponding saturated and monounsaturated as well as the di-and polyunsaturated fatty acid esters. From the polar liquid phase column the polyunsaturated fatty acid esters are eluted last and suffer some losses. As a result, the contribution of the arachidonoyl species is slightly underestimated. In addition to the peaks for cholesteryl palmitate, oleate, linoleate, and arachidonate, there are readily detectable peaks corresponding to campesteryl plus dihydrobrassicasteryl, and sitosteryl palmitate, oleate, and linoleate. There are also discernible elevations in the baseline with retention times expected for the plant steryl arachidonates. The recovery of the polyunsaturates may be improved by GLC on polarizable liquid phases in combination with temperature programming (87), which provides resolution based on molecular weight and degree of unsaturation; however, instability of the polyunsaturated esters still presents a problem.

Table 3 gives the quantitative estimates for the various steryl esters in total plasma and isolated lipoprotein fractions of a patient with sitosterolemia (37). There are minor differences in the composition of the steryl esters between the two lipoprotein fractions. The VLDL fraction contains more esterified cholesterol than plant sterol in comparison with the combined LDL + HDL fraction.

Effective separations of cholesteryl and plant steryl esters on short nonpolar capillary GLC columns have been also obtained by Evershed et al. (88,89), who reported detailed protocols for the analysis of a steryl ester mixtures employing argentation chromatography, HPLC, GC, and GC/MS. Preparative TLC was used for a preliminary separation of synthetic steryl esters into short-chain (C_2–C_8) and long-chain (C_{10}–C_{22}) species. The shorter chain species were further resolved by adsorption HPLC and the longer chain species by reversed-phase HPLC before GLC on short capillary columns. The longer chain steryl esters were also resolved into saturated, mono-, di-, tri-, and polyene acyl types by preparative HPLC. Each of the steryl ester fractions was subjected to GC/MS under electron impact ionization.

Evershed et al. (90) have since used steryl palmitates, varied in the nature of the steryl moiety, as model compounds for investigation of the mass spectrometric behavior of steryl long-chain fatty acyl esters. Compounds were chosen to be representative of biochemically important steryl esters, and included variations in the position and degree of unsaturation of the steroid nucleus and C_{17} side chain and position and degree of methylation. M^+ ions were generally weak or absent, and the major high-mass ions arose from characteristic fragmentations of the steroid nucleus following loss of the acyl moiety ($[M-RCO_2H]^+$). Fragment ions characteristic of the acyl moiety were lacking. Negative ion chemical ionization using ammonia as reagent gas afforded spectra containing characteristic fragment ions $[RCO_2]$, RCO_2-18]$^-$,

Table 3 Quantitative Composition of Steryl Esters of the VLDL and LDL + HDL Fractions of Plasma from a Patient with Phytosterolemia as Estimated by GLC on Polar Capillary Columns (mol %)

Peak no.	Steryl esters	Total	Lipoproteins VLDL	LDL + HDL
1	Chol 14:0	0.8	0.6	0.9
2	Chol 16:0	12.9	11.8	13.6
3	Chol 16:1(n-9)		0.5	0.6
4	Chol 16:1 (n-7)	4.6	5.0	4.9
5	Camp 16:0	0.7	0.6	1.1
6	Unknown		0.3	0.4
7	Sito 16:0 + Chol 18:0	1.9	1.0	2.2
8	Chol 18:1	23.3	25.3	19.3
9	Chol 18:2	41.6		
10	Camp 18:1		44.0	36.5
11	Unknown			
12	Sito 18:1	1.3	1.3	3.1
13	Camp 18:2	2.0	1.6	3.8
14	Sito 18:2	4.2	1.8	5.7
15	Chol 20:4	3.2	2.9	4.9
16	Chol 20:37		0.3	1.0
17	Camp 20:4	Trace	0.3	0.5
18	Sito 20:4	Trace	Undetermined	0.4
Other			2.7	1.0

[a] Steryl esters were resolved intact on a polar capillary GLC column as described in legend to Fig. 8.
Source: Ref. 37.

and $[RCO_2\text{-}19]^-$ from which the nature of the fatty acyl moiety can be readily deduced.

Fenner and Parks (91) reported GC analysis of intact steryl esters in wild-type *Saccharomyces cerevisiae* and in an ester-accumulating mutant. The yeast steryl esters were resolved on a nonpolar capillary column (bonded methyl silicone, Supelco, 20 m × 0.32 mm i.d.) with temperature programming and hydrogen as carrier gas. A compound tentatively identified as zymosteryl palmitate was the most prevalent ester in wild-type log phase cells; ergosta-5,7-dienoyl palmitate and ergosta-5,7-dienyl palmitoleate were the major esters in stationary cells. In the mutant strain, ergosteryl esters of palmitate, palmitoleate, oleate, and stearate were the major ester components throughout the culture cycle.

Kuksis et al. (92) reported the detection and quantification of intact cholesteryl esters in total plasma lipid extract by GC/MS with positive chemical ionization. The GLC separations were obtained on a polarizable capillary column (OV-22, New Haven, CT) using the same chromatography conditions as those previously described for plasma total lipid profiling separation with polarizable capillary columns (87).

More recently, nonpolar capillary GC of intact steryl esters has been applied to the separation and quantification of plant steryl esters in crude vegetable oils and foods. Miller and Gordon (42) obtained partial identification of the plant steryl esters based on GC retention times reported in the literature and authentic steryl esters, and by the principal fragment ions produced by GC/MS with electron impact ionization. Cholesteryl palmitate was used as internal standard for GC quantification. The major steryl esters were found to be the oleates, linoleates, and linolenates of campesterol and sitosterol. Kamm et al. (93) used on-line LC/GC for the separation of plant steryl esters. The identities of the compounds were confirmed by nonpolar capillary column GC/MS investigation of the collected HPLC fractions and by comparison of the mass spectra (chemical ionization using ammonia as ionization gas) to those of synthesized reference compounds. Cholesteryl laurate was used as internal standard for GC quantification. The major steryl esters were the palmitates, oleates, and linoleates of campesterol, sitosterol, and stigmasterol.

Kalo and Kuuranne (51) briefly described the analysis of free and esterified sterols in fats and oils by flash chromatography, GC, and ESI/MS/MS. The steryl esters were identified and quantified by precursor ion MS of intact steryl esters. Cholesteryl esters were measured in lipase modified butterfat, while plant steryl esters were measured in rapeseed oil and in sunflower oil.

Jover et al. (94) have described complete characterization of lanolin steryl esters by subambient pressure GC/MS in the electron impact and chemical ionization modes. Steryl esters with different sterol (cholesterol, lanosterol, and dihydrolanosterol) and acid moieties either according to carbon number (C_{10}–C_{20}) or isomeric forms (normal, iso-, and anteiso-) were resolved, identified, and quantified. Chemical ionization in either positive or negative ion modes using methane, isobutene, and ammonia gases was utilized. Steryl ester isomers were identified based on chromatographic retention parameters and by the fatty acid profile of the lanolin.

Earlier, Lohninger et al. (95) had shown that total plasma lipid extracts can be injected at low column temperatures and the interference from pyrolysis of phospholipids avoided without removing them prior to GLC. Kuksis et al. (53) have recommended removal of phospholipids by hydrolysis with phospholipase C or by flash chromatography with zeolite columns.

B. HPLC and LC/MS

1. Resolution of Free Sterols

Rees et al. (96) first demonstrated that reversed-phase HPLC is well suited for the resolution of mixtures of nonesterified sterol, and their method has been widely utilized by other workers (17). Child and Kuksis (64,65) employed reversed-phase LC/UV for the study of cell membrane uptake following conversion to 7-dehydro and calciferol derivatives of cholesterol, campesterol, and sitosterol. Figure 9 shows reversed-phase HPLC profiles of the 7-dehydrosterols following incubation of brush-border and red blood cell membranes with a mixture of equimolar amounts of 7-dehydrocholesterol, 7-dehydrocampesterol, and 7-dehydrositosterol. Table 4 compares the relative uptake of $\Delta^{5,7}$-and Δ^5-sterols by rat erythrocytes. The plant sterols not only were less effectively absorbed than cholesterol, but they also inhibited cholesterol absorption. However, ergosterol is assimilated to a considerable extent (64) and is administered to humans as a source of ergocalciferol (vitamin D_2).

Hidaka et al. (97) have reported an HPLC system that can also distinguish between cholesterol, cholestanol, and plant sterols. For this purpose, 0.1 ml of plasma with 10 μg of 5β-cholestane-3α-ol (as an internal standard) was treated with 1 M KOH and extracted twice with hexane. The sterols in the extracts were converted to their benzoate derivatives with benzoyl chloride reagent that was freshly prepared for each assay. The benzoate derivatives were reextracted with 1,2-dichloroethane and dissolved again in 250 μl of acetonitrile-dichloroethane (2:1, v/v) after evaporation under a stream of nitrogen. The separation of the sterols was performed on a reversed-phase column (SBC-ODS, 150 × 2.5 mm, Shimadzu, Kyoto, Japan) installed in a Shimadzu LC-6A system and maintained at 50°C and monitored at 228 nm. The solvent used for the elution was acetonitrile-water-acetic acid (97:3:0.2, by vol) at a flow rate of 0.5 ml/min.

Plank and Lorbeer (81,82) have also described an on-line LC/GC system for the analysis of free and esterified sterols in vegetable oil methyl esters used as diesel fuel substitutes. Qualitative and quantitative information about these minor components is provided without saponifcation and off-line preseparation. Prior to analysis the free stero ls are silylated with methyl-N-trimethylsilyltrifluoroacetamide. Betulinol is used as an internal standard. Using concurrent eluent evaporation with a loop-type interface for eluent transfer, transfer temperatures and transfer efficiency were carefully optimized. The concentration of the free sterols as well as their qualitative and quantitative composition and the concentration of the sterol esters are determined in five different types of vegetable oil methyl esters. Recovery

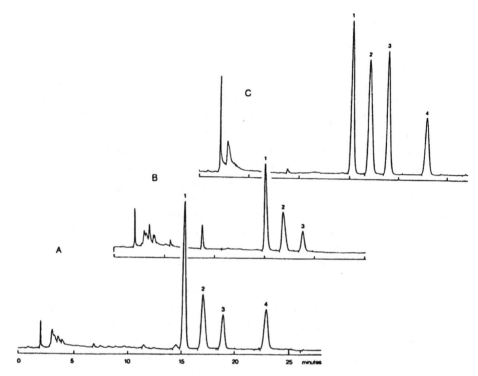

Figure 9 HPLC profiles of total lipid extracts of brush-border and erythrocyte membranes following incubation with 7-dehydrosterols in a mixed micellar medium. A, brush border extract; B, erythrocyte extract; C, extract of medium prior to incubation. Peak identification: 1, 7-dehydrocholesterol; 2, 7-dehydrocampesterol; 3, 7-dehydrositosterol; 4, 7-dehydrocholesterol acetate (internal standard). Brush borders and erythrocytes were incubated for 30 min at 37°C in a medium containing 6.6 mM sodium taurocholate, 0.6 mM egg lecithin, and 30 μM each of the 7-dehydrosterols. HPLC conditions: Hewlett-Packard model 5880 liquid chromatograph equipped with a variable-wavelength UV detector and a reversed-phase Supelcosil LC-18 column (5 μm particle size, 25 × 0.4 cm i.d., Supelco). The sterols were eluted with isocratic methanol-acetonitrile (1:1, v/v) solution at a flow rate of 1 ml/min and an oven temperature of 30°C. The sterol peaks were monitored at 265 nm. Sample: total lipid extracts in methanol-acetonitrile. (From Ref. 64.)

Table 4 A Comparison of the Uptake of $\Delta^{5,7}$-and Δ^5-Sterols by Rat Erythrocytes

Experiment no.	Cellular uptake[a]					
	$\Delta^{5,7}$-C$_{27}$[a]	$\Delta^{5,7}$-C$_{28}$[b]	$\Delta^{5,7}$-C$_{29}$[b]	Δ^5-C$_{27}$	Δ^5-C$_{28}$	Δ^5-C$_{29}$
1					30.2 ± 3.9[c] (30)	16.7 ± 3.2 (30)
2		25.5 ± 5.0 (35)	12.5 ± 2.3 (33)		ND	13.8 ± 1.1 (33)
3	49.0 ± 6.3 (30)	24.7 ± 3.2 (30)	9.9 ± 1.8 (30)			
4	58.0 ± 12.6 (31)	38.0 ± 7.9 (37)	17.1 ± 3.7 (36)			
5	55.6 ± 9.7 (32)	35.4 ± 6.4 (38)	16.4 ± 3.7 (34)	ND[d] (50)		

[a] Rat erythrocytes were incubated for 30 min at 37°C in media containing 6.6 mM taurocholate, 0.6 mM egg yolk phospholipid, and the sterols in the concentrations indicated in the table. Cellular uptake is expressed as nmol sterol/ml packed erythrocytes. Each value represents the mean ± S.D. of three to six replicates from separate cell preparations. The media concentrations of each sterol (µM) are listed in parentheses immediately below the uptake value.
[b] Sterols assayed by UV absorbance during resolution by HPLC.
[c] Sterols assayed as the TMS ethers by GLC.
[d] Not determined.
[e] Assayed by liquid scintillation counting (using 22,23-[^{3}H]-sitosterol tracer) following resolution of the sterols by HPLC.
Source: Ref. 65.

of the HPLC/GLC procedure and reproducibility of the quantitative results were evaluated.

Careri et al. (20,21) recently validated the use of reversed-phase HPLC for quantification of plant sterols in vegetable oils and food products. Solid-phase extraction on silica cartridges was used for purification of the unsaponifiable matter, and determination of plant sterols was carried out using HPLC on a narrow-bore HPLC column and UV detection. Careri et al. (20) also explored the applicability of APCI for the characterization of sterol fraction in a soybean oil sample. The chromatographic separation was carried out using a C8 narrow-bore column (150 × 2.1 mm, 5 µm) (Supelco, Bellefonte, PA) under isocratic conditions with acetonitrile-water (86:14, v/v) as the mobile phase at a flow rate of 0.3 ml/min. The operative wavelength was set at 208 nm. LC/MS determinations were performed by operating the mass spectrometer (Quattro LC triple quadrupole, Micromass, Manchester, UK) in the positive ion mode. Full-scan mass spectra range of 300–400 mass units using a step size of 0.1 mass unit and a scan time of 0.5 s; the resolution of quadrupole was tuned to unit resolution. The HPLC column was operated under the above conditions. Cholesterol, campesterol, sitosterol, and 7-ketocholesterol were provisionally identified by their retention times compared with commercially available standards. Their identities were confirmed using GC/MS (20,21). Helium was used as carrier gas (2 ml/min flow rate). The ion source was held at 80°C and operated in the electron impact mode. Samples were applied via an on-column injector, with oven temperature 210–290°C and ramping at 2°C/min. Sterols were identified by their fragmentation pattern while scanning the mass range m/z 200–680.

HPLC provides an excellent resolution of cholestanol from the corresponding stenols and allother sterols. However, these methods are not fully adequate for dealing with the minor sterols, which are sensitive to peroxidation and degradation, and require modification of conventional methods of analysis (2,17,66).

2. Resolution of Steryl Esters

HPLC provides an alternative method for the resolution of intact steryl esters. While separations of the esters of a single sterol moiety offer marginal advantages over analyses of the released fatty acids, the resolution of intact esters is indispensable for the determination of the fatty acid esters of mixed sterols.

Kuksis et al. (98) and Marai et al. (99) used a reversed-phase HPLC system with on-line chemical ionization MS for separation of the fatty acid esters of cholesterol in normal plasma and of campesterol and sitosterol found in vegetable oils and Intralipid. The steryl esters, which overlapped with the triacylglycerols during chromatography on the reversed-phase column using a linear gradient of 30–90% propionitrile in acetonitrile, were identified and quantified on basis of single-ion mass chromatograms extracted from the total ion current and the relative retention times of the nonesterified and esterified sterols. The ion at m/z 383 identifies the nonesterified campesterol, campesteryl linoleate, campesteryl oleate, and campesteryl palmitate in order of increasing retention time. Similarly, the ion at m/z 397 identifies nonesterified sitosterol, sitosteryl linoleate, sitosteryl oleate, and sitosteryl palmitate in order of increasing retention time (Fig. 10). The ions m/z 383 and 397 represent the $[MH-H_2O]^+$ and $[MH-RCOOH]^+$ fragments of the nonesterified and esterified forms of campesterol and sitosterol, respectively. In the selected ion mode, the mass spectrometer permits the detection of nonesterified and esterified plant sterols in a total lipid extract without prior separation from the triacylglycerols.

At about the same time, Billheimer et al. (100) reported reversed-phase separation of 30 synthetic steryl esters made up of 10 different fatty acids

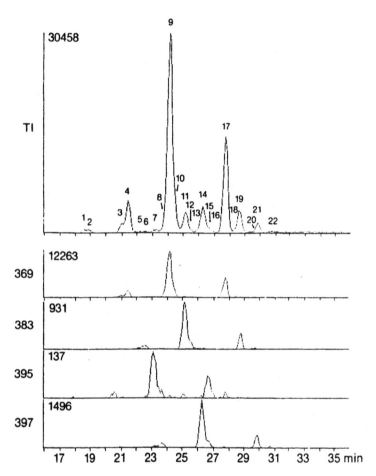

Figure 10 LC/MS elution profile of steryl esters of the LDL + HDL fraction of plasma from a patient with phytosterolemia. TI, total ion current; 369, 383, 395, and 397, fragment ions representing the steroid nuclei of cholesterol, campesterol, stigmasterol and avenasterol, and sitosterol. Peak identification: 1, cholesteryl docosahexaenoate; 2, cholesteryl eicosapentaenoate; 3, cholesteryl linolenate; 4, cholesteryl arachidonate; 5, unknown; 6, campesteryl arachidonate; 7, avenasteryl linoleate; 8, sitosteryl arachidonate; 9, cholesteryl linoleate; 10, cholesteryl palmitoleate, 11, campesteryl linoleate; 12 and 13, unknown; 14, sitosteryl linoleate; 15, avenasteryl oleate; 16, unknown; 17, cholesteryl oleate; 18, unknown; 19, cholesteryl palmitate; 19, campesteryl oleate; 20, unknown; 21, sitosteryl oleate; 22, unknown LC/MS conditions: Hewlett-Packard model 1084 liquid chromatograph interfaced with Hewlett-Packard model 5985B mass spectrometer via a Hewlett-Packard direct liquid inlet interface; column, reversed-phase Supelcosil C_{18} (25 × 0.26 cm i.d.); eluent, a linear gradient of 30–90% propionitrile in acetonitrile. (From Ref. 37.)

and 14 sterols using acetonitrile/isopropanol (60:40, v/v) as a mobile phase and UV monitoring at 205 nm. The steryl esters were characterized by their relative retention times. The method was used to determine the steryl ester composition of corn oil, which was found to contain stigmasteryl linoleate, campesteryl linoleate, sitosteryl linoleate, stigmasteryl oleate, campesteryl oleate, and sitosteryl oleate as major components.

Figure 10 shows an LC/MS elution profile of steryl esters of the LDL + HDL fraction of plasma from a patient with phytosterolemia (see Figure 8 for GLC profile). In this instance the signal from the steryl esters could be increased by performing the analysis in presence of chloride ions, which amplified the ion current (101). The total ion current and the single-ion mass chromatograms extracted from the total ion current are shown in relation to the corresponding steroid nuclei of cholesterol (m/z 369), campesterol (m/z 383), stigmasterol (m/z 395), and avenasterol and sitosterol (m/z 397). The steryl ester peaks were identified by their relative retention times when compared to synthetic standards. The total ion current represents the sum of the $[MH-RCOOH]^+$ ion currents of the individual steryl esters.

Cheng and Kowal (102) reported excellent separation and quantification of rat adrenal cholesteryl esters (CE 18:0–22:6) using reversed-phase HPLC (5 μm Ultrasphere ODS, Beckman, 25 cm × 4.6 mm i.d.) with acetonitrile-isopropanol (1:1, v/v) and UV detection at 210 nm.

Ferrari et al. (103) resolved steryl esters of fatty acids from vegetable oils by rapid chromatography on a short column of silica gel and quantitatively determined by HPLC on reversed-phase columns using an evaporative light scattering detector that was calibrated to sitosteryl oleate as external standard. The areas of all individual peaks in the range of steryl esters were summed and the total steryl ester content was calculated by using sitosteryl oleate as external standard.

Mezine et al. (22) observed that the separation of the steryl esters on a hexylphenyl phase could be achieved using isocratic elution with 90% acetonitrile in water; however, there was noticeable broadening of the late-eluted peaks. Therefore, a gradient was used. The position of free sterols was established by injecting a plant sterol mixture and confirming the characteristic spectra. The HPLC elution profile of steryl esters reflects differences both in the sterol moieties and in the fatty acid composition. The pair of cholesteryl palmitate and cholesteryl oleate can be resolved under isocratic conditons (90% acetonitrile). Assignment of the molecular species of the steryl esters was done on the basis of the elution profile of cholesteryl esters and the known composition of fatty acids derived from canola oil. Ion extraction corresponding to m/z 350–410 $[M-OCOR]^+$ of TIC of the steryl ester chromatogram revealed the presence of four major peaks of esters of sitosterol (m/z 397–398) and campesterol (m/z 383–384). Ion extraction in the m/z 395–396 range revealed a more complex picture, presumably due to the separation of stigmasterol and avenasterol esters.

Mezine et al. (22) used APCI/MS for the detection of steryl esters. The mass selective detector settings were optimized using flow injection analysis. It was set to scan in the m/z range of 200–900. It was found that steryl esters at the optimal APCI conditions produce two types of major fragment ions described as $[M-OCOR]^+$ and $[M-OCOR+H]^+$. The method was applied to determine the steryl ester composition of certain plant sterol–enriched margarines. A setting of the mass spectrometer detector to scan in m/z 350–410 range allowed specific detection of steryl ester peaks. A total ion current chromatogram, obtained in the scan mode in the mass range of m/z 350–410, as well as the ion extraction chromatograms were shown in Fig. 10 in stacked line-up of mass chromatograms. The presence of either natural or hydrogenated steryl esters was shown in the two margarines investigated. Sitostanol (m/z 399, [M-OCOR], m/z 400 protonated form) and campestanol (m/z 385, [M-OCOR], m/z 386, protonated form), as well as sitosterol esters, were seen.

The positive-ion APCI mass spectra of phytosterols were characterized by the protonated molecular ions of the analytes and an abundant signal corresponding to fragmentation due to loss of a water molecule. In the sitosterol mass spectrum an interfering signal at m/z 383 was also detected, attributable to the $[M+H-H_2O]^+$ ion of campesterol, as an impurity in the standard powder of sitosterol. The presence of campesterol and sitosterol in the soybean oil extract was confirmed by LC/APCI/MS analysis carried out on the same sample. The total ion current profile was compared to the extracted ion chromatograms of sitosterol, stigmasterol, and campesterol.

Combining HPLC with MS provides a method for detection and identification of incompletely resolved plant sterols in the presence of triacylglycerols or steryl esters. Early workers (37) described LC/MS of total sterol fraction, isolated from plasma lipoproteins by preparative TLC. An application of this approach to the plasma phytosterolemia patients led to the detection of several campesteryl, sitosteryl, and avenasteryl esters (37,62). The authors also noted the need for background subtraction to compensate for solvent impurities associated with propionitrile. Mezine et al. (22) have developed a simple, reliable method capable of analyzing the content of intact plant steryl esters in food products, and have investigated the chromatography of steryl esters on an HPLC column packed with hexylphenyl media, which is less hydrophobic and allows elution of steryl esters by commonly used water-organic mixtures. Moreover, the hexylphenyl phase, besides providing pure hydrophobic interactions, is capable of π-π interactions with the double bonds present in the steryl esters, implying higher selectivity in comparison to C18 phases. Mezine et al. (22) developed the LC/APCI/MS combination in a positive-ion mode for an application to analysis of plant sterol and stanol esters in food and beverage products, but it is obvious that the method can be readily adapted for work with biological samples.

The LC/MS system also allows an examination of each steryl ester peak for the presence of the corresponding 5α-stanyl esters. The presence of cholestanol (m/z 371) and stigmastanol (m/z 399) could be discerned for each of the major ester peaks and its characteristic retention time. The fragment ions of the stanols (possess the same mass as the parent ion (M + 2 ion) of their unsaturated homologues and must be clearly distinguished from them on the basis of the retention times of the steryl ester peaks. The LC/MS results confirmed the presence of both saturated and unsaturated plant sterols in the various fatty acid ester classes as observed from GLC analyses of the sterols in the AgNO$_3$ TLC fractions of the intact steryl esters (see above). In control plasma only small amounts of the 5α-saturated stanols were found (0.4% of total). In contrast, substantial amounts of the 5α-saturated stanols were present in the phytosterolemia plasma. The LC/MS approach using a single quadrupole did not allow differentiation between the fatty acid esters of campesterol and its isomer 22,23-dihydrobrassicasterol, which were eluted with similar retention times and possessed very similar chemical ionization mass spectra.

V. QUANTIFICATION

Careri et al. (20,21) described a validation procedure for the detection limit (L_D) and the quantification limit (L_Q), expressed as signals based on the mean blank (x_b) and the standard deviation (s_b) of the blank responses, as follows:

$$L_D = x_b + 2ts_b \qquad L_Q = xb + 10s_b$$

Where t is a constant of the t Student distribution (one sided) dependent on the confidence level and the degree of freedom ($v = n - 1$; n = number of measurements). Ten blank measurements were performed to calculate xb and s_b. L_D and L_Q wer converted from signal domain to concentration domain [limits of detection (LOD) and quantification (LOQ), respectively] using a calibration function calculated in the 0.8–2.8 µg/ml concentration range (18). In order to satisfy basic requirements such as homoscedasticity and linearity, the Bartlett test and linearity tests [the lack-of-fit test and the Mandel fitting test (104)] were performed at the 95% significance level.

Linearity was studied over two orders of magnitude of concentrations in the 0.7–70 µg/ml range. Six equispaced concentration levels were chosen and three replicated injections were performed at each point. As in the case of LOD and LOQ calculation, the homoscedasticity test, the lack-of-fit test, and the Mandel fitting test were run.

Precision was calculated in terms of intraday repeatability and inter-day reproducibility. The intraday repeatability was calculated in terms of RSD% ($n = 5$) to two concentration levels (2 and 100 µg/ml) for each analyte. The interday reproducibility was checked on three different days at the same concentration levels as for intraday repeatability; for this purpose, a homoscedasticity test and analysis of variance were performed on the replicated measurements ($n = 15$) at the 95% significance level.

Careri et al. (20,21) also calculated the matrix effect. For this purpose, the calibration function of the fundamental analytical procedure was first determined:

$$Y = a_c + b_c x_c$$

Analytical calibration procedure was peformed on the unspiked and spiked samples following the standard addition method. The analytical results x_f were then calculated using the found signal values y_f and the analysis function, i.e., the calibration function solved for x:

$$x_f = y_f - a_c / b_c$$

By plotting the "found concentrations" (x_f) vs. the original calibrations (x_c), the recovery curve was calculated, which is mathematically described by the recovery function (linear regression line):

$$X_f = a_f + b_f x_c$$

In the ideal case, the recovery function results in a line with intercept $a_f = 0$ and the slope $b_f = 1$ as well as a residual standard deviation that corresponds to the standard process deviation of the fundamental analytical procedure (104).

In order to calculate the sterol ester composition derived by LC/MS, Mezine et al. (22) established the response factors for the sterols. First, the percentage of corresponding ion current in TIC was calculated:

$$S_i = I_i / I_{TIC} \times 100$$

where S_i is percentage of the ion current of the sterol, I_i, the integrated area of extracted ion current in m/z window corresponding to ($[\text{M-OCOR}]^+$ and $[\text{M-OCOR} + \text{H}]^+$), and I_{TIC}, the integrated area of total ion current. The response factor was established by comparing the calculated values with data obtained by GC/FID. The calculated composition of molecular species of sterol esters in the analyzed mixture was obtained as follows:

$$SE_i = \frac{I_i}{ITLC} \times K \times 100$$

where K is the response factor.

Mezine et al. (22) determined the linearity, precision, and accuracy of the method of steryl ester quantification by LC/MS. The linear calibration was performed in the range 250–5000 ng of total sterol ester mixture per injection. The limit of detection (S/N ratio 4:1) was found to be 25 ng of total steryl ester mixture, and the limit of quantification (SD < 20%) was 150 ng of ASE per injection. The calibration curve was constructed in coordinates: ratio of concentrations ("sterol esters"/internal standard)- ratio of areas ("sterol esters"/internal standard), where "sterol esters" refers to both peaks of sterol ester and free sterols present in the commercial preparation of steryl esters. The following parameters were obtained: intercept −0.2492, slope 0.6287, and correlation coefficient 0.9985, which indicate good linearity of the method. The method also has good precision—intraday grand CV% ($n = 15$) was 6.6 and accuracy (RE%) was 3.5.

VI. SIGNIFICANCE

The main purpose of the above studies was the assessment of model compounds, which will allow a more rational approach for isolation of natural hypocholesterolemic sterol analogues or development of synthetic analogues. Increased plasma cholesterol is generally accepted as a marker for hyperlipoproteinemia and atherosclerosis. The increases in the uncommon plasma sterols, while highly significant, nevertheless may represent only minor components of plasma sterols, for whose detection, identification, and quantification highly sophisticated methodology is needed. The practical progress to date is summarized below.

A. Differential Sterol Absorption

Early work by Kuksis and Huang (54,55) showed that campesterol (24α-methyl, also known as γ-sitosterol) was absorbed into dog lymph at a rate between that of cholesterol and sitosterol (24β-ethyl), whereas Vahouny et al. (105) later showed that plant sterols with 24-methyl and methylidine substitutions were absorbed more efficiently, but less so than cholesterol. Vahouny et al. (105) also showed a significant lymphatic absorption of brassicasterol (Δ^{22}, 24β-methyl), which is a significant component of bivalve shellfish sterols, at about 9% when assessed after administration of mixed sterols.

    Several studies support discrimination between cholesterol and sitosterol at the brush-border membrane. Child and Kuksis (64,65), using the 7-dehydro analogues of the two sterols in micellar solution, reported a four- to fivefold uptake differential by rat brush-border membranes in vitro. Figure 11 shows the capillary GLC elution patterns of the plant sterols of menhaden oil as the total unsaponifiable matter (bottom), the sterols of the intestinal brush-border membrane (middle), and lymph sterols (top) during menhaden oil absorption in the rat (Kuksis, Yang, and Myher, 1992, unpublished work). In

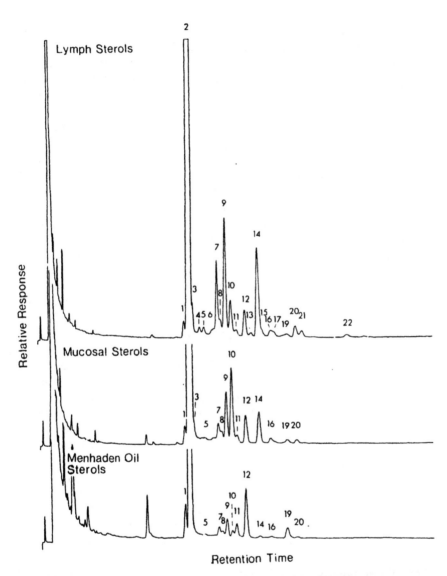

Figure 11 GLC elution profile of the sterols in the unsaponifiable matter of men-hade oil (bottom), intestinal brush-border (middle), and intestinal lymph (top) sterols of rats receiving menhaden oil in the diet. Peak identification is as follows: 1, cholestanol; 2, cholesterol; 4, brassicastanol; 5, brassicasterol; 7, Δ^5-cholestenol (latho-sterol); 8, campestanol; 9, campesterol; 10, desmosterol; 12, 24-methylenecholesterol; 13, sitostanol; 14, sitosterol; 15, Δ^7-ergosterol; 16, $\Delta^{5,7}$-ergostadienol; 18, $\Delta^{7,24}$-methylenecholesterol; 19, Δ^5-avenasterol; 21, Δ^7-stigmasterol; 22, Δ^7-avenasterol; 23, unknown. GLC conditions as given in Fig. 4. Sample: sterol acetates. (From Kuksis, Yang, Myher, 1992. Unpublished results.)

comparison to the original unsaponifiable matter, the brush-border sterols are enriched in campesterol and Δ^7-cholestenol (lathosterol) and sitosterol, specifically discriminating against 24-methylene cholesterol. In contrast to the brush-border sterols, the lymph sterols have become enriched in desmosterol, while also transferring efficiently the 24-methylene cholesterol that had entered the brush border. However, the transfer of the Δ^7-cholestenol to lymph was strongly impaired, although its transfer from the lumen to the brush border had been favored. This study shows that the sterols of menhaden oil were subject to three levels of selection: first, at the luminal solubilization; second, at the brush-border uptake; and third, at the lymphatic transfer. Because of insufficient concentration in the unsaponifiable matter of menhaden oil, the lymphatic absorption of brassicasterol was not established.

Early work suggested several potential sites for discrimination, including the level of mucosal sterol esterification (106) and luminal solubilization (107,108). Two major sites appeared to be involved, namely, differential uptake at the brush border and differential affinity for bile salt in micelles. The possibility that absorbable and unabsorbable sterols possess different affinities for the micelle as suggested by Armstrong and Carey (108) could result in a differential delivery of the sterols to the cell surface. Ikeda and Sugano (109) reported that the intestinal uptake of sitosterol intubated into the stomach of rats was about one-fifth that of cholesterol.

Chemical saturation of the Δ^5 double bond leads to the introduction of a 5α position of the hydrogen atom (campestanol and sitostanol), whereas enzymatic transformation by bacteria in the intestine leads to the 5β position (methyl and ethyl coprostanols). Several studies have indicated that sitostanol is not absorbed at all (110–113), and feeding of this saturated plant sterol seems to have a more pronounced effect on reduction of serum cholesterol than does sitosterol.

Ikeda et al. (107) have presented data to suggest that the mechanism that accounts for inhibition of cholesterol absorption is distinct from that which discriminates between cholesterol and plant sterols for absorption. Specifically, sitosterol (112) and fucosterol (107) displaced cholesterol from micellar solution, accounting for the inhibition of its absorption, but micellar solubilization does not ensure absorption because sitosterol is not absorbed even when fully solubilized (107). Several potential sites for discrimination have been suggested. One is at the level of mucosal sterol esterification and is based on the repeated observations that, in contrast to cholesterol, sitosterol appears largely unesterified in the lymph of experimental animals. This observation is supported by in vitro studies demonstrating that sitosterol is esterified less well (113,114) than cholesterol by the intestinal esterification enzymes cholesterol esterase (107) and acyl coenzyme A acyltransferase (115).

Several laboratories have attempted to effect a complete or partial replacement of cholesterol with plant sterols in mammalian cells (116–118), or with 7-dehydro-, des-AB,8-one, or calciferol derivatives (64,65). From solutions containing egg yolk phospholipid, a preference for cholesterol over plant sterol uptake by red cells was observed, increasing with time from a cholesterol/sitosterol ratio of unity to a maximum of 2 after 60 min incubation (118). Table 4 demonstrates that the relative rates of uptake of the 7-dehydrosterols parallel those of parent sterols. Comparable rates of uptake for sitosterol and its 7-dehydro derivative were obtained when the cells were incubated in the presence of campsterol, sitosterol, and their 7-dehydro analogues. Mammalian cell lines have now been isolated that are sterol auxotrophs. Rujanavech and Silbert (119) have used one of these mutants to identify sterols that can replace cholesterol as a growth factor without metabolism. These workers isolated cholesterol along with the plant sterols by saponification and analyzed them by GLC in the free from on a 3% OV-17 column at 250°C. The sterols cholesterol, campesterol, brassicasterol, stigmasterol, and ergosterol, along with saturated and unsaturated derivatives of cholesterol, were identified on the basis of retention times established under these conditions for the different analogues. Quantification was achieved by addition of known amounts of coprostanol or of stigmasterol to samples to serve as an internal standard. Derivatives with a methyl group on the side chain or with small changes in the nucleus, such as reduction of the Δ^5 bond, which makes the nucleus less planar, support growth only at reduced rates. More marked alterations in the side chain structure, such as introduction of an ethyl group, or an alkyl group plus a trans double bond, or of a *cis*-ethylenic bond (all of which increase the membrane volume required to accommodate the side chain), resulted in sterols that cannot support growth of LM cells as the sole supplement.

Buttke and Bloch (120) had earlier used GC/MS to demonstrate that a yeast cell auxotroph grows with a wide variety of sterols, including ergosterol (the natural yeast sterol), sitosterol, stigmasterol, cholesterol, 7-dehydrocholesterol, 19-norcholesterol, and cycloartenol, a 9, 19-cyclopropane analogue of lanosterol. Thus, yeast can use a wider spectrum of sterols than the mammalian cell.

Suarez et al. (121) have recently used HPLC to determine the incorporation of ergosterol by HL-60 cells. The HL-60 cells (1.5×106 per ml) were incubated in DCCM-1 medium supplemented with 15 μg/ml ergosterol or placebo (0.44% ethanol), for 10 h, at 37°C in 5% CO_2 atmosphere. Then the cells were washed with ice-cold $NaCI/P_i$ and resuspended in 0.5 ml of 10% KOH. The lipid extract was further subfractionated into the saponifiable and nonsaponifiable fractions (96), and the latter was analyzed by reversed phase

HPLC with the use of a Luna-C18 column (250 × 4.6 mm, 5 µm; Phenomenex, Torrance, CA). Sterols were eluted with acetonitrile-water (95:5, v/v) at a flow rate of 1.2 ml/min at 45°C, and monitored by UV (variable-wavelength detector, Beckman Instruments, Palo Alto, CA). For this study, both promyelocytic HL-60 and lymphoblastoid MOLT-4 human cell lines were cultured in cholesterol-free medium and treated with SKF 104976 to inhibit cholesterol synthesis, and the effects of ergosterol (as compared to cholesterol) on DNA synthesis, cell cycle progression, and Cdk1 activity were determined. The results were interpreted on the basis of the differential action of these sterols, with ergosterol contributing to cell membrane formation and cholesterol required for cyclin-dependent kinase activation.

B. Inhibition of Cholesterol Absorption

Peterson (3) reported for the first time that the increase of plasma cholesterol levels in chickens caused by cholesterol feeding can be prevented by including 1% soybean sterols in the diet. Since then, numerous studies have confirmed a hypocholesterolemic action of plant sterols, especially sitosterol (4,5,122–125).

Currently, a great renewal has occurred in the use of sitosterols for the inhibition of cholesterol absorption. The novel development was to convert plant sterols to the corresponding stanols and esterify them to a fat-soluble form (125). Miettinen and Gylling (reviewed in 126) have developed a sitostanol ester margarine that includes sitostanol in soluble ester form. The significance of this development is that the margarine can replace a small amount of normal dietary fat. The effect of the sitostanol margarine on the plasma sterols was determined (assessed) at several levels, including total, free and esterified cholesterol.

Campesterol and sitosterol, plant sterols that may be indicators of dietary cholesterol absorption (127), can be measured within the same analytical run. Measurement of the concentrations of phytosterols may be useful for the study of cholesterol metabolism particularly in large-scale epidemiological studies (127). Concentrations of campesterol and phytosterols are similar to published levels using similar methods, both in whole serum (128,129) and in lipoproteins (130).

Miettinen and coworkers (127,131,132) have shown that the levels of campesterol and sitosterol are proportional to the total and fractional dietary absorption of cholesterol, determined by continuous [^{14}C]cholesterol/[^{3}H] sitosterol feeding. Determination of cholesterol absorption by the latter method is not ideal for large-scale studies because it is time consuming and involves the administration of radioactive isotopes. Therefore, the measurement of phytosterols may provide a simple and convenient means of measuring cholesterol absorption. Recovery of cholesterol from the lipoprotein fractions varied from 71% to 74% as determined by [^{3}H]cholesterol. Linearity was excellent using known amounts of sterol standard added to patient samples, over the range 0.15–20 µmol/L for 7-ketocholesterol, campesterol, and sitosterol. The recovery of 0.4 µg of 7-ketocholesterol, campesterol, and sitosterol added to patient sample was 95–99%. The coefficient of variation depended on the concentration of sterol and varied from 6.6 to 16 in the concentration range 2–4.5 µmol/L.

It could be shown that the "campestanol" peak found in serum samples was made up of two sterols: $\Delta^{8,24}$-dimethylsterol precursor of cholesterol synthesis and avenasterol, with traces of sitostanol being found only during sitostanol feeding. Cholestanol and other plant sterol peaks obtained by GLC from sterols in serum appeared to include only the respective sterol. In addition, squalene and noncholesterol sterols, including Δ^8-cholestenol, desmosterol, and lathosterol, plant sterols (campesterol and sitosterol), and cholestanol, were determined by GLC. The resulting decrease of hepatic cholesterol was apparently balanced by enhanced LDL receptor activity and up-regulation of cholesterol synthesis. The relative decrease of the serum campesterol proportion was shown to reduce cholesterol absorption efficiency by more than 60% in familial hypercholesterolemia (FHC) subjects (127).

Vahouny et al. (133) had earlier reported the comparative absorption of sitosterol, stigmasterol, and fucosterol and the differential inhibition of cholesterol absorption. The test sterols were extracted by the Folch method from brown algae. After saponification with alcoholic KOH, the nonsaponifiable lipids were extracted into hexane. Sterols were precipitated by addition of methanol and recrystallized twice from methanol. Purity was verified by GC/MS with packed SE-30 column. The column separated cholesterol, campesterol, and stigmasterol but did not resolve sitosterol and fucosterol. These sterols were identified and quantified by an SP-1000 column. Analysis of 24-h lymph collections by GC/MS demonstrated that all three sterols were poorly absorbed to the extent of only 3–4% of the administered dose of 50 mg. In contrast, cholesterol absorption under similar conditions was about 42% of the administered dose. Administration of either sitosterol or stigmasterol resulted in an equally effective inhibition of cholesterol absorption (54%). Under identical conditions, fucosterol had no effect on absorption of luminal cholesterol.

From a structural standpoint, nonabsorbability is most clearly related to substitutions in position C-24 [54,64,65,105,109,133] on the sterol side chain with methyl or ethyl groups or their unsaturated counterparts as exemplified by campesterol (24-methyl), sitosterol (24-ethyl), stigmasterol (Δ^{22}, 24-ethyl), and fucosterol (24-ethylidine). Each of these is poorly absorbed and inhibits cholesterol absorption.

C. Sitosterolemia and Xanthomatosis

The association of high plasma cholesterol concentrations and atherosclerosis is best demonstrated in the inherited lipid disorder familial hypercholesterolemia, where absent or reduced functional tissue LDL receptors result in extraordinarily high plasma LDL cholesterol levels. As a consequence, these individuals show extensive tendon and tuberous xanthomatosis and develop atherosclerosis rapidly, with death often occurring because of myocardial infarction at a young age. Increased level of plasma cholesterol is generally accepted as a marker for hyperlipoproteinemia and atherosclerosis. There is good evidence that increased levels of plasma sterols other than cholesterol can serve as markers for abnormalities in lipid metabolism associated with clinical disease (30,60). Thus, premature atherosclerosis and extensive xantomatosis have been described in two rare lipid storage diseases, *cerebrotendinous xanthomatosis* and *sitosterolemia with xanthomatosis*. In *cerebrotendinous xanthomatosis*, severe neurological dysfunction, cataracts, pulmonary abnormalities, and endocrine hypofunction are found, whereas in sitosterolemia with xanthomatosis, arthritis and episodic erythrocyte hemolysis are noted occasionally. Specifically, in *cerebrotendinous xanthomatosis*, cholestanol, the 5α-dihydro derivative of cholesterol, is present in all tissues with particularly high amounts in brain, nerve, xanthomas, and bile. In distinction, large quantities of plant sterols (campesterol and sitosterol) along with 5α-saturated stanols, cholestanol, 5α-campestanol, and 5α-sitostanol accumulate in plasma of subjects with sitosterolemia with xanthomatosis. Although the mechanism for the enhanced sterol accumulation is not known, abnormal lipoprotein sterol transport has been suggested to have a role during sterol absorption and tissue distribution.

Total plasma lipid profiling provides a simple and highly effective means for demonstrating excessive amounts of plant sterols in plasma in sitosterolemia. Figure 12 shows the total plasma GLC profile of a patient with phytosterolemia and of two healthy family members (37). This particular separation was obtained following conversion of the plasma phospholipids into diacylglycerols and ceramides and trimethylsilylation of the sample. The figure shows clearly resolved peaks for free cholesterol, campesterol, and sitosterol along with those of the corresponding palmitoyl, oleoyl, linoleoyl, and arachidonoyl esters of cholesterol and of the plant sterols, although the latter ones are not as clearly resolved.

Salen et al. (60) recently reported the percent absorption, turnover, and distribution of campestanol in a sitosterolemic homozygote, her obligate heterozygous mother, and three healthy human control subjects. The campestanol, like other noncholesterol sterols, was hyperabsorbed and retained in sitosterolemic homozygotes. However, campestanol absorption was only slightly increased in the sitosterolemic heterozygote, and removal was as rapid as in control subjects.

Lutjohann et al. (134) studied the sterol absorption and sterol balance in phytosterolemia by deuterium-labeled sterols with special attention to sitostanol treatment. The results showed that the absolute difference in absorption rate of different sterols between the patients and healthy volunteers was about the same. Thus, patients with phytosterolemia seem to have generally increased absorption of sterols rather than a loss of a specific discriminatory mechanism. Jones et al. (135) have also examined the modulation of plasma lipid levels and cholesterol kinetics by phytosterol vs. phytostanol esters. It was shown that cholesterol turnover, as measured by deuterium labeling, was not influenced by diet. The data further indicated that plant sterol and stanol esters differentially lowered circulating total and LDL cholesterol levels by suppression of cholesterol absorption in hypercholesterolemic subjects.

In the absence of an established mechanism of cholesterol absorption, it is difficult to speculate about the mechanism of plant sterol absorption and especially about the plant sterol inhibition of cholesterol absorption. Nevertheless, the speculation continues unabated. Based on recent work with ABC transport proteins by Berge et al. (136) and on other recent studies, Allayee et al. (137) have provided a rationalization for the differential absorption of cholesterol and plant sterols by the intestine. Dietary sterols passively enter intestinal cells, and a proportion of them are actively pumped back into gut lumen by the ABC transporter proteins. Berge et al. (136) used a combination of mapping information and functional data to identify the genes that are defective in sitosterolemia. Kuwabara and Labousesse (138) recently discussed the protein families possessing sterol-sensing domains and speculated about their role in cholesterol homeostasis, cell signaling, and cytokinesis, the dietary control of which, if any, is unknown.

D. Cancer Prevention

In addition to reduction of serum or plasma total and LDL cholesterol (13,139), dietary plant sterols have been reported to prevent tumor growth in experimental animals and cell cultures. Rats fed a diet supplemented with 0.3% sitosterol had a significantly lower incidence of chemically induced tumors than controls (140). Janezic and Rao (141) reported a reduction in cholic acid–induced hyperproliferation of colonic epithelium with feeding phytosterols to mice, and Awad et al. (142) reported similar results in rats. In tissue cultures, Awad et al. (143) observed that sitosterol inhibited HT-29 human colon cancer cell growth and altrered membrane lipids by decreasing membrane sphingomyelin. Sitosterol was also found to be effective in inhibiting the growth of LNCaP cells, a human prostate cancer cell line (144), and MBA-MD-231, a human breast cancer cell line (145). Furthermore, epidemiological studies have suggested that populations consuming diets high in phytosterols might be at decreased risk for colon cancer (146,147). In this connection, Li et al. (148) have proposed the measurement of plasma sitosterol and campesterol as a new biomarker for cancer prevention. The measurements were made by GC of the free sterols prepared by saponification of the total plasma sample in the presence of 5α-cholestane. Sterols, including cholesterol, were extracted with hexane and determined by GC without derivatization (143). The gas-chromatograph was equipped with a 30-foot nonpolar capillary column and the analyses were performed at 265°C with nitrogen as the carrier gas. The authors reported high reproducibility and good overtime reliability of plasma sitosterol and campesterol measurements, which were thought to be suitable for potential clinical and population-based studies on cancer prevention. However, it should be pointed out that GC analysis of underivatized sterols is not generally recommended, although it has been used in the past to resolve saturated and unsaturated sterols, which cannot be readily achieved following derivatization (see above).

E. Other Metabolic Effects

The significance of the metabolic transformations of the plant sterols in the animal and human tissues remains largely unknown. It is generally believed

---

Figure 12 Gas chromatographic profiles of total lipids of two healthy family members (controls) (A and B) and of a patient with phytosterolemia (C). Peak identification: 16 and 18, free fatty acids with 16 and 18 carbon atoms; 20, 22 and 24, monoacylglycerols with 16, 18 and 20 acyl carbons, respectively; 27, free cholesterol; 28, free campesterol (arrow); 29, free sitosterol (arrow); 30, tridecanoylglycerol (internal standard); 32–42, diacylglycerol and ceramide moieties with 30–40 fatty chain carbons; 43–49, steryl esters with 43–49 total carbon atoms (46 is C_{18} fatty acid ester of campesterol and the descending shoulder of peak 47 is the C_{18} fatty acid ester of sitosterol); 50–56, triacylglycerols with 50–56 acyl carbons. GC conditions: Hewlett-Packard model 5880A gas chromatograph equipped with an on-column capillary injector and a flame ionization detector; column, flexible quartz capillary (8 m × 0.30 mm i.d.) coated with a permanently bonded nonpolar SE-54 liquid phase (Hewlett-Packard); Oven temperature programmed from 40 to 150°C at 30°C/min, then to 230°C at 20°C/min, to 280°C at 10°C/min and to 340°C at 5°C/min; injector and detector temperatures were 100°C and 360°C, respectively. Carrier gas, H_2 at 55.1 kPA. (From Ref. 37.)

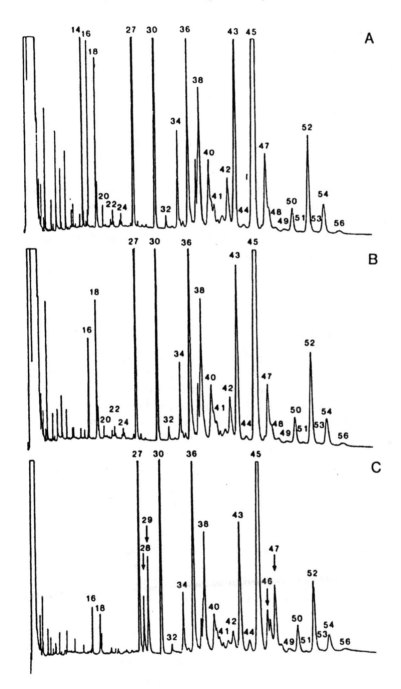

that the plant sterols do not interfere with the metabolism of cholesterol. However, certain specific effects of plant sterols have also been recorded. Field et al. (149) have examined the effect of micellar sitosterol on cholesterol metabolism in Caco-2 cells and have noted several specific effects, which they have reviewed in the light of previous investigations. However, none of the effects on cholesterol metabolism were attributed to specific sitosterol transformation products. It must therefore be concluded that the metabolic transformation products of sitosterol appear to be of little or no importance to the well-being of the animal or to the sitosterolemia patient (see above).

The plant sterols possess steroid ring structures identical to those of cholesterol and are known to undergo esterification to fatty acid esters presumably via acyl coenzyme A:cholesterol acyltransferase and lecithin cholesterol acyltransferase, although specific demonstrations of these transformations are lacking. Likewise, the plant sterols are believed to be subject to reduction of the double bond with the formation of 5α-stanols. However, these transformations take place at rates much lower than those affecting cholesterol.

1. Conversion to 5β-Sterols

Under the influence of gut bacteia, the plant sterols along with cholesterol become subject to 5β transformation. In early studies, Kuksis and Huang (55) and Miettinen et al. (56) demonstrated that GC/FID can be used for an effective separation and quantification of the bacterial reduction products of cholesterol and plant sterols. Kuksis (83) showed the separation of the fecal sterols from rats fed plant sterols for the purpose of inhibition of cholesterol absorption (see also Fig. 2). The 5β forms of cholesterol, campesterol, and sitosterol are clearly seen to precede the elution of the corresponding 5α forms.

More detailed assessment of plant sterol metabolism is readily accomplished by a combined application of TLC and capillary GLC. TLC greatly simplifies the identification and quantification of the sterols in the fecal unsaponifiable matter. Miettinen (150) has used Florisil plates, which were developed with diethyl ether-heptane (55:45, v/v) and sprayed with rhodamine G. Viewing under UV light revealed three bands (top to bottom): fraction 1, 3-ketocoprostanone and 3-keto derivatives of plant sterols; fraction II, coprostanol and ring-saturated 5β derivatives of plant sterols; fraction III, cholesterol, campesterol, stigmasterol, sitosterol, and corresponding ring-saturated 5α-sterols. The individual TLC bands were examined by GLC using a 46-m-long glass capillary SE-30 column that was temperature programmed from 170°C to 265°C at 20°C/min. The sterols originating from cholesterol (coprostanol, epicoprostanol, coprostanone, cholesterol, and cholestanol) are quantitatively separated from the plant sterols and their bacterial conversion products. Bacterial transformation products of plant sterols are consistently present in large amounts in every fecal sterol sample. On the capillary column, cholesterol and methylcoprostanol are clearly separated, allowing quantification even of cholestanol peak.

2. Conversion to Unnatural Bile Acids

There is evidence that an ethyl group at C−24 obstructs the conversion of sitosterol to the common C_{24} bile acids (151–153). Several studies have shown that sitosterol is converted to polar compounds in the bile acid fraction of bile (154,155). Subbiah and Kuksis (151) observed that the acidic steroids produced from ^{14}C-labeled sitosterol were relatively more polar than the common C_{24} bile acids, while Kritchevsky et al. (152) tentatively identified the acidic steroids in feces of monkeys fed sitosterol as C_{27} bile acids. Muri-Boberg et al. (153) and Lund et al. (156) have identified most of the polar products. It was shown that that the polar products resulting from (4) sitosterol were di-and trihydroxylated C_{21} bile acids. Using mass spectrometry, nuclear magnetic resonance, stereospecific dehydrogenases, and reagents, the major trihydroxylated C_{21} bile acids were identified as 5β-pregnan-3α,11β,15β-triol-21-oic acid, 5β-pregnan-3α,11β,15α-21-oic acid, and 5β-pregnan-3α, 12α-diol-21-oic acid. Considerably less C_{21} bile acid was formed from labeled sitosterol in male than in female Wistar rats. The C_{21} bile acids formed in male rats did not contain a 15-hydroxyl group. Since Lund et al. (157) have shown that such C_{21} bile acids are also formed from labeled cholesterol, the possibility exists that the C_{21} bile acids are formed from both cholesterol and plant sterols. By analysis of purified most polar products of [4-^{14}C]sitosterol by radio-GC and the same products of 7α,7β-[[2]H2]sitosterol by GC/MS, Muri-Boberg et al. (158) were able to identify two major metabolites as C_{21} bile acids. One metabolite had three hydroxyl groups (3α, 15, and unknown), and one had two hydroxyl groups (3α, 15) and one keto group. A possible explanation for the formation of C_{21} bile acids from sitosterol in rat liver is a hypothetical mechanism.

After intravenous injection of deuterium-labeled sitosterol into bile fistula of female Wistar rats, the isolated C_{21} bile acids were found to contain very little isotope. This is in contrast to early studies by Salen et al. (6) that claimed efficient formation of cholic and chenodeoxycholic acid from intravenously administered [22,23-^{3}H]sitosterol. In a reinvestigation with [4-^{14}C] sitosterol, Muri-Boberg et al. (153) could not find any significant conversion to labeled C_{24}-bile acids in two healthy human subjects. The results suggested that healthy human subjects, like other mammalian species studied, have little or no capacity to convert sitosterol into the normal C_{24} bile acids. A conversion of plant sterols to C_{24} bile acids could theoretically occur following dealkylation of the sitosterol to cholesterol before conversion to bile acids. Such a dealkylation occurs in some worms and crabs (159), but attempts to demonstrate such pathways in mammals have failed.

3. Conversion to Steroid Hormones

Muri-Boberg et al. (153) and Lund et al. (157) showed that conversion of labeled sitosterol into C_{21} bile acids also occurred in adrenalectomized and overiectomized rats, indicating that endocrine tissues were not involved. In this connection it may be noted that Subbiah and Kuksis (160,161) demonstrated the conversion in rat testes of [4-^{14}C]sitosterol into progesterone, pregnenolone, 17α-progesterone, and other polar steroids at rates comparable to those observed with [4-^{14}C]cholesterol. It was concluded that the plant sterol was converted to steroid hormones by a mitochondrial enzyme system of the rat testes that has been shown to cleave the side chain of cholesterol at $C-20$ to yield pregnenolone and isocaproic aldehyde.

Werbin et al. [43] had earlier demonstrated the conversion of sitosterol to urinary cortisol in the guinea pig.

4. Oxygenation of Sterols and Steryl Esters

Plat et al. (162) used GC/MS to analyze the oxophytosterol concentrations in serum from patients with phytosterolemia or cerebrotendinous xanthomatosis, in a pooled serum and in two lipid emulsions. 7-Ketositosterol, 7β-hydroxysitosterol, 5α,6α-epoxysitosterol, 3β,5α,6β-sitostanetriol, and probably also 7α-hydroxysitosterol were present in markedly elevated concentrations in serum from phytosterolemic patients only. In addition, campesterol oxidation products, such as 7α-hydroxycamapesterol and 7β-hydroxycampesterol, were found. The same oxyphytosterols were also found in two lipid emulsions in which the ratio of oxidized sitosterol to sitosterol varied between 0.038 and 0.041. For this purpose, Plat et al. (162) synthesized oxyphytosterols from a mixture consisting of sitosterol, campesterol, and traces of stigmasterol based on previously described methods. All plant sterols were acetylated prior to use for synthesis of all oxyphytosterols, except for the triols. For use as an internal sandard; Plat et al. (162) synthesized deuterated trihydroxy phytosterols and deuterated 7-keto phytosterols from a mixture of d_5-2,2,4,4,6-sitosterol, campesterol, and stigmasterol. The oxyphytosterols were isolated from blood or serum in the presence of the internal standards and the samples saponified in a closed tube under nitrogen for 2 h at room temperature by adding 10 ml of 0.35 M ethanolic KOH solution. Prior to saponification, the samples were extensively saturated with nitrogen for removal of oxygen to minimize autoxidation. After saponification, phosphoric acid 50% in water (v/v) was added to neutralize the solution, followed by addition of 6 ml NaCl solution in water (9 mg/ml), according to procedures previously described by Lujohann et al. (134). After extraction with dichloromethane, the organic solvent was evaporated in a round-bottomed flask and evaporated to dryness. The sample was then redissolved in 1 ml toluene for chromatography.

According to Plat et al. (162) GC/MS analysis was performed with the TMS derivatives of oxophytosterols. For this, 2 μl of the TMS derivatives in cyclohexane was injected via a Trace GC2000 autosampler (Thermoquest CE Instruments Austin, Texas, U.S.A.) equipped with an aRTX5MS column (30 m × 0.25 mm i.d., 0.25 μm film thickness) coupled to a Gas Chrom Q-plus ion trap (Thermoquest). Analysis was carried out in single-ion monitoring mode, making m/z the primary resolving parameter other than retention time. Helium was used as carrier gas. The injector temperature was 280°C. Table 5 gives some plant sterol and oxidized sitosterol concentrations in three samples of phytosterolemic patients vs. two patients and pool serum.

Mezine et al. [22] also investigated the formation of steryl ester oxidation products. For this purpose, the steryl esters were subjected to thermally induced oxidation by air and the obtained mixture was subjected to HPLC-APCI-MS analysis. The oxidation led to more polar, earlier emerging compounds. The scanning of the eluate in the range m/z range 200–600 μm revealed that most of he newly formed compounds gave rise to ions corresponding to a sterol nucleus $[M-OCOR]^+$ and $[M-OCOR+H]^+$, implying that oxidation occurred at the fatty acid moieties. There were also detected ions, which could be ascribed as monooxygenated species of the sterol nucleus

Table 5   Serum Plant Sterol and Oxidized Sitosterol Concentrations in Three Samples of a Phytosterolemic Patient vs. CTX Patients and Pool Serum[a]

	Control serum	Phytosterolemia	CTX
Plant sterols (mg/dl)[b]	1.3	44.0 ± 3.5	1.2 and 1.4
Oxidized sitosterol forms (μg/ml)			
7-Keto-	ND	0.92 ± 0.09	0.04 and ND
7β-Hydroxy-	ND	0.88 ± 0.08	0.01 and ND
5α,6α-Epoxy-	ND	2.33 ± 0.44	ND and ND
3β,5α,6β-Trihydroxy-	ND	0.54 ± 0.73	ND and ND
7α-Hydroxycholesterol	76.3	83.3 ± 36.9	1221 and 671

[a] Values are mean ± SD. ND, not detectable.
[b] Plant sterols are calculated as the sum of sitosterol and campesterol.
  CTX, cerebrotendinous xanthomatosis.
Source: Ref. 161.

[M-OCOR + O]$^+$ and their protonated species. Ions corresponding to dioxygenated sterol nucleus were present at intensities below 10–15% compared to monooxygenated species. Surprisingly, the products of steryl ester oxidation were not detected in any of the food products.

VII. SUMMARY AND CONCLUSIONS

Much progress has been made in the chromatographic analysis of phytosterols. The early gas chromatographic separations obtained on packed columns have been extended to high-resolution capillary columns, which, in combination with mass spectrometry, have resulted in virtually complete resolution and identification of all molecular species. High-performance liquid chromatography in combination with mass spectrometry has provided reliable resolution and quantification of the thermally unstable phytosterols and the fatty acid esters of phytosterols. These separations have clearly resolved the unesterified plant sterols and cholesterol as well as their derivatives.

The distinction between plant sterols and cholesterol has been less obvious in regard to their metabolism. Although phytosterols are absorbed by the intestine much less effectively than cholesterol and are known to inhibit cholesterol absorption, the metabolic basis of the differential absorption and the inhibition of cholesterol absorption by plant sterols remains to be established. Likewise, progress in chromatographic analysis has not led to improved understanding of the endogenous metabolism of the absorbed phytosterols. The early claims of dealkylation of sitosterol to cholesterol and conversion to natural bile acids have not been substantiated. Instead, sitosterol has been demonstrated to yield unnatural bile acids of both higher and lower molecular weight. There is evidence, however, that the metabolic pathways of cholesterol and sitosterol may converge following cleavage of the sterol side chain at C–20. Direct analytical and biochemical evidence is lacking for the effect, if any, of absorbed sitosterol on cholesterol metabolism, although lowering of plasma cholesterol has been postulated. Knowledge at the molecular level of the sterol carrier proteins and enzymes involved in cholesterol metabolism promises to be a step toward better understanding of the sterol structure–function relationship, which analytical methodology alone has not been able to provide.

REFERENCES

1. T Itoh, T Tamura, T Matsumoto. Sterol composition of 19 vegetable oils. J Am Oil Chem Soc 50:122–125, 1973.

2. A Kuksis. Plasma non-cholesterol sterols. J Chromatogr A 935:203–236, 2001.
3. DW Peterson. Effect of soybean sterols in the diet on plasma and liver cholesterol in chicks. Proc Soc Exptl Biol Med 78:143–148, 1951.
4. OJ Pollak. Successful prevention of experimental hypercholesterolemia and cholesterol atherosclerosis in the rabbit. Circulation 2:696–701, 1953.
5. JMR Beveridge, W Ford Connel, GA Mayer, HL Haust, M White. Plant sterols, degree of unsaturation, and hypercholesterolemic action of certain fats. Can J Biochem Physiol 36:895–911, 1958.
6. G Salen, EH Ahrens Jr, SM Grundy. Metabolism of β-sitosterol in man. J Clin Invest 49:952–967, 1970.
7. AH Lichtenstein, RJ Deckelbaum. Stanol/sterol ester–containing foods and blood cholesterol levels. Circulation 103:1177–1179, 2001.
8. KB Hicks, RA Moreau. Phytosterols and phytostanols: functional food cholesterol busters. Food Technology 55:63–67, 2001.
9. RH Frolich, M Kunze, I Kiefer. Cancer preventive value of natural, non-nutritive food constituents. Acta Med Austr 24:108–113, 1997.
10. AB Awad, CS Fink. Phytosterols as anticancer dietary compounds: evidence and mechanism of action. J Nutr 130:2127–2130, 2000.
11. PJ Bouic, JH Lamprecht. Plant sterols and sterolins: a review of their immune-modulating properties. Altern Med Rev 4:170–177, 1999.
12. T Akihisa, K Yasukawa, M Yamaura, M Ukiya, Y Kimura, N Shimizu, K Arai. Triterpene alcohol and sterol ferrulates from rice bran and their anti-inflammatory effects. J Agric Food Chem 48:1319–2313, 2000.
13. WH Ling, PJH Jones. Dietary phytosterols: A review of metabolism, benefits and side effects. Life Sci 57:195–206, 1995.
14. B Yang, KO Kalimo, RL Tahvonen, LM Mattila, JK Katajisto, HP Kallio. Effects of dietary supplementation with sea buckthorn (*Hippophae rhamnoides*) seed and pulp oils on fatty acid composition of skin glycerophospholipids of patients with atopic dermatitis. J Nutr Biochem 11:338–340, 2000.
15. AA Jekel, HAM Vaessen, RC Schothorst. Capillary gas chromatographic method for determining underivatized sterols—some results of analyzing the duplicate 24-h-diet samples collected in 1999. Fresenius J Anal Chem 360:595–600, 1998.
16. AK Johansson, H Korte, B Yang, JC Stanley, H Kallio. Sea buckthorn berry oil inhibits platelet aggregation. J Nutr Biochem 11:4911–495, 2000.
17. SL Abidi. Chromatographic analysis of plant sterols in foods and vegetable oils. J Chromatogr A 935:173–201, 2001.
18. KM Phillips, DM Ruggio, JA Bailey. Precise quantitative analysis of phytosterols, stanols, and cholesterol metabolites in human serum by capillary gas chromatography. J Chromatogr B 732:17–29, 1999.
19. GSMJE Duchateau, CG Bauer-Plank, AJH Louter, M van der Ham, JA Boerma, JJM Boerma, PA Zandbelt. Fast and accurate method for total 4-desmethyl sterol(s) content in spreads, fat-blends, and raw materials. J AM Oil Chem Soc 79:273–278, 2002.
20. M Careri, L Elviri, A Mangia. Liquid chromatography–UV determination and liquid chromatography–atmospheric pressure chemical ionization mass

spectrometric characterization of sitosterol and stigmasterol in soybean oil. J Chromatogr A 935:249–257, 2001.

21. M Careri, F Bianchi, C Corradini. Recent advances in the application of mass spectrometry in food-related analysis. J Chromatogr A 970:3–64, 2002.

22. I Mezine, C Macku, H Zhang, R Lijana. Analysis of plant sterol and stanol esters in cholesterol-lowering food products using high performance liquid chromatography–atmospheric pressure chemical ionization–mass spectrometry. J Chromatogr A (2002) In press.

23. A Babiker, U Diczfalusy. Transport of side-chain oxidized oxysterols in the human circulation. Biochim Biophys Acta 1392:333–339, 1998.

24. S Prattes, G Hori, A Hammer, A Blaschitz, WF Graier, W Sattler, R Zech, E Streyrer. Intracellular distribution and mobilization of unesterified cholesterol in adipocytes: triglyceride droplets are surrounded by cholesterol-rich ER-like surface layers structures. J Cell Sci 113:2977–2989, 2000.

25. L Liscum, NJ Munn. Intracellular cholesterol transport. Biochim Biophys Acta 1438:19–37, 1999.

26. RA Hagerman, MJ Anthony, RA Willis. Solid-phase extraction of lipid from *Saccharomyces cerevisiae* followed b high-performance liquid chromatographic analysis of coenzyme Q content. Anal Biochem 296:141–143, 2001.

27. J Folch, M Lees, GH Stanley. A simple method for the isolation and purification of total lipid from animal tissues. J Biol Chem 226:497–509, 1957.

28. EG Bligh, WJ Dyer. A rapid method of total lipid extraction and purification. Can J Biochem 37:911–917, 1959.

29. EA Podrez, EE Poliakiv, Z She, R Zhang, Y Deng, PJ Finton, L Shan, M Febbraio, DP Febbraio, RL Silverstein, HF Hoff, RG Solomon, SL Hazen. A novel family of atherogenc oxidized phospholipids promotes macrophage foam cell formation via the scavenger receptor CD36 and is enriched in atherosclerotic lesions. J Biol Chem 277:38517–38523, 2002.

30. G Salen, S Shefer, L Nguyen, GC Ness, GS Tint, V Shore. Sitosterolemia. J Lipid Res 33:945–955, 1992.

31. FH Mattson, SM Grundy, JR Crouse. Optimizing the effect of plant sterols on cholesterol absorption in man. Am J Clin Nutr 35:697–700, 1982.

32. THJ Beveridge, TSC Li, JCG Drover. Phytosterol content in American ginseng seed oil. J Agric Food Chem 50:744–750, 2002.

33. SW Park, PB Addis. Identification and quantitative estimation of oxidized cholesterol derivatives in heated tallow. J Agric Food Chem 34:653–659, 1986.

34. KO Esuoso, H Lutz, E Bayer, M Kutubuddin. Unsaponifiable lipid constituents of some underutilized tropical seed oils. J Agric Food Chem 48:21–234, 2000.

35. E Stahl. Thin Layer Chromatography. Springer-Verlag: New York, 1969, p. 887.

36. JJ Myher, A Kuksis, WC Breckenridge, JA Little. Studies of triacylglycerol structure of very low density lipoprotein of normolipemic subjects and patients with type III and type IV hyperlipoproteinemia. Lipids 19:683–691, 1984.

37. A Kuksis, JJ Myher, L Marai, JA Little, RG McArthur, DAK Roncari. Fatty

acid composition of individual plasma steryl esters in phytosterolemia and xanthomatosis. Lipids 21:371–377, 1986.

38. S Popov, RMK Carson, C Djerassi. Occurrence and seasonal variation of 19-norcholest-4-en-3-one and 3β-monohydroxy sterols in the californian giorginian, *Muricea californica*. Steroids 41:537–548, 1983.
39. T Akihisa, S Thakur, FU Rosenstein, T Matsumoto. Sterols of Cucurbitaceae: the configurations at C-24 pf 24-alkyl-Δ^5-, Δ^7- and Δ^8-sterols. Lipids 21:39–47, 1986.
40. B Ham, B Butler, P Thionville. Evaluating the isolation and quantification of sterols in seed oils by solid-phase extraction and capillary gas-liquid chromatography. LC-GC 18:1174–1181, 2000.
41. DJ McNamara, CA Bloch, A Botha, D Mendelsohn. Sterol synthesis in vitro in freshly isolated blood mononuclear leukocytes from familial hypercholesterolemia patients treated with probacol. Biochim Biophys Acta 833:412–416, 1985.
42. LAD Miller, MH Gordon. Use of lipolysis in the isolation of sterol esters. Food Chem 56:55–59.
43. H Werbin, IL Chaikoff, EE Jones. The metabolism of ^{3}H-β-sitosterol in the guinea pig: its conversion to urinary cortisol. J Biol Chem 235:1629–1633, 1960.
44. HL Haust, A Kuksis, JMR Beveridge. Quantitative precipitation of various 3β-hydroxysterols with digitonin. Can J Biochem 44:119–128, 1966.
45. JJ Myher, A Kuksis. Facile gas-liquid chromatographic resolution of saturated and unsaturated sterols using a polar capillary column. J Biochem Biophys Meth 15:111–122, 1987.
46. N Frega, F Bocci, G Lercker. Direct gas chromatographic analysis of the unsaponifiable fraction of different oils with a polar capillary column. J Am Oil Chem Soc 69:447–450, 1992.
47. RC Heupel, Y Sauvaire, PH Le, EJ Paris, WD Nes. Sterol composition and biosynthesis in sorghum: importance to developmental regulation. Lipids 21:69–75, 1986.
48. WD Nes, RC Heupel. Physiological requirements for biosynthesis of 24β-methyl sterols in *Gliterella figikuroi*. Arch Biochem Biophys 244:211–217, 1986.
49. G Salen, P Kwiterovich Jr, S Shefer, GS Tint, I Horak, V Shore, B Dayal, E Horak. Increased plasma cholestanol and 5α-saturated plant sterol derivatives in subjects with sitosterolemia and xanthomatosis. J Lipid Res 26:203–209, 1985.
50. B Dayal, GS Tint, AK Batta, J Speck, AK Khachadurian, S Shefer, G Salen. Identification of 5α-stanols in patients with sitosterolemia and xanthomatosis: stereochemistry of the protonolysis of steroidal organoboranes. Steroids 40:233–243, 1982.
51. P Kalo, T Kuuranne. Analysis of free and esterified sterols in fats and oils by flash chromatography, gas chromatography and electrospray tandem mass spectrometry. J Chromatogr 935:237–248, 2001.

52. Myher JJ, Kuksis A. General strategies in chromatographic analysis of lipids. J Chromatogr B 671:3–33.
53. A Kuksis, L Marai, DA Gornall. Direct gas chromatographic examination of total lipid extracts. J Lipid Res 8:352–358, 1967.
54. A Kuksis, TC Huang. Differential absorption of plant sterols in the dog. Can J Biochem Physiol 40:1493–1504, 1962.
55. A Kuksis, TC Huang. A gas chromatographic study of the in vivo conversion of plant sterols to derivatives of coprostanol. Fed Proc 23:553, 1964.
56. TA Miettinen, SM Grundy, EH, Ahrens Jr. Quantitative isolation and gas–liquid chromatographic analysis of total dietary and fecal neutral steroids. J Lipid Res 6:411–424, 1965.
57. TT Ishikawa, JB Brazier, LE Steward, RW Fallat, CJ Glueck. Direct quantitation of cholestanol in plasma by gas–liquid chromatography. J Lab Clin Med 87:345–353, 1976.
58. G Salen, V Berginer, V Shore, I Horak, E Horak, GS Tint, S Shefer. Increased concentrations of cholestanol and apolipoprotein B in the cerebrospinal fluid of patients with cerebrotendinous xanthomatosis. N Engl J Med 316:1233–1238, 1987.
59. S Serizawa, Y Sayama, H Otsuka, T Kasama, T Yamakawa. Simplified determination of cholestanol in serum by gas liquid chromatography: biochemical diagnosis of cerebrotendinous xanthomatosis. J Biochem 90:17–21, 1981.
60. G Salen, G Xu, GS Tint, AK Batta, S Shefer. Hyperabsorption and retention of campesterol in a sitosterolemic homozygote: comparison with her mother and three control subjects. J Lipid Res 41:1883–1889, 2000.
61. N Gerst, B Ruan, J Pang, WK Wilson, GJ, Schroepfer Jr. An updated look at the analysis of unsaturated C_{27} sterols by gas chromatography and mass spectrometry. J Lipid Res 38:1685–1701, 1997.
62. RG Dyer, C Hetherington, KGMM Albeti, MF Laker. Simultaneous measurement of phytosterols (campesterol and β-sitosterol) and 7-ketocholesterol in human lipoproteins by capillary column gas chromatography. J Chromatogr 663:1–7, 1995.
63. RH Thompson Jr, G Patterson, MJ Thompson, HT Slover. Separation of pairs of C-24 epimeric sterols by glass capillary gas liquid chromatography. Lipids 16:694–699, 1981.
64. P Child, A Kuksis. Critical role of ring structure in the differential uptake of cholesterol and plant sterols by membrane preparations in vitro. J Lipid Res 24:1196–1209, 1983.
65. Child P, Kuksis A. Uptake of 7-dehydro derivatives of cholesterol, campesterol and β-sitosterol by rat erythrocytes, jejunal villus cells, and brush border membranes. J Lipid Res 24:552–565.
66. WR Nes. A comparison of methods for the identification of sterols. Methods Enzymol 111:3–37, 1985.
67. VK Garg, W Nes. Studies on the C-24 configurations of Δ^7-sterols in the seeds of *Cucurbita maxima*. Phytochemistry 23:2919–2923, 1984.
68. VK Garg, WR Nes. Occurrene of Δ^5-sterols in plants producing predominantly

Δ^7-sterols: studies on the sterol composition of six Cucurbitaceae seeds. Phytochemistry 25:2591–2597, 1986.

69. A Kamal-Eldin, K Maatta, J Toivo, A Lampi, V Piironen. Acid catalyzed isomerization of fucosterol and Δ^5-avenasterol. Lipids 33:1073–1077, 1998.

70. A Rahier, P Benveniste. Mass spectral identification of phytosterols. In: Analysis of Sterols and Other Biologically Significant Steroids. WD Nes and EJ Parrish, Eds. Academic Press: San Diego, pp. 223–250, 1989.

71. A Kamal-Eldin, LA Appelqvist, G Yosif, GM Iskander. Seed lipids of *Sesamum indivum* and related wild species in Sudan The sterols. J Sci Food Agric 59:327–334, 1992.

72. PC Dutta, L Normen. Capillary column gas–liquid chromatographic separation of delta5-unsaturated and saturated phytosterols. J Chromatogr A 816: 177–184, 1998.

73. T Itoh, H Tani, K Fukushima, T Tamura, T Matsumoto. Structure-retention relationship of sterols and tripterpine alcohols in gas chromatography on a glass capillary column. J Chromatogr 234:65–76, 1982.

74. RJ Reina, KD White, EGE Jahngen. Validated method for quantitation and identification of 4,4-desmethylsterols and triterpene diols in plant oils by thin-layer chromatography-high resolution gas chromatography-mass spectrometry. J AOAC Int 80:1272–1280, 1997.

75. VK Garg, WR Ness. Codisterol and other Δ^5-sterols in the seeds of *Cucurbita maxima*. Phytochemistry 23:2925–2929, 1984.

76. M Kalinowska, WR Nes, FG Crumley, WD Ness. Stereochemical differences in the anatomical distribution of C-24 alkylated sterols in *Kalanchoe daigremontiana*. Phytochemistry 29:3427–3434, 1990.

77. T Matsumoto, T Akihisa, S Soma, M Takido, S Takahasi. Composition of unsaponifiable lipid from seed oils of *Panax ginseng* and *P quiquefolium*. J Am Oil Chem Soc 63:544–546, 1986.

78. T Itoh, T Tamura, T Matsumoto. Sterols, methylsterols, and triterpene alcohols in three theaceae and some other vegetable oils. Lipids 9:173–184, 1974.

79. SK Stankovic, MB Bastic, JA Jovanovic. Composition of the triterpine alcohol fraction of horse chestnut seed. Phytochemistry 24:119–121, 1985.

80. B Yang, RM Karlsson, PH Oksman, HP Kallio. Phytosterols in sea buckthorn (*Hippophae rhamnoides* L.) berries: identification and effects of different origins and harvesting times. J Agr Food Chem 49:5620–5629, 2002.

81. C Plank, E Lorbeer. Analysis of free and esterified sterols in vegetable oil methyl esters by capillary GC. J High Resol Chromatogr 16:483–487, 1993.

82. C Plank, E Lorbeer. On-line liquid chromatography-gas chromatography for the analysis of free and esterified sterols in vegetable oil methyl esters used as diesel fuel substitutes. J Chromatogr A 683:95–104, 1994.

83. A Kuksis. Gas chromatography of sterols, steryl esters and steroid glycosides. Fette Seifen Anstrichmitt 75:420–433, 1973.

84. A Kuksis. Lipoprotein analysis. In: Analysis of Sterols and Biologically Significant Steroids (WD Nes, EJ Parish, eds.) Academic Press: New York, 1989, pp.151–202.

85. A Kuksis. Examination of total lipid extracts by direct gas chromatography. Fette Seifen Anstrichmitt 75:517–533, 1973.
86. A Kuksis, L Marai, JJ Myher, K Geher. Identification of plant sterols in plasma and red blood cells of man and experimental animals. Lipids 11:581–586, 1976.
87. A Kuksis, JJ Myher, P Sandra. Gas–liquid chromatographic profiling of plasma lipids using high temperature–polarizable capillary columns. J Chromatogr 500:427–441, 1990.
88. RP Evershed, VL Male, J Goad. Strategy for the analysis of steryl esters from plant and animal tissues. J Chromatogr 400:187–205, 1987.
89. RP Evershed, LJ Goad. Capillary gas chromatography/mass spectrometry of cholesteryl esters with negative ammonia chemical ionization. Biomed Environ Mass Spectrom 14:131–140, 1987.
90. RP Evershed, M Prescott, N Spooner, LJ Goad. Negative ion ammonia chemical ionization and electron impact ionization mass spectrometric analysis of steryl fatty acyl esters. Steroids 53:285–309, 1989.
91. GP Fenner, LW Parks. Gas chromatographic analysis of intact steryl esters in wild type *Saccharomyces cerevisiae* and in an ester accumulating mutant. Lipids 24:625–629, 1989.
92. A Kuksis, JJ Myher, K Geher. Quantification of plasma lipids by gas–liquid chromatography on high temperature-polarizable capillary columns. J Lipid Res 34:1029–1038, 1993.
93. W Kamm, F Dionisi, L-B Fay, C Hischenhuber, H-G Schmarr, K-H Engel. Analysis of steryl esters in cocoa butter by on-line liquid chromatography–gas chromatography. J Chromatogr 918:341–349, 2001.
94. E Jover, Z Moldovan, JM Bayona. Complete characterization of lanolin steryl esters by sub-ambient pressure gas chromatography–mass spectrometry in the electron impact and chemical ionization modes. J Chromatogr A 970:249–258, 2002.
95. A Lohninger, P Preis, L Linhart, SV Sommoggy, M Landau, E Kaiser. Determination of plasma free fatty acids, free cholesterol, cholesteryl esters, and triacylglycereols directly from total lipid extract by capillary gas chromatography. Anal Biochem 186:243–250, 1990.
96. HH Rees, PL Donnahey, TW Goodwin. Separation of C27, C28, C29 sterols by reversed-phase high performance liquid chromatography on small particles. J Chromatogr 116:281–291, 1976.
97. HT Hidaka, T Nakamura, A Aoki, H Kojima, Y Nakajima, K Kosugi, I Harada, M Harada, A Kobayashi, A Tamura, T Fuji, Y Shigeta. J Lipid Res 31:881–888, 1990.
98. A Kuksis, L Marai, JJ Myher. Strategy of glycerolipid separation and quantification by complementary analytical techniques. J Chromatogr 273:43–66, 1983.
99. L Marai, JJ Myher, A Kuksis. Analysis of triacylglycerols by reversed-phase high pressure liquid chromatography with direct liquid inlet mass spectrometry. Can J Biochem Cell Biol 61:840–849, 1983.

100.   JT Billheimer, S Avart, B Milani. Separation of steryl esters by reversed-phase liquid chromatography. J Lipid Res 24:1646–1651, 1983.
101.   A Kuksis, L Marai, JJ Myher. Plasma lipid profiling by liquid chromatography with chloride attachment mass spectrometry. Lipids 26:240–246, 1991.
102.   B Cheng, J Kowal. Analysis of adrenal cholesteryl esters by revered phase high performance liquid chromatography. J Lipid Res 35:1115–1121, 1994.
103.   RA Ferrari, W Esteves, KD Mukherjee, E Schulte. Alteration of sterols and steryl esters in vegetable oils during industrial refining. J Agric Food Chem 45: 4753–4757, 1997.
104.   W Funk, V Dammann, G Donnevert. Quality Assurance in Analytical Chemistry, VCH: Weinheim, 1995.
105.   GV Vahouny, WE Connor, S Subramaniam, DS Lin, LL Gallo. Comparative lymphatic absorption of sitosterol, stigmasterol, and fucosterol and differential inhibition of cholesterol absorption. Am J Clin Nutr 37:805–809, 1983.
106.   KE Suckling, EF Stange. Role of acyl-CoA:cholesterol acyltransferase in cellular cholesterol metabolism. J Lipid Res 26:647–671, 1985.
107.   I Ikeda, K Tanaka, M Sugano, GV Vahouny, LL Gallo. Inhibition of cholesterol absorption in rats by plant sterols. J Lipid Res 29:1573–1582, 1988.
108.   MJ Armstrong, MC Carey. Thermodynamic and molecular determinants of sterol solubilities in bile salt micelles. J Lipid Res 28:1144–1155, 1987.
109.   I Ikeda, M Sugano. Some aspects of mechanism of inhibition of cholesterol absorption by β-sitosterol. Biochim Biophys Acta 732:651–658, 1983.
110.   AS Hassan, AJ Rampone. Intestinal absorption and lymphatic transport of cholesterol and β-sitosterol in the rat. J Lipid Res 20:646–653, 1979.
111.   M Sugano, M Morioka, I Ikeda. A comparison of hypocholesterolemic activity of β-sitosterol and β-sitostanol in rats. J Nutr 107:2011–2019, 1977.
112.   T Heinemann, O Leiss, K von Bergmann. Effect of low dose sitostanol on serum cholesterol in patients with hypercholesterolemia. Atherosclerosis 61:219–223, 1986.
113.   T Slota, NA Kozlov, HV Ammon. Comparison of cholesterol and bet5a-sitosterol: effects on jejunal fluid secretion induced by oleate, and absorption from mixed micellar solutions. Gut 24:653–658, 1983.
114.   LL Gallo, S Myers, GV Vahouny. Rat intestinal co-enzyme A:cholesterol acyltransferase properties and localization. Proc Soc Exp Biol Med 177:188–196, 1984.
115.   AK Bhattacharya. Uptake and esterification of plant sterols by rat small intestine. Am J Physiol 240:G50–G55, 1981.
116.   KR Bruckdorfer, RA Demel, J De Gier, LLM Van Deenen. The effect of partial replacements of membrane cholesterol by other steroids on the osmotic fragility and glycerol permeability of erythrocytes. Biochim Biophys Acta 183:334–345, 1969.
117.   HW Kircher, FU Rosenstein, A Kandutsch. Lethal effect of cis- but not tans-22-dehydrocholesterol on mouse fibroblast cells. Lipids 16:943–945, 1982.
118.   P Child, A Kuksis. Different uptake of cholesterol and plant sterols by erythrocytes in vitro. Lipids 17:748–754, 1982.

119. C Rujanavech, DF Silbert. LM cell growth and membrane lipid adaptation to sterol structure. J Biol Chem 261:7196–7203, 1986.
120. TM Buttke, K Bloch. Utilization and metabolism of methyl-sterol derivatives in the yeast mutant strain GL7. Biochemistry 20:3267–3272, 1981.
121. Y Suarez, C Fernandez, B Ledo, AJ Ferruelo, M Martin, MA Vega, D Gomez-Coronado, MA Lasuncion. Differential effects of ergosterol and cholesterol on Cdk1 activation and SRE-driven transcription. Eur J Biochem 269: 1761–1771, 2002.
122. JMR Beveridge, WF Connell, GA Mayer. The nature of the substances in dietary fat affecting the level of plasma cholesterol in humans. Can J Biochem Physiol 35:257–270, 1957.
123. AM Lees, HYI Mok, RS Lees, MA McCluskey, SM Grundy. Plant sterols as cholesterol-lowering agents: clinical trials in patients with hypercholesterolemia and studies of sterol balance. Atherosclerosis 28:325–338, 1977.
124. HT Vanhanen, TA Miettinen. Effects of unsaturated and saturated dietary plant sterols on their serum contents. Clin Chim Acta 205:97–107, 1992.
125. H Gylling, P Puska, E Vartiainen, TA Miettinen. Serum sterols during stanol ester feeding in a mildly hypercholesterolemic population. J Lipid Res 40:593–600, 1999.
126. TA Miettinen, H Gylling. Regulation of cholesterol metabolism by dietary plant sterols. Curr Opin Lipidol 10:9–14, 1999.
127. H Gylling, M Siimes, TA Miettinen. Sitostanol ester margarine in dietary treatment of children with familial hypercholesterolemia. J Lipid Res 36:1807–1812, 1995.
128. TA Miettinen, RS Tilvis, YA Kesaniemi. Serum plant sterols and cholesterol precursors reflect cholesterol absorption and synthesis in volunteers of a randomly selected male population. Am J Epidemiol 131:20–31, 1990.
129. HJM Kempen, JFC Glatz, JA givers Leuven, HA van der Voort, MB Katan. Serum lathosterol concentration is an indicator of whole body cholesterol synthesis in humans. J Lipid Res 29:1149–1155, 1988.
130. WH Sutherland, RS Scott, CJ Lintott, MC Robertson, SA Stapely, C Cox. Plasma non-cholesterol sterols in patients with non-insulin dependent diabetes mellitus. Horm Metab Res 24:172–175, 1992.
131. RS Tilvis, TA Miettinen. Serum plant sterols and their relation to cholesterol absorption. Am J Cli Nutr 43:92–97, 1986.
132. HT Vanhanen, J Kajander, H Lehtovirta, TA Miettinen. Serum levels, absorption efficiency, faecal elimination and synthesis of cholesterol during increasing dose of dietary sitostanol esters in hypercholesterolaemic subjects. Clin Sci 87:61–67, 1994.
133. GV Vahouny, D Kritchevsky. Plant and marine sterols and cholesterol metabolism. In: G Spiller, ed. Nutritional Pharmacology. New York: Alan R. Liss, 1981, pp. 32–72.
134. D Lutjohann, I Bjorkhem, UF Beil, K von Bergmann. Sterol absorption and sterol balance in phytosterolemia evaluated by deuterium-labeled sterols: effect of sitostanol treatment. J Lipid Res 36:1763–1773, 1995.
135. PJ Jones, M Raeini-Sarjaz, FY Ntanios, CA Vanstone, JY Feng, WE Parsons.

Modulation of plasma lipid levels and cholesterol kinetics by phytosterol versus phytostanol esters. J Lipid Res 41:697–705, 2000.

136. KE Berge, H Tian, GFA Graf, L Yu, NV Grishin, J Schultz, P Kwitterovich, B Barnes, R Barnes, HH Hobs. Accumulation of dietary cholesterol in sitosterolemia caused by mutations in adjacent ABC transporters. Science 290: 1771–1775, 2000.

137. H Allayee, BA Lafitte, AJ Lusis. An absorbing study of cholesterol. Science 290:1709–1711, 2000.

138. PA Kuwabara, M Labouesse. The sterol-sensuing domain: multiple families, a unique role? Trends Genet 18:193–220, 2002.

139. PJH Jones, DE MacDougall, F Ntanios, CA Vanstone. Dietary phytosterols as cholesterol-lowering agents in humans. Can J Physiol Pharmacol 75:217–227, 1997.

140. RFB Raicht, BI Cohen, EP Fazzini, AN Sarwal, M Takahashi. Protective effects of plant sterols against chemically induced colon tumours in rats. Cancer Res 40:403–405, 1980.

141. SA Janezic, VA Rao. Dose-dependent effects of dietary phytosterol on epithelial cell proliferation of the murine colon. Food Chem Toxicol 30:611–616, 1992.

142. AB Awad, AYT Hernandez, CS Fink, SL Mendel. Effect of dietary phytosterols on cell proliferation and protein kinase C activity in rat colonic mucosa. Nutr Cancer 27:210–215, 1997.

143. AB Awad, YC Chen, SC Fink, T Hennessey. β-Sitosterol inhibits growth of HT-29 human colon cancer cell growth and alter membrane lipids. Anticancer Res 16:2797–2804, 1996.

144. RL VonHoltz, CS Fink, AB Awab. β-Sitosterol activates the sphingomyelin cycle and induces apoptosis in LNCaP human prostate cancer cells. Nutr Cancer 32:8–12, 1998.

145. A Dawnie, CS Fink, AB Awad. Effect of phytosterols on MDA-MB-231 human breast cancer cell growth. FASEB J 13:O433–O437, 1999.

146. PP Nair, N Turjman, G Kessie, B Calkins, GT Goodman, H Davidovitz, G Nimmagadda. Diet, nutrition intake, and metabolism in populations at high and low risk for colon cancer—dietary cholesterol, β-sitosterol, and stigmasterol. Am J Clin Nutr 40:927–930, 1984.

147. BM Calkins, DJ Whittaker, AA Rider, N Turjman. Diet, nutrition intake, and metabolism in populations at high risk and low risk for colon cancer—population: demographic and anthropomorphic characteristics. Am J Clin Nutr 40: 887–895, 1984.

148. JH Li, AB Awad, CS Fink, YW Wu, M Terevisan, M Muti. Measurement variability of plasma β-sitosterol and campesterol, two new biomarkers for cancer prevention. Eur J Cancer Prev 10:245–249, 2001.

149. FJ Field, E Born, SN Mathur. Effect of micellar β-sitosterol on cholesterol metabolism in Caco-2 cells. J Lipid Res 38:348–360, 1997.

150. FA Miettinen. Gas–liquid chromatographic determination of fecal neutral sterols using a capillary column. Clin Chim Acta 124:245–248, 1982.

151. MTR Subbiah, A Kuksis. Differences in metabolism of cholesterol and sito-

sterol following intravenous injection in rats. Biochim Biophys Acta 306:95–105, 1973.

152. D Kritchevsky, LM Davidson, EH Mosbach, BI Cohen. Identification of acidic steroids in feces of monkeys fed β-sitosterol. Lipids 16:77–78, 1981.

153. KM Boberg, K Einarsson, I Bjorkhem. Apparent lack of conversion of sitosterol into C-24 bile acids in humans. J Lipid Res 31:1083–1088, 1990.

154. L Swell, CR Treadwell. Metabolic fate of injected ^{14}C-phytosterols. Proc Soc Exp Biol Med 108:810–813, 1961.

155. B Skrede, I Bjorkhem, O Bergesen, HJ Kayden, S Skrede. Studies on 5α-sitostanol: its presence in the serum of a patient with phytosterolemia and its biosynthesis from plant sterols in bile fistula rats. Biochim Bophys Acta 836: 368–375, 1985.

156. E Lund, O Anderson, J Zhan, A Babiker, G Ahlborg, U Diczfalusy, K Sjovall, I Sjovall, I Bjorkhem. Arterioscler Thromb Vasc Biol 16:208, 1996.

157. E Lund, M Boberg, S Bytsrom, K Carlstrom, J Olund, I Bjorkhem. Formation of novel C_{21}-bile acids from cholesterol in the rat. J Biol Chem 266:4929–4937, 1990.

158. KM Boberg, E Lund, J Olund, I Bjorkhem. Formation of C21 ble acids from plant sterols in the rat. J Biol Chem 265:7967–7975, 1990.

159. I Bjorkhem. Mechanism of degradation of the steroid side chain in the formation of bile acids. J Lipid Res 33:455–471, 1992.

160. MTR Subbiah, A Kuksis. Conversion of β-sitosterol to steroid hormones by rat testes in vitro. Experientia 31:763–764, 1973.

161. MTR Subbiah. Dietary plant sterols: current status in human and animal sterol metabolism. Am J Clin Nutr 26:219–225, 1973.

162. J Plat, H Brzezinka, D Jutjohann, RP Mensink, K Von Bergmann. Oxidized plant sterols in human serum and lipid infusions as measured by combined gas-liquid chromatography–mass spectrometry. J Lipid Res 42:2030–2038, 2001.

5

Does Phytosterol Intake Affect the Development of Cancer?

Lena Normén
*Healthy Heart Program, University of British Columbia,
Vancouver, British Columbia, Canada*

Susan W. Andersson
Sahlgrenska Academy at Göteborg University, Göteborg, Sweden

I. INTRODUCTION

Phytosterols are well known for their cholesterol lowering properties (11), but increasing evidence suggests that they also may have effects on cancer development (12). Generally speaking, cancer is still one of the largest killers of men and women living in Western societies (13). It affects the quality of life (14) and contributes to unnecessary morbidity and societal costs considering that many cancers can be prevented (15).

Cancer encompasses a wide category of conditions, which are all characterized by an abnormally regulated growth and spread of diseased cells (16). External factors, such as chemicals, radiation, and viruses, are implicated in the causal pathways, but cancer is also very often a result of internal factors, such as hormones, immune conditions, and genetics (16). The American Cancer Society estimated for 2001 that 1,268,000 Americans would be diagnosed with cancer and 553,400 would die of the disease. It is estimated that about one-third of the deaths are related to nutrition, physical activity, and other lifestyle factors, which is similar to the 31% (172,000) of deaths due to smoking (16). Therefore, many cancers are preventable through alteration of dietary habits and smoking cessation, constituting a challenge for public health programs and the clinical sector. Demographic changes worldwide may lead to as much as a 30% increase in cancer deaths in the period 1990–2010, as estimated by Pisani et al (17). Today more than 70% of cancers occur in low- and middle-income countries, but despite this, the relative risk of cancer in a developed country like the United States is higher (13).

In the year 2000, the National Institutes of Health (NIH) estimated that $180 billion was spent on cancer treatment in the United States (16). The costs are high, and substantial amounts could be saved if cancer prevention programs would be successful in facilitating behavioral change including diet and lifestyle changes at an early age. However, the high morbidity and mortality of cancers demonstrate that prevention programs are presently less than successful and that identification of preventive dietary agents could play an important role in reducing cancer incidence. The identification of preventive agents would therefore be beneficial to society, should they be acceptable and adopted by a large part of the population.

There have been several animal studies addressing a hypothesis of cancer prevention with phytosterol supplementation (18–23). Cell studies have explained potential mechanisms of this preventive action of phytosterols (23–28) and have laid the ground for a hypothesis of additional physiological effects other than that of lowering of blood lipids.

The cancer types that are most strongly related to dietary factors are colon, breast, and prostate cancer. The two latter are endocrine-dependent cancers (29), which explains the interest in potential effects of phytosterols on hormones such as testosterone and estrogen. Due to the link between immune function and cancer (30), phytosterol studies addressing effects on the immune system will also be discussed. Phytosterols are part of a normal diet and have therefore to be evaluated as such. The issue of diet and cancer in general will be discussed with particular emphasis on a context in which phytosterol supplementation may prove useful as part of primary prevention.

High doses of phytosterols appear to prevent cancer in animal studies, which is supported by the evidence of mechanisms from cell studies. At this time, however, there is little information about such effects based on epidemiological studies and human intervention studies, which limits the potential to translate these observations to human conditions. Average phytosterol intake in Western populations with high blood lipid levels is likely to increase in the next decade due to the fact that phytosterols are frequently used as functional ingredients in food products to lower low-density lipoprotein (LDL) concentrations (31,32). Therefore, the effect on cancer needs further extensive research, as there are too few studies to support a conclusion about the overall effect of phytosterols on cancer. Against available scientific literature, however, the question "Does phytosterol intake affect cancer development?" will be addressed.

II. CANCER EPIDEMIOLOGY AND ETIOLOGY

The most common cancers in the United States in males and females are presented in Figure 1. In men, lung and bronchus cancer is the number one killer, followed by prostate and colorectal cancers (16). In women, lung and bronchus cancer accounts for most of the cancer mortality, closely followed by breast and colorectal cancers (16). Cancer incidence (new emerging cases/ year) differs slightly from the mortality rates for both men and women. The lung and bronchus may be the major cause of cancer mortality in both

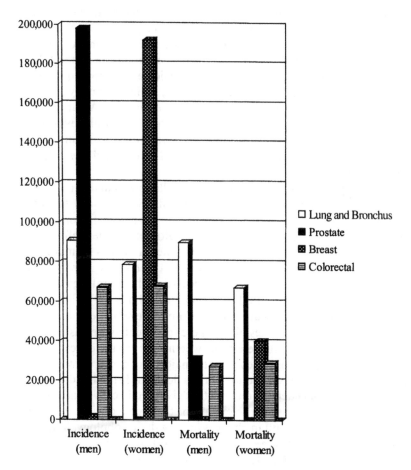

Figure 1   Estimated incidence and total mortality from cancer in the United States 2001 (16).

genders, but the incidence of prostate and breast cancer are each higher, demonstrating that survival of the two latter cancers together with colon and rectal cancer is higher than that of lung and bronchus (16).

In general, the most common cancers are classified as epithelial cancers, as they are initiated from the epithelium of the lung, colon, breast, and prostate (33,34). Carcinogenesis results from the accumulation of multiple sequential mutations and alterations in nuclear and cytoplasmic molecules, which ultimately culminates in invasive neoplasms (35). There are two types of neoplasms or tumors—benign and malignant—and the common name for the malignancy is *cancer* (36).

Cancer etiology is divided into three separate phases: initiation, promotion, and progression (2). During the initiation phase, DNA damage occurs due to radiation or through a genotoxic element, which binds to DNA. Both events are rapid and lead to a so-called mutation of the DNA. During the promotion phase, damaged cells replicate as a consequence of malfunctioning proteins, often due to mutations, which normally leads to abnormal cell proliferation and apoptosis (i.e., cell death). A cancer may require as many as 10 or more mutations to reach its full devastating malignant character, taking into account all events during initiation and promotion (36). The cancer cells after phases I and II are termed *premalignant* or *preneoplastic* (10).

During the progression phase, the cancer cells spread and cause minor or major changes in human physiology, which finally may lead to organ malfunction and, ultimately, death. Exposure to the causal factor, with a following stepwise cancer development, may be initiated decades before there are clinical signs of tumor development (33). The hallmarks of malignant transformation are the capabilities of invasion and metastasis. For the latter processes to proceed, tumor cells must be able to detach, migrate, gain access to blood or lymphatic vessels, and disseminate in the body (37).

During all of the evolutionary stages of tumor development, a complicated multifactorial scenario takes place where treatment as well as environmental and genetic components determine whether or not the cancer will be fatal (38). There is compelling evidence that a majority of cancers are clonal in the beginning and that new subgroups of cells develop through a darwinian intercellular selection of genetic variations that give certain cells an advantage in terms of growth (36). The stepwise mutations that occur during the cancer development process may result in changes in growth morphology, hormone dependence, enzyme and cytokine production, as well as expression of surface antigens. This may ultimately lead to cancer cells escaping homeostatic controls, which allows the abnormal cells to escape the body's defense system, with a malignant tumor as a result (36).

A. Colorectal Cancer

In 2001, 135,400 American men and women were diagnosed with colorectal cancer (16). Internationally,* approximately 945,000 cases of colorectal cancer occurred in 2000 (most recent year of compared international statistics), accounting for 9.5% of all cancers (13). Colon and rectal cancer are thereby not only common in the United States, but are among the most common causes of cancer morbidity and mortality worldwide. The incidence varies 20-fold in different parts of the world, with the highest incidence found in developed countries, while countries like China and India still have low incidences (13). Age, race, education, smoking, exercise, alcohol, family history of colon cancer, and obesity are all general risk factors for colorectal cancer (39). Insulin resistance has been identified as a potential risk factor for colon cancer development (40–42), which is quite worrisome taking into account that 22% of the U.S. population suffers from the metabolic syndrome whereof insulin resistance is a central component (43).

The majority of colorectal cancers arise in preexisting adenomas, which are initiated by mutations in a tumor suppressor gene (44). An adenoma is a benign epithelial tumor with a glandular organization (1). Screening programs show that the occurrence of adenoma–carcinoma has a ratio of 30:1, which indicates that not all adenomas develop into tumors (45). Instead, certain molecular pathways affect and trigger development of an adenoma into a cancer. Well-defined molecular pathways have been identified for sporadic colorectal cancer, familial adenomatous polyposis, hereditary nonpolyposis colorectal cancer, cancer development in inflammatory bowel diseases, and familial colorectal cancer (2). The effects on physiology by the five mentioned pathways differ significantly, and the existence of different cancer etiologic factors raises the point that different treatments and preventive agents are needed to combat colorectal cancer. The benefits of surgical removal of adenomas in the prevention of colorectal cancer has been tested in the context of three population-based screening programs, but only one has shown a weak significant preventive effect by this procedure (46–48). This indicates that there may be more potent targets for colorectal cancer prevention than surgery and that knowledge about the mechanisms of agents at a molecular level is of utmost importance for chemoprevention.

---

*The term "internationally" refers here to available cancer statistics from WHO. There is a lack of statistics from many developing countries, which most likely results in underestimation of total cancer morbidity and mortality.

B. Breast Cancer

During 2001, an estimated 192,200 new invasive cases and 46,400 new cases of in situ breast cancer were expected to occur among women in the United States (16). Internationally, 1,017,217 breast cancer cases occurred during 2000, and 371,680 women died from the disease (13). Environment and lifestyle also play a strong role in the development of breast cancer, as demonstrated by migration studies. The incidence rates of breast cancer among Japanese women are high in Los Angeles County compared with those in Japan (49). However, the timing of immigration to the United States is important in determining breast cancer risk. When migration occurs later in life, rates for breast cancer are substantially lower than when migration occurs early, although still much higher than in Japan. Environmental factors in early life may therefore be important in the etiology of breast cancer (49). General risk factors are age, family history of breast cancer, long menstrual history, obesity after menopause, recent use of oral contraceptives, and late parity (first child after age 30) (16).

The breast as an organ consists of 70–90% fat. Functional elements such as the mammary glands are dispersed in this lipid matrix and aligned with epithelial cells. It is from these latter cells that the carcinomas arise (34). Breast cancer has a genetic component in that 5% of breast cancer cases are attributed to inheritance of high penetrance susceptibility genes, such as *BRCA1* and *BRCA2*. Still, 95% of cases are not due to a hereditary gene, and suggested initiating factors have been radiation and fat-soluble genotoxins from the environment, including the diet (34).

As reviewed by Stoll (50), estrogen receptors (ER) and insulin-like growth factor-I receptor (IGF-IR) are found in breast cancer specimens, where they enhance proliferative activity in normal and malignant human mammary epithelial cells in culture. Activation of their respective receptors may induce synergistic stimulation of mammary carcinogenesis (50), which demonstrates that estrogen concentrations are linked to breast cancer.

Prevention programs today are mainly based on mammography as well as on DNA testing of susceptibility genes in families with a strong history of premature breast cancer. National screening programs reduce breast cancer mortality by approximately 30% (51). Though the numbers are impressive, there is still much room for improvement in breast cancer prevention.

C. Prostate Cancer

Although lung cancer incidence has reached a plateau, prostate cancer has increased dramatically and has become one of the most common cancers among men in the United States (16). The World Health Organization (WHO) estimated that there would be 536,279 new cases worldwide during

2000 and that 202,201 men would die from the disease in the same year (13). In the United States, 198,100 new cases of prostate cancer were expected to occur in 2001 (16).

The prostate is susceptible to pathological abnormalities more than any other human male organ (52). Benign hyperplasia and adenocarcinomas originate both in the peripheral zone and in the normally small anteromedial transition zone (53). Environmental factors are of utmost importance for the occurrence of prostate cancer, as demonstrated by the large differences in incidence between different regions of the world (13). For Japanese men who have emigrated to the United States, the incidence rate of prostate cancer is substantially higher in Los Angeles County than for men remaining in Japan (49). However, the rates are similar to those of U.S.-born Hispanic and Japanese individuals, regardless of age at immigration (49). Therefore, risk of developing prostate cancer seems not to depend on timing of the exposure to risk or protective factors. Risk factors include age, high testosterone concentrations, insulin resistance, history of prostatitis, syphilis, and gonorrhea (54–57). Hereditary causes, such as susceptibility genes, can explain only a small portion of prostate cancer cases (58). In the United States, community-based prostate-specific antigen screening programs are sometimes undertaken as a useful way to detect prostate cancer but are not as widespread as the breast cancer screening programs (59).

III. CHEMOPREVENTION WITH PHYTOSTEROLS

Cancer preventive agents work either by diminishing mutations or by reducing the growth or replication of cancer cells at a later stage. Central to cancer development are built-in regulatory proteins, which control cell proliferation and growth, where cancer-promoting factors may work together simultaneously or act in a sequence. As reviewed by Gescher et al. (35), these also constitute the molecular targets for phytochemicals, where these dietary constituents may work alone or in combination with other components to prevent adverse events. During the two slow phases of promotion and progression, exposure to preventive agents may be beneficial in inhibiting the promotion and/or progression of the cancer cells (35). The most important targets for chemoprevention are:

1. Hormones, growth factor, tumor promoters (extracellularly)
2. Detector proteins (intracellularly)
3. Transducer proteins (intracellularly)
4. Modulator proteins (intracellularly)
5. Secondary signaling proteins (intracellularly)
6. Transcription factors (intracellularly)

Based on the described background of different initiating and promoting factors of cancer, it seems obvious that cancer prevention programs would benefit by being multifactorial, working on several parallel targets simultaneously.

The U.S. National Cancer Institute has presented a science-based approach to evaluate cancer-preventive effects of dietary constituents. Here it is stressed that as human cancers can take numerous years to develop, cancer preventive effects of dietary constituents need to be tested using animal and cell models (60). For phytosterols, it is clear that there is a shortage of human data, which is a common problem shared with many other potential cancer-preventive agents (60).

A. Bioavailability of Phytosterols

For dietary constituents to prevent cancer through interactions with the listed targets, they must first be absorbed from the diet. Thereafter endogenous bioavailability in terms of stimulating/inhibiting effects on one or more of the six identified targets must be demonstrated. Phytosterols exist in all foods of plant origin, as monomers, glycosides, esters, or glucosylated esters (see Chapter 5 for a comprehensive review on food sources and chemical properties of phytosterols). The different chemical forms exist in different compartments of the plant cell, i.e., free phytosterols are mainly found in the plant membrane wall to give structural properties, while phytosterol glucosides and esters mainly are found in the cytosol and endoplasmic reticulum (61).

Phytosterols have limited fat and micellar solubility (62), but are nevertheless absorbed to a low extent (63). The fractional absorption depends on their physicochemical properties (i.e., decreasing absorption with increasing hydrophobicity) through variation in the structure of both the sterol nucleus and side chain (63). In Table 1 an overview of clinical trials or mg based on phytosterol and phytostanol supplementation is given. Only studies that present phytosterol concentrations in plasma or serum as micromoles per liter are used. There are several studies presenting the relative change of serum phytosterols as a factor of total cholesterol concentration (64–71). However, as a majority of these are based on saturated phytosterols, we chose to use the studies using absolute concentrations of sitosterol and campesterol in plasma. Most of the latter are based on supplementation of the unsaturated phytosterols, which have a known absorption (63).

Before absorption takes place, the esters are split in the human duodenum to the free form by human lipases (72). When humans consume daily doses of 0.8–3.2 g of unsaturated phytosterols, the plasma concentrations of campesterol and sitosterol increase with 15% per gram PS consumed (SD

Table 1 Clinical Trials with Phytosterol-Enriched Spreads and the Absolute and Relative Change in Phytosterol Concentrations of Plasma

Population	Phytosterol dose (g/d)	Plasma campesterol (μM)			Plasma sitosterol (μM)			Ref.
		Baseline	Post-intervention	Relative change[f] (%)	Baseline	Post-intervention	Relative change (%)	
HC[a], n=155	1.5 UPS[c]	7.5	9.6	−25	3.0	7.1	+133	73
	3.0 UPS	6.8	7.7	−40	2.8	6.3	+68	
NC[b], n=84	3.0 UPS	14.9	21.3	+41	6.1	8.3	+11	74
	6.0 UPS	14.8	17.0	+31	5.9	6.6	+12	
	9.0 UPS	12.7	19.2	+57	4.9	6.0	+21	
HC, n=155	3.0 SPS	17.8	10.5	−73	8.9	6.8	−30	79
HC, n=22	0.8 SPS[d]	18.6	13.8	−34	8.5	6.3	−36	78
	1.6 SPS		11.1	−67		5.2	−64	
	2.4 SPS		10.8	−72		5.2	−67	
	3.2 SPS		9.9	−87		4.7	−82	
NC, n=88	1.6 UPS	8.0	31.0	+74	9.4	19.4	+52	75
HC, n=32	1.7 UPS	22.1	27.5	0	5.4	4.4	0	76
HC, n=15	1.8 UPS	14.1	24.3	+99	8.6	9.6	+39	77
	1.8 SPS	17.3	8.5	−28	9.2	4.8	−24	
	1.8 Mix[e]	13.8	17.7	+55	7.2	7.4	+22	
Mean (SD)	2.7 (2.0)		Treatment arms with UPS supplementation: +34%(42%)			Treatment arms with UPS supplementation: +39% (37%)		
			Treatment arms with SPS supplementation: −60% (24%)			Treatment arms with SPS supplementation: −50% (24%)		

a Hypercholesterolemic.
b Normocholesterolemic.
c Unsaturated phytosterols.
d SPS, saturated phytosterols.
e 50% unsaturated and 50% saturated.
f Change relative to control group.

22% per gram) and 20% per gram (SD 25% per gram), respectively (73–77). In contrast, when saturated phytosterols are consumed, the plasma concentrations of campesterol and sitosterol decrease with 30% per gram PS consumed (SD 11% per gram) and 27% per gram (SD 14% per gram), respectively (77–79). The overview identifies one important property difference between certain phytosterols: unsaturated phytosterol esters may inhibit cholesterol absorption equally effective as saturated phytosterols in the human duodenum (80), but saturated phytosterols also inhibit phytosterol absorption from occurring, as reflected by decreased plasma/serum concentrations. The first step of bioavailability of phytosterols is therefore fulfilled only by the unsaturated phytosterols; the latter are absorbed and may therefore have a direct or indirect effect on one or several targets for endogenous cancer prevention. The saturated phytosterols may have indirect effects on cancer by actually lowering the normal plasma concentrations of phytosterols, but the biological effects of this still remain unknown.

The one form of cancer that may be less dependent on phytosterol absorption is colorectal cancer. Phytosterols that are excreted into the large bowel may also have a theoretical local effect if they are incorporated into the cell membranes of the colon or rectum.

Following absorption, phytosterols are distributed in tissue after being transported with the lipoproteins (see Chapter 7 for overview of blood transport) (22). This has been demonstrated in an animal study where male Sprague-Dawley rats were fed *ad libitum* for 21 days and sacrificed thereafter for analysis of the following tissues: epidymal fat pads, liver, testis, kidney, and prostate. Phytosterol incorporation in the diet increased plasma concentrations of sitosterol and campesterol 5.6- and 4.6-fold, respectively (22). The livers of intervention animals were found to accumulate campesterol at a level 19 times that of control animals, whereas sitosterol did not appear to accumulate at all. In testis, campesterol incorporation increased 6-fold, whereas sitosterol increased 64% (22). The kidneys and prostate were devoid of sitosterol, whereas adipose tissue had the highest accumulation of all the sites (22). Phytosterol incorporation also had secondary effects on cholesterol levels in certain tissues and fatty acid composition of cell membranes, but what this means in terms of physiology and cancer is not known.

B. In Vivo Evidence of Anticancer Properties

1. Colon Cancer

Seventh Day Adventists experience lower rates of colon cancer and have at the same time higher dietary intakes of phytosterols than the general population (81). This observation led to the first hypothesis of a preventive effect of phytosterols on colon cancer. As bile acids are well-known tumor promoters (18), the lower incidence observed in Seventh Day Adventists was hypothesized to be due to a decreased bile acid excretion following phytosterol intakes (82). Observations in ileostomy subjects in studies describing the effects of diets rich in different dietary fibers or vegetables oils on sterol metabolism lend support to this (83–86). In these studies, a relative increase in phytosterol intake often seemed to be followed by a decrease in bile acid excretion. However, the dietary composition varies in more aspects than phytosterol content, which limits the use of these findings for supporting that phytosterols diminish bile acid excretion.

Bile acid excretion in earlier studies of phytosterols in humans and animals show conflicting results (Table 2), probably due to small group size and variations in fecal output (66,72,80,87–94). Different phytosterols may also have different effects on bile acid excretion. A study of pure stigmasterol in rats demonstrated an increased bile acid excretion (89), whereas sitosterol supplementation in rats gave no change (87) or even reduced bile acid excretion (88). Sterol excretion in response to phytosterol intervention could also be a question of phytosterol and cholesterol dose (95). Studies have been performed with phytosterol/cholesterol ratios of 2:250. A 2 × 2 factorial comparison of different doses of a phytosterol mixture (high vs. low dose of 60% sitosterol, 35% campesterol, and 5% stigmasterol) and cholesterol (high vs. low dose) showed that low phytosterol doses in combination with low dietary cholesterol decreased total fecal acidic steroid excretion by 43% in comparison with a high dietary cholesterol intake (90). In contrast, high phytosterol doses resulted in a fecal bile acid excretion 113% higher with low dietary cholesterol than with high dietary cholesterol intakes. Addition of mixed phytosterols to the low-cholesterol diet produced a nearly twofold increase, whereas such an addition to the high-cholesterol diet produced a significant decrease by about 53% in the total fecal acidic steroid excretion.

Studies in normal human subjects show inconclusive results (66,91, 93,94) and are often difficult to interpret due to the large variation in bowel transit time. Instead, ileostomy studies provide a more reliable technique as the lack of the large bowel in these subjects decreases bowel passage time from 2–3 days to 6–8 h, thus eliminating some of the large interindividual variation of daily sterol excretion (96). The results from a study in seven ileostomists showed that there was a nonsignificant 17% reduction in chenodeoxycholic acid excretion induced by the soy sterol ester but no difference with sitostanol (80). An alternative approach to the ileostomy model would be to study the rate-limiting enzyme in bile acid synthesis, cholesterol-7α-hydroxylase, which is inhibited by β-sitosterol in cell cultures (97). A decreased activity of the enzyme in response to β-sitosterol treatment could result in a lower bile acid production followed by decreased bile acid excretion.

Table 2 Overview of Studies of Sterol Excretion in Response to Phytosterol and Cholesterol Feeding

Species	Phytosterols (mg/d)		Cholesterol (mg/d)	Bile acid excretion	R
Rat	Sitosterol	200	7.5	No change	
Rat	Sitosterol	55	—	↓CA[a], ↓CCAC[b]↓DCA[c]	
Rat	Stigmasterol	2	—	↑LCA[d]	
		4		↑LCA	
		11		↑CCAC, ↑LCA	
		52		↑CCAC, ↑LCA, ↑DCA, and ↑HDCA[e]	
Monkey	Phytosterol mix	3000	12	↑DCA, ↑TBA[f]	
			90	↓LCA, ↓TBA	
Human	Phytosterol mix	3000	400–600	↑TBA	
Human	Phytosterol mix	3000	700	↓TBA	
		9000		No change	
Human	Sitosterol	6000	163	No change	
	Sitostanol	1500	185	↑TBA	
Human	Sitosterol	1005	326 ± 28	No change	
	Sitostanol	1040		No change	
	Sitopstanol ester	1219		No change	
Human	Phytosterol mix (esterified)	8600	—	↓Secondary BA[g]	
Human	Sitostanol ester	2000	775	No change	
Human	Phytosterol mix (esterified)	1500		No change	
	Sitostanol ester	1500	775		

[a] Cholic acid.
[b] Chenodeoxycholic acid.
[c] Deoxycholic acid.
[d] Lithocholic acid.
[e] Hyodexoxycholic acid.
[f] Total bile acids.
[g] Bile acids.

In general, however, the effect on bile acid excretion of phytosterols seems quite limited in magnitude, suggesting that alternative cancer-preventing mechanisms may be involved. More than two decades ago, the first rat study was published that addressed the hypothesis that a decreased intake of phytosterols, which occurs with the high intake of refined food items in westernized societies, might be partly responsible for an increased incidence of colon cancer (19). The study was performed in male Wistar rats, which were randomized into four groups: (a) Control diet, (b) Sitosterol (0.2%), (c) control + carcinogen to stimulate cancer development, and (d) sitosterol (0.2%) + carcinogen to stimulate cancer development. The carcinogen that was given was the frequently used N-methyl-N-nitrosurea (MNU). The dose of MNU was chosen to produce about 50% tumor incidence in the two latter groups. In the noncarcinogen-supplemented animals no tumors occurred during the experimental period of 28 days. In the two groups of animals that received MNU, tumors were detected in 54% of animals that were not given a sitosterol supplement, compared to 33% of animals that were. The frequency was 1.1 tumors/animal on MNU without simultaneous phytosterol supplementation compared to 0.44 tumors per animal when sitosterol was also added to the test diet (19). The study was the first to demonstrate that phytosterols retard colonic tumor formation, when cancer was initiated through a carcinogen. A later study of rats induced with MNU demonstrated that feeding of 0.3% sitosterol decreased the rate of colonic cell proliferation and compressed the crypt's proliferative compartment, thereby suppressing expression of the altered and damaged DNA (98).

Both studies revealed interesting results, but it was not clear if the carcinogen-induced DNA damage corresponds to the cancer etiology under normal conditions. As agents/factors other than carcinogens may cause damage to the genome, a similar study was repeated, but this time without an added carcinogen (21). Instead a bile acid, cholic acid, was chosen for the intervention, as it is believed to cause hyperproliferation by way of its cytotoxic bacterial metabolite deoxycholic acid (99). This time female C57B1/6J mice were randomized into five groups: (a) control; (b) supplementation of cholic acid; (c) 0.3% phytosterol mixure (60% sitosterol, 30% campesterol, and 5% stigmasterol); (d) 1% phytosterol mixure; and (e) 2% phytosterol mixture (21). The results showed that the prevention of hyperproliferation with phytosterols that took place with administration of MNU also took place with cholic acid. The effect was dose dependent and also entailed an expansion of aberrant crypts that previously had only been demonstrated with MNU. Awad et al. confirmed the protective effect on enhanced cell proliferation (20).

Despite the evidence of a decreased proliferation rate, a French group did not find supportive evidence when repeating a similar design with MNU in Wistar rats addressing the question of whether or not the fat content of the diet had an effect on the outcome (100). The authors performed a 2×2 factorial design with four groups: (a) control—low saturated fat (8%); (b) high saturated fat; (c) low saturated fat (8%) + 24 mg phytosterol mixture (55% sitosterol, 41% campesterol, and 4% stigmasterol); and (d) high saturated fat + 24 mg phytosterol mixture. There were no significant differences in number of tumors between control and phytosterol-treated animals with MNU. There was also no difference in benign-appearing colonic glands in the area of the lymphoid follicle (CGLF). Interesting to note, however, was that the CGLF per animal was 0.6 with the high-fat diet, but only 0.1 in each of the other three groups. However, there was no statistical difference (100).

2. Breast Cancer

To date there is only one published study on the effects of phytosterols on breast cancer (25). In this study, female SCID mice were inoculated in the right inguinal mammary fat pad with cultured breast cancer cells (MDA-MB-231 type). The animals were randomized into two groups: (a) control diet + 0.2% cholic acid + 2% cholesterol, and (b) control diet + 0.2% cholic acid + 2% phytosterols (56% sitosterol, 28% campesterol, 10% stigmasterol, and 6% dihydrobrassicasterol). Animals that did not develop cancers were excluded (one per group). After 8 weeks of tumor growth, the tumor area and weight was 33% smaller in the phytosterol group compared to control. Another interesting aspect of this study was that metastases in the lungs and lymph nodes were only found in 52% of the phytosterol-treated animals compared to 71% of the control group. This study provided evidence that phytosterol supplementation in animals inhibits breast cancer tumor growth and leads to a lower frequency of metastases (25).

3. Prostate Cancer

The effect of phytosterols on prostate cancer development has been investigated in a few animal models. In a study based on a mouse model, a similar intervention protocol to the one described above for breast cancer was applied (23). Here SCID mice (male) had cultured prostate cancer cells (PC-3 type) inoculated into the right flank. After 8 weeks, animals were sacrificed and histological examinations of sections of lymph nodes, lung, and liver were performed. The phytosterol-treated animals had 28% smaller tumors 6 weeks following the inoculation with cancer cells, and this increased to 40–43% in the last week of the experiment. Animals fed with phytosterols also had only about one-half the rate of metastasis (37% of phytosterol treated animals compared to 73% of controls). There was a 62% reduction of tumor metastasis to the lung and a 33% reduction to the lymph nodes. Phytosterol-treated animals had no metastases in the liver compared to 18% of controls (23).

C. Cancer-Preventive Mechanisms of Phytosterols

In 2000, Awad and Fink reviewed the evidence of anticancer properties of phytosterols, based on their extensive work in this area, and proposed some potential mechanisms of action (101). That review remains the most extensive summary of potential pathways that may explain a variety of favorable effects of phytosterols. The central hypothesis is that phytosterol supplementation inhibits growth of organs/cells through a stimulation of apoptosis (101). In vitro studies of tumor cell lines of colon (102), prostate (23), and breast cancer cells (28) all showed that incubation with sitosterol in physiological concentrations reduced tumor growth. The effect was more potent in colon cancer cells than in prostate cancer cells (101), and with sitosterol, whereas cholesterol and campesterol showed no effect (28).

The phytosterol trigger mechanism that has received most attention with implications for cancer prevention is the stimulatory effect on the sphingomyelin cycle (28). Sphingomyelin is concentrated in the outer layer of the plasma membrane and provides a barrier to the extracellular environment (103). It was assumed only to have a structural function until the existence of a sphingomyelin-based signaling pathway was proposed (104,105). This pathway can be activated by receptor-mediated mechanisms (106) for which ceramide is a second messenger (107). Sphingolipid metabolism has proved to be a dynamic process; besides ceramide, other sphingolipid metabolites, such as sphingosine and sphingosine 1-phosphate, are now also recognized as messengers, playing essential roles in cell growth, survival, and death (108,109).

To exert its effects on the sphingomyelin cycle (Fig. 2), sitosterol has to be incorporated into the cell wall (24). An in vitro study showed that sphingomyelin content was reduced in cell membranes by 26% and ceramide production increased by 45% from colon cancer cells following incubation with sitosterol in physiological concentrations (24). Similar conditions also reduced cell growth by 55% compared with that with cholesterol (24). A fourfold increase of apoptosis in prostate cancer cells when incubated with sitosterol has also been reported (110). Sitosterol exerts its effects without affecting protein kinase C; the latter is a receptor for certain tumor promoters and also plays a crucial role in events related to tumor progression (111).

Two antagonistic enzymes suggested to be involved in the action of ceramide on cell growth are protein phosphate 2A (PP2A) and phospholipase D (PLD). PPA2 inhibits cell growth (8), whereas PLD stimulates growth (9). Incubation of sitosterol with prostate cancer cells confirms a substantial activation of PPA2 but also shows that PDL activity increases simultaneously in a significant though modest way (26). It is speculated that this increase is a secondary effect to incorporation of sitosterol into the cell membrane, which also hosts PLD due to alterations of membrane fluidity (26). The net effect of this shift of enzyme activity nevertheless results in reduced cellular growth and even reduced potency of metastasis (23).

Another potential mechanism that has been presented is a hypothesized indirect effect on prostate cancer and possibly breast cancer (Fig. 2), mediated through changes in testosterone concentrations (101). Testosterone can

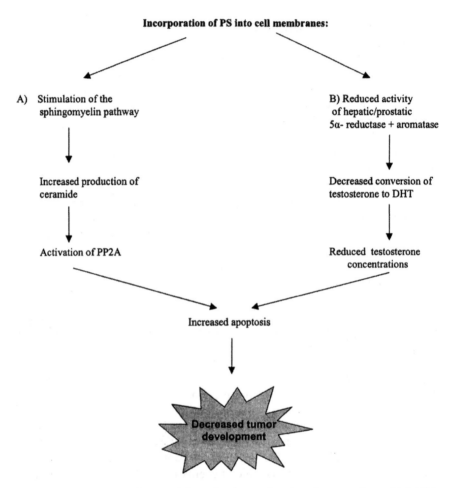

Figure 2 Overview of potential mechanisms whereby phytosterols could inhibit cancer development (101).

be activated by reduction of dihydrotestosterone and other reduced metabolites (with 5α-reductase) or by conversion to estrogen by aromatization (with aromatase) (3,112). Most of the testosterone is synthesized by the testis and metabolized by the liver, whereas prostate is the target organ (3). Feeding with 2% of sitosterols reduced testosterone concentrations in plasma by 33% in rats (113). The activity of 5α-reductase was decreased by 41–44% in the liver and by 33% in the prostate. No effect was observed on the testis levels of the same enzyme. Aromatase activity was reduced by 55% with the phytosterol feeding (113). These results suggest that testosterone metabolism may be involved in the protective effects that sitosterol is hypothesized to have on cancer development. However, the hypothesis is not supported by a clinical study in men on the effects of a spread enriched in unsaturated phytosterols, which showed no significant differences in total or free testosterone concentrations during one year (75). However, the fact that phytosterol supplementation has no effect on testosterone concentrations in plasma does not necessarily mean that there is no local effect in the prostate (75). A clinical trial of the effect of 60 mg of sitosterol per day in men with benign prostatic hyperplasia showed that there was a significant improvement in symptoms and urinary flow parameters (114) that lasted for a period of up to 18 months after treatment initiation (115). This was despite a relevant reduction of prostatic volume (114). A lack of reduction of prostate size, however, also has been seen during more controlled conditions in rats, where prostate size was not significantly different even though doses as high as 8.6% of phytosterols were incorporated in the feed (116).

If compared with normal intakes, phytosterol doses used in animal trials are substantially higher and correspond to even higher intakes than that from phytosterol-enriched food products (31,32). The doses in animal trials are difficult to translate directly into human conditions, but compared as dose/kg body weight (BW), the protective effects on growth and spread of prostate cancer were seen with a dose of 2.6 g/kg BW^{-1} d^{-1} (calculation based on 2% phytosterols, 2.57 g of food per day, BW = 20 g), which corresponds to that of 200 g of phytosterols per day for a 75-kg person (23). The phytosterol dose in the benign prostatic hyperplasia trial (114) was relatively small and would only correspond to a 20–50% increased dietary intake of phytosterols (117,118). Functional food products with cholesterol-lowering properties increase the intake as much as 7–20 times (32), which suggests that these could have a stronger physiological effect.

D. Phytosterols Are Not Phytoestrogens

Phytosterols are sometimes classified with well-described phytoestrogens (7,119) and have been suggested to exert estrogenic activity in a number of studies that were published before phytosterol-enriched food products were introduced into the food market (120–127). However, most of these had design flaws in that they were often based on crude plant extracts where purity and exact composition was not always presented. The early trials were performed to investigate whether or not sitosterol could affect fertility due to direct effects on estrogen receptors. Sitosterol is part of waste products from the wood industry, found in pulp and paper mill effluents. When wood industry waste has been dumped into freshwater, a subsequent infertility in fish has been observed (125). This led to the testing of sitosterol and other phytosterols as possible compounds responsible for the observed effect. Due to the flaws of the earlier studies, a new series of studies was performed as part of a safety evaluation before the new phytosterol-enriched products were allowed on the food market (94,116,128–132). Three studies (116,128,130) dealt directly with the suggested estrogenic activities of phytosterols. The study by Baker et al. was a three-step assay evaluation of estrogenic activity (128). The first step assessed whether or not sitosterol binds to estrogen receptors that were obtained from the uteri of 10-week-old female Wistar rats. Phytosterols [a well-characterized mixture of sitosterol (48%), campesterol (29%), and stigmasterol (23%) tested as both monomers and esters] were negative in the assay and did not compete with estradiol at the tested concentrations. Estradiol, on the other hand, did bind to the estrogen receptor. In the second step, a recombinant yeast strain was used to assess the ability of phytosterols to directly stimulate the transcriptional activity of human estrogen receptors. This also was negative for phytosterols, which did not affect the transcription level, while a stimulated production was seen with both estradiol and coumestrol—the latter a well-characterized phytoestrogen (133). In the third part of the study, phytosterols and phytosterol esters also had no effect on uterine weight in vivo, whereas both estradiol and coumestrol gave increased weight in a dose-dependent manner (128).

A two-generation reproduction study showed that phytosterol esters had no effect on reproduction (116). Measures of pup mortality, precoital time, mating, male and female fertility, gestation, number of females with stillborn pups, pup development, and sexual maturation were all unaffected. The only human study of estrogenic effects in women was based on a small sample size: five women in the intervention group were compared with a control group of six individuals (130). The study demonstrated a lack of significant effect on concentrations of the following blood variables: follicle-stimulating hormone, luteinizing hormone, progesterone, estradiol, sex hormone–binding globulin, and estrone. Due to the limited sample size and the many methodological problems that may arise when studying female sex hormones, this study seems to have little weight in completely ruling out a possible reproductive effect. In consideration of the lack of evidence of estrogenic activity from the available results from animal trials, however, it seems as if the fears about effects on reproduction have little support. Accordingly, the decreased fertility of trout observed in freshwater where wood pulp waste has been dumped (125) may depend on something else in the wood pulp rather than the phytosterols themselves.

E.   Phytosterol Intake in Relation to Immune Function

Cell-mediated immunity (CMI) is a manifestation of the immune system mediated by antigen-sensitized T lymphocytes via lymphokines or direct cytotoxicity in the absence of circulating antibody (or where an antibody has a subordinate role) (1). CMI is suppressed in virtually all malignant disease, including colorectal and prostate cancer (134). As the malignant process develops, the cancer cells evolve to subvert CMI response. Moreover, the reduced CMI seen in colorectal cancer patients is completely reversed following curative surgery, which strongly supports the hypothesis that colorectal cancer can suppress the systemic immune response (134). Suppressed CMI occurs simultaneously with angiogenesis and reduced apoptosis, which hypothetically may provide the ideal environment for serial mutations that is required for the development of malignant disease.

Though the role of phytosterols in modulation of the immune system has not been extensively evaluated, several studies suggest that phytosterols could have a T-cell-mediated effect (135). A mixture of free sitosterol and sitosterol glucoside was tested in three different populations in South Africa: (a) healthy marathon runners (136), (b) patients with tuberculosis (137), and (c) patients with human immunodeficiency virus (HIV) (138). The two first studies were parallel double-blind randomized studies, but based on relatively small study samples. They all suggested that a daily dose of 60 mg of sitosterol had effects on the immune system. The trial in healthy marathon runners showed that the supplemented individuals had a significantly increased number of total white blood cells and neutrophils, and that the percentage of lymphocytes remained the same. Similar findings were seen in phytosterol-supplemented patients with pulmonary tuberculosis. The study in HIV patients showed that the phytosterol-supplemented patients had higher counts of CD4 cells (138)—the classical clinical sign of HIV is a lack of these cells, which in the end leads to the development of acquired immune deficiency syndrome (AIDS) (139). The study has, however, limited use for support of an effect on the immune system, as it was an open-label study without control group.

The immune studies of phytosterols have all been performed in groups of persons who are immune suppressed, and phytosterols may lack an effect in individuals with normal immune function. In short-term clinical studies (4 weeks), C-reactive protein (inflammatory marker) and white blood cells were not affected in healthy normolipidemic men and women when they consumed sitostanol (67). On the other hand, sitostanol has a very limited absorption, which may limit its effects on immune function (63).

Mechanistically, the South African findings could be explained by a several-fold increased proliferation of lymphocytes (30), which is in contrast to what is found for colon cancer cells where phytosterols decrease the proliferation (21). In vitro experiments of the addition of sitosterol to T cells led to an enhanced secretion of interleukin-2 (IL-2) and γ-interferon (INF-γ) (137). As IL-2 therapy has been shown to have antitumor effects (140), it could be hypothesized that phytosterol supplementation would increase

IL-2 expression, thereby having a favorable effect on cancer. Along with breast and ovarian carcinoma, trials have also focused on metastatic melanoma, acute myelogenous leukemia, and metastatic renal cell carcinoma (140). Though a hypothesis linking immune modulating effects of phytosterols to cancer prevention is attractive, it seems less plausible due to the fact that IL-2 therapy often needs to be based on high-dose regimens to demonstrate a meaningful clinical effect (140). However, the theory cannot be ruled out at this point due to the lack of clinical studies but should be tested in people with cancer and a suppressed cell-mediated immune response.

IV. POTENTIAL ADVERSE EFFECTS OF PHYTOSTEROLS ON CANCER

A. Effect of Lowering Plasma Levels of Carotenoids on Cancer Risk

Lipid-standardized levels of blood concentrations of carotenoids decrease with consumption of phytosterol esters (32,78,141–149). The 10–20% reduction that has been demonstrated in clinical trials seems to be of little physiological importance but has still resulted in some debate (Fig. 3). Theoretically, it could be argued that carotenoids are involved in antioxidative pathways with implications for cancer. Glutathione S-transferases (GSTs), for example, are enzymes with antioxidant properties, that catalyze the addition of aliphatic, aromatic, or heterocyclic radicals as well as epoxides and arene oxides to glutathione. They have been shown to inhibit adduction of activated amines to DNA in cell-free systems. In humans, silencing* of the gene GST pi (*GSTP1*) through hypermethylation is found in nearly all prostate carcinomas and is believed to be an early event in prostate carcinogenesis (5). Loss of *GSTP1* expression in human prostate may, therefore, enhance the susceptibility of the prostate to carcinogenic insult by certain compounds. Conversely, induction of GSTs in early-stage prostate carcinogenesis may be a useful protective strategy that may take place through dietary supplementation with beta-carotene (150). It is uncertain at this point if the slight reduction of carotenoids, in particular beta-carotene, has an effect on the activity of GSTs. One way to counteract the reduction of blood carotenoids is to follow the dietary advice to increase the intake of fruit and vegetables. With an increase of one daily serving of high-carotenoid vegetables or fruit, beta-carotene concentrations of subjects who consumed no phytosterols did

---

*The term "silencing through hypermethylation" entails a technique to make the gene nonfunctional.

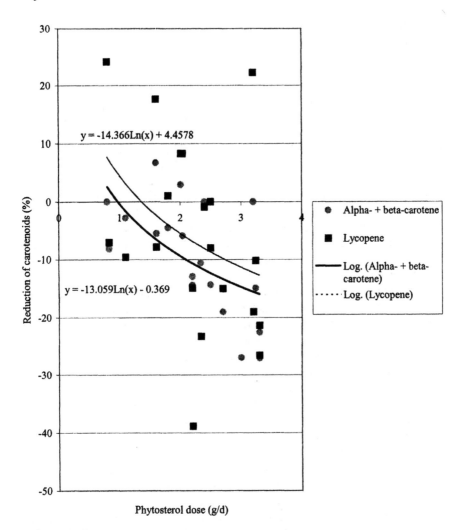

Figure 3 Relative reduction of carotenoids during phytosterol/stanol ester consumption.

not differ significantly from those of subjects who consumed phytosterol esters (151). A 10–20% decrease in plasma carotenoids falls well within the seasonal variation observed in individuals (100). Current knowledge of the physiological functions of carotenoids does not indicate any major health risk associated with the slight decrease in blood levels following intake of phytosterols.

B. Sterol Fermentation, Mutagenic Activity, Oxidation, and Genotoxicity

Phytosterol supplementation has been suggested to have an unfavorable effect on bacterial activity and could theoretically alter the sterol structure during fermentation (see word list) in the gut. An increased excretion of coprostanol was found in a rat study following phytosterol supplementation (152). The authors argued that previous research shows that cholesterol in itself can act as a promoter of colon cancer. This could be adverse for the development of colon cancer (153). An increased excretion of cholesterol and coprostanol has been found to be associated with colon tumors in animal research (153), but the theory has not attracted much interest in recent years. Coprostanol is formed from cholesterol with the aid of 7α-hydroxylase and has been found in increased levels in individuals at high risk for colon cancer and in people consuming high-fat diets (154,155). This does not, however, prove that cholesterol and coprostanol are promoters that have a direct effect on the development of colon cancer, and the evidence seems more circumstantial than directly linked to the effect. People on high-fat diets often consume diets with a lower quality, i.e., less fruits and vegetables, and less dietary fiber, and it seems unlikely that one small factor like increased excretion of coprostanol would increase the cancer risk in a meaningful way.

Mutagenic potential of phytosterols and phytosterol esters has been assessed in a bacterial mutation assay and an in vitro chromosome aberration assay (132). In addition, an in vitro mammalian cell gene mutation assay and two in vivo mutagenicity studies on rat bone marrow micronucleus and liver unscheduled DNA synthesis were also conducted with phytosterol esters (132). Phytosterols and phytosterol esters were not mutagenic in any of these assays. Breakdown products of cholesterol like 4-cholesten-3-one and 5β-cholestan-3-one also showed no evidence of mutagenic activity in these assays (132), which speaks against phytosterols and phytosterol esters as colon cancer promotors.

Unsaturated forms of phytosterols have been suggested to be more oxidation prone than the saturated ones and therefore potentially harmful. Kinetic studies of sitosterol and sitosterol ester show that sitosterol ester is oxidized to a larger extent in the beginning of an autoxidative reaction (156). During the next step, peroxides of both sitosterol and its ester have been established together with 7-OH and 7-keto derivatives, which diminishes the difference in autoxidation between the two compounds. It has been suggested that sitosterol peroxides have an accelerating influence on the autoxidation but are also weaker initiators than the hydroperoxides of fatty acids (157). It should be noted, however, that peroxides are formed under experimental conditions that do not resemble a real-life situation. Still, assays of genotoxicity and mutagenic activity could shed light on some of the adverse effects of the oxides, which can exist in small amounts after long storage, especially if the foods (i.e., vegetables oils) have not been stored in darkness. There is little available information on this kind of biological activity of phytosterols. Nonoxidized stanol esters from wood and vegetables oils have been shown to have no genotoxic effects in bacterial assays (158).

V. EPIDEMIOLOGICAL EVIDENCE OF PHYTOSTEROL EFFECTS ON CANCER

There are few epidemiological studies of the relationship between phytosterol intake per se and cancer. Phytosterol intake does not normally occur in isolation, and therefore a first step would be to examine phytosterols as part of complex diets.

A. Phytosterol Intake in Relation to Foods and Nutrients

There are three major sources of phytosterols from the diet: (a) fats and oils with plant origin; (b) bread and cereals; and (c) fruits and vegetables (159). In one of the few studies evaluating sources of phytosterols, the Netherlands Cohort Study of Diet and Cancer, intake of five phytosterols (sitosterol, campesterol, stigmasterol, sitostanol, and campestanol) was measured. The major source of the five mentioned was bread (mainly rye and whole wheat), closely followed by vegetable fats (mainly 80% margarines), and fruits and vegetables (118).

Several nutrients have been identified as associated with cancer of the colon, breast, and prostate. As reviewed by Walter Willett, energy intake, dietary fat (especially from animal sources), meat, and alcohol have been suggested to have a positive association, whereas dietary fiber, fruits and vegetables, and certain dietary supplements have been suggested to have inverse relationships (160). Phytosterol intake correlates well with energy intake, total fat, polyunsaturated fat, monounsaturated fat, linoleic acid, dietary fiber (especially cereal fiber), whereas it has a low correlation with antioxidant intake (118). This means that phytosterols share the same food sources of the mentioned nutrients to a large extent.

B. Phytosterol Intake and Cancer

The first epidemiological study of phytosterols and cancer was based on a dietary assessment with regard to phytosterol intake in individuals of a low-risk population, the Seventh Day Adventists (81). The study showed that the phytosterol intake of lacto-ovo-vegetarians and nonvegetarians among the

Seventh Day Adventists was higher than that of the normal population and the Seventh Day Adventists vegans. The study has frequently been referred to, although it does not directly assess the relationship between cancer and phytosterol intake. There are only a few studies of the direct association between dietary phytosterols and risk of cancer: to date three case-control studies from Uruguay (161–163), one case-control study from the United States (119), and one prospective study from the Netherlands (118). The case-control studies assessed the relationship between phytosterol intake and cancer of the breast (163), stomach (161), and lung (164). They all found strong inverse associations between phytosterol intake and cancer development. The American study by Strom et al. assessed the effect of phytosterols on prostate cancer and found the opposite, i.e., an increased risk with higher intakes (119). There may be methodological problems in that the largest dietary source was black tea. It appears as if the dry weight of tea has been used for wet weight and phytosterol intake has thus been overestimated in the applied database (7). This makes it difficult to assess the study, as tea drinkers would have very high estimates of phytosterol intake compared to non–tea drinkers.

The only published prospective study does not support a protective effect of phytosterols on colon cancer (118). Instead positive associations were observed, especially for rectal cancers, but due to the lack of animal data supporting a role of phytosterols in the causal pathway, no conclusion could be drawn from these findings.

VI. EPIDEMIOLOGICAL STUDIES OF DIET AND CANCER IN GENERAL: WHAT IS THE ROLE OF PHYTOSTEROLS?

A. Diet and Colorectal Cancer

When ethnic groups migrate from one part of the world to the other, their incidence of colorectal cancer starts to resemble that of the country of residence, which points out the very important role of diet and environment in colon cancer development. The most well-studied example are Japanese men and women who attain a similar risk to that of Americans of developing colon, breast, or prostate cancer when emigrating to the United States, despite having an initially low risk (49,165,166). The American Cancer Society suggests that as much as 90% of colon cancer deaths could be prevented through a healthy diet and a more active lifestyle (16). Since the 1950s, there has been a 29% decrease of colon cancer cases in American Caucasian men and women, and there is evidence that dietary modifications toward a healthier style (less fat, more dietary fiber, more fruits and vegetables) has contributed to this decrease (167).

Case-control studies on diet and colorectal cancer with special regard to dietary fiber (168) and fruit and vegetable intake (169) suggest protective effects of high-fiber and/or fruit and vegetables. Prospective studies of diet and cancer show less convincing effects of the same components. Table 3 shows some major international prospective studies addressing nutrient/food intake and colon cancer (170–198). As can be seen in the table, only two of seven studies support a positive association for fat, one of five studies supports an inverse association for dietary fiber, and four out of eight studies support an inverse association for fruit and vegetables. Only one study addresses the effect of fat quality, as different fatty acids may have different effects. This study found no difference between saturated, polyunsaturated, and monounsaturated fatty acids on colon cancer risk (193). It is uncertain if the content of phytosterols in some way affects the outcome, and more studies would be needed to address to what extent phytosterol intake could be a confounder of those results.

The present outlook on diet and colorectal cancers is shifting from a strong focus on single nutrients to diet and lifestyle as a whole. One study showed that subjects with a high red meat intake, a low legume intake, and a high body mass experienced a more than threefold elevation in risk relative to all other patterns based on these variables (40–42). In individuals with an unhealthy diet, an increased intake of protective agents could play a role. This reasoning leads us to the second major focus of potential cancer-causing mechanisms: a loss of endothelial barrier function due to a failure of terminal differentiation of cells, ultimately leading to increased expression of COX-2, oxidative stress, increased proliferation, and mutagenesis (199). These events can be diminished with such anti-inflammatory agents as aspirin (200). It may be that compounds like phytosterols could work in a similar positive fashion in a high-risk population when given in pharmacological levels if proven effective in increasing apoptosis in humans. In an environment where low physical activity leads to increased problems with overweight and obesity, future foods may need increased enrichments of protective agents alone or in combination to combat the risks to which our lifestyle exposes us in terms of cancer and heart disease. An interesting study of colorectal cancer risk in Swedish women indicates that a 55% lower incidence was found in the group with the healthiest dietary pattern (201). In light of the complex composition of diet and multifactorial etiology of cancer, this study exemplifies that single substances may possibly have limited effects on prevention. Combination therapies of different nutrients/bioactive substances alone or with anti-inflammatory compounds (natural or synthetic) may be a more optimal approach.

B. Breast Cancer and Diet in General

Agents initiating human mammary and prostatic cancers remain largely unidentified (34). It has been suggested that the two diseases may share

Table 3 Prospective Studies on Diet and Cancer: Main Results and Conclusions

Study	Association with colorectal cancer	Association with breast cancer	Association prostate ca
Seventh Adventists Health Study			
Dietary fat	—	None (171)	—
Fruit and vegetables	Inverse[a] (170)	—	Inverse (17
Alpha-Tocopherol, Beta-Carotene Cancer Prevention Study			
Dietary fat	None (173)	—	—
Dietary fiber	None (173)	—	—
Fruit & vegetables	None (173)	—	—
Canadian National Breast Screening Study			
Dietary fat	—	Positive (174)	—
Dietary fiber	—	Inverse (175)	—
Health Professionals Follow-up Study			
Dietary fat	Positive[b] (176)	—	Positive[d] (1
Dietary fiber	Inverse[c] (177)	—	—
Fruit & vegetables	None (178)	—	Inverse[e] (18
Iowa Women's Health Study			
Fruit & vegetables	Inverse (30)	—	—
Netherlands Cohort Study (NLCS)			
Dietary fat	No association (182)	None (184)	None (186)
Dietary fiber	—	None (185)	—
Fruit & vegetables	No association[f] (183)	None (185)	None (36)
New York State Cohort Study			
Dietary fat	—	None (228)	—
Dietary fiber	—	None (228)	—
New York University Women's Health Study			
Dietary fat	None (188)		—
Dietary fiber	None (188)	—	—
Fruit & vegetables	None (188)	—	—
Nurse's Health Study			
Dietary fat	Positive[g] (189)	None (191,192)	—
Dietary fiber	None (190)	None (192)	—
Fruit & vegetables	None (178)	—	—
Sweden Mammography Study			
Dietary fat	None (193)	None/inverse/positive[i] (195)	—
Dietary fiber	None (194)	—	—
Fruit and vegetables	Inverse[h] (194)	Inverse[j] (196)	—
Pooled analysis			
Dietary fat	—	No association (197,229)	—

Table 3 Continued

Study	Association with colorectal cancer	Association with breast cancer	Association with prostate cancer
Dietary fiber	—	—	—
Fruit & vegetables	—	No association (198)	—

[a] The inverse association in this study was only described for legumes.
[b] Saturated fat was positively associated with risk of colorectal adenoma.
[c] Modest reduced risk of distal colon adenoma with increasing intake of fiber from fruit, but not cereals or vegetables. Soluble fiber, but not insoluble fiber, was inversely associated with distal colon adenoma.
[d] Animal fat, especially fat from red meat, was associated with an elevated risk of advanced prostate cancer.
[e] Combined intake of tomatoes, tomato sauce, tomato juice, and pizza (which accounted for 82% of lycopene intake) was inversely associated with risk of prostate cancer and advanced prostate cancers.
[f] Colon and rectum treated separately in the NLCS. An exception from the main result was an inverse association for colon cancer in women. Brassica vegetables and cooked leafy vegetables showed inverse associations for both men and women.
[g] Only animal fat was positively associated with the risk of colon cancer. No association found with fat of plant origin.
[h] Subanalyses showed that this association was due largely to fruit consumption. Total, cereal, and fruit fiber had no separate effects.
[i] No association with saturated fat, an inverse association with monounsaturated fat, and a positive association with polyunsaturated fat.
[j] Only brassica vegetables were studied.

a common cause, with changes in diet, rather than in environment, being responsible for the migration-related increases in cancer incidence observed (34,49,166,202,203).

Breast cancer risk rises with increasing alcohol intake, which widely is assumed to be mediated by changes in estrogen concentrations (204). Alcohol is thereby one of the most important promoting factors for breast cancer development especially in combination with hyperinsulinemia. Alcohol can stimulate growth of precancerous breast lesions in the years leading up to the menopause and may increase the risk of breast cancer manifesting after the menopause (204).

Energy restriction is well known to reduce the development of mammary tumors (205,206). Studies of energy intake in relation to cancer are however complex matters, as energy intake is also a marker of physical activity which is known to prevent breast cancer as well as other types of cancers (colon and prostate) (207). In contrast, epidemiological evidence on the relationship of dietary fat to breast cancer from cohort and case-control studies has been inconsistent. As can be seen from Table 3, only one of six studies supports a positive association for total fat, one of four studies supports an inverse association for dietary fiber, and one of three studies supports an inverse association of fruit and vegetables. Several of the major prospective studies have been pooled in an attempt to increase the power even further during the statistical analysis. The two papers that have been published so far deal with dietary fat and fruit and vegetable consumption (197,198). Despite the increased power, they showed that none of these factors was associated with breast cancer development.

Another factor that has been evaluated is the consumption of red meat. Cooked meat contains varying amounts of heterocyclic amines, which have mutagenic activity (34). A pooled analysis of several prospective studies suggests that there is no association with red meats (208). However, the fact that the content of heterocyclic amines depends on the cooking conditions (time and temperature) as well as method (grilling, boiling) has made it difficult to evaluate the effect.

The role of overall diet has been evaluated with regard to development of breast cancer (209). There is no apparent association between a "Western" dietary pattern (characterized by foods such as red and processed meats, refined grains, fat, and sweets) or a "healthy" dietary pattern (fruit and vegetables, fish and poultry, low-fat dairy, and whole grains) and breast cancer risk. As reviewed and debated by Falkenberry and Legare, it is established that endogenous factors such as hormones influence breast cancer development (210), yet a difficult challenge remains to identify the environmental risk factors that predispose to breast cancer. As an effect of diet seems to be poorly supported by present research, it is almost impossible to discuss what role phytosterols could play as protective agents. However, it could be argued that if breast cancer develops mainly from mutagenic activity caused by heterocyclic amines in cooked meat, phytosterols could prove a useful complement to other agents that reduce the formation of heterocyclic amines. This reasoning builds on an assumption that phytosterols would be shown to have an effect on apoptosis also in humans.

C. Prostate Cancer and Diet in General

Alcohol also works as a promotor for prostate cancer, predisposing individuals at risk (211). As reviewed by Kolonel, ecological* and case-control studies implicate fats and meats as risk factors for prostate cancer (212); however, there is presently no consensus on whether dietary fat affects pros-

---

*Studies of the association of two factors, such as a nutrient and colon cancer, at a national or international level. Correlation studies is one example of this methodology.

tate cancer. Fewer studies have evaluated the relationship between diet and prostate in a prospective way.

Epidemiological research has suggested that the initiating factors are dietary carcinogens (found in particular in fried meat), which are metabolized by prostatic epithelial cells and thereby activated (213). One dietary carcinogen is the compound 2-amino-1-methyl-6-phenylimidazo[4,5-b]pyridine (PhIP) (6). Large differences in PhIP intake between different U.S. ethnic groups may at least partly explain why prostate cancer kills approximately twofold more African-American than Caucasian men (214). Experimental data indicate that PhIP mutates prostate DNA and causes prostate tumors in rats (6).

As can be seen from Table 3, only one of the two available studies supports a positive association for total fat and two of three studies support an inverse association with fruit and vegetables. In one of the latter, an inverse association is actually observed with a more frequent consumption of tomato products (215). Similarly, the carotenoid lycopene was also associated with reduced risk of prostate cancer, and intake of tomato sauce, the primary source of bioavailable lycopene, was associated with an even greater reduction in prostate cancer risk (215). However, another study demonstrated a lack of effect of lycopene, but instead showed a positive association between prostate cancer risk and intake of β-cryptoxanthin (216). In the same study, inverse associations for retinol, alpha-carotene, and beta-carotene were demonstrated among nondrinkers; which suggests that certain vitamins and antioxidants may have effects that are masked by alcohol consumption. The role of phytosterols is still unknown, but future prospective studies could shed some light on their association with prostate cancer.

VII. METHODOLOGICAL ISSUES OF PHYTOSTEROLS IN NUTRITIONAL EPIDEMIOLOGY

A. Phytosterol Databases

Associations between phytosterol intake and cancer risk have been difficult to assess due to the lack of suitable databases covering phytosterol concentrations in a sufficient number of food items. In 1978, the U.S. Department of Agriculture published a report on lipid content of foods (159). One of these summarized literature values of phytosterol concentrations from the 1960s and onward. In total, the report contains phytosterol concentrations of 51 vegetables, 22 fruits, 9 cereals, 16 margarines, 56 oils, 12 seeds and nuts, 35 dry spices, 10 legumes, and 9 miscellaneous items. It is an extensive report, set as a basis for the evaluation of mixed diets. The report contained only a few samples of cereal items but can serve as an indication of phytosterol content of different foods. However, lack of consistency in the analytical methodology limits its use in epidemiological studies, as dietary concentration determination must be reliable to give proper estimates of intake in relation to risk for disease development. This is exemplified in a study by Strom et al. (119), who have used a modified version of Dietsys, the nutrient analysis program associated with the National Cancer Institute Health Habits and History Questionnaire. Phytosterol values from the report of Weihrauch and Gardner (159) along with four other reports (7) on phytosterol concentrations of foods were added to the database for later use in epidemiological studies of plant compounds such as phytosterols. The database has been applied in a case-control study (119), though the validity of the database can be questioned due to the methodological concerns just mentioned.

Due to increasing interest in phytochemicals, an expanded listing of analytical values of bioactive substances is now available. Ethnobotanical databases such as the Phytochemical Database from the U.S. Agricultural Service can be directly accessed through the internet (http://www.ars-grin.gov/duke/). This gives information based on the name of the botanical plant, and concentrations of phytosterols are estimated in parts per millions. Again, the validity could be questioned regarding use epidemiological studies, as these are also summaries of several studies using different techniques, which might induce a large variation in phytosterol estimates.

Two new databases have recently been developed in Scandinavia based on a large number of food samples (217,218). The databases are based on extensive work performed with one and the same methodology for use in epidemiological studies. This approach is less likely to give bias, and as a majority of the phytosterol values have been or will be published, this provides an alternative source of phytosterol information for epidemiological applications. However, several factors that may affect the accuracy and representativity of phytosterol values require consideration. Examples of some including the accuracy of databases are (a) phytosterol composition may depend on plant variety, geographic region, growing season, shipping, storage, and preparation; (b) different brands are sold in different countries; and (c) there may be a change of ingredient composition over time (219). Therefore, databases must be used judiciously in epidemiological studies of phytosterols and cancer.

B. Methodological Concerns

Familiarity with different methodologies applied when evaluating evidence of dietary intake in general or of phytosterols in particular is of utmost importance. In general, ecological studies and case-control studies show strong associations, while the same factors seem to have weak to moderate associations in prospective epidemiological studies of diet and cancer (220). This is mainly due to methodological reasons, as ecological studies do not control for all confounding factors, and case-control studies are believed to be heavily affected by recall bias, i.e., the interviewed persons under- or overestimate their food intake (221). Methodological strengths and limitations of different study designs are presented in Table 4 to further the understanding of the discussion about dietary intake and cancer. In general, ecological studies and case-control studies generate a hypothesis, whereas prospective and intervention studies to a larger extent are used to draw an evidence-based conclusion. The majority of phytosterol studies have been case-control studies and have often found support for a cancer preventive effect. This can in part depend on recall bias, where so-called healthy items get underreported in a conscious or subconscious way. Therefore, it is important that future studies assess the effect of phytosterols using a prospective study design. It may be that phytosterol intake correlates strongly with other nutrients and phytochemicals, which limits the power of the study to detect an effect. It is also possible that phytosterol intake from naturally occurring dietary sources may be too low to have a beneficial effect on cancer. Instead, intervention studies in high-risk populations with high phytosterol doses would need to be performed, as these would give substantially stronger evidence of a cancer-preventive effect of phytosterols.

Other methodological concerns that need to be taken into account when evaluating the effects of dietary phytosterol intake on cancer are confounding factors, the lack of food frequency questionnaires that have been validated specifically for phytosterol intake, and underreporting. Confounding factors for phytosterol intake are other nutrients and food components that affect the outcome. Total intakes of energy, vitamin E, and dietary fiber are potential confounding factors for colon cancer risk (118). Food frequency questionnaires need to be validated with regard to the investigated ingredients. Most questionnaires are validated with regard to energy, macronutrients, and antioxidants (222). However, whether or not they at the same time give reasonable estimates of phytosterol intake needs to be assessed. Examples of other methodological issues are the facts that food frequency questionnaires often include questions about aggregated food items to limit the number of questions and certain phytosterol-rich foods may not be included. Subjects are also asked to recall information from a period of up to a year, which may result in imprecise and biased intake estimates (219).

Another methodological problem is the frequency of underreporting. This is known to always occur but seems to be worse in groups that are overweight (223). By finding a reliable biomarker (see Sect. VII.C), the effect of phytosterol intake on cancer development could be assessed in larger population-based studies with the help of a blood sample, which could be one

Table 4 Descriptions of Epidemiological Study Designs Including Strength and Limitations (230)

Study design	Description	Strength	Limitation	Weight and type of evidence
Ecological/correlation study	Examination of per capita consumption of a dietary factor and the prevalence/incidence/mortality of a cancer in a population.	Simple assessment that rapidly can suggest evidence of that the studied dietary factor is directly/indirectly related to the cancer.	Does not take into account confounding factors[a], and is likely to give false positive results.	Light Indicative, not conclusive.
Case-control studies	Comparisons of consumption of dietary factor by subjects with cancer with matched control subject.	Confounders can be taken into account when dealing with the results statistically. Relatively small sample size can be used, which decreases the cost of the study.	High likeliness of recall and selection bias.	Light to moderate. Suggestive, not conclusive.

	Description	Advantages	Disadvantages	Strength of evidence
Prospective studies	Assessment of diets in a large group of healthy people with long-term follow-up over a period when cancers will be developed by the studied sample.	Confounders can be taken into account when dealing with the results statistically. Less likeliness of selection or recording bias than case-control.	Large sample size is needed, which increases study costs. Follow-up is often too short, and dietary intake at baseline may be wrongly assumed to reflect the intake during the whole study period.	Moderate. Can be used together with other types of studies (mechanistic and intervention) to draw conclusion.
Intervention studies	Randomized study where nutritional intervention is undertaken.	If the study entails a placebo-controlled and double-blind regimen of a dietary supplement, it gives hard evidence of an effect on cancer.	Expensive, too short follow-up and small sample size, poor compliance, high drop-out, and use of nonoptimal biomarkers.	Strong. Should ultimately be performed to find support for the conclusion.

[a] Confounding factors are factors that can cause or prevent the outcome of interest, are not intermediate variables, and are not associated with the factor(s) under investigation. They give rise to situations in which the effects of two processes are not separated, or the contribution of causal factors cannot be separated, or the measure of the effect of exposure or risk is distorted because of its association with other factors influencing the outcome of the study (definition taken from the PubMed MESH terms). Example: Number of TVs and cancer may be correlated. The confounding factor may be differences in diet, smoking, or other factor that actually causes the result, which then would the causal factor, and not the TV factor alone.

way to overcome problems with underreporting, especially in overweight populations. In light of the accelerated rate of obesity development in most Western countries (224), this could be a more optimal approach.

C. Potential Methodology: Biomarkers of Phytosterol Intake

To evaluate the agreement between consumption and estimate of intake, a reliable biomarker needs to be assessed in relation to intake. Li et al. suggested that blood concentrations (i.e., plasma or serum) of sitosterol and campesterol would be useful in this assessment as they present high reproducibility and reliability over time (225). Tilvis and Miettinen, on the other hand, suggested that blood concentrations of phytosterols also are positively associated with cholesterol absorption (226). They also showed that concentrations of phytosterols in whole serum and in each lipoprotein fraction were significantly correlated with the percentage of absorbed dietary cholesterol but were independent of the amount of dietary cholesterol and plant sterols, which is evidence against using blood concentrations of phytosterols as biomarkers. However, the study was not based on a controlled diet and presents a theoretical problem that would rule against using phytosterols as biomarkers, as blood concentrations of phytosterols are also dependent on the rate of cholesterol absorption. To assess the possibility of using blood concentrations of phytosterols, a controlled study with 2×2 factorial design would need to be performed, where the effect of a low and a high cholesterol intake with a simultaneously low or high phytosterol intake is examined. If phytosterol intake always predicts plasma concentrations, they may even be an excellent alternative to assess intake directly, though this needs to be evaluated.

VIII. FUTURE PERSPECTIVES ON PHYTOSTEROL RESEARCH IN RELATION TO CANCER

Phytosterols have an important function in the human body due to their cholesterol lowering effect, which has led to the introduction of phytosterol-enriched spreads in more than 20 countries. Their role in cancer prevention is, however, still under investigation, and considerable work must be done before the effects on human cancer risk can be judged. Should there be a proven effect of phytosterols, consumers of high amounts of phytosterols may have an added benefit—that of cancer prevention. Even though animal and cell studies support an effect of sitosterol, there is a need for demonstrating the effects also in humans. Different approaches can be taken to do this. One approach is to investigate the effects of phytosterol supplementation on biomarkers of cancer. An alternative could be to carry out studies in high-risk populations (persons with intestinal polyps) or populations who already have developed cancer to assess polyp or tumor development, though this may entail some ethical dilemmas. According to the animal trials, phytosterol supplementation could lead to reduced metastasis and tumor growth of certain cancers (25,102), which may be applicable in the human population as well.

This chapter uncovers the fact that there is still a lot of work to be done in relation to the potential effects of phytosterols on cancer. We have instead tried to describe the background against which phytochemicals could play a role in cancer prevention. Cancer is on the rise worldwide due to changes in lifestyle (17). Also, in the United States, lifestyle-related cancers are on the rise with a simultaneous increase in the incidence of obesity and the metabolic syndrome—two predicting factors of cancer development and survival (39,42,227). Primary prevention programs need to be improved in general. Whether or not phytosterols can play a role in such programs still needs to be determined.

GLOSSARY

Adenocarcinoma	A malignant epithelial tumor with a glandular organization (1).
Apoptosis	One of the two mechanisms by which cell death occurs. It is the mechanism responsible for the physiological deletion of cells and appears to be intrinsically programmed. Characterized by distinctive morphological changes in the nucleus and cytoplasm, chromatin cleavage at regularly spaced sites, and the endonucleolytic cleavage of genomic DNA at internucleosomal sites. Serves as a balance to mitosis in regulating the size of animal tissues and in mediating pathological processes associated with tumor growth (1).
Benign hyperplasia	An increase in the number of cells in a tissue or organ, not due to nonmalignant tumor formation (1).
Carcinogens	Substances that increase the risk of neoplasms in humans or animals. Both genotoxic chemicals, which affect DNA directly, and nongenotoxic

	chemicals, which induce neoplasms by other mechanism, are included (1).
Cancer promotion	Spread of cancer cells, which cause minor or major changes in human physiology that finally may lead to organ malfunctioning (2).
Cancer proliferation	Damaged cells replicate as a consequence of malfunctioning proteins, which normally regulate cell proliferation and apoptosis (i.e., cell death) (2).
Cancer initiation	Mutation of the DNA occurs through radiation or through a genotoxic element (2).
Chemoprevention	Use of chemotherapeutic agents as the means of preventing the development of a specific disease (1).
CMI	Cell-mediated immunity.
CRP	C-reactive protein.
DHT	Dihydrotestosterone, a biologically active metabolite of testosterone (3).
Epidemiology	Study of how often a disease occurs and why (4).
Etiology	Study of disease development.
Fermentation	An enzyme-induced chemical change in organic compounds that occurs in the absence of oxygen (1).
GST	Glutathione S-transferase (5).
GSTP1	GST pi (5).
INF-γ	γ-Interferon.
HIV	Human immunodeficiency virus.
Incidence	The number of new cases of a given disease during a given period in a specified population. It also is used for the rate at which new events occur in a defined population (1).
IL-2	Interleukin-2.
Malignancy	Cancer.
Neoplasm	New abnormal growth of tissue. Malignant neoplasms show a greater degree of anaplasia and have the properties of invasion and metastasis, compared to benign neoplasms (1).
PhIP	2-Amino-1-methyl-6-phenylimidazo[4,5-b]pyridine (6).
Phytoestrogens	Plant constituents that have a weak estrogenic activity (bind to the estrogen receptor) (7).

PLD/PP2A	Protein phosphate 2A and phospholipase D: two antagonistic enzymes involved in the action of ceramide on cell growth. PPA2 inhibits cell growth (8), whereas PLD stimulates growth (9).
Premalignant cells	Cancer cells after the initation and promotion steps (10).
Preneoplastic cells	Cancer cells after the initation and promotion steps (10).
Prevalence	The total number of cases of a given disease in a specified population at a designated time (1).

REFERENCES

1. PUBMED Definition of MESH terms. National Library of Medicine at [http://www.ncbi.nlm.nih.gov/PubMed/] 2002.
2. M Ponz de Leon, A Percesepe. Pathogenesis of colorectal cancer. Dig Liver Dis 32(9):807–821, 2000.
3. JE Montie, KJ Pienta. Review of the role of androgenic hormones in the epidemiology of benign prostatic hyperplasia and prostate cancer. Urology 43(6):892–899, 1994.
4. C Coggon, G Rose, D Barker. What is epidemiology? In: Epidemiology for the Uninitiated, 3rd ed. London: British Medical Journal (BMJ) Publishing Group, 1993, pp. 1–5.
5. CP Nelson, LC Kidd, J Sauvageot, WB Isaacs, AM De Marzo, JD Groopman, WG Nelson, TW Kensler. Protection against 2-hydroxyamino-1-methyl-6-phenylimidazo[4,5-b]pyridine cytotoxicity and DNA adduct formation in human prostate by glutathione S-transferase P1. Cancer Res 61(1):103–109, 2001.
6. T Shirai, M Sano, S Tamano, S Takahashi, M Hirose, M Futakuchi, R Hasegawa, K Imaida, K Matsumoto, K Wakabayashi, T Sugimura, N Ito. The prostate: a target for carcinogenicity of 2-amino-1-methyl-6-phenylimidazo [4,5-b]pyridine (PhIP) derived from cooked foods. Cancer Res 57(2):195–198, 1997.
7. PC Pillow, CM Duphorne, S Chang, JH Contois, SS Strom, MR Spitz, SD Hursting. Development of a database for assessing dietary phytoestrogen intake. Nutr Cancer 33(1):3–19, 1999.
8. RA Wolff, RT Dobrowsky, A Bielawska, LM Obeid, YA Hannun. Role of ceramide-activated protein phosphatase in ceramide-mediated signal transduction. J Biol Chem 269(30):19605–19609, 1994.
9. ME Venable, GC Blobe, LM Obeid. Identification of a defect in the phospholipase D/diacylglycerol pathway in cellular senescence. J Biol Chem 269 (42):26040–26044, 1994.
10. R Cotran, V Kumar, S Robbins. Neoplasia: pathologic Basis for Disease. Philadelphia: W.B. Saunders, 1994.

11. PJ Jones, DE MacDougall, F Ntanios, CA Vanstone. Dietary phytosterols as cholesterol-lowering agents in humans. Can J Physiol Pharmacol 75(3):217–227, 1997.
12. AB Awad, LA Begdache, CS Fink. Effect of sterols and fatty acids on growth and triglyceride accumulation in 3T3-L1 cells. J Nutr Biochem 11(3):153–158, 2000.
13. C Mathers, C Boschi-Pinto, A Lopez, C Murray. Cancer incidence, mortality and survival by site for 14 regions of the world. Global Programme on Evidence for Health Policy Discussion Paper No. 13. Geneva: WHO, 2001.
14. MK Soni, D Cella. Quality of life and symptom measures in oncology: an overview. Am J Manag Care 8(suppl 18):S560–573, 2002.
15. R Doll, R Peto. The causes of cancer: quantitative estimates of avoidable risks of cancer in the United States today. J Natl Cancer Inst 66(6):1191–1308, 1981.
16. ACS. Cancer Facts and Figures. Atlanta, GA: American Cancer Society, 2001.
17. P Pisani, DM Parkin, F Bray, J Ferlay. Estimates of the worldwide mortality from 25 cancers in 1990. Int J Cancer 83(1):18–29, 1999.
18. PP Nair. Role of bile acids and neutral sterols in carcinogenesis. Am J Clin Nutr 48(suppl 3):768–774, 1988.
19. RF Raicht, BI Cohen, EP Fazzini, AN Sarwal, M Takahashi. Protective effect of plant sterols against chemically induced colon tumors in rats. Cancer Res 40(2):403–405, 1980.
20. AB Awad, AY Hernandez, CS Fink, SL Mendel. Effect of dietary phytosterols on cell proliferation and protein kinase C activity in rat colonic mucosa. Nutr Cancer 27(2):210–215, 1997.
21. SA Janezic, AV Rao. Dose-dependent effects of dietary phytosterol on epithelial cell proliferation of the murine colon. Food Chem Toxicol 30(7):611–616, 1992.
22. AB Awad, MD Garcia, CS Fink. Effect of dietary phytosterols on rat tissue lipids. Nutr Cancer 29(3):212–216, 1997.
23. AB Awad, CS Fink, H Williams, U Kim. In vitro and in vivo (SCID mice) effects of phytosterols on the growth and dissemination of human prostate cancer PC-3 cells. Eur J Cancer Prev 10(6):507–513, 2001.
24. AB Awad, RL von Holtz, JP Cone, CS Fink, YC Chen. beta-Sitosterol inhibits growth of HT-29 human colon cancer cells by activating the sphingomyelin cycle. Anticancer Res 18(1A):471–473, 1998.
25. AB Awad, A Downie, CS Fink, U Kim. Dietary phytosterol inhibits the growth and metastasis of MDA-MB-231 human breast cancer cells grown in SCID mice. Anticancer Res 20(2A):821–824, 2000.
26. AB Awad, Y Gan, CS Fink. Effect of beta-sitosterol, a plant sterol, on growth, protein phosphatase 2A, and phospholipase D in LNCaP cells. Nutr Cancer 36(1):74–78, 2000.
27. AB Awad, AC Downie, CS Fink. Inhibition of growth and stimulation of apoptosis by beta-sitosterol treatment of MDA-MB-231 human breast cancer cells in culture. Int J Mol Med 5(5):541–545, 2000.

28. AB Awad, H Williams, CS Fink. Phytosterols reduce in vitro metastatic ability of MDA-MB-231 human breast cancer cells. Nutr Cancer 40(2):157–164, 2001.
29. CJ Portier. Endocrine dismodulation and cancer. Neuroendocrinol Lett 23 (suppl 2):43–47, 2002.
30. PJ Bouic, S Etsebeth, RW Liebenberg, CF Albrecht, K Pegel, PP Van Jaars-veld. beta-Sitosterol and beta-sitosterol glucoside stimulate human peripheral blood lymphocyte proliferation: implications for their use as an immunomo-dulatory vitamin combination. Int J Immunopharmacol 18(12): 693–700, 1996.
31. TA Miettinen, P Puska, H Gylling, H Vanhanen, E Vartiainen. Reduction of serum cholesterol with sitostanol-ester margarine in a mildly hypercholester-olemic population. N Engl J Med 333(20):1308–1312, 1995.
32. JA Weststrate, GW Meijer. Plant sterol-enriched margarines and reduction of plasma total- and LDL-cholesterol concentrations in normocholesterolaemic and mildly hypercholesterolaemic subjects. Eur J Clin Nutr 52(5):334–343, 1998.
33. T Muto, HJ Bussey, BC Morson. The evolution of cancer of the colon and rectum. Cancer 36(6):2251–2270, 1975.
34. PL Grover, FL Martin. The initiation of breast and prostate cancer. Carcino-genesis 23(7):1095–1102, 2002.
35. AJ Gescher, RA Sharma, WP Steward. Cancer chemoprevention by dietary constituents: a tale of failure and promise. Lancet Oncol 2(6):371–379, 2001.
36. H Schreiber. Tumor immunology. In: W Paul, ed. Fundamental Immunology, 4th ed. Philadelphia, New York: Lippincott–Raven, 1999, pp. 1237–1270.
37. EC Woodhouse, RF Chuaqui, LA Liotta. General mechanisms of metastasis. Cancer 80(suppl 8):1529–1537, 1997.
38. M McCredie. Cancer epidemiology in migrant populations. Rec Results Can-cer Res 154:298–305, 1998.
39. TK Murphy, EE Calle, C Rodriguez, HS Kahn, MJ Thun. Body mass index and colon cancer mortality in a large prospective study. Am J Epidemiol 152(9):847–854, 2000.
40. PN Singh, GE Fraser. Dietary risk factors for colon cancer in a low-risk popu-lation. Am J Epidemiol 148(8):761–774, 1998.
41. WR Bruce, TM Wolever, A Giacca. Mechanisms linking diet and colorectal cancer: the possible role of insulin resistance. Nutr Cancer 37(1):19–26, 2000.
42. E Giovannucci. Insulin, insulin-like growth factors and colon cancer: a review of the evidence. J Nutr 131(suppl 11):3109S–3120S, 2001.
43. ES Ford, WH Giles, WH Dietz. Prevalence of the metabolic syndrome among US adults: findings from the third National Health and Nutrition Exam-ination survey. JAMA 287(3):356–359, 2002.
44. H Lamlum, A Papadopoulou, M Ilyas, A Rowan, C Gillet, A Hanby, I Talbot, W Bodmer, I Tomlinson. APC mutations are sufficient for the growth of early colorectal adenomas. Proc Natl Acad Sci U S A 97(5):2225–2228, 2000.
45. AM Pollock, P Quirke. Adenoma screening and colorectal cancer. BMJ 303 (6811):1202, 1991.

46. O Kronborg, C Fenger, J Olsen, OD Jorgensen, O Sondergaard. Randomised study of screening for colorectal cancer with faecal-occult-blood test. Lancet 348(9040):1467–1471, 1996.

47. JD Hardcastle, JO Chamberlain, MH Robinson, SM Moss, SS Amar, TW Balfour, PD James, CM Mangham. Randomised controlled trial of faecal-occult-blood screening for colorectal cancer. Lancet 348(9040):1472–1477, 1996.

48. JS Mandel, TR Church, JH Bond, F Ederer, MS Geisser, SJ Mongin, DC Snover, LM Schuman. The effect of fecal occult-blood screening on the incidence of colorectal cancer. N Engl J Med 343(22):1603–1607, 2000.

49. H Shimizu, RK Ross, L Bernstein, R Yatani, BE Henderson, TM Mack. Cancers of the prostate and breast among Japanese and white immigrants in Los Angeles County. Br J Cancer 63(6):963–966, 1991.

50. BA Stoll. Oestrogen/insulin-like growth factor-I receptor interaction in early breast cancer: clinical implications. Ann Oncol 13(2):191–196, 2002.

51. SW Duffy, L Tabar, HH Chen, M Holmqvist, MF Yen, S Abdsalah, B Epstein, E Frodis, E Ljungberg, C Hedborg-Melander, A Sundbom, M Tholin, M Wiege, A Akerlund, HM Wu, TS Tung, YH Chiu, CP Chiu, CC Huang, RA Smith, M Rosen, M Stenbeck, L Holmberg. The impact of organized mammography service screening on breast carcinoma mortality in seven Swedish counties. Cancer 95(3):458–469, 2002.

52. JE McNeal. The prostate gland: morphology and pathophysiology. Monogr Urol 9:36–63, 1988.

53. JE McNeal. Cancer volume and site of origin of adenocarcinoma in the prostate: relationship to local and distant spread. Hum Pathol 23(3):258–266, 1992.

54. LK Dennis, CF Lynch, JC Torner. Epidemiologic association between prostatitis and prostate cancer. Urology 60(1):78–83, 2002.

55. LK Dennis, DV Dawson. Meta-analysis of measures of sexual activity and prostate cancer. Epidemiology 13(1):72–79, 2002.

56. R Shi, HJ Berkel, H Yu. Insulin-like growth factor-I and prostate cancer: a meta-analysis. Br J Cancer 85(7):991–996, 2001.

57. T Shaneyfelt, R Husein, G Bubley, CS Mantzoros. Hormonal predictors of prostate cancer: a meta-analysis. J Clin Oncol 18(4):847–853, 2000.

58. SA Gayther, KA de Foy, P Harrington, P Pharoah, WD Dunsmuir, SM Edwards, C Gillett, A Ardern-Jones, DP Dearnaley, DF Easton, D Ford, RJ Shearer, RS Kirby, AL Dowe, J Kelly, MR Stratton, BA Ponder, D Barnes, RA Eeles. The frequency of germ-line mutations in the breast cancer predisposition genes BRCA1 and BRCA2 in familial prostate cancer. The Cancer Research Campaign/British Prostate Group United Kingdom Familial Prostate Cancer Study Collaborators. Cancer Res 60(16):4513–4518, 2000.

59. KL Schwartz, TY Kau, RK Severson, RY Demers. Prostate-specific antigen in a community screening program. J Fam Pract 41(2):163–168, 1995.

60. GJ Kelloff, JA Crowell, VE Steele, RA Lubet, WA Malone, CW Boone, L Kopelovich, ET Hawk, R Lieberman, JA Lawrence, I Ali, JL Viner, CC

Sigman. Progress in cancer chemoprevention: development of diet-derived chemopreventive agents. J Nutr 130(suppl 2S):467S–471S, 2000.

61. W Nes. Multiple roles for phytosterols. In: P Stumpf, ed. The Metabolism, Structure and Function of Plant Lipids. New York: Plenum Press, 1987, pp. 3–9.

62. l Ikeda, K Tanaka, M Sugano, GV Vahouny, LL Gallo. Inhibition of cholesterol absorption in rats by plant sterols. J Lipid Res 29(12):1573–1582, 1988.

63. RE Ostlund Jr, JB McGill, CM Zeng, DF Covey, J Stearns, WF Stenson, CA Spilburg. Gastrointestinal absorption and plasma kinetics of soy delta(5)- phytosterols and phytostanols in humans. Am J Physiol Endocrinol Metab 282(4):E911–916, 2002.

64. H Gylling, TA Miettinen. Cholesterol reduction by different plant stanol mixtures and with variable fat intake. Metabolism 48(5):575–580, 1999.

65. H Gylling, TA Miettinen. Serum cholesterol and cholesterol and lipoprotein metabolism in hypercholesterolaemic NIDDM patients before and during sitostanol ester-margarine treatment. Diabetologia 37(8):773–780, 1994.

66. TA Miettinen, H Vanhanen. Dietary sitostanol related to absorption, synthesis and serum level of cholesterol in different apolipoprotein E phenotypes. Atherosclerosis 105(2):217–226, 1994.

67. RP Mensink, S Ebbing, M Lindhout, J Plat, MM van Heugten. Effects of plant stanol esters supplied in low-fat yoghurt on serum lipids and lipoproteins, non-cholesterol sterols and fat soluble antioxidant concentrations. Atherosclerosis 160(1):205–213, 2002.

68. P Nestel, M Cehun, S Pomeroy, M Abbey, G Weldon. Cholesterol-lowering effects of plant sterol esters and non-esterified stanols in margarine, butter and low-fat foods. Eur J Clin Nutr 55(12):1084–1090, 2001.

69. HT Vanhanen, S Blomqvist, C Ehnholm, M Hyvonen, M Jauhiainen, I Torstila, TA Miettinen. Serum cholesterol, cholesterol precursors, and plant sterols in hypercholesterolemic subjects with different apoE phenotypes during dietary sitostanol ester treatment. J Lipid Res 34(9):1535–1544, 1993.

70. H Vanhanen. Cholesterol malabsorption caused by sitostanol ester feeding and neomycin in pravastatin-treated hypercholesterolaemic patients. Eur J Clin Pharmacol 47(2):169–176, 1994.

71. HT Vanhanen, J Kajander, H Lehtovirta, TA Miettinen. Serum levels, absorption efficiency, faecal elimination and synthesis of cholesterol during increasing doses of dietary sitostanol esters in hypercholesterolaemic subjects. Clin Sci (Lond) 87(1):61–67, 1994.

72. TA Miettinen, M Vuoristo, M Nissinen, HJ Jarvinen, H Gylling. Serum, biliary, and fecal cholesterol and plant sterols in colectomized patients before and during consumption of stanol ester margarine. Am J Clin Nutr 71(5):1095–1102, 2000.

73. LI Christiansen, PL Lahteenmaki, MR Mannelin, TE Seppanen-Laakso, RV Hiltunen, JK Yliruusi. Cholesterol-lowering effect of spreads enriched with microcrystalline plant sterols in hypercholesterolemic subjects. Eur J Nutr 40(2):66–73, 2001.

74. MH Davidson, KC Maki, DM Umporowicz, KA Ingram, MR Dicklin, E

Schaefer, RW Lane, JR McNamara, JD Ribaya-Mercado, G Perrone, SJ Franke, WC Franke. Safety and tolerability of esterified phytosterols administered in reduced-fat spread and salad dressing to healthy adult men and women. J Am Coll Nutr 20(4):307–319, 2001.

75. HF Hendriks, EJ Brink, GW Meijer, HM Princen, FY Ntanios. Safety of long-term consumption of plant sterol esters-enriched spread. Eur J Clin Nutr 57:681–692, 2003.

76. PJ Jones, FY Ntanios, M Raeini-Sarjaz, CA Vanstone. Cholesterol-lowering efficacy of a sitostanol-containing phytosterol mixture with a prudent diet in hyperlipidemic men. Am J Clin Nutr 69(6):1144–1150, 1999.

77. CA Vanstone, M Raeini-Sarjaz, WE Parsons, PJ Jones. Unesterified plant sterols and stanols lower LDL-cholesterol concentrations equivalently in hypercholesterolemic persons. Am J Clin Nutr 76(6):1272–1278, 2002.

78. MA Hallikainen, ES Sarkkinen, MI Uusitupa. Plant stanol esters affect serum cholesterol concentrations of hypercholesterolemic men and women in a dose-dependent manner. J Nutr 130(4):767–776, 2000.

79. H Gylling, P Puska, E Vartiainen, TA Miettinen. Serum sterols during stanol ester feeding in a mildly hypercholesterolemic population. J Lipid Res 40(4):593–600, 1999.

80. L Normén, P Dutta, A Lia, H Andersson. Soy sterol esters and beta-sitostanol ester as inhibitors of cholesterol absorption in human small bowel. Am J Clin Nutr 71(4):908–913, 2000.

81. PP Nair, N Turjman, G Kessie, B Calkins, GT Goodman, H Davidovitz, G Nimmagadda. Diet, nutrition intake, and metabolism in populations at high and low risk for colon cancer. Dietary cholesterol, beta-sitosterol, and stigmasterol. Am J Clin Nutr 40(4 Suppl):927–930, 1984.

82. AV Rao, SA Janezic. The role of dietary phytosterols in colon carcinogenesis. Nutr Cancer 18(1):43–52, 1992.

83. A Lia, G Hallmans, AS Sandberg, B Sundberg, P Aman, H Andersson. Oat beta-glucan increases bile acid excretion and a fiber-rich barley fraction increases cholesterol excretion in ileostomy subjects. Am J Clin Nutr 62(6):1245–1251, 1995.

84. AS Sandberg, H Andersson, I Bosaeus, NG Carlsson, K Hasselblad, M Harrod. Alginate, small bowel sterol excretion, and absorption of nutrients in ileostomy subjects. Am J Clin Nutr 60(5):751–756, 1994.

85. JX Zhang, E Lundin, H Andersson, I Bosaeus, S Dahlgren, G Hallmans, R Stenling, P Aman. Brewer's spent grain, serum lipids and fecal sterol excretion in human subjects with ileostomies. J Nutr 121(6):778–784, 1991.

86. AM Langkilde, H Andersson, I Bosaeus. Sugar-beet fibre increases cholesterol and reduces bile acid excretion from the small bowel. Br J Nutr 70(3):757–766, 1993.

87. RF Raicht, BI Cohen, S Shefer, EH Mosbach. Sterol balance studies in the rat. Effects of dietary cholesterol and beta-sitosterol on sterol balance and rate-limiting enzymes of sterol metabolism. Biochim Biophys Acta 388(3):374–384, 1975.

88. K Uchida, H Takase, Y Nomura, K Takeda, N Takeuchi, Y Ishikawa. Changes in biliary and fecal bile acids in mice after treatments with diosgenin and beta-sitosterol. J Lipid Res 25(3):236–245, 1984.

89. R Andriamiarina, L Laraki, X Pelletier, G Debry. Effects of stigmasterol-supplemented diets on fecal neutral sterols and bile acid excretion in rats. Ann Nutr Metab 33(5):297–303, 1989.

90. AK Bhattacharyya, DA Eggen. Effects of feeding cholesterol and mixed plant sterols on the fecal excretion of acidic steroids in rhesus monkeys. Atherosclerosis 53(3):225–232, 1984.

91. B Kudchodkar, L Horlick, H Sodhi. Effects of plant sterols on cholesterol metabolism in man. Atherosclerosis 23:239–248, 1976.

92. AM Lees, HY Mok, RS Lees, MA McCluskey, SM Grundy. Plant sterols as cholesterol-lowering agents: clinical trials in patients with hypercholesterolemia and studies of sterol balance. Atherosclerosis 28(3):325–338, 1977.

93. M Becker, D Staab, K Von Bergmann. Treatment of severe familial hypercholesterolemia in childhood with sitosterol and sitostanol. J Pediatr 122(2): 292–296, 1993.

94. JA Weststrate, R Ayesh, C Bauer-Plank, PN Drewitt. Safety evaluation of phytosterol esters. Part 4. Faecal concentrations of bile acids and neutral sterols in healthy normolipidaemic volunteers consuming a controlled diet either with or without a phytosterol ester–enriched margarine. Food Chem Toxicol 37(11):1063–1071, 1999.

95. S Shefer, G Salen, J Bullock, LB Nguyen, GC Ness, Z Vhao, PF Belamarich, I Chowdhary, S Lerner, AK Batta, et al. The effect of increased hepatic sitosterol on the regulation of 3-hydroxy-3-methylglutaryl-coenzyme A reductase and cholesterol 7 alpha-hydroxylase in the rat and sitosterolemic homozygotes. Hepatology 20(1 pt 1):213–219, 1994.

96. H Andersson, I Bosaeus. Sterol balance studies in man. A critical review. Eur J Clin Nutr 47(3):153–159, 1993.

97. KM Boberg, JE Akerlund, I Bjorkhem. Effect of sitosterol on the rate-limiting enzymes in cholesterol synthesis and degradation. Lipids 24(1):9–12, 1989.

98. EE Deschner, BI Cohen, RF Raicht. The kinetics of the protective effect of beta-sitosterol against MNU-induced colonic neoplasia. J Cancer Res Clin Oncol 103(1):49–54, 1982.

99. PA Craven, J Pfanstiel, R Saito, FR DeRubertis. Relationship between loss of rat colonic surface epithelium induced by deoxycholate and initiation of the subsequent proliferative response. Cancer Res 46(11):5754–5759, 1986.

100. RV Cooney, AA Franke, JH Hankin, LJ Custer, LR Wilkens, PJ Harwood, L Le Marchand. Seasonal variations in plasma micronutrients and antioxidants. Cancer Epidemiol Biomarkers Prev 4(3):207–215, 1995.

101. AB Awad, CS Fink. Phytosterols as anticancer dietary components: evidence and mechanism of action. J Nutr 130(9):2127–2130, 2000.

102. AB Awad, YC Chen, CS Fink, T Hennessey. beta-Sitosterol inhibits HT-29 human colon cancer cell growth and alters membrane lipids. Anticancer Res 16(5A):2797–2804, 1996.

103. RN Kolesnick. Sphingomyelin and derivatives as cellular signals. Prog Lipid Res 30(1):1–38, 1991.

104. RN Kolesnick. Sphingomyelinase action inhibits phorbol ester-induced differentiation of human promyelocytic leukemic (HL-60) cells. J Biol Chem 264(13):7617–7623, 1989.

105. RN Kolesnick, S Clegg. 1,2-Diacylglycerols, but not phorbol esters, activate a potential inhibitory pathway for protein kinase C in GH3 pituitary cells. Evidence for involvement of a sphingomyelinase. J Biol Chem 263(14):6534–6537, 1988.

106. T Okazaki, RM Bell, YA Hannun. Sphingomyelin turnover induced by vitamin D3 in HL-60 cells. Role in cell differentiation. J Biol Chem 264(32):19076–19080, 1989.

107. T Okazaki, A Bielawska, RM Bell, YA Hannun. Role of ceramide as a lipid mediator of 1 alpha,25-dihydroxyvitamin D3-induced HL-60 cell differentiation. J Biol Chem 265(26):15823–15831, 1990.

108. YA Hannun, C Luberto, KM Argraves. Enzymes of sphingolipid metabolism: from modular to integrative signaling. Biochemistry 40(16):4893–4903, 2001.

109. S Mathias, LA Pena, RN Kolesnick. Signal transduction of stress via ceramide. Biochem J 335(pt 3):465–480, 1998.

110. RL von Holtz, CS Fink, AB Awad. Beta-sitosterol activates the sphingomyelin cycle and induces apoptosis in LNCaP human prostate cancer cells. Nutr Cancer 32(1):8–12, 1998.

111. R Gopalakrishna, U Gundimeda. Protein kinase C as a molecular target for cancer prevention by selenocompounds. Nutr Cancer 40(1):55–63, 2001.

112. NN Stone, VP Laudone, WR Fair, J Fishman. Aromatization of androstenedione to estrogen by benign prostatic hyperplasia, prostate cancer and expressed prostatic secretions. Urol Res 15(3):165–167, 1987.

113. AB Awad, MS Hartati, CS Fink. Phytosterol feeding induces alteration in testosterone metabolism in rat tissues. J Nutr Biochem 9:712–717, 1998.

114. RR Berges, J Windeler, HJ Trampisch, T Senge. Randomised, placebo-controlled, double-blind clinical trial of beta-sitosterol in patients with benign prostatic hyperplasia. Beta-sitosterol Study Group. Lancet 345(8964):1529–1532, 1995.

115. RR Berges, A Kassen, T Senge. Treatment of symptomatic benign prostatic hyperplasia with beta-sitosterol: an 18-month follow-up. BJU Int 85(7):842–846, 2000.

116. DH Waalkens-Berendsen, AP Wolterbeek, MV Wijnands, M Richold, PA Hepburn. Safety evaluation of phytosterol esters. Part 3. Two-generation reproduction study in rats with phytosterol esters—a novel functional food. Food Chem Toxicol 37(7):683–696, 1999.

117. G Morton, S Lee, D Buss, P Lawrance. Intakes and major dietary sources of cholesterol and phytosterols in the British diet. J Hum Nutr Diet 8:429–440, 1995.

118. AL Normen, HA Brants, LE Voorrips, HA Andersson, PA van den Brandt,

RA Goldbohm. Plant sterol intakes and colorectal cancer risk in the Netherlands Cohort Study on Diet and Cancer. Am J Clin Nutr 74(1):141–148, 2001.

119. SS Strom, Y Yamamura, CM Duphorne, MR Spitz, RJ Babaian, PC Pillow, SD Hursting. Phytoestrogen intake and prostate cancer: a case-control study using a new database. Nutr Cancer 33(1):20–25, 1999.

120. MI Elghamry, R Hansel. Activity and isolated phytoestrogen of shrub palmetto fruits (*Serenoa repens* Small), a new estrogenic plant. Experientia 25(8): 828–829, 1969.

121. T Malini, G Vanithakumari. Effect of beta-sitosterol on uterine biochemistry: a comparative study with estradiol and progesterone. Biochem Mol Biol Int 31(4):659–668, 1993.

122. T Malini, G Vanithakumari. Comparative study of the effects of beta-sitosterol, estradiol and progesterone on selected biochemical parameters of the uterus of ovariectomised rats. J Ethnopharmacol 36(1):51–55, 1992.

123. T Malini, G Vanithakumari. Antifertility effects of beta-sitosterol in male albino rats. J Ethnopharmacol 35(2):149–153, 1991.

124. T Malini, G Vanithakumari. Rat toxicity studies with beta-sitosterol. J Ethnopharmacol 28(2):221–234, 1990.

125. P Mellanen, T Petanen, J Lehtimaki, S Makela, G Bylund, B Holmbom, E Mannila, A Oikari, R Santti. Wood-derived estrogens: studies in vitro with breast cancer cell lines and in vivo in trout. Toxicol Appl Pharmacol 136(2): 381–388, 1996.

126. FA El Samannoudy, AM Shareha, SA Ghannudi, GA Gillaly, SA El Mougy. Adverse effects of phytoestrogens-7. Effect of beta-sitosterol treatment on follicular development, ovarian structure and uterus in the immature female sheep. Cell Mol Biol 26(3):255–266, 1980.

127. D MacLatchy, Gvd Kraak. The phytoestrogen beta-sitosterol alters the reproductive endocrine status of goldfish. Toxicol Appl Pharmacol 134:305–312, 1995.

128. VA Baker, PA Hepburn, SJ Kennedy, PA Jones, LJ Lea, JP Sumpter, J Ashby. Safety evaluation of phytosterol esters. Part 1. Assessment of oestrogenicity using a combination of in vivo and in vitro assays. Food Chem Toxicol 37(1): 13–22, 1999.

129. PA Hepburn, SA Horner, M Smith. Safety evaluation of phytosterol esters. Part 2. Subchronic 90-day oral toxicity study on phytosterol esters—a novel functional food. Food Chem Toxicol 37(5):521–532, 1999.

130. R Ayesh, JA Weststrate, PN Drewitt, PA Hepburn. Safety evaluation of phytosterol esters. Part 5. Faecal short-chain fatty acid and microflora content, faecal bacterial enzyme activity and serum female sex hormones in healthy normolipidaemic volunteers consuming a controlled diet either with or without a phytosterol ester-enriched margarine. Food Chem Toxicol 37(12):1127–1138, 1999.

131. DJ Sanders, HJ Minter, D Howes, PA Hepburn. The safety evaluation of phytosterol esters. Part 6. The comparative absorption and tissue distribution of phytosterols in the rat. Food Chem Toxicol 38(6):485–491, 2000.

132. AM Wolfreys, PA Hepburn. Safety evaluation of phytosterol esters. Part 7. Assessment of mutagenic activity of phytosterols, phytosterol esters and the cholesterol derivative, 4-cholesten-3-one. Food Chem Toxicol 40(4):461–470, 2002.

133. W Jefferson, E Padilla-Banks, G Clark, R Newbold. Assessing estrogenic activity of phytochemicals using transcriptional activation and immature mouse uterotrophic responses. J Chromatogr B Anal Technol Biomed Life Sci 777(1–2):179, 2002.

134. AG Dalgleish, KJ O'Byrne. Chronic immune activation and inflammation in the pathogenesis of AIDS and cancer. Adv Cancer Res 84:231–276, 2002.

135. PJ Bouic. The role of phytosterols and phytosterolins in immune modulation: a review of the past 10 years. Curr Opin Clin Nutr Metab Care 4(6):471–475, 2001.

136. PJ Bouic, A Clark, J Lamprecht, M Freestone, EJ Pool, RW Liebenberg, D Kotze, PP van Jaarsveld. The effects of B-sitosterol (BSS) and B-sitosterol glucoside (BSSG) mixture on selected immune parameters of marathon runners: inhibition of post marathon immune suppression and inflammation. Int J Sports Med 20(4):258–262, 1999.

137. PR Donald, JH Lamprecht, M Freestone, CF Albrecht, PJ Bouic, D Kotze, PP van Jaarsveld. A randomised placebo-controlled trial of the efficacy of beta-sitosterol and its glucoside as adjuvants in the treatment of pulmonary tuberculosis. Int J Tuberc Lung Dis 1(6):518–522, 1997.

138. PJ Bouic, A Clark, W Brittle, JH Lamprecht, M Freestone, RW Liebenberg. Plant sterol/sterolin supplement use in a cohort of South African HIV-infected patients—effects on immunological and virological surrogate markers. S Afr Med J 91(10):848–850, 2001.

139. RS Hogg, B Yip, C Kully, KJ Craib, MV O'Shaughnessy, MT Schechter, JS Montaner. Improved survival among HIV-infected patients after initiation of triple-drug antiretroviral regimens. CMAJ 160(5):659–665, 1999.

140. MB Atkins. Interleukin-2: clinical applications. Semin Oncol 29(3 suppl 7): 12–17, 2002.

141. H Gylling, P Puska, E Vartiainen, TA Miettinen. Retinol, vitamin D, carotenes and alpha-tocopherol in serum of a moderately hypercholesterolemic population consuming sitostanol ester margarine. Atherosclerosis 145(2):279–285, 1999.

142. MA Hallikainen, ES Sarkkinen, MI Uusitupa. Effects of low-fat stanol ester enriched margarines on concentrations of serum carotenoids in subjects with elevated serum cholesterol concentrations. Eur J Clin Nutr 53(12):966–969, 1999.

143. HF Hendriks, JA Weststrate, T van Vliet, GW Meijer. Spreads enriched with three different levels of vegetable oil sterols and the degree of cholesterol lowering in normocholesterolaemic and mildly hypercholesterolaemic subjects. Eur J Clin Nutr 53(4):319–327, 1999.

144. J Plat, EN van Onselen, MM van Heugten, RP Mensink. Effects on serum lipids, lipoproteins and fat soluble antioxidant concentrations of consumption

frequency of margarines and shortenings enriched with plant stanol esters. Eur J Clin Nutr 54(9):671–677, 2000.

145. KC Maki, MH Davidson, DM Umporowicz, EJ Schaefer, MR Dicklin, KA Ingram, S Chen, JR McNamara, BW Gebhart, JD Ribaya-Mercado, G Robins, SJ Robins, WC Franke. Lipid responses to plant-sterol-enriched reduced-fat spreads incorporated into a National Cholesterol Education Program Step I diet. Am J Clin Nutr 74(1):33–43, 2001.

146. AL Amundsen, L Ose, MS Nenseter, FY Ntanios. Plant sterol ester-enriched spread lowers plasma total and LDL cholesterol in children with familial hypercholesterolemia. Am J Clin Nutr 76(2):338–344, 2002.

147. A Sierksma, JA Weststrate, GW Meijer. Spreads enriched with plant sterols, either esterified 4,4-dimethylsterols or free 4-desmethylsterols, and plasma total- and LDL-cholesterol concentrations. Br J Nutr 82(4):273–282, 1999.

148. JT Judd, DJ Baer, SC Chen, BA Clevidence, RA Muesing, M Kramer, GW Meijer. Plant sterol esters lower plasma lipids and most carotenoids in mildly hypercholesterolemic adults. Lipids 37(1):33–42, 2002.

149. FY Ntanios, GS Duchateau. A healthy diet rich in carotenoids is effective in maintaining normal blood carotenoid levels during the daily use of plant sterol-enriched spreads. Int J Vitam Nutr Res 72(1):32–39, 2002.

150. A Sarkar, B Mukherjee, M Chatterjee. Inhibition of 3'-methyl-4-dimethyla-minoazobenzene-induced hepatocarcinogenesis in rat by dietary beta-caro-tene: changes in hepatic anti-oxidant defense enzyme levels. Int J Cancer 61(6): 799–805, 1995.

151. M Noakes, P Clifton, F Ntanios, W Shrapnel, I Record, J McInerney. An increase in dietary carotenoids when consuming plant sterols or stanols is effective in maintaining plasma carotenoid concentrations. Am J Clin Nutr 75(1):79–86, 2002.

152. D Quilliot, F Boman, C Creton, X Pelletier, J Floquet, G Debry. Phytosterols have an unfavourable effect on bacterial activity and no evident protective effect on colon carcinogenesis. Eur J Cancer Prev 10(3):237–243, 2001.

153. JP Cruse, MR Lewin, GP Ferulano, CG Clark. Co-carcinogenic effects of dietary cholesterol in experimental colon cancer. Nature 276(5690):822–825, 1978.

154. JS Crowther, BS Drasar, MJ Hill, R Maclennan, D Magnin, S Peach, CH Teoh-chan. Faecal steroids and bacteria and large bowel cancer in Hong Kong by socio-economic groups. Br J Cancer 34(2):191–198, 1976.

155. BS Reddy, K Watanabe, JH Weisburger. Effect of high-fat diet on colon carcinogenesis in F344 rats treated with 1,2-dimethylhydrazine, methylazox-ymethanol acetate, or methylnitrosourea. Cancer Res 37(11):4156–4159, 1977.

156. N Yanishileva, E Marinova. Autooxidation of sitosterol. I. Kinetic studies on free and esterified sitosterol. La Rivistina Italiana delle Sostanze Grasse 58: 477–480, 1980.

157. N Yanishlieva, E Marinova, H Schiller, A Seher. Comparison of sitosterol autooxidation in free form, as fatty acid, and in triacylglycerol solution. Kinetics of the process and structure of the products formed. 16th ISF conference. Budapest, Hungary: Fat Science, 1983.

158. D Turnbull, VH Frankos, JH van Delft, N DeVogel. Genotoxicity evaluation of wood-derived and vegetable oil-derived stanol esters. Regul Toxicol Pharmacol 29(2 pt 1):205–210, 1999.
159. JL Weihrauch, JM Gardner. Sterol content of foods of plant origin. J Am Diet Assoc 73(1):39–47, 1978.
160. WC Willett. Diet and cancer. Oncologist 5(5):393–404, 2000.
161. E De Stefani, P Boffetta, AL Ronco, P Brennan, H Deneo-Pellegrini, JC Carzoglio, M Mendilaharsu. Plant sterols and risk of stomach cancer: a case-control study in Uruguay. Nutr Cancer 37(2):140–144, 2000.
162. M Mendilaharsu, E De Stefani, H Deneo-Pellegrini, J Carzoglio, A Ronco. Phytosterols and risk of lung cancer: a case-control study in Uruguay. Lung Cancer 21(1):37–45, 1998.
163. A Ronco, E De Stefani, P Boffetta, H Deneo-Pellegrini, M Mendilaharsu, F Leborgne. Vegetables, fruits, and related nutrients and risk of breast cancer: a case-control study in Uruguay. Nutr Cancer 35(2):111–119, 1999.
164. E De Stefani, P Brennan, P Boffetta, M Mendilaharsu, H Deneo-Pellegrini, A Ronco, L Olivera, H Kasdorf. Diet and adenocarcinoma of the lung: a case-control study in Uruguay. Lung Cancer 35(1):43–51, 2002.
165. DM Flood, NS Weiss, LS Cook, JC Emerson, SM Schwartz, JD Potter. Colorectal cancer incidence in Asian migrants to the United States and their descendants. Cancer Causes Control 11(5):403–411, 2000.
166. RG Ziegler, RN Hoover, MC Pike, A Hildesheim, AM Nomura, DW West, AH Wu-Williams, LN Kolonel, PL Horn-Ross, JF Rosenthal, et al. Migration patterns and breast cancer risk in Asian-American women. J Natl Cancer Inst 85(22):1819–1827, 1993.
167. KC Chu, RE Tarone, WH Chow, BF Hankey, LA Ries. Temporal patterns in colorectal cancer incidence, survival, and mortality from 1950 through 1990. J Natl Cancer Inst 86(13):997–1006, 1994.
168. B Trock, E Lanza, P Greenwald. Dietary fiber, vegetables, and colon cancer: critical review and meta-analyses of the epidemiologic evidence. J Natl Cancer Inst 82(8):650–661, 1990.
169. KA Steinmetz, JD Potter. Vegetables, fruit, and cancer prevention: a review. J Am Diet Assoc 96(10):1027–1039, 1996.
170. GE Fraser. Associations between diet and cancer, ischemic heart disease, and all-cause mortality in non-Hispanic white California Seventh-day Adventists. Am J Clin Nutr 70(3 Suppl):532S–538S, 1999.
171. PK Mills, WL Beeson, RL Phillips, GE Fraser. Dietary habits and breast cancer incidence among Seventh-Day Adventists. Cancer 64(3):582–590, 1989.
172. PK Mills, WL Beeson, RL Phillips, GE Fraser. Cohort study of diet, lifestyle, and prostate cancer in Adventist men. Cancer 64(3):598–604, 1989.
173. P Pietinen, N Malila, M Virtanen, TJ Hartman, JA Tangrea, D Albanes, J Virtamo. Diet and risk of colorectal cancer in a cohort of Finnish men. Cancer Causes Control 10(5):387–396, 1999.
174. GR Howe, CM Friedenreich, M Jain, AB Miller. A cohort study of fat intake and risk of breast cancer. J Natl Cancer Inst 83(5):336–340, 1991.
175. TE Rohan, GR Howe, CM Friedenreich, M Jain, AB Miller. Dietary fiber,

vitamins A, C, and E, and risk of breast cancer: a cohort study. Cancer Causes Control 4(1):29–37, 1993.

176. E Giovannucci, MJ Stampfer, G Colditz, EB Rimm, WC Willett. Relationship of diet to risk of colorectal adenoma in men. J Natl Cancer Inst 84(2):91–98, 1992.

177. EA Platz, E Giovannucci, EB Rimm, HR Rockett, MJ Stampfer, GA Colditz, WC Willett. Dietary fiber and distal colorectal adenoma in men. Cancer Epidemiol Biomarkers Prev 6(9):661–670, 1997.

178. KB Michels, G Edward, KJ Joshipura, BA Rosner, MJ Stampfer, CS Fuchs, GA Colditz, FE Speizer, WC Willett. Prospective study of fruit and vegetable consumption and incidence of colon and rectal cancers. J Natl Cancer Inst 92(21):1740–1752, 2000.

179. E Giovannucci, EB Rimm, GA Colditz, MJ Stampfer, A Ascherio, CC Chute, WC Willett. A prospective study of dietary fat and risk of prostate cancer. J Natl Cancer Inst 85(19):1571–1579, 1993.

180. E Giovannucci, A Ascherio, EB Rimm, MJ Stampfer, GA Colditz, WC Willett. Intake of carotenoids and retinol in relation to risk of prostate cancer. J Natl Cancer Inst 87(23):1767–1776, 1995.

181. KA Steinmetz, LH Kushi, RM Bostick, AR Folsom, JD Potter. Vegetables, fruit, and colon cancer in the Iowa Women's Health Study. Am J Epidemiol 139(1):1–15, 1994.

182. RA Goldbohm, PA van den Brandt, P van't Veer, HA Brants, E Dorant, F Sturmans, RJ Hermus. A prospective cohort study on the relation between meat consumption and the risk of colon cancer. Cancer Res 54(3):718–723, 1994.

183. LE Voorrips, RA Goldbohm, G van Poppel, F Sturmans, RJ Hermus, PA van den Brandt. Vegetable and fruit consumption and risks of colon and rectal cancer in a prospective cohort study: the Netherlands Cohort Study on Diet and Cancer. Am J Epidemiol 152(11):1081–1092, 2000.

184. PA van den Brandt, P van't Veer, RA Goldbohm, E Dorant, A Volovics, RJ Hermus, F Sturmans. A prospective cohort study on dietary fat and the risk of postmenopausal breast cancer. Cancer Res 53(1):75–82, 1993.

185. DT Verhoeven, N Assen, RA Goldbohm, E Dorant, P van 't Veer, F Sturmans, RJ Hermus, PA van den Brandt. Vitamins C and E, retinol, beta-carotene and dietary fibre in relation to breast cancer risk: a prospective cohort study. Br J Cancer 75(1):149–155, 1997.

186. AG Schuurman, PA van den Brandt, E Dorant, HA Brants, RA Goldbohm. Association of energy and fat intake with prostate carcinoma risk: results from The Netherlands Cohort Study. Cancer 86(6):1019–1027, 1999.

187. AG Schuurman, RA Goldbohm, E Dorant, PA van den Brandt. Vegetable and fruit consumption and prostate cancer risk: a cohort study in The Netherlands. Cancer Epidemiol Biomarkers Prev 7(8):673–680, 1998.

188. I Kato, A Akhmedkhanov, K Koenig, PG Toniolo, RE Shore, E Riboli. Prospective study of diet and female colorectal cancer: the New York University Women's Health Study. Nutr Cancer 28(3):276–281, 1997.

189. WC Willett, MJ Stampfer, GA Colditz, BA Rosner, FE Speizer. Relation of

meat, fat, and fiber intake to the risk of colon cancer in a prospective study among women. N Engl J Med 323(24):1664–1672, 1990.

190. CS Fuchs, EL Giovannucci, GA Colditz, DJ Hunter, MJ Stampfer, B Rosner, FE Speizer, WC Willett. Dietary fiber and the risk of colorectal cancer and adenoma in women. N Engl J Med 340(3):169–176, 1999.

191. WC Willett, DJ Hunter, MJ Stampfer, G Colditz, JE Manson, D Spiegelman, B Rosner, CH Hennekens, FE Speizer. Dietary fat and fiber in relation to risk of breast cancer. An 8-year follow-up. Jama 268(15):2037–2044, 1992.

192. C Byrne, H Rockett, MD Holmes. Dietary fat, fat subtypes, and breast cancer risk: lack of an association among postmenopausal women with no history of benign breast disease. Cancer Epidémiol Biomarkers Prev 11(3):261–265, 2002.

193. P Terry, L Bergkvist, L Holmberg, A Wolk. No association between fat and fatty acids intake and risk of colorectal cancer. Cancer Epidemiol Biomarkers Prev 10(8):913–914, 2001.

194. P Terry, E Giovannucci, KB Michels, L Bergkvist, H Hansen, L Holmberg, A Wolk. Fruit, vegetables, dietary fiber, and risk of colorectal cancer. J Natl Cancer Inst 93(7):525–533, 2001.

195. A Wolk, R Bergstrom, D Hunter, W Willett, H Ljung, L Holmberg, L Berg-kvist, A Bruce, HO Adami. A prospective study of association of mono-unsaturated fat and other types of fat with risk of breast cancer. Arch Intern Med 158(1): 41–45, 1998.

196. P Terry, A Wolk, I Persson, C Magnusson. Brassica vegetables and breast cancer risk. Jama 285(23):2975–2977, 2001.

197. DJ Hunter, D Spiegelman, HO Adami, L Beeson, PA van den Brandt, AR Folsom, GE Fraser, RA Goldbohm, S Graham, GR Howe, et al. Cohort studies of fat intake and the risk of breast cancer—a pooled analysis. N Engl J Med 334(6):356–361, 1996.

198. SA Smith-Warner, D Spiegelman, SS Yaun, HO Adami, WL Beeson, PA van den Brandt, AR Folsom, GE Fraser, JL Freudenheim, RA Goldbohm, S Graham, AB Miller, JD Potter, TE Rohan, FE Speizer, P Toniolo, WC Wolk, A Wolk, A Zeleniuch-Jacquotte, DJ Hunter. Intake of fruits and vege-tables and risk of breast cancer: a pooled analysis of cohort studies. Jama 285(6):769–776, 2001.

199. WR Bruce, A Giacca, A Medline. Possible mechanisms relating diet and risk of colon cancer. Cancer Epidemiol Biomarkers Prev 9(12):1271–1279, 2000.

200. HG Yu, JA Huang, YN Yang, H Huang, HS Luo, JP Yu, JJ Meier, H Schrader, A Bastian, WE Schmidt, F Schmitz. The effects of acetylsalicylic acid on proliferation, apoptosis, and invasion of cyclooxygenase-2 negative colon cancer cells. Eur J Clin Invest 32(11):838–846, 2002.

201. P Terry, FB Hu, H Hansen, A Wolk. Prospective study of major dietary patterns and colorectal cancer risk in women. Am J Epidemiol 154(12):1143–1149, 2001.

202. AS Whittemore, LN Kolonel, AH Wu, EM John, RP Gallagher, GR Howe, JD Burch, J Hankin, DM Dreon, DW West, et al. Prostate cancer in relation

to diet, physical activity, and body size in blacks, whites, and Asians in the United States and Canada. J Natl Cancer Inst 87(9):652–661, 1995.

203. G Ursin, AH Wu, RN Hoover, DW West, AM Nomura, LN Kolonel, MC Pike, RG Ziegler. Breast cancer and oral contraceptive use in Asian-American women. Am J Epidemiol 150(6):561–567, 1999.

204. BA Stoll. Alcohol intake and late-stage promotion of breast cancer. Eur J Cancer 35(12):1653–1658, 1999.

205. A Tannenbaum, H Silverstone. Nutrition in relation to cancer. Adv Cancer Res 1:451–501, 1953.

206. A Tannenbaum. The genesis and growth of tumors. III. Effects of a high fat diet. Cancer Res 2:468–475, 1942.

207. CM Friedenreich, MR Orenstein. Physical activity and cancer prevention: etiologic evidence and biological mechanisms. J Nutr 132(11 Suppl):3456S–3464S, 2002.

208. SA Missmer, SA Smith-Warner, D Spiegelman, SS Yaun, HO Adami, WL Beeson, PA van den Brandt, GE Fraser, JL Freudenheim, RA Goldbohm, S Graham, LH Kushi, AB Miller, JD Potter, TE Rohan, FE Speizer, P Toniolo, WC Willett, A Wolk, A Zeleniuch-Jacquotte, DJ Hunter. Meat and dairy food consumption and breast cancer: a pooled analysis of cohort studies. Int J Epidemiol 31(1):78–85, 2002.

209. P Terry, R Suzuki, FB Hu, A Wolk. A prospective study of major dietary patterns and the risk of breast cancer. Cancer Epidemiol Biomarkers Prev 10(12):1281–1285, 2001.

210. SS Falkenberry, RD Legare. Risk factors for breast cancer. Obstet Gynecol Clin North Am 29(1):159–172, 2002.

211. AG Schuurman, RA Goldbohm, PA van den Brandt. A prospective cohort study on consumption of alcoholic beverages in relation to prostate cancer incidence (The Netherlands). Cancer Causes Control 10(6):597–605, 1999.

212. LN Kolonel. Fat, meat, and prostate cancer. Epidemiol Rev 23(1):72–81, 2001.

213. WHO. The WHO Health Report. Geneva, Switzerland: World Health Organization, 1997.

214. KT Bogen, GA Keating. US dietary exposures to heterocyclic amines. J Expo Anal Environ Epidemiol 11(3):155–168, 2001.

215. E Giovannucci, EB Rimm, Y Liu, MJ Stampfer, WC Willett. A prospective study of tomato products, lycopene, and prostate cancer risk. J Natl Cancer Inst 94(5):391–398, 2002.

216. AG Schuurman, RA Goldbohm, HA Brants, PA van den Brandt. A prospective cohort study on intake of retinol, vitamins C and E, and carotenoids and prostate cancer risk (Netherlands). Cancer Causes Control 13(6):573–582, 2002.

217. V Piironen, J Toivo, A-M Lampi. Natural sources of dietary plant sterols. J Food Comp Analys 13(4):619–624, 2000.

218. L Normen, M Johnsson, H Andersson, Y van Gameren, P Dutta. Plant sterols in vegetables and fruits commonly consumed in Sweden. Eur J Nutr 38(2):84–89, 1999.

219. RG Ziegler. The future of phytochemical databases. Am J Clin Nutr 74(1):4–5, 2001.
220. K Young-In. Nutrition and Cancer. In: B Bowman, R Russel, eds. Present Knowledge in Nutrition, 8th ed. Washington, DC: International Life Sciences Institute, 2001, pp. 573–589.
221. L Kohlmeier. Problems and pitfalls of food-to-nutrient conversion. In: Food and Health Data. Their Use in Nutrition Policy Making, 34th ed. WHO Regional Publications, European Series, Copenhagen, 1991.
222. M Nelson. The validation of dietary questionnaires. In: B Margetts, M Nelson, eds. Design Concepts in Nutritional Epidemiology. Oxford: Oxford University Press, 1991, pp. 266–291.
223. GR Goldberg, AE Black, SA Jebb, TJ Cole, PR Murgatroyd, WA Coward, AM Prentice. Critical evaluation of energy intake data using fundamental principles of energy physiology: 1. Derivation of cut-off limits to identify underrecording. Eur J Clin Nutr 45(12):569–581, 1991.
224. AH Mokdad, MK Serdula, WH Dietz, BA Bowman, JS Marks, JP Koplan. The continuing epidemic of obesity in the United States. JAMA 284(13):1650–1651, 2000.
225. JH Li, AB Awad, CS Fink, YW Wu, M Trevisan, P Muti. Measurement variability of plasma beta-sitosterol and campesterol, two new biomarkers for cancer prevention. Eur J Cancer Prev 10(3):245–249, 2001.
226. RS Tilvis, TA Miettinen. Serum plant sterols and their relation to cholesterol absorption. Am J Clin Nutr 43(1):92–97, 1986.
227. S Zhang, AR Folsom, TA Sellers, LH Kushi, JD Potter. Better breast cancer survival for postmenopausal women who are less overweight and eat less fat. The Iowa Women's Health Study. Cancer 76(2):275–283, 1995.
228. S Graham, M Zielezny, J Marshall, R Priore, J Freudenheim, J Brasure, B Haughey, P Nasca, M Zdeb. Diet in the epidemiology of postmenopausal breast cancer in the New York State Cohort. Am J Epidemiol 136(11):1327–1337, 1992.
229. SA Smith-Warner, D Spiegelman, HO Adami, WL Beeson, PA van den Brandt, AR Folsom, GE Fraser, JL Freudenheim, RA Goldbohm, S Graham, LH Kushi, AB Miller, TE Rohan, FE Speizer, P Toniolo, WC Willett, A Wolk, A Zeleniuch-Jacquotte, DJ Hunter. Types of dietary fat and breast cancer: a pooled analysis of cohort studies. Int J Cancer 92(5):767–774, 2001.
230. W Willett. Overview of nutritional epidemiology. In: Nutritional Epidemiology, 2nd ed. New York: Oxford University Press, 1998, pp. 1–17.

6

Role of Plant Sterols in Cholesterol Lowering

Lena Normén and Jiri Frohlich
Healthy Heart Program, University of British Columbia,
Vancouver, British Columbia, Canada

Elke Trautwein
Unilever Research and Development, Vlaardingen, The Netherlands

I. INTRODUCTION

There is substantial evidence that an elevated concentration of low-density lipoprotein cholesterol (LDL-C) is one of several important risk factors for the development of coronary heart disease (CHD) (1–7). The increasing public awareness of LDL-C as a risk factor for CHD has created an opportunity for use of phytosterol-enriched products in the marketplace. In the last years, consumers in the United States, Australia, and many European countries have been introduced to phytosterol-enriched food products as novel means to reduce their low-density LDL-C concentrations. The way to communicate the effects of phytosterol-enriched products on health is in part conventional marketing, but also food labeling with "health claims" or "structure function claims" (see Chapter 8 for full explanation) opens up for easy ways to increase public awareness. The commercials and advertisements or more specific claims of food labeling about the beneficial effects of phytosterols on LDL-C and thus on cardiovascular health are based on a large number of clinical and experimental studies.

  In this chapter we will review the evidence for phytosterols as effective LDL-C reducing agents, their metabolism, their potential side effects in relation to cholesterol lowering, as well as their overall implications for prevention of CHD. Implications of phytosterol use in public health programs, as well as primary and secondary prevention, will also be discussed. Information on current trends in CHD mortality and morbidity provides a background against which the effects of phytosterols will be discussed.

II. TRENDS IN CORONARY HEART DISEASE

A. Mortality and Morbidity

The spectrum of CHD includes stable and unstable angina pectoris, major coronary events such as myocardial infarction (MI), and coronary death (8). It is one of the leading causes of death in many developed and developing countries (9,10). Large differences exist between various regions (Fig. 1), with Southwest and Central Asia and Eastern Europe having the highest prevalence of CHD with the more westernized countries being intermediate and the lowest prevalence occurring in Japan (11). In 1999, 12.6 million Americans had CHD; a majority were diagnosed with MI (12). CHD caused 529,659 deaths in the same year, which corresponds to 1 of 5 deaths. In 2002, 1.1 million Americans were estimated to have a new or recurrent coronary event (MI or fatal CHD) (12). During the last decades, mortality from CHD has decreased in many industrialized countries (13–16). In the United States,

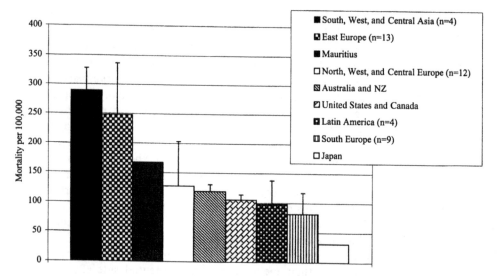

Figure 1 Average age-standardized mortality of ischemic heart disease for men and women combined (11).

death rates have declined approximately 25% during 1996–1998, compared to 1986–1988 (12). In contrast, CHD rates in urban regions of Eastern Europe, Southeast Asia, and India are still on the rise (15,17–19). The trend toward a decreased mortality that has been observed in many industrialized countries is expected to continue into the 21st century (10). With the lower mortality rates, also a decreased incidence of CHD would be expected. Data from the World Health Organization (WHO) Monica study (15), a prospective study in 16 hospitals in Massachusetts (20), the Minnesota Heart Study (21), and the Nurses Health Study (16) all report a reduction in nonfatal events or incidence of CHD, although there was an increase in CHD morbidity in the Athero-sclerosis Risk in Communities Study (13). Improvements in medical care and in secondary intervention may explain the decreased CHD mortality rates, while preventive measures such as changes in lifestyle and dietary pattern as well as increased use of lipid-lowering medications could explain the reduced incidence. The reductions of CHD mortality and incidence should not be interpreted as indicating that CHD will soon be a problem of the past. In fact, the prevalence of CHD is expected to increase due to the fact that our population is aging. Moreover, generally healthier dietary patterns and a reduction in the number of smokers can explain some of the reductions in CHD mortality and morbidity, but this is counteracted by an increase in the prevalence of obesity (16). With the obesity still on the increase among adults and children in both developed and developing countries (22–24), a growing fraction of the world's population is likely to suffer from the metabolic syndrome (syndrome X) and/or diabetes, which substantially increases the risk of CHD. The United States and many other countries will have to deal with the prevention of CHD and other chronic diseases during the next decades in a cost-effective way. Otherwise these disorders will have a major effect on society in terms of costs (12) and lower quality of life (25). The potential of lifestyle and dietary measures to reduce the risk of CHD incidence and mortality is substantial and needs to be considered, as drug treatment may not be applicable from a primary prevention perspective. Health care costs are already very high, and drug treatment of a common and preventable condition in the public seems not to be an optimal way to go.

B. Atherosclerosis

Atherosclerosis—a complex multifactorial, possibly chronic inflammatory condition with multiple steps—is the major cause of CHD. Examples of mod-ifiable risk factors (Table 1) are dyslipidemia (elevated LDL-C or triglyceride concentrations) alone or in combination with low high-density lipoprotein (HDL) concentrations, cigarette smoking, and diabetes mellitus, while pre-disposing risk factors are age, overweight/obesity, physical inactivity, male

Table 1 Categories of Risk Factors to Identify Patients with High Risk of Coronary Heart Disease

Type of risk factor	Expression	Possible to influence with advice
Causative		
	Elevated total cholesterol or LDL cholesterol[a]	x
	Alternative: elevated apoB[a]	x
	Elevated blood pressure	x
	Cigarette smoking	x
	Low HDL[a]	x
	Diabetes mellitus	x
Conditional[b]		
	Triglycerides[a]	x
	Small LDL particles	x
	Lp[a][a]	x
	Homocysteine[a]	x
	Coagulation factors:	
	Plasminogen activating factor inhibitor-1[a]	x
	Fibrinogen[a]	
Predisposing		
	Overweight and obesity[c]	x
	Physical inactivity[c]	x
	Male sex	—
	Family history of premature CHD	—
	Socioeconomic factors	—
	Behavioral factors (e.g., mental depression)	—
	Insulin resistance	x
Susceptibility		
	Left ventricular hyperthrophy	—

[a] Concentrations in blood/serum or plasma.
[b] Conditional risk factors are considered as risk factors only when concentrations in blood are abnormally high.
[c] Considered as major risk factors by the American Heart Association.
Source: Ref. 8.

gender, family history of CHD, socioeconomic status, and insulin resistance (8). Several of these factors are modifiable by a healthy lifestyle and diet (26).

One of the theories concerning the etiology of atherosclerosis has been the "oxidation hypothesis." This suggests LDL entrapment in the extracellular matrix of the subendothelial space of the artery wall results in a cascade of events that ultimately leads to uptake of oxidized LDL by macrophages and the formation of lipid-containing foam cells (27). The oxidation hypothesis has lost some support over the last few years, due to several clinical trials with antioxidant supplementation (e.g., β-carotene, vitamin E, vitamin C), which have not been able to demonstrate a preventive effect on CHD in populations with low or high risk of CHD (28–31). However, it could be argued that antioxidants could prevent CHD from occurring if consumed throughout life, and not only at an older age, but the lack of support from the clinical trials has stimulated a debate in which alternative hypotheses have been promoted. Instead, a growing interest in low-grade inflammation of the endothelium has emerged. This process involves white blood cells (32), surface-selective adhesion molecules (32), and leukocytes (33,34). The subclinical inflammation seen in early atheromas in humans is likely the result of damage to endothelial cells caused by dyslipidemia (33). The fact that impaired endogenous atheroprotective mechanisms occur at branch points in arteries, where endothelial cells are exposed to a disturbed flow (33), suggests that homeostatic processes involving nitrogen oxide are also involved in the pathogenesis of atherosclerosis.

Atherosclerosis consists of two different phenomena: atherosis and fibrosis (35). Atherosis in infants has been described as yellow dots at the root of the aorta, whereas at puberty it consists of more extensive yellow streaks in many parts of the aorta and in the coronary arteries. Microscopically, early dots and streaks consist mostly of intracellular but also some extracellular lipids in the intima. They differ from adult lesions by the absence of fibrosis. In adults, atherosis and fibrosis form fibrous plaques, representing a more advanced stage of atherosclerosis. Production of extracellular fibrous material is a reparative mechanism in response to the increasing lipid accumulation and rearrangement of the structure of the intima (35,36). Some of these lesions later get calcified, and some develop into lipid-rich unstable lesions with a fibrous cap, which cause the majority of acute events of CHD.

Populations predisposed to CHD, i.e., subjects with familial hypercholesterolemia, adult smokers, and male body builders taking anabolic-androgenic steroids (37–40) all show signs of endothelial dysfunction, which is an early functional abnormality in the preclinical stage of atherosclerosis. A number of factors can cause endothelial dysfunction, but there is still some controversy regarding the mechanism. In long-term studies of statin treatment, endothelial function improved (41,42). This implies either that lowering of LDL concentrations improves endothelial function or that statins have a beneficial effect on the endothelium. The latter is supported by an in vitro observation that atorvastatin simulates NO production in cell lines in the absence of LDL (43). However, LDL reduces nitric oxide synthase through the modulation of caveolin in endothelial cells in a dose-dependent manner (44), which supports the possibility that LDL also has a direct adverse effect on endothelial function. In contrast, elevated HDL ameliorates abnormal vasoconstriction at any given extent of atherosclerotic wall thickening (45).

Atherosclerosis and CHD have long been regarded as diseases of middle-aged and older populations. Most studies have been done in older individuals who already show signs of CHD, However, the problems start early in life. The goal of therapy later in life is to stabilize the plaques, whereas the main goal in early adulthood is to prevent the formation of plaques (26).

C. LDL: A Causal Risk Factor for Coronary Heart Disease

Epidemiological evidence shows that LDL-C is closely linked to the incidence of CHD. In populations with very low LDL-C concentrations (<100 mg/dl or <2.6 mM), CHD is nearly nonexistent (46). Populations with higher LDL-C concentrations have more CHD than those with lower concentrations (47, 48). When people migrate from regions where LDL-C concentrations are generally low to areas with average higher levels, they adapt to a more Western lifestyle, which increases their cholesterol concentrations and subsequently their CHD risk (49,50). However, the ultimate proof of a connection between LDL-C and CHD is found in individuals with the genetic forms of hypercholesterolemia (i.e., familial hypercholesterolemia) and grossly elevated LDL-C concentrations, who have an accelerated rate of atherosclerosis and premature CHD (51).

In longitudinal population studies, total cholesterol has been used as a surrogate marker of LDL-C concentrations. The Lipid Research Clinics Trial (1,2), the Multiple Risk Factor Trial (52), and the Framingham Heart Study (53) all demonstrated an association between total cholesterol or LDL-C and incidence of new cases of CHD. The same pattern is seen with recurrent coronary events in people with established CHD with an increasing frequency of repeat events in those with higher LDL-C concentrations.

The best proven link between LDL-C lowering and CHD are the statin trials, which demonstrate that LDL-C lowering reduces the incidence of CHD and total mortality, MI, as well as related conditions (3,54–58). Based on these trials and other available studies, concentrations of total cholesterol and LDL-C according to level of risk have been identified (Table 2). Lowering of LDL-C both in people with and without CHD is therefore the most important strategy for primary and secondary prevention.

D. Effects of Diet on LDL-C and Coronary Heart Disease

A healthy lifestyle affects LDL-C concentrations in a favorable way and includes healthy dietary habits, adequate physical exercise, and a body weight within the recommended range. Even small individual differences in LDL-C

Table 2 Definitions of Total Cholesterol and LDL-C Concentrations According to the Adult Treatment Panel III

	Total cholesterol			LDL-C (mM)	
	(mg/dl)	(mM)		(mg/dl)	(mM)
			Optimal	< 100	< 2.6
Desirable total cholesterol	< 200	< 5.2	Near or above optimal	100–129	2.6–3.3
Borderline high	200–239	5.2–6.19	Borderline high	130–159	3.31–4.1
High	≥240	≥6.2	High	160–189	4.1–4.9
Very high	—	—	Very high	≥190	≥4.9

According to the ATP III, no or little atherosclerosis occurs at LDL concentrations below 100 mg/dl, or 2.6 mM.

lead to substantial differences in CHD risk at a population level, and early changes have large implications for quality of life at an older age. More specifically, and as estimated from observational and intervention studies, for every percentage point of LDL-C reduction at a population level, the risk of CHD decreases by 1–2% (59). The major LDL-C-raising dietary constituents are saturated fatty acids, *trans* fatty acids, and, to a minor extent, dietary cholesterol (60,61). For every percentage point of increase in the total energy from saturated fatty acids, serum LDL-C increases by about 2% (62). When a typical American diet with 34% of total energy from fat (i.e., energy percent, E%) and with 15% representing saturated fat was compared to a diet with less fat (29 E% of total fat, 9 E% of saturated fat), LDL-C was reduced by 7% (63). A recent meta-analysis of the effects of dietary cholesterol shows that an increased intake of 100 mg/day would elevate total cholesterol by 0.056 mmol/L (2.2 mg/dl) (64).

   trans-Fatty acids have effects that are similar to that of saturated fatty acids, albeit even more adversely on serum HDL (65). Replacement of saturated fat with unsaturated fat [monounsaturated fatty acids (MUFAs) and polyunsaturated fatty acids (PUFAs)] has been shown to lower LDL-C (62). Substitution of fat with carbohydrates will also lower LDL-C but raises TG concentrations (66). All three classes of fatty acids (saturated, PUFAs and MUFAs) elevate HDL concentrations, when they replace carbohydrates (62), though the effect is stronger with MUFAs than PUFAs at a relatively high intake (>10 E%) (62).

   Recommended diets for CHD prevention differ slightly in their distribution of nutrients, but most of them are similar to that advocated by the American National Cholesterol Education Program (26). According to the most recent version of these guidelines, total and saturated fat should be limited to 25–35 E%, and <7 E%, respectively. Polyunsaturated and mono-unsaturated fatty acids can be consumed up to 10 and 20 E%, respectively (26). A meta-analysis of dietary intervention studies based on the recommendations from the NCEP suggests that if people follow the NCEP diet, LDL-C concentrations can be lowered by 16% (67). A pooled analysis of 19 randomized trials investigating the dietary effect on cholesterol concentrations showed that the percentage reduction in blood total cholesterol attributable to dietary advice (after at least 6 months of intervention) only was 5.3% (68). Including both short- and long-term studies, the effect was 8.5% at 3 months and 5.5% at 12 months, which suggests that compliance to dietary advice diminishes with time, unless the intervention is repeated frequently. A meta-analysis of primary and secondary CHD prevention trials showed that total mortality and CHD incidence was reduced by 6% and 11%, respectively, with dietary intervention (69).

Dietary advice has proven to be useful compared to the insulin-sensitizing drug metformin in preventing diabetes when patients are followed frequently by dietitians (70). Still, it seems to have a more limited role in primary prevention, which normally is based on a limited number of events of dietary counseling, public health campaigns, or other less individualized approaches. Changes in food habits are difficult to achieve without persistent and repeated advice; rather they occur over the course of decades in response to impulses other than nutrition recommendations. Phytosterol-enriched food products have been shown to improve the effect of diet on LDL-C concentrations irrespective the quality of the habitual diet (71). Nevertheless, such products should always be consumed in the context of a heart-healthy diet based on nutrition recommendations.

III. METABOLISM AND PHYSIOLOGICAL EFFECTS OF PHYTOSTEROLS

The cholesterol-lowering properties of phytosterols (PSs) were discovered already in 1951, when Peterson fed soybean sterols to chickens on a high-cholesterol diet and found that the expected elevation of blood cholesterol concentrations did not occur (72). Since then many studies addressing physiological effects in both humans and animals have been performed. In the early trials, the dominant phytosterol was the unsaturated sitosterol, and therefore we will focus on this compound. Furthermore, campesterol and stigmasterol, as well as the saturated phytosterols sitostanol and campestanol, will also be considered. For a full understanding of the effects of phytosterols on plasma cholesterol, some basic information on the mechanism of absorption of cholesterol and phytosterols will also be addressed.

A. Lipoprotein Metabolism

Cholesterol is a steroid synthesized by all human tissues in a process via two pathways (Fig. 2) (73). It is a precursor for bile acids and steroid hormones (73), and has important structural functions in the cell membranes of most human and animal cells (74). Humans once relied heavily on their endogenous cholesterol production, as a majority of prehistoric diets were likely to be low in saturated fat and dietary cholesterol (< 50 mg/day) (74). Today the situation is different: dietary cholesterol decreases the need for the relative endogenous production. Despite this, the endogenous production still represents the major source of cholesterol in the human body pool, which is estimated to be around 75 g (75). The total turnover is 1200 mg/day, with the endogenous production in the whole body representing 60–75% and dietary intake 25–40% (75). The average American's dietary cholesterol intake is within the range of 180–290 mg/day for children (76) and 230–340 mg/day for adults (77).

As reviewed by Lichtenstein and Jones (78) and Schaefer (79), lipoprotein metabolism consists of an exogenous and an endogenous pathway. The exogenous pathway consists of lipid absorption and lipid transport from the intestine to the liver. Following emulsification and hydrolysis by lipases (80), long-chain fatty acids and dietary cholesterol are absorbed by the enterocytes. Intestinal fatty acid binding protein (FABP2) and microsomal triglyceride transfer protein (MTP) are essential for the uptake, intracellular metabolism, and assembly and secretion of chylomicrons. The main function of chylomicrons is to transport dietary triglycerides, cholesterol, and other lipophilic compounds from the intestine to tissue where these nutrients are metabolized. Apolipoprotein B48 (apoB48) is a hydrophobic protein that is incorporated into chylomicrons before secretion. After being released from mucosal cells, chylomicrons enter the lymphatic system, reach the *superior vena cava*, and then reach the peripheral circulation. The enzyme lipoprotein lipase (LPL) located on the capillary walls (activated by apoC-II), catalyzes the hydrolysis of triacylglycerols (TGs) into fatty acids to allow for uptake of the hydrolyzed fat. Adipose tissue and muscle are the principal sources of LPL and take up the vast majority of exogenous fat. Following the uptake of fatty acids, the cholesterol-enriched chylomicron remnants are subsequently taken up by the lipoprotein receptors in the liver. There the cholesterol may be incorporated in very-low-density lipoproteins (VLDLs) and excreted into the circulation. Apolipoprotein B48 and apoE are involved in the receptor uptake of chylomicron remnants of the liver. The endogenous pathway consists of the cycle of lipids between the liver and the peripheral tissues. MTP aids in the production of VLDL from fatty acids synthesized in the liver or taken up from the bloodstream, hepatic TG stores, and lipoprotein remnants. After entering the bloodstream, the TGs in VLDL are catalyzed by LPL, which results in the

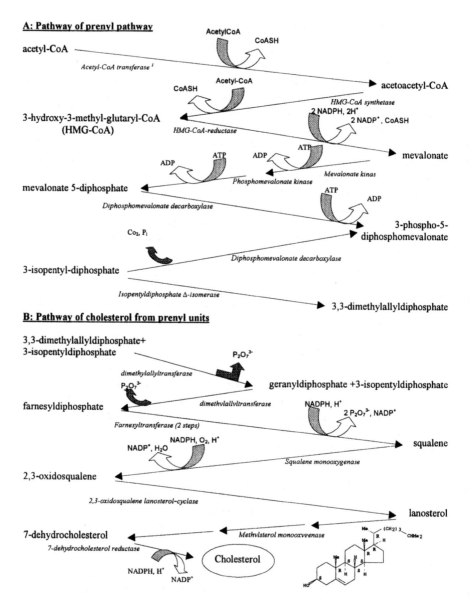

Figure 2 Pathways of endogenous cholesterol synthesis. (From Ref. 73. Reproduced with permission from John Wiley and Sons.)

formation of intermediate-density lipoprotein (IDL). IDL is either taken up by the liver, or transformed to LDL by LPL or hepatic lipase. LDL is taken up by peripheral or liver cells through the aid of apoB-100. Cholesterol is transported from peripheral tissues via the reverse cholesterol transport pathway in which nascent high-density lipoprotein (HDL) particles mature and are taken up by the liver. For more detailed descriptions, consult the excellent reviews by Lichtenstein and Jones (77) or Ernst J. Schaefer (79) on lipoprotein metabolism in relation to nutrition and CHD.

B. Sterol Absorption Mechanisms and Mutations

Originally, cholesterol absorption was thought to only occur by passive diffusion aided by the concentration gradient of free cholesterol created between the extra- and intracellular space or by cholesterol in bile. However, several lines of evidence indicate that cholesterol absorption to a certain extent is a protein-mediated process involving several receptors and transporters, which facilitate cholesterol influx into enterocytes or flux into the enterocyte/efflux back to the intestinal lumen. However, passive diffusion is not completely ruled out, as there may be some cholesterol entering the enterocyte this way, especially following a fatty meal (81).

The dietary cholesterol intake of the American population (77) corresponds only to a small amount compared to the endogenous secretion from bile, which is within the range of 800–1200 mg/day (82). Measurements with stable isotopes indicate that 29–80% of the cholesterol is absorbed (83), where the absorption declines with increased cholesterol intake (84).

Two enterocyte membrane proteins have been suggested to facilitate intestinal cholesterol absorption: scavenger receptor class B type 1 (SR-B1) and cell cluster determinant antigen 36 (CD36). The first protein is a receptor for HDL and modified LDL (85,86). Some cell and animal studies suggest that SR-B1 can mediate cholesterol absorption (87,88), but studies with SR-B1 knockout mice suggest that SR-B1 is not essential for intestinal cholesterol absorption (89). The second protein, CD36, has been suggested to have a similar effect to SR-B1 in mediating the cholesterol absorption (90). In general, the data on SR-B1 and CD36 are not conclusive, but it appears as if they both may contribute to cholesterol absorption rather than controlling it.

Cholesterol absorption may also depend on proteins for reverse cholesterol transport back the lumen after uptake of the enterocyte. Earlier work on HDL metabolism has revealed that the adenosine triphosphate binding cassette A1 (ABCA1) protein is a transporter mediating cholesterol efflux from peripheral tissues (91). Subsequent studies have pointed out that the ABCA1 transporter may also be active in the small bowel (92,93). Nuclear hormone receptors such as oxysterol receptors (LXRs) and the bile acid receptor (FXR) form obligate heterodimers with retinoid X receptors (RXR) and are activated by RXR agonists such as rexinoids. Animals treated with rexinoids exhibit inhibition of cholesterol absorption, a fact that supports the hypothesis of ABCA1 transporter involvement in intestinal cells (92,93). Knockout mice with the gene lacking for ABCA1 had a 62% increase in cholesterol absorption compared to the wild strain (94). The recent finding that ABCG5 and ABCG8 cooperate to limit the intestinal absorption of sterols and to promote biliary excretion of sterols (95) indicates that other ABC transporters also may be involved in sterol absorption. Cholesterol feeding increases the relative mRNA level for ABCA1, ABCG5, and ABCG8 in the jejunum (96), which is one of the main sites for cholesterol absorption (81). Their relative importance in the human gut remains to be determined.

The cholesterol portion, which is taken up by the enterocyte but escapes reverse cholesterol transport, is esterified by acyl CoA: cholesterol acyltransferase 2 (ACAT2) (97). The existence of SR-B1, ABC transporters, and ACAT2 suggests that the intestinal cholesterol absorption is a highly regulated process with implications for overall sterol homeostasis in the human body.

Sitosterolemia is a very rare autosomal recessive disorder clinically characterized by tendon and tuberous xanthomas, premature atherosclerosis, and premature death (98–101). Biochemical abnormalities include elevated concentrations of sterols, particularly phytosterols, suggesting an increased phytosterol absorption. Sitosterolemia is caused by mutations in two adjacent genes encoding coordinately regulated ATP-binding cassette (ABC) half transporters (ABCG5 and ABCG8) (102), and in the homozygote form the clearance of phytosterol is impaired (98). The existence of the mutation indicates that both cholesterol and sitosterol in normal subjects undergo reverse transport (efflux) after being taken up by the gastrointestinal cell.

C. Mechanism of Action by Phytosterols

The main physiological response to ingestion of phytosterols is suppressed cholesterol absorption. Several possible mechanistic steps have been proposed, one being the competition for micellar space in the small bowel between phytosterols and cholesterol (103,104), and others focusing on effects at the epithelial cell level (105).

For cholesterol to be absorbed, it needs to be incorporated into micelles. Phytosterols reduce cholesterol micellar incorporation when added simultaneously with cholesterol and thereby reduce its availability for absorption (106). Mixed bile salt micelles solubilize both cholesterol and β-sitosterol to a comparable extent (107). Partition of free cholesterol and sitosterol between the oil and the micellar phase is approximately the same when the system is not saturated (107), and suggests a competition for space in the micelle.

When the lipid phase gets saturated before or during micellar incorporation, crystallization may take place, and differences between the sterols may explain the reduced absorption further. Sitosterol is known to crystallize and coprecipitate with cholesterol, whereas sitostanol only forms eutectic mixtures (108). Due to the lack of difference in efficacy following intake of unsaturated and saturated phytosterols (109,110), these in vitro observations seem to have little importance for in vivo conditions.

Plat and Mensink (105), on the other hand, recently showed that there is an increased expression of ABCA1 in intestinal cells following absorption of phytostanols. This indicates stimulation of cholesterol efflux by phytosterols. Cholesterol absorption is a slow process, and Beaumier-Gallon showed that most of the dietary cholesterol only enters the bloodstream following a second meal rather than being absorbed directly in the postprandial state (111). This finding suggest that there is a time span following uptake of the enterocyte, when phytosterols can diminish cholesterol absorption further due to an upregulation of ABCA1 and subsequent increased sterol efflux.

Caco-2 cells have been used to examine the effect of β-sitosterol on cholesterol trafficking, cholesterol metabolism, and apoB secretion (112). Cells incubated with micelles containing only cholesterol showed an increase in the influx of plasma membrane cholesterol to the endoplasmic reticulum and increased secretion of cholesteryl esters derived from the plasma membrane. β-Sitosterol did not alter cholesterol trafficking or cholesteryl ester secretion. Inclusion of β-sitosterol in the micelle together with cholesterol attenuated the influx of plasma membrane cholesterol and prevented the secretion of cholesteryl esters derived from the plasma membrane. Among several phytosterols, only campesterol did not promote an influx of plasma membrane cholesterol. This study reported that inclusion of β-sitosterol in the micelle with cholesterol reduced the uptake of cholesterol by 60% (112).

D. Effect of Phytosterols on Cholesterol Absorption

The normal phytosterol intake with Western-type diets is in the range of 100–400 mg/day (113–119) but is normally considered to be too small to have a meaningful effect on cholesterol absorption. However, when studying the variation of phytosterol content of controlled dietary interventions in ileostomy subjects, there is a relationship between the phytosterol content of foods and cholesterol absorption (120). The inhibiting effect on cholesterol absorption by the phytosterol content of an edible oil was demonstrated by Ostlund et al, who compared a phytosterol-free corn oil with one in which the phytosterols had been added back to the oil. This led to a substantial reduction of cholesterol absorption (121). Corn oil has a high concentration of phytosterol (>1 g/100 g oil), and the tested doses (150 and 300 mg) in the mentioned study (121) can be found in 1–3 tablespoons of corn oil per day (122). The high phytosterol content of corn oil also lead to a 6% lower LDL-C concentration compared to olive oil (121), which has a lower phytosterol content (123).

The reduction of cholesterol absorption and subsequently LDL concentrations can, however, be enhanced by higher doses. The effect of supplemented phytosterols on cholesterol absorption has been evaluated by several groups (98,109,110,124–142) and is determined by the dose and the physical state of the phytosterol. The amount of dietary cholesterol and the amount of biliary cholesterol secreted in the gut could theoretically also affect the reduction. The reduction in fractional cholesterol absorption occurs both with a high (> 0.5 g/day) (109,125,126), moderate (0.3–0.5 mg/day) (132,134) and low dietary cholesterol intakes (<0.3 g/day) (127–129). Figure 3 gives an overview of 13 studies in which cholesterol absorption has been measured during controlled conditions (109,110,125–132,134,135,139). Only studies with supplemented phytosterols, five or more subjects, and a control group have been included. Infusion studies have not been included as the technique in itself promotes peristalsis, which increases gut passage and therefore may effect the absorption (143). As can be seen from the summary of the results, the efficacy partly depends on the physicochemical properties, which are determined by whether or not the phytosterols are ingested in a free form, as esters, or solubilized in phospholipid micelles. The design of the studies varies together with the extent of control of sterol intake, but the figure can be used for an evaluation of the efficacy on cholesterol absorption. In general, it appears as if doses of free and esterified phytosterols of less than 1 g have less effect. The largest fractional reductions were observed with doses of 3 g of esterified phytosterols (expressed as sterol equivalents) (132,134). The micellar phytosterols have large effects in small doses, however, 0.6, 0.8, and 3.0 g/day gave a similar reduction, though in different studies (129–131).

The data used for Figure 3 indicate that free sterols have effects similar to those of esterified phytosterols when comparing phytosterol doses less than 2.0 g/day. However, this topic has been a matter of debate and determined by whether or not the free sterols are solubilized in the gut or not. In an early direct comparison between free phytosterols with the esterified ones, the free form was more efficient (126). A later comparison of low doses found that free sitosterol and sitostanol ester inhibited equally well, whereas subjects did not respond well to free sitostanol (128). However, free phytosterols normally have a low solubility, which limits their use, as crystallization of free phy-

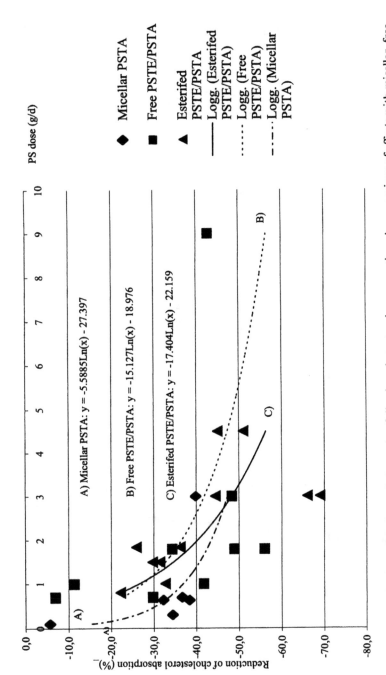

Figure 3 Reduction of cholesterol absorption following phytosterol consumption. A comparison of effects with micellar, free, and esterified phytosterols.

Table 3 Absolute and Fractional Sterol Absorption

Subjects	n	Absorption method	PS intake (mg)	PS absorption		Ref.
				(%)	(mg)[a]	
Healthy	7	Intubation study[b] (mixed PS in liquid formula diet)	1. 0.075 cholesterol	24/h	0.36/h	148
			2. 0.075 sitosterol	16/h	0.24/h	
			3. 0.0375 cholesterol + 0.0375 sitosterol	16/h	0.012/h	
				16/h	0.012/h	
Healthy	10	Intubation study (mixed PS in liquid formula diet)	Sitosterol 21.1/h	4.2/h	0.91/h	149
			Stigmasterol 1.6/h	4.8/h	0.08/h	
			Campesterol 4.4/h	9.6/h	0.46/h	
			Campestanol 0.5/h	12.5/h	0.07/h	
Healthy	4	Recovery of sitosterol isotopes in plasma	Sitosterol 330/d	2.8/d	8.4/d	150
Healthy	1	Plasma dual-isotope ratio[c] (PS from diet)	Sitosterol 1909/d	0.6/d	11.5/d	98
Healthy	1		Sitosterol 250/d	4/d	10.0/d	
Sitosterolemic	2	(PS from diet)	Sitosterol 250/d	28/d	70.0/d	
			Sitosterol 250/d	63/d	158/d	
Healthy	3	Fecal dual-isotope ratio[d] (PS from diet)	Sitosterol 145/d	6/d	8.2/d	137
			Campesterol 27/d	19/d	5.0/d	
Sitosterolemic	3		Sitosterol 145/d	16/d	23.7/d	
			Campesterol 27/d	27/d	7.2/d	

Group	N	Method	Sterol	Intake	Absorption		Ref.
Healthy	3	Plasma dual-isotope ratio	Campestanol	540/d	0.1/d[e]	0.70/d	151
Sitosterolemic[f]	1	(PS from diet in the homozygote sitosterolemic, and from supplements in the rest)	Campestanol	2/d	31/d	0.67/d	
Sitosterolemic[g]	1		Campestanol	540/d	0.3/d	2.10/d	
Healthy	10	Plasma dual-isotope ratio	Sitosterol	236	0.51	1.20/d	152
	10	(Comparing two test meals with 600 mg supplemented sterols or stanols)	Campesterol	161	1.90	3.04/d	
			Sitostanol	436	0.044	0.19/d	
			Campestanol	142	0.16	0.22/d	

[a] Calculated from the phytosterol intakes and measured absorption rates.
[b] Intestinal perfusion technique, where samples are aspirated in the gut. Absorption calculated with the ratio between administered sterols to sitostanol as the nonabsorbable marker.
[c] Builds on the use of two differently marked isotopes (preferably stable), where one is consumed and the other injected in plasma. Absorption is calculated as the (ratio of injected/ingested tracers)/administered dose × 100.
[d] Phytosterols marked with isotopes given together with sitostanol as a nonabsorbable marker. Absorption calculated from the intestinal disappearance of phytosterols vs. the nonabsorbable marker ([²H₄] sitostanol).
[e] In this particular study, max absorption (at a single time) was also measured. This was 5.5% for the healthy subjects, 80% for the homozygote, and 14.3% for the heterozygote. This indicates that daily absorption rate is lower than if absorption is measured in a short time interval.
[f] Homozygotes.
[g] Heterozygotes.
PS, phytosterol.

tosterols may be a problem with higher doses. For example, in the study by Lees *et al.*, a dose of 9 g/day was used, giving a slightly lower reduction of cholesterol absorption than the 3-g dose (125). Newer formats are, however, built on microcrystallization, which improves solubilization and gives products LDL-lowering efficacy (144). Low doses of free phytosterols have similar efficacy as free phytostanols on cholesterol absorption (128,139), and esterified phytosterols have the same efficacy as esterified phytostanols (109,110). Micellar phytosterols seem to have the highest efficacy when compared with free phytosterols and phytosterol esters. Phytosterol micelles prepared with phospholipid gave reductions of cholesterol absorption in the range 13–56%/g (129–131).

E. Quantification of Phytosterol Absorption

Early studies by Schönheimer suggested that phytosterols were not absorbed in mammals (145,146). However, recent evidence of increased plasma concentrations following phytosterol supplementation in clinical trials indicates at least a partial absorption (110,147). However, normal subjects show, less than a doubling of the plasma phytosterol concentrations despite daily intakes that have been increased to doses of 3–18 g/day (125,147). The absolute magnitude of the phytosterol absorption has been controversial. There are only a few studies providing quantitative information about the phytosterol absorption in humans: all are based on different methods, phytosterol sources, as well as tested doses, supporting a wide range of fractional phytosterol absorption (98,137,148–152). Estimates of the fractional phytosterol absorption in healthy individuals range between 0.6% to 28% for absorption of unsaturated and 0.044% to 12.5% for the saturated ones (Table 3). The absorption rate is often expressed as a percentage of the intake, which creates an impression of a relatively high absorption when the dietary intake of phytosterols is low. By changing the percentages into absolute amounts of phytosterols (mg), the absorption range becomes narrower. In healthy subjects where phytosterols have been provided over the course of the day, absorption is usually less than 10 mg/day depending on the method and daily dose. Available intubation studies of phytosterol absorption (148,149) have little relevance as the used doses of phytosterols have been low and far below the solubility threshold. Therefore, these experiments do not resemble physiological conditions where phytosterols are part of a meal ingested in a bolus. Currently, studies using isotope techniques are preferred to estimate fractional and absolute phytosterol absorption. The most recent study of phytosterol absorption detected sitosterol absorption to be 0.5% and campesterol to 1.9% (152)—much lower than previous estimates. The authors suggest that the use of radioactive isotopes (98,151) tends to result in overestimation of the fractional and sometimes also the absolute absorption of phytosterols. This is because the activity of the radioactive label in plasma is assumed to be structurally identical to the supplemented isotope molecule. Taking into account that radioactive tracers are seldom pure, either before or after labeling, there is a risk of misclassification of the sterols when analyzed. However, advances in analytical procedures now allow exact characterization of tracer material (152). By calculating the absolute phytosterol absorption using the rate measured by Ostlund et al. and using the dietary intake of phytosterols as described in the Netherlands Cohort Study on Diet and Cancer (116), the summarized amount of sitosterol, campesterol, stigmasterol (assuming it has the same absorption as sitosterol), sitostanol, and campestanol would be approximately 2.1 mg/day. Taken together, these data suggest that absorption in healthy individuals is very low, even when the absorption of phytosterols is improved by emulsification with lecithin (152) or bile acids (148).

After uptake by the enterocyte, both phytosterols and cholesterol are esterified via (ACAT). The reaction depends on prior CoA activation of fatty acids (153). ACAT is up-regulated by diets high in cholesterol (154). The ACAT esterification rate of phytosterols is about 60 times lower than that of cholesterol (155), thus partly explaining the relatively low intestinal uptake of phytosterols by enterocytes. The absorbed sterols are exported into intestinal lymph and can enter the circulation as chylomicrons (quantitatively most important), very-low-density lipoprotein (VLDL), or nascent HDL (156).

The fractional and absolute absorption of phytosterols in sitosterolemia shows large variation from case to case (98,137,151), but is in general substantially greater in homozygotes. Fractional absorption rates of 14–68% has been observed, though results are difficult to interpret as most homozygotes consume phytosterol-restricted diets and therefore the absolute amounts absorbed are small. The increased phytosterol absorption in sitosterolemia leads to an exaggerated body pool of phytosterols. The total body pool of sitosterol has been estimated by Salen et al. to be in the range of 70–290 mg for normal subjects during average phytosterol consumption (98,150), whereas in two sitosterolemic subjects the pool was 3500 mg and 4800 mg, respectively (98). Following ingestion of a phytosterol-enriched spread (540 mg campestanol per day), the campestanol pool contained only 9.4 ± 7.6 mg in three control subjects, compared with 264 mg in a homozygous sitosterolemic subject (151). In the last case, the subject's heterozypous mother had a campestanol pool similar to the control subjects (27 mg).

F. Postabsorptive Fate of Sterols

Following their absorption, sterol-containing chylomicrons and nascent HDL enter the bloodstream (157). Cholesterol concentrations in apoB-48-containing chylomicrons are moderate after a first meal, reach a maximum after the second meal, and are still elevated after a third meal (111). A marked and stepwise increase in the concentration of plasma total cholesterol occurs during the first 6 h after a meal. LDL and HDL enrichment of cholesterol parallels that of plasma but is threefold and plateaus after 24 h (111). Data on the distribution rate of phytosterols are scarce, but when comparing the distribution of sitosterol and campesterol between different lipoprotein subclasses in a fasting state following a normal consumption of phytosterols, 69–70% are found in LDL, 24% in HDL, and 6% in VLDL (158). Approximately 70% of the sterols in LDL are esterified (sitosterol, campesterol, and cholesterol ester), whereas 80% of the sterols in HDL are esterified compared to 50% of the sterols in VLDL.

Similar to cholesterol, phytosterols are excreted via the route of bile (159), with a minor fraction excreted through the skin (160,161). In contrast to that of cholesterol, it has yet to be determined whether or not transformation of phytosterols to bile acids occurs, as data are conflicting. Salen and Grundy (150) suggested that phytosterols could be converted to bile acids, and found that approximately 20% of phytosterols had been directly converted to primary bile acids, as indicated by measurement of radioactivity in the bile acids fraction. This finding was later contradicted by Boberg et al., who administered [4-^{14}C]sitosterol intravenously to two healthy subjects (162). They found no labeled C24 bile acid products appearing in bile and concluded that humans have little or no capacity to convert sitosterol to C24 bile acids.

Sterols have varying half-lives, reflecting the different types of body pools (163). The short one, $t_{1/2\alpha}$, reflects the distribution into the various body tissue with a rapid excretion, while $t_{1/2\beta}$ reflects elimination from tissue with storage function, i.e., body fat (164). In general, the total body pool of phytosterols is small in comparison wish cholesterol due to the relatively fast turnover through the lower absorption and the faster excretion via bile (150,165–167); phytostanols are more rapidly excreted than phytosterols and cholesterol (152,168). The $t_{1/2\alpha}$ for campesterol in humans is estimated at 4.1±0.3 days, compared to 2.9±0.2 days for sitosterol, 1.8±0.2 days for sitostanol, and 1.7±0.11 for campestanol (152). The $t_{1/2\alpha}$ for sitosterol described by Ostlund et al. (152) is slightly shorter than that described by Salen in 1970 (3.8 days), but this may depend on analytical methods. The slowly mixing pool consists of approximately 35% of the total sitosterol pool (152) and has a $t_{1/2\beta}$ of 12.7 days (150). The slower clearance of campesterol is verified by the finding that campesterol injected in humans is cleared at a slower rate than sitosterol and stigmasterol (169).

Tissue deposition and conversion of fecal bacteria are events that take place after intestinal sterol absorption or secretion into the large bowel. This will be discussed in Chapter 6 because these events play more of a role in cancer etiology.

G. Effects of Phytosterols on Lipoproteins and Other Risk Factors for Coronary Heart Disease

The reduction of cholesterol absorption by phytosterols leads to a number of physiological effects that ultimately decrease plasma LDL concentrations. Nevertheless, a reduction in LDL takes place because of up-regulation of LDL receptor synthesis in response to the decreased cholesterol pool (170). This results in lower LDL concentrations in plasma (171). The reduction of plasma LDL-C is positively correlated with the magnitude of the expression of hepatic LDL receptors, demonstrating that the more efficient up-regulation in response to phytosterol feeding, the larger the magnitude of the drop in LDL (171). There is in general no effect on plasma HDL or TG concentrations when low or high doses of phytosterols are administered (172,173). A study of phytosterols in relation to the different subfractions of lipoproteins showed that total and esterified cholesterol in plasma as well as free and esterified cholesterol in LDL decrease with sitostanol ester intake, and the decrease was paralleled by a decrease of apoB (174). The reduction of apoB indicates that the number of LDL particles is reduced. However, subjects with a more atherogenic LDL particle size (pattern B: <25.5 nm in diameter) seem not to change their LDL particle size to the less atherogenic type (pattern A: ≥25.5 nm in diameter) in response to phytosterol consumption (175).

A compensatory response to the reduction of absorption and total body pool of cholesterol is an increase in cholesterol synthesis. This results in increased concentrations of cholesterol precursors in plasma (127,128,132,135, 176–179) as measured by using sterol balance technique (127,128,132) or deuterated water method (110,127,128,132,135,176–178). When concentrations of the cholesterol precursor Δ7-lathosterol were assessed in a dose-response study, the effect was not dose related (180). This inconsistency may be explained by the fact that the rate of endogenous cholesterol synthesis differs among individuals (181). Phytosterols, in conjunction with an inhibitor of the endogenous cholesterol synthesis, i.e., 3-hyroxy-3-Methylglutaryl-CoA reductase inhibitors (statins), have been shown to enhance LDL-C reduction (182,183).

Other CHD risk factors, such as coagulation factors and fibrinolytic markers, do not change in response to phytosterol treatment (184,185). Similarly, in a short-term clinical study (4 weeks), C-reactive protein (CRP) and white blood cell count were not affected in healthy normolipidemic men and women (179). Studies with a longer duration are needed to assess changes in inflammatory markers, such as CRP. From the above data, it appears that the main known CHD protective effect comes from the reduction in LDL-C concentrations.

H. Effects of Phytosterols on Absorption of Lipophilic Compounds and Metabolism

As lipophilic compounds share the same absorption pathway, it has been suggested that the absorption of a number of lipophilic compounds may be impaired by phytosterols. Examples are fat-soluble vitamins like vitamin A (retinol), D, E (tocopherol), and K, as well as some carotenoids (α- and β-carotenes, lycophene, lutein, β-cryptoxanthin, and ubiquinol-10). According to the predictive rules from a theoretical model (i.e., the log-compartment model, which predicts the absorption of compounds based on their physicochemical properties) (186), interactions occur between components with similar lipophilicity (187). The physicochemical properties determine how the lipophilic components solubilize and subsequently orient themselves in the core and surface of fat droplets in emulsions and lipoproteins (80), which ultimately affects their absorption and bioavailability. Phytosterol/stanol esters have high lipophilicity, which may impair the absorption of other highly lipophilic compounds. However, clinical studies in which retinol, 25-hydroxyvitamin D, vitamin K, and ubiquinol-10 have been included as measurements have suggested that their plasma concentrations are not reduced by phytosterol supplementation (188–193). Weststrate and Meijer were the first to show that participants in a clinical trial consuming phytosterol or phytostanol ester–enriched spreads at doses of 2–3 g/day had a reduced plasma concentration of α- and β-carotenes, as well as lycopene (147). Reduction of carrier proteins, namely, VLDL and LDL, was thought to be the reason for this change. The total lipid-standardized (i.e., correction taking into account that the concentrations of total lipids, here mainly LDL-C, is also reduced) concentrations of α- and β-carotenes as well as lycopene decreased by approximately 20%, and remained significantly lower for all carotenoids except lycopene in the phytosterol ester period of the trial. Plat et al. showed that the decrease of blood carotenoid concentrations following phytosterol ester consumption depended on the carotenoid lipophilicity (193). The hydrocarbon carotenoids (β-carotene, α-carotene, and lycopene) decreased the most, followed by the less lipophilic oxygenated cartenoids (lutein/zeaxanthin and β-cryptoxanthin) and the tocopherols. The reductions were associated with the reduction of LDL, which carries most of these carotenoids. There was an association between the decrease in the hydrocarbon carotenoids and a decrease in cholesterol absorption, which suggests that they compete for space in the lipid phase during fat absorption. LDL-C-standardized carotenoid concentrations were unchanged, except for β-carotene, which was nonsignificantly lowered by about 10% (193). These results are supported by the findings of Borel et al., who showed that the most lipophilic carotenoids distribute in the core of lipid particles, whereas the more polar ones solubilize in the surface (194). Absorption of polar carotenoids requires TG lipolysis so that they can be transported through the gut before uptake. The fact that phytosterol esters are located in the core of lipid droplets (195) suggests that phytosterol esters and the apolar carotenoids use the same transport route before absorption in the gut, which makes them compete for space. In contrast, a postprandial study suggested that the absorption of α-tocopherol, retinol, and β-carotene was not impaired (196). However, these results had a large interindividual difference and small study size, which may explain these findings.

Reduced concentrations of β-carotenes in lipoproteins may theoretically affect the antioxidant capacity of blood, which is believed to be protective against oxidative damage. As mentioned earlier, the effects of antioxidants on CHD are still debated. Therefore, the relevance of the reductions of carotenoids following phytosterol supplementation is of questionable clinical importance. Furthermore, and despite these facts, the normal concentrations of carotenoids in plasma and tissues do not correlate with clinical markers of antioxidant activity and oxidative stress (197). The phytosterol-induced decreased concentration of carotenoids can be countered by consumption of more fruit and vegetables. Subjects who were advised to eat the daily recommended five servings of vegetables and fruit, whereof at least one serving was rich in carotenoids, had a 13% increase in plasma β-carotene during the control period. This did not differ significantly from the plasma β-carotene concentration during phytosterol ester or phytostanol ester consumption (198).

I. Effect of Phytosterols on Atherosclerosis in Animal Studies

To date, there are no studies dealing with the effect of phytosterols on the progression of atherosclerosis in humans. With access of surrogate markers such as intima media thickness of carotid arteries, flow-mediated dilation of the brachial artery, angiography and magnetic resonance imaging in coronary arteries (199), this could be measured in controlled human trials. However, such studies generally require a large number of study subjects and a minimal duration of at least a year, which therefore entails a high study cost. Instead reports about animal studies on the effects of phytosterols and atherosclerosis development have to be evaluated. Below we assess these animal studies for evidence that treatment with phytosterol not only lowers LDL-C but also prevents atherosclerotic plaques.

Peterson was the first to demonstrate the antiatherosclerotic affects by phytosterols in chickens. He fed soy sterols to the animals when they were on an atherogenic diet high in cholesterol and showed that the unsaturated soy sterols protected against atherosclerosis (72). Later, both unsaturated and saturated forms of phytosterols were shown to be antiatherogenic in animal trials using apoE-deficient mice (200–202) or rabbits (203) on an atherogenic diet. There were no differences between plant stanols derived from vegetable oils or wood on the effect on atherosclerotic lesion size and the severity of intimal fatty streaks/mild plaques (202). Plant stanol esters also did not increase the adherence of monocytes to the vessel wall (202). In the comparison of unsaturated and saturated phytosterols fed to rabbits, the largest effect was seen with stanols, whereas two groups with unsaturated sterols showed no change in atherosclerotic lesions (203). The mice studies, however, support an antiatherogenic effect of unsaturated phytosterols (200,201): phytosterol treatment resulted in approximately 50% reductions in the atheroma area. Moreover, a recent dose-response study in hamsters demonstrated that lipid-filled foam cell areas were 70–100% smaller in hamsters fed unsaturated sterols in a dose from 0.24% to 2.84% compared to controls (204). Phytosterol treatment does not seem to give regression of advanced atherosclerotic plaques (205). It does, however, slow the rate of progression.

It has been suggested that unsaturated phytosterols may be atherogenic (189) due to the following: (a) Monocytes, the precursor of foam cells, can accumulate phytosterols. (b) In a phytosterolemic patient who died prematurely of sudden cardiac death, phytosterols were present to a larger extent than cholesterol found in the coronary arteries. (c) Treatment with bile acid-binding resins lowers plasma phytosterol levels, which may supposedly improve cardiac symptoms (206). In regards to the accumulation of phytosterols seems logical as monocytes also accumulate other lipids, including cholesterol, a process that ultimately leads to foam cell formation (35). Among the important protective mechanisms against hypercholesterolemia and atherosclerosis is decreased cholesterol absorption and increased cholesterol excretion via bile with higher intakes of cholesterol (207). A specific mechanism mediated via ABCG5 and ABCG8 allows the body to distinguish among sterols (102). In sitosterolemia, this mechanism is abolished due to a mutation, and phytosterols are absorbed and accumulated, while cholesterol absorption and LDL concentration remain normal (98). This in turn could explain why sitosterol was found in higher concentrations than cholesterol in atherosclerotic tissue of a sitosterolemic individual (189).

Glueck et al. suggested, based on an epidemiological study, that high plasma concentrations of phytosterols were associated with an increased risk of CHD independent of total serum cholesterol concentrations (208). In subjects with elevated blood cholesterol an association between plasma concentrations of phytosterols and premature CHD was found. Among 506 individuals who replied to questions about personal and family history of

CHD, premature CHD was twice as common in the 10% with highest stigmasterol concentrations (208). It is interesting to note that there was also a positive correlation between plasma phytosterol concentrations and LDL concentrations ($r = 0.34$), although the higher risk of premature CHD with higher stigmasterol concentrations was independent of plasma cholesterol (208). There were also no significant differences in plasma phytosterol concentrations of all cases of CHD compared to noncases with an exception of campesterol, which was marginally increased but weakened when controlled for plasma cholesterol concentration (208). Therefore, the data do not show a consistent significant positive association between serum phytosterol levels and risk of CHD, independent of blood total cholesterol. The study design is not the best one to answer the question of whether or not phytosterols are more atherogenic than cholesterol. For example, the studied population have cholesterol levels in the top quintile from the original screening. Therefore, it is not possible to make assessments about subjects with low cholesterol values from this study or to make judgments on the effects of plant sterols in the total population. The statistics also raise some questions, as serum cholesterol is not mentioned as a covariate despite the finding of a statistically significant correlation between serum cholesterol and serum phytosterols. Instead, this was done in a recent study by Sudhop et al. (209), in which elevated serum phytosterols were linked with a family history of CHD independently of sex, age, TC, LDL-C, HDL, and TG. The statistics are based on a small sample but confirm that serum phytosterols had a strong influence on the results. Nevertheless, it is still unclear whether modestly elevated serum plant sterols are a risk factor or if they are a surrogate marker for other mechanisms, which could result in increased risk for CHD. However, it seems unlikely that a small increase in serum phytosterol concentrations would be offset by a significant reduction in LDL-C, which is linked to a reduced risk of CHD.

Another interesting observation from a cell model is that unsaturated phytosterols have been shown to inhibit vascular smooth cell hyperproliferation (210). Demonstration of this mechanism speaks against the point that unsaturated phytosterols would give atherosclerosis, but it remains to be verified under in vivo conditions.

IV. PHYTOSTEROLS AS CHOLESTEROL-LOWERING AGENTS

Since 1951, numerous studies have been performed addressing the hypocholesterolemic effects of phytosterols in normolipidemic, hyperlipidemic, and diabetic men and women, or children with familiar hypercholesterolemia. The interest in pharmacological treatment with phytosterols reached its peak during the 1970s, but declined later as there was a general belief that high doses were needed to achieve a significant effect on the serum cholesterol concentration. This was due to the low efficacy of many of the commercial preparations, which were generally formulated powders with limited solubility. The interest in phytosterols as hypocholesterolemic agents started again in the late 1980s, when animal and human studies suggested that β-sitostanol had a stronger effect on serum cholesterol and cholesterol absorption than β-sitosterol (104,141,149,168). As mentioned in the introduction, esterification of phytosterols proved to be a useful way of increasing their solubility for incorporation into foods without influencing food texture. Miettinen et al. subsequently described that esterified phytosterols, i.e., β-sitostanol ester from tall oil, enriched in a canola oil–based spread had a significant serum cholesterol lowering effect in humans (170).

Here, clinical trials from the mid-1980s until present will be reviewed and discussed. Earlier trials have been reviewed elsewhere (211,212) and therefore will not be discussed here. Many of these earlier studies have struggled with problems of low efficacy due to low solubility of the phytosterol preparations used and/or lack of control in the study design.

Since the mid-1980s, a number of clinical trials and case descriptions of phytosterol supplementation, in different adult populations have been published. The plasma cholesterol lowering effect of phytosterols has been supported by most of these studies. Table 4 gives an overview of 57 clinical studies and case descriptions with 98 treatment arms in 3284 adults and 183 children. Esterified phytosterols or phytostanols were used in 44 studies, free sterols in 13 studies, and micellar phytosterols in 2 studies (some studies have encompassed more than one format, as can be seen in the table). The used carriers have been spreads, dressings, ground beef, yogurt, mayonnaise, lemonade, pudding, chocolate as well as capsules. Doses varying from 0.7 to 8.6 g/day of phytosterols (expressed as free sterol equivalents) reduced LDL-C concentrations in the range of 1.1 to 29%. This wide range reflects a number of factors that may contribute to the overall phytosterol effect. These factors will be discussed in the following sections.

A. The LDL-C-Reducing Effect of Phytosterols Ingested as Natural Food Ingredients

As mentioned, most studies on the hypocholesterolemic effect of phytosterols were based on the use of purified phytosterol mixtures or enriched functional food products rather than studying effects on the dietary phytosterol intake. The usual dietary content of phytosterols represents only about 10–30% of that given with an enriched functional food; thus, the impact on serum cholesterol is expectedly substantially smaller (123,213). The classic equations of Keys and Hegsted take into account fatty acid and cholesterol intake and their effect on total cholesterol concentration (60,61). These equations are useful tools for predicting the change of cholesterol concentrations in lipoprotein fractions in response to changes in dietary fat intake, but some variations cannot be fully explained by fat quality alone. In part, this variation in response may depend on the phytosterol intake with these fats, which is considered neither in the classical formulas nor in their updated versions. Subsequently, a later meta-analysis by Mensink and Katan of 27 clinical trials with different fat intakes did not include phytosterols in the predicting formula (62).

So far only a few studies so far have dealt with the effects of the phytosterol contents of foods and their effects on serum cholesterol. In one experimental study, 16 normolipidemic subjects consumed a controlled diet containing corn oil, olive oil, or olive oil enriched with phytosterols (123). Corn oil has a high content of PUFAs and a higher content of phytosterols than olive oil (corn oil about 850 mg/100 g and olive oil about 110 mg/100 g). Olive oil is in contrast, a rich source of MUFAs. Based on the predictions by Keys (61), Hegsted (60), and Mensink and Katan (62), both oils should have a similar effect on TC and LDL-C concentrations. In this study, however, LDL concentrations were 9% higher on the olive oil–enriched diet than on the diet with corn oil (123). The supplementation of olive oil with phytosterols had an effect on LDL-C similar to that of the olive oil diet, which may be explained by low phytosterol solubility. In a comparison of sunflower oil (high PUFA + high phytosterol content) with olive oil (high MUFA + low phytosterol content), the diet rich in olive oil resulted in 10% higher LDL and apoB concentrations than the diet rich in sunflower oil (213). In contrast, canola oil (high MUFA + high phytosterol content) had an effect similar to sunflower oil on plasma lipids and apolipoprotein concentrations (214). The observed differences in LDL concentrations can also be explained by the large differences in squalene concentration of these oils. For example, olive oil has a 17 times higher squalene concentration than sunflower and canola oil (213). As can be seen in Figure 2 (see page 252), squalene is a precursor of cholesterol in the endogenous cholesterol synthesis. Squalene feeding is known to elevate cholesterol synthesis, though the effects on LDL-C concentrations are more limited (215). For instance, addition of 1 g purified squalene to a canola oil–rich diet for 9 weeks caused a net increase in serum total, VLDL, IDL, and LDL cholesterol concentrations by 12%, 34%, 28%, and 12%, respectively (216). Various plant oils would be expected to cause only minor differences in LDL-C concentrations. Instead significant differences in LDL-C concentrations were observed, which supports the assumption that minor components such as phytosterols and squalene in these oils could explain some of the

Table 4 Clinical Trials of the LDL-Lowering Effect of Phytosterols

Study design/ref.	N (M/F)	Diet	Duration (weeks)	PS source	PS carrier	Chemical form	PS dose (g/d)	ΔLDL-C (%)
Normo- and mildly hypercholestrolemic adults								
Parallel (292)	24 (12/12)	Controlled, typical British	4	Veg. oil	Spread	Sterol esters	8.6	−23[a]
Parallel (293)	84 (46/38)	Ad libitum, normal	8	Veg. oil	Spread Dressing	Sterol esters	3.0	−3.7[NS]
							6.0	−1.5[NS]
							9.0	−7.7[NS]
Crossover (249)	88	Ad libitum,	3	Veg. oil	Spread	Sterol esters	3.2	−11[a]
Crossover (135)	22 (−/22)	Ad libitum, heart diet	7	Wood	Spread	Stanol ester	3.0	−15[a]
Latin square (173)	100 (42/58)	Ad libitum, normal	3.5	Veg. oil	Spread	Sterol ester	0.8	−6.7[a]
							1.6	−8.5[a]
							3.2	−9.9[a]
Parallel (294)	185 (90/95)	Ad libitum, normal	52	Veg. oil	Spread	Sterol ester	1.6	−5.9[a]
Parallel (248)	96 (34/62)	Ad libitum, normal	4	Veg. oil	Spread	Sterol ester	2.0: apoE3	−9.2[a]
							apoE4	−11.0[a]
							3.0: apoE3	−8.7[a]
							apoE4	−6.4[NS]
Crossover (295)	22	Controlled, typical Canadian	1.4	Wood	Spread	Free sterols	1.6	−9.1[a]
Crossover (71)	23	Controlled, typical American diet	3	Veg. oil	Spread + dressing	Sterol esters	3.3	−14.3[a]
		Controlled, NCEP step I diet						−16.8[a]

Design (ref)	N (M/F)	Diet	Dur.	Source	Vehicle	Type	Dose	% change
Crossover (296)	56 (28/28)	Controlled, typical American diet	3	Veg. oil	Dressing	Sterol esters	2.2	-9.7^a
Parallel (175)	34 (34/0)	Normal diet, partly controlled	4	Beef	Soy bean oil	1/3 free sterols + 2/3 sterol esters	2.7	-13.4^a
Parallel (179)	60 (16/44)	Partly controlled, ad libitum, normal	4	Veg. oil	Yogurt	Stanol ester	3	-13.7^a
Before and after (138)	11	Ad libitum, normal	1	Wood	Spread	Stanol ester	2	-14.8^b
Parallel (297)	24 (8/16)	Ad libitum, normal	5	Wood	Spread	Stanol ester	2.2	$-13^{a,c}$
Crossover (298)	12 (12/–)	Controlled, typical French	4	Soybean oil	Butter	Free sterols	0.7	-15^a
Crossover (192)	39 (11/28)	Partly controlled, ad libitum, normal	4	Veg. oil + wood	Spread Shortenings	Stanol ester	2.5 (in 1 dose/d) 2.5 (in 3 doses/d)	-9.4^a -10.4^a
Parallel (184)	112 (41/71)	Partly controlled, ad libitum, normal	8	Veg. oil Wood	Spread Shortenings	Stanol ester Stanol ester	3.8 4.0	-14.6^a -12.8^a
Crossover (299)	76 (39/37)	Ad libitum, normal	3	Soy oil	Spread	Free sterols	0.8	-6^a
Parallel (131)	24 (16/8)	Ad libitum, hypocholesterolemic	10	Soy	Lemonade	Micellar stanols	1.9	-14.3^a
Crossover (300)	42 (22/20)	Ad libitum, normal	4	Veg. oil	Spread	Sterol ester	2.1	-10^a
Crossover (223)	60 (28/32)	Ad libitum, normal	3	Rice bran oil	Spread	Free sterols	2.1	-8.5^a
Before and after (190)	(a) 16 (3/13) (b) 12 (5/7)	Ad libitum, hypocholesterolemic	12 6	Wood	Spread	Stanol ester	2.2	-12.4^b -19.9^b
Latin square[d] (147)	96	Ad libitum, normal	3.5	Soybean oil Rice bran oil Wood	Spread	Sterol ester Sterol ester Stanol ester	3.2 1.7 2.7	-13^a -1.5^{NS} -13^a

Table 4 Continued

Study design/ref.	N (M/F)	Diet	Duration (weeks)	PS source	PS carrier	Chemical form	PS dose (g/d)	ΔLDL-C (%)
Hypercholesterolemic adults								
Parallel (301)	61 (28/33)	Controlled, hypocholesterolemic	8	Wood	Spread	Stanol ester	2.0	−7[a]
Parallel (254)	167 (100/67)	*Ad libitum*, normal	8	Canola oil	Spread	Stanol ester	2.0	−12[a]
						Stanol ester	3.0	−10
Parallel (108)	134	*Ad libitum*, normal	26	Wood	Spread	Free sterols	1.5	−11.3[a]
							3.0	−10.6[a]
Parallel (302)	62 (30/32)	*Ad libitum*, hypocholesterolemic	4	Wood	Chocolate	Free sterols	1.8	−11.1[a]
Before and after (252)	33 (33/0)	*Ad libitum*, hypocholesterolemic	3	Capsules	Wood	Free stanols	3.0	−3.6[NS]
Parallel (177)	23 (0/23)	*Ad libitum*, normal	6	Wood	Spread	Stanol ester	3.2	−8[b]
				Wood	Butter	Stanol ester	3.2	−10[b]
				Veg. oil	Spread	Stanol ester	3.2	−12[a]
Parallel (222)	55 (20/35)	Partly controlled, hypocholesterolemic	8.	Wood	Spread	Stanol ester	2.2	−8.6[a]
Crossover (172)	22 (8/14)	Controlled, typical Finnish	4	Wood	Spread	Stanol ester	2.3	−13.7[a]
				Wood			0.8	−1.7[NS]
Latin square (218)	34	Partly controlled, hypocholesterolemic	4				1.6	−5.6[a]
							2.4	−9.7[a]
							3.2	−10.4[a]
				Veg. oil	Spread	Sterol ester	2.0	−12.7[a]
				Wood		Stanol ester	2.0	−10.4[a]
Before and after (140)	6 (3/3)	*Ad libitum*, hypocholesterolemic	4	Wood	Capsules	Free stanol	1.5	−15[a]
Parallel (303)	32 (32/0)	Controlled, hypocholesterolemic	4.3	Wood	Spread	1/5 free stanols + 4/5 free sterols	1.7	−15.5[a]

Study (ref)	n (M/F)	Diet	Wk	Oil	Vehicle	Type	Dose	ΔLDL %
Latin square (110)	15 (15/0)	Controlled, hypocholesterolemic	3	Veg oil	Spread	Sterol ester	1.8	−13.2[a]
				Veg oil	Spread	Stanol ester		−6.4[a]
Parallel (251)	224 (101/123)	Ad libitum, hypocholesterolemic	5	Soybean oil	Spread	Sterol ester	1.1	−7.6[a]
							2.2	−8.1[a]
Parallel (128)	31 (22/9)	Ad libitum, normal	9	—	Mayonnaise	Free sterols	0.7	−6.2[NS]
						Free stanols	0.7	−2.6[NS]
						Stanol ester	0.8	−7.7[a,b]
Parallel (170)	141 (−/−)	Ad libitum, normal	26	Wood	Spread	Stanol ester	1.8	−8.7[a]
			52					
Parallel (255)	62 (26/36)	Ad libitum, normal	8	Veg oil	Spread	Sterol ester	2.6	−13.0[a]
				Veg oil	Spread	Sterol ester	2.5	−10.0[a]
Study 1. Before and after + crossover	22 (18/4)	Ad libitum, normal	4	Wood	Spread	Stanol ester	2.4	−13.6[a]
							2.4	−8.3[a]
Study 2. Crossover (219)			4	Veg oil	PS spread vs butter	Sterol ester	2.4	−12.2[a]
Parallel (304)	318	Ad libitum, normal	8	—	Spread	Stanol ester	3.0	−10[a]
					Spread	Sterol esters	3.0	−5.3[a]
					Spread	Sterol esters	3.0	−10.2[a]
					Spread	Sterol esters	2.0	−4.2[a]
Crossover (185)	53 (22/31)	Ad libitum, hypocholesterolemic diet	8	Veg oil	Spread	Sterol ester	1.6	−7.5[a]
Crossover (198)	(a) 46 (20/26)	Ad libitum, carotenoid-rich diet	3	Veg oil	Spread	Sterol ester	2.3	−7.7[a]
				Wood		Stanol ester	2.5	−9.5[a]
Parallel (183)	(b) 35 (20/15)	Ad libitum, hypocholesterolemic diet	4	Veg oil	Spread	Sterol ester	2.0	−9.6[a]
	152 (74/78)			Veg oil	Spread	Sterol esters	2.0	−10.2[a]
						Sterol esters + cerivastatin		−6.1[a]
Parallel (178)	67 (47/20)	Ad libitum, normal	6	Wood	Mayonnaise	Stanol ester	3.4	−9[a]
Parallel, Before and after (127)	15 (11/4)	Ad libitum, normal	6	Wood	Mayonnaise	Stanol ester	0.8	−7.7[a]
							2.0	−15.0[a]

Table 4 Continued

Study design/ref.	N (M/F)	Diet	Duration (weeks)	PS source	PS carrier	Chemical form	PS dose (g/d)	ΔLDL-C (%)
Parallel (182)	14 (5/9)	*Ad libitum,* hypocholesterolemic	6	Wood	Mayonnaise	Stanol ester	1.5	−1.1[NS]
Crossover (139)	16 (10/6)	Controlled, typical Canadian	3	Soy	Butter	Free sterols	1.8	−11.3[a]
						Free stanols 50:50 mix of each		−13.4[a] −16.0[a]
Before and after (190)	4 (2/2)	*Ad libitum,* hypocholesterolemic	12	Wood	Spread	Stanol ester	2.2	−11.4[a]
Diabetic adults								
Crossover (132)	11 (11/0)	*Ad libitum,* normal	6	Wood	Spread	Stanol ester	3.0	−9[a]
Crossover (134)	7(7/0)	*Ad libitum,* normal	7	Wood	Spread	Stanol ester	3.0	−14[a]
Parallel (231)	81	*Ad libitum,* normal	4	Veg. oil	Spread	Sterol ester	1.6	−6.8[a]
Healthy children								
Crossover (233)	19 (8/11)	*Ad libitum,* normal	4	Wood	Spread	Stanol ester	3.0	−15.5[a,e]
Children with familial hypercholesterolemia								
Crossover (237)	38 (19/19)	*Ad libitum,* hypocholesterolemic	8	Veg. oil	Spread	Sterol ester	1.6	−10.2[a]

Before and after (256)	7 (4/3)	*Ad libitum,* hypocholesterolemic	12	—	Pastilles	Free sterols	6.0	−17[b]
Before and after (305)	9 (6/3)	*Ad libitum,* hypocholesterolemic	12	—	Pastilles	Free sterols	6.0	−19.5[b]
Before and after (306)	14 (7/7)	*Ad libitum,* hypocholesterolemic	28	—		Free stanols	1.5	−29.2[b]
			6	Wood	Spread	Stanol ester	3.0	−15[b]
Crossover (191)	72 (40/32)	*Ad libitum,* hypocholesterolemic	12	Wood	Spread	Stanol ester	1.5	−7.5[a]
Before and after (190)	24 (8/16)	*Ad libitum,* hypocholesterolemic	12	Wood	Spread	Stanol ester	2.2	−17.9[b]
Heterozygote sitosterolemics								
Before and after (217)	2 (1/1)	*Ad libitum,* normal	4	Veg. oil	Spread	Sterol ester	3.2	−10.8[f]

[a] Significant different from control.
[b] Statistically different from baseline (no control product used).
[c] (Non-HDL).
[d] Latin square with five periods and four treatments.
[e] Reduction from baseline as the control group entailed a wheat bran intervention, which changed the dietary intake during the trial.
[f] No statistics performed due to too small power of study.
Note: Populations are classified according to their average LDL-C cholesterol concentrations.
PS, phytosterol.

variations in LDL-C concentrations. Although this suggests that the dietary intake of phytosterols from plant oils may not be substantial enough to give a substantial LDL-C reduction, the small impact on lowering of total and LDL-C can have an impact in terms of reducing the risk of cardiovascular disease.

B. LDL-C-Lowering Efficacy of Phytosterol-Enriched Foods in Adult Populations

The average LDL-C reduction from the clinical trials summarized in Table 4 was calculated. An average daily dose of 2.8 g/day (SD 1.7 g/day) phytosterols in studies of normo- and mildly hypercholesterolemic subjects with a mean total cholesterol concentration of 5.2 $\pm$ 0.5 mM (201 $\pm$ 19 mg/dL) gave an overall average LDL-C reduction of 10.9% (SD 4.6). The LDL reduction in hypercholesterolemic populations with a mean total cholesterol concentration of 6.9 $\pm$ 1.1 mM (267 $\pm$ 43 mg/dL) was similar at 9.5% (SD 3.4%). When all the trials were summarized, including the three trials with diabetic patients, the LDL reduction was 10.1% (SD 4.0%). Exclusion of trials with non-optimal design such as before-and-after treatment or lack of a control group did not affect the average LDL reduction substantially, as it remained at 9.9% (SD 3.9%).

One trial tested the effects of phytosterols in two subjects with hetero-zygous sitosterolemia. The small case description showed that a phytosterol-enriched spread reduced LDL-C concentrations without increasing the plasma concentrations of phytosterols more than that in normal adults (217). In contrast, a woman with the homozygous form of sitosterolemia who was fed a phytosterol-enriched spread got substantial increases of plasma phytosterols (Connor and Frohlich, unpublished results, 2003).

The effect of the phytosterol dose can roughly be evaluated by dividing the dose used by the LDL reduction. When this is done for all randomized trials with a control group, the average LDL reduction per gram of phytos-terol is 4.9% (SD 2.8%) per gram. This number does not, take into account the physicochemical properties of the phytosterols used (free vs. esterified or micellar form) or that the effect might plateau at higher doses. To visualize this, the relative reduction of LDL concentrations has been plotted against the used dose of phytosterols (Fig. 4). Studies using free phytosterols are plotted separately from those using esterified phytosterols. The results in different populations have all been plotted together, as the relative response is similar in both normo- and hypercholesterolemic populations. As can be seen in the figure, higher doses of phytosterols do not contribute to a substantially larger effect, and the dose-response effect is not linear. Instead it plateaus with

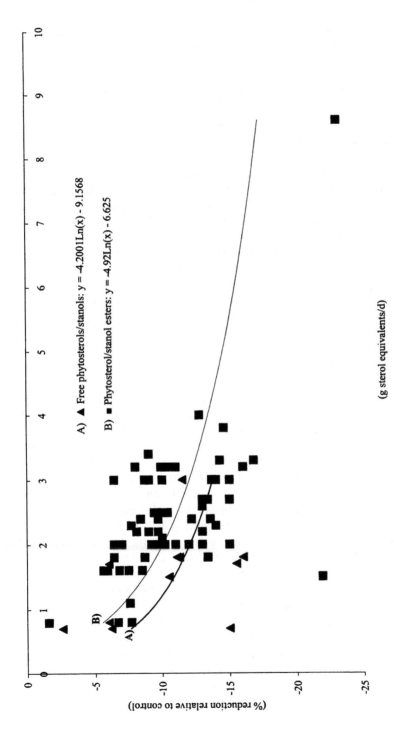

A) ▲ Free phytosterols/stanols: $y = -4.2001\mathrm{Ln}(x) - 9.1568$

B) ■ Phytosterol/stanol esters: $y = -4.92\mathrm{Ln}(x) - 6.625$

(g sterol equivalents/d)

(% reduction relative to control)

Figure 4 A comparison of LDL-reducing effects of free and esterified phytosterols.

larger intakes. The two well-controlled dose-response studies that exist for phytosterol and phytostanol esters both showed that a dose of 1.6 g/day resulted in a significant reduction of −8.5% and −5.6%, respectively, and that larger doses only increase this effect modestly (172,173).

The trend lines depicted in Figure 4 show that free phytosterols have a similar effect to esterified phytosterols in doses lower than 3 g (data on higher doses are scarce of both free and esterified phytosterols). This is expected, as free sterols are less soluble and more difficult to incorporate into food products in a soluble form. The study by Christiansen *et al.* used a new technique, called microcrystallization, which enables up to 30 weight % of free phytosterols to be added to dietary fats and oils without the use of emulsification agents. In this particular study, the reduction of LDL-C by 10.5% with 1.5 g/day was similar to the 8.5% decrease with 1.6 g of phytosterol ester and even higher than the 5.6% lowering with phytostanol ester, as assessed in the two mentioned dose-response studies (172,173). The physicochemical properties are therefore very important.

Several questions have been raised about efficacy of different formulas. One of them is whether or not phytostanol esters are more efficient than phytosterol esters. There are several published well-controlled randomized comparisons of the two types of esters (110,147,198,218,219). All of them show that there is no difference in efficacy between unsaturated and saturated phytosterols when they are esterified. This is also the rationale behind plotting them together in Figure 4.

Another factor that has attracted interest is whether or not the phytosterol composition determines the efficacy of LDL lowering. The source decides the phytosterol profile, as the latter in one way is to be seen as the "fingerprint" of the plant from which the phytosterols come. The materials that have been used as a phytosterol source for the products in the clinical trials are different vegetable oils and wood pulp. In general, when phytosterols are extracted from vegetable oils, their major component is sitosterol (approximately 60%), followed by campesterol (30%), and the rest is a mix of brassicasterol, avenasterol, and stigmasterol (122,220). One exception would be olive oil, sitosterols content is normally more than 90%, which is quite unique for this oil. Plant oils such as canola, sunflower, and soy are more commonly used as a source of phytosterol-enriched products (221). In wood, the profile is similar to that of olive oil, with sitosterol dominating the content (85–90%) followed by campesterol and sitostanol (221). Two studies have compared esterified phytostanols from vegetable oils and wood (184,222). Despite a difference in profile of the 4-desmethyl sterols, there were no significant differences between the two treatments. In rice bran oil, the phytosterols consist not only of the dominating 4-desmethyl sterols, but also of a substantial amount of 4,4-dimethyl sterols, esterifed to ferrulic acid. In clinical trials, phytosterols from rice bran oil have only yielded a modest reduction in LDL-C concentrations (147,223), and it seems as if the 4-desmethyl sterols found in most other vegetable oils and wood have a greater effect.

C. LDL-Lowering Efficacy of Phytosterol-Enriched Foods in Subjects with Diabetes

Diabetes puts individuals at a high risk for CHD and treatment to reduce LDL-C concentrations is crucial (224). The typical subject with type II diabetes or the metabolic syndrome is overweight with a high-risk profile for atherosclerosis (26), where an increased in the number of LDL particles often coexists with other risk factors for heart disease, i.e., hypertension, insulin resistance, low HDL and elevated triglyceride concentrations (225, 226).

Overweight or obesity in the metabolic syndrome and diabetes promote an abnormal sterol metabolism (227). Contrary to what can be assumed, the obese and insulin-resistant individual has relatively low cholesterol absorption and relies more on an endogenous cholesterol synthesis, which is increased (227–229). These abnormalities improve with weight loss and intensified insulin regimen (228,230). Despite the shift in sterol metabolism that occurs with insulin resistance, phytosterol therapy lowers LDL in diabetics to a similar extent as in normal individuals (132,134,231). In a slightly overweight population with good-to-moderate glycemic control, sitostanol ester intake lowered LDL-C with 9% (132). In addition, cholesterol was mostly reduced in the smaller and the more dense LDL particles (132), considered to be the most atherogenic (232). In another study in individuals with well-controlled diabetes, phytosterol ester–enriched spreads lowered LDL by 6.8% after 4 weeks of treatment, though limited compliance gave less of an effect after 8 and 12 weeks (231). Phytosterols could therefore be an important adjunct to diabetes treatment.

D. LDL-Lowering Efficacy of Phytosterol-Enriched Foods in Children

In most countries, phytosterols are not recommended for general use by children because children do not normally have elevated cholesterol; hence, cholesterol management is not usually a priority unless blood lipids are elevated. Furthermore, phytosterol intake may interfere with carotenoid absorption, which normally is not seen as a concern, but can be in subgroups who have higher demand for vitamin A. The LDL-C-lowering effect of phytosterols is similar in children and adults (191). There is only one controlled clinical trial in healthy children, which shows that the LDL-C-lowering effect was 15%, without any adverse short-term effects (233).

One young population who would benefit from phytosterol treatment are children with familial hypercholesterolemia. Due to mutations in the genes coding for the LDL receptor, their blood lipids are grossly elevated, and they are predisposed to premature CHD (234). The current standard treatment for children is with bile acid absorption inhibitors (resins), but compliance is generally low due to the gastrointestinal side effects that these cause (235). Statins are generally not recommended, though it is probable that this is about to change due to the recent trial of simvastatin demonstrating lack of adverse effect on growth and pubertal development with doses up to 40 mg/day (236). Six clinical trials using phytosterols have been performed in children with familial hypercholesterolemia, but only two of these were randomized and based on a controlled design (191,237). In these two trials, 1.5 g of phytostanol ester and 1.6 g of phytosterol ester per day reduced LDL-C concentrations on average by 7.5% and 10.2% (191,237). Resins were not used in these trials, and the effects of phytosterol treatment in combination with resins are not known. Also, in these trials there were no short-term side effects. However, as in many adult studies, serum concentrations of some of the carotenoids were modestly reduced (191,237).

Bile acid sequestrants lower LDL-C concentrations within the range of 15–20% depending on the dose and whether it is used in combination with other drugs that further improve the effect, such as niacin (238). Phytosterol in combination would be a useful tool for the treatment of children with familial hypercholesterolemia as it can be used in combination with drugs, and has no known gastrointestinal side effects. It could also help to lower the required doses of drugs in children with familial hypercholesterolemia.

E. ApoE Genotype Does Not Have a Major Influence on Phytosterol Efficacy

ApoE is a protein constituent of plasma VLDL, chylomicrons, and a subfraction of HDL, that plays an important role in the uptake of remnants from the blood (239). In mice, apoE deficiency leads to elevated LDL-C concentrations and atherosclerosis due to an accumulation of VLDL and chylomicron remnants (240–242). Phytosterol treatment in apoE-deficient mice reduced atherosclerosis and LDL-C concentrations in blood (201,202) as well as hepatic VLDL-C secretion (243). This finding shows that the presence of apoE is not conditional for the antiatherosclerotic properties of phytosterols.

ApoE has a natural variation of its DNA, i.e. a *polymorphism*, with three different so-called genotypes. ApoE3 is the most common genotype, whereas apoE4 resulting from an arginine for cysteine substitution at residue 112 and the apoE2 resulting from a cysteine for arginine substitution at residue 158 are less common (79). In the Framingham study, the distribution of ApoE4, E3, and E2 was 21%, 65%, and 14%, respectively (244,245). In Europe, there is a clear north–south gradient of the genotype, with more E4 in the north (17% and 15% in Finland and Ireland, respectively) compared to countries in the south (8.5% in Greece) (246). Different genotypes of apoE impact the intestinal cholesterol absorption and cholesterol synthesis differently. Subjects who are either heterozygous or homozygous for the E2 allele absorb less and synthesize more cholesterol than those with phenotype E4/3 and E4/4, with the individuals with the most common phenotype E3/3 in between (247). This finding, together with the observation that apoE deficiency limits the hepatic up-regulation of LDL receptors, suggests that polymorphisms of apoE may affect phytosterol efficacy of LDL-C lowering. Individuals with the apoE4 allele had a greater reduction of their LDL-C concentrations in response to phytosterol consumption in two studies (128,178). However, there was no direct statistical comparison of the groups with different apoE phenotypes. Moreover, the sample size in the two studies was small ($n < 100$), which is a low level to assess subtle genetic differences. Three newer studies include direct statistical comparisons of all three apoE genotypes and do not support a difference in the LDL-C-reducing efficacy of phytosterol treatment (184,248,249). However, two of these studies, had limited power to assess differences in genotypes [(study 1, $n = 112$ (184)/study 2, $n = 96$ (248) split up in a two group and a three group parallel design, respectively]. In the third, people were selected based on their phenotype for a direct comparison of the LDL-C-lowering effect (249). Also, this study lacked support for the assumption that subjects with apoE4 would have a stronger response to phytosterol ester–enriched foods. In conclusions there is limited support for the hypothesis that polymorphism of apoE has an effect on the LDL-C-lowering efficacy of phytosterols.

F. Relation of Phytosterol Intake Frequency and Efficacy

Most studies have been performed with phytosterols consumed in two or three daily doses. This has been based on the assumption that giving phytosterols with the major meals would optimize their effect, as the phytosterols were to impair micelle formation during lipid digestion in the postprandial state. The only study that has compared a single dose to several doses was carried out by Plat et al., where 2.5 g of esterified plant stanols was either consumed in one dose at lunch or split into three doses given with the daily meals (192). The relative distribution was 0.42 g at breakfast, 0.84 g at lunch, and 1.25 g at dinner (proportional to dietary cholesterol intake). The study showed almost identical effects on LDL-C concentrations of the two regimens. This finding is in line with the belief that cholesterol absorption has a long duration, where inhibition can take place over a certain time span rather than directly after specific meals. Plat et al. recently showed that there is an increased expression of ABCA1 in intestinal cells following absorption of plant stanols (105). This indicates stimulation of cholesterol efflux by phytosterols. Both cholesterol and phytosterols may be secreted back into the intestinal lumen following absorption into the enterocyte via one of the ABC receptors (ABCA1, ABCG5, or ABCG8). It appears as if the reducing effect of phytosterols on cholesterol absorption depends both on a reduced incorporation of cholesterol into micelles when phytosterols are present in the gut and on increased sterol efflux, though this needs to be verified in animal studies. Frequency of phytosterol intake may have less importance; phytosterols divided into several doses or given as a single dose seem to have similar effects.

G. The LDL-Reducing Effect by Phytosterols: Effect of Background Diets

As can be seen from Table 4, the LDL-reducing effect of phytosterols is similar whether or not the dietary regimen of the study was controlled or *ad libitum*, and if it was typical for the country or based on nutritional guidelines. The effect of phytosterols is additive to that of a healthy diet, as shown in two clinical trials (250,251). For instance, Hallikainen et al. showed that by switching from a controlled high-saturated-fat diet to a low-fat diet, there was a 10% lowering of LDL-C concentration (250). Compared to baseline, addition of 2.3 g of esterified phytosterols (expressed as sterol equivalents) or 2.2 g of esterified phytostanols reduced LDL-C concentrations by 24% and 18% respectively, demonstrating an additive effect. In contrast, another study with hypercholesterolemic individuals on a low-fat low-cholesterol diet showed no significant further reductions of LDL-C by phytosterols (252). Here it was argued that limiting cholesterol intake would decrease the phytosterol effect, but the above-mentioned studies by Hallikainen et al. and Maki et al. contradict this view. Other plausible explanations for the lack of effect could be: (a) the fact that the phytosterols were used in capsules and thus not properly solubilized (this is known to limit their effect); (b) the subjects had an inappropriate type of lipid disorder; or (c) the lack of sufficient power, as the reduction in fact was 4%, but not statistically significant.

The most illustrative study compared phytosterol spreads in a randomized trial, where a typical American diet (TAD) was consumed during period 1, TAD + phytosterol spreads in period 2, an NCEP Step I diet in period 3, and an NCEP Step I + phytosterol spread diet in period 4. The study showed that phytosterol esters lowered LDL-C by 14% when consumed with the

TAD, and 11% on top of the Step I diet (71). The results were not significantly different, indicating that the background diet does not affect the efficacy of phytosterols. However, background diet can give an additive reducing effect, as can be seen from the results in this study (71). When LDL-C concentrations with a typical American diet were compared with the NCEP Step I with phytosterols added diet the difference in LDL-C was 17%.

Phytosterols can be consumed with any kind of diet, as their effectiveness is not dependent on the quality of diet per se. However, phytosterol supplementation should be part of a heart-healthy diet, as many more risk factors than LDL-C concentrations can be beneficially affected by the diet composition. Consumption of unsaturated dietary fatty acids, vegetable (soy) protein, dietary fiber of different types, vitamins, and antioxidants from plant foods, as well as fish consumption, together with restriction of simple carbohydrates, saturated and *trans* fatty acids, and salt, have been shown to be particularly useful in lowering cholesterol and triglyceride concentrations, blood pressure, and other relevant CHD risk factors and thus contribute in preventing CHD (26). Another positive booster of the LDL-C-lowering effect may be contributed by the carrier of the phytosterols. Most commercially available phytosterol products are either low-fat (35–40 E%) or full-fat (60–80 E%) spreads rich in poly- or monounsaturated fatty acids. With the recommended intake of 20 g/day of phytosterol spread, a low-fat spread would contribute 5–6 g fat rich in unsaturated fatty acids, and a high-fat alternative would contain 10–12 g. Butter has 80% fat, and the same portion size contains 3 g of unsaturated fat. If energy expenditure is assumed to be 10.5 MJ (2500 kcal) and butter would be replaced by one of these two spreads, the difference in intake of 2–9 g/day of unsaturated fat would give a LDL-C reduction of 1–4%. Spreads reduce LDL-C concentrations more if they replace butter. The energy from 20 g of low-fat or high-fat spreads is 260–300 kJ and 527–600 kJ, respectively (60–70 and 125–140 kcal, respectively). The extra energy obtained with a low-fat spread alternative seems therefore quite small and is not expected to have any adverse effects on the weight of individuals.

H. Potential Existence of Nonresponders

Though there is little information on nonresponders, Gylling et al. raised the possibility that some people respond less well than others to phytosterols. In the first long-term study on phytosterol treatment, plant stanol ester intake lowered serum cholesterol in about 88% of subject (176). There is no similar assessment of phytosterol ester treatment, but it is likely that the existence of nonresponders depend on how much the studied individuals depend on their hepatic cholesterol synthesis in comparison to their intestinal uptake.

In a subanalysis of the 4S trial, Miettinen et al. suggested that the subjects with high cholesterol synthesis benefited by a larger LDL-C reduction by taking statins than cholesterol absorption blockers (181). This would support the theory that some people will respond less to phytosterols due to a high endogenous cholesterol synthesis. In contrast, people who have been reported not to respond well to statins (253) may respond better to phytosterol-enriched foods.

I.   Phytosterol-Enriched Foods in Combination with Drug Treatment

Phytosterols have been used in conjunction with statins (135,182,190,254, 255), fibrates (185,256), and insulin sensitizers (132,257). In all these trials, phytosterol therapy had additive or independent effects. Statin therapy inhibits the enzyme HMG-CoA reductase, which catalyzes one of the steps in the endogenous cholesterol synthesis (Fig. 2). Combination therapy with phytosterols and statins has an additive effect on reducing LDL-C concentrations, as the two types of components act on different metabolic pathways (183). Moreover, the increased cholesterol synthesis in response to the cholesterol absorption caused by supplemented phytosterols is counteracted by statins, which means that their combination could be beneficial. In a clinical trial of statin-treated subjects, cerivastatin reduced LDL-C by 32%, and phytosterols on top of this caused an 8% extra reduction (183).

A reduction of 10% with phytosterols plus a reduction of 20% with pravastatin decreases LDL-C by 30% (182), which is quite substantial. Combination treatment with phytosterols could be preferred over doubling the statin dose as usually only 6% additional reduction occurs with the latter method (258). Vanhanen et al. showed that simultaneous inhibition of cholesterol synthesis and absorption effectively reduces LDL-C concentration (182). Recent studies with the new drug ezetimibe (in US Zetia™) 10 mg/day suggest that coadministration with 10 mg/day of statins yields a comparable LDL-C reduction to that of the highest dose of statins (80 mg/day) (259). The reduction of LDL-C concentrations with 10 mg/day of ezetimibe alone was 18%, which is similar to that achieved with phytosterols. Although it remains to be investigated, it seems likely that phytosterols could also reduce the need for higher statin doses. Thus, the risk of potential side effects of statins may be decreased with combination therapy with phytosterols (260).

Fibrates are often used in subjects with the metabolic syndrome or non-insulin-dependent diabetes mellitus as they lower blood concentrations of triglycerides and increase HDL-C (261). They bind to the nuclear hormone receptor superfamily, the peroxisome proliferator–activated receptors, which results in an enhanced catabolism of triglyceride-rich particles and reduced secretion of VLDL (262). Combining phytosterols with fibrates may be particularly beneficial for the above-mentioned patient categories, as there are large benefits of also managing moderately increased LDL concentrations in diabetics (286).

V. USE OF PHYTOSTEROLS IN STRATEGIC POPULATION PREVENTION OF CORONARY HEART DISEASE

A. Phytosterols as Part of Public Health Programs

The aim of modern nutrition is not only to prevent malnutrition, death, and disease but also to prevent mental and physical disability (263) and thus improve the quality of life. The most common approach to public health improvement is through national dietary guidelines with the aim of preventing chronic disorders such as obesity, CHD, diabetes, cancer, osteoporosis, etc. These guidelines are mainly implemented through public health campaigns as well as through specific meal planning in institutes such as schools and hospitals. Most of today's guidelines stress a reduced intake of dietary cholesterol, saturated fat and *trans* fatty acids, together with an increased intake of fruits and vegetables, unsaturated fatty acids, whole grain foods, and dietary fiber. However, no general guidelines suggest that phytosterol-rich foods can be used in conjunction with a healthy diet to further lower LDL-C concentrations, perhaps because the national guidelines do not specifically deal with CHD prevention. The *Dietary Guidelines for Americans* and *Canada's Food Guide to Healthy Eating* are examples of guidelines of food intake for the general population (264,265). There is a need of more knowledge about the overall biological effects of phytochemicals before they can be generally recommended in national guidelines.

In 1985, the National Heart, Lung, and Blood Institute (NHLBI) launched the National Cholesterol Education Program (NCEP). The purpose of the NCEP is to contribute to reducing illness and death from CHD by reducing the number of Americans with high blood cholesterol. The efforts, directed to health professionals and the public, are aimed at (a) raising awareness and understanding about high blood cholesterol as a risk factor for CHD and the benefits of lowering cholesterol levels as a mean of preventing CHD, and (b) making suggestions about a heart-healthy diet. The NCEP guidelines comprise both a population approach and a clinical program. While the clinical program focuses on individuals with high risk of developing CHD, the population strategy is directed to the whole population. ATP III has a special section directed to public health programs (for the general public and not specifically the high-risk population), which focuses on a message that blood cholesterol can be reduced by reducing intakes of saturated fat and cholesterol, but does not mention phytosterols (26). The American Heart Association (AHA) has similar dietary recommendations to that of the NCEP (266), but has chosen a somewhat more conservative approach with phytosterols, as it discusses their use under "issues that merit further research." Simultaneously, the clinical guidelines classify phytosterols as a potential lipid-lowering tool for adults with hypercholesterolemia or as part of secondary prevention.

There are several prerequisites for working out successful national food and nutrition policies. In addition to knowledge about dietary pattern, nutrient and food recommendations, and general fortification and supplementation, food labeling is an important part of building a strategy (267). Information about ingredients and nutrient contents on the packed food products, as well as physiological functionality, can be useful for spreading knowledge about specific nutrients and phytochemicals. Recommendations from health authorities regarding health claims make it easier for consumers to select the appropriate food products (267). The U.S. Food and Drug Administration (FDA) allows certain health claims for specific ingredients if there is significant scientific evidence acknowledged by agreed among qualified experts that such a claim in valid. Currently, 14 such claims have been authorized (268). Unilever Bestfoods/North America and McNeil Consumer Healthcare submitted a health claim petition in February 2000, and in September of the same year, an interim ruling for a health claim about phytosterol and phytostanol esters was granted in the United States (268).

The FDA assessment of the clinical studies led to a different ruling concerning recommended doses (268). Most studies of phytosterol esters have used relatively lower doses than the ones based on phytostanol esters (Table 4). The document stated that "daily dietary intake levels of plant sterol and stanol esters that have been associated with reduced risks are (a) 1.3 g or more of plant sterol esters and (b) 3.4 g or more of plant stanol esters. This conclusion is based on a lack of clinical studies testing lower doses of phytostanol esters and should not be interpreted as indicating that phytostanol esters are less effective for risk reduction. The FDA also recommends that phytosterol esters be consumed in a minimum of two portions over the day, which then should contain either 0.65 g of phytosterol esters or 1.7 g of phytostanol esters. The health claim can be used for spreads, dressings, snack bars, and dietary supplements in a soft-gel form. Together with existing guidelines, the health claims can favor a more frequent use of phytosterols through more awareness of their effects.

B. Phytosterols in Primary Prevention of Coronary Heart Disease

The NCEP guidelines point out that clinical professionals are a link between clinical management and population strategy (26). There is, however, some controversy about priorities for primary prevention within the medical sector. The American College of Physicians states that cholesterol screening should be focused on people with symptoms of CHD (269), whereas NCEP supports broader screening for risk factors (26). NCEP guidelines suggest that in addition to those with a short-term elevated risk of CHD, people with an elevated long-term risk (> 10 years) should also be targeted. Current European and U.S. studies show that treatment and lifestyle advice do not always reach the high-risk populations (270,271). The implementation of guidelines is largely limited by the high societal costs for screening and treatment.

The NCEP guidelines (26) together with recommendations from the Spanish Cardiology Society (272), the Association of Clinical and Public Health Nutritionists in Finland (273), and the National Heart Foundation of Australia (274) have all identified phytosterols as an important additional dietary option of the clinical treatment. In a first step in the NCEP guidelines (26), the physician is urged to explain the basis for the "therapeutic lifestyle changes" (TLC) diet (full description in Table 5). The focus during the first clinical visit following a diagnosis of high total cholesterol or LDL-C concentration is to reduce saturate fat and dietary cholesterol, encourage an increase in physical activity, and to refer the patient to a dietitian. After 6 weeks of treatment with the TLC diet, blood lipids are reevaluated. If goals are not met, phytosterols-enriched foods (2 g sterols/day) are recommended together with dietary fiber (10–25 g/day) to further boost the TLC diet. After 3 months totally, and if goals are not met with the dietary changes, lipid med-

Table 5 Dietary Advice According to National Cholesterol Education Program

---

Build a healthy base
► Let the food pyramid guide your food choices.
► Choose a variety of grains daily, especially whole grains.
► Choose a variety of fruits and vegetables daily.
► Keep foods safe to eat.
Choose sensibly
► Choose a diet that is low in saturated fat and cholesterol and moderate in total fat.
► Choose beverages and foods to moderate your intake of sugars.
► Choose and prepare foods with less salt.
► If you drink alcoholic beverages, do so in moderation.

Source: Ref. 26.

ication should be prescribed. The recommended scenario does not always correspond to the real-life situation as many patients get drugs prescribed immediately. However, as can be seen from the following reasoning, dietary changes including phytosterol consumption would save health insurance costs or tax money (depending on country), as drug treatment is expensive. Using a population-based treat-to-target analysis, statin treatment costs U.S.$456–5344 per person-year (275). If the whole U.S. population with elevated blood lipids would be treated with statins, this would be an unrealistic cost.

C.   Phytosterols as Part of Secondary Prevention

Guidelines for LDL reduction in secondary prevention emphasize a more aggressive approach (276). The compelling evidence from a number of large clinical trials suggests that significant CHD risk reduction occurs at all LDL concentrations, even in the range of 2.6–3.2 mM (100–129 mg/dl or 2.6–3.2 mM). To slow down or even reverse atherosclerosis, a multifaceted approach may be needed where diet, physical activity, and drugs all play important roles.

It was long thought that diet and lifestyle interventions only had substantial effects in primary prevention interventions, but several studies have shown positive effects also during secondary prevention (277–279). Dietary interventions based on low-fat diets (277,278) or a Mediterranean diet (279,280) provide evidence that the risk of CHD, in particular MI, can be reduced through dietary means and other lifestyle changes. Compliance with dietary advice is higher in motivated people with established CHD, as exemplified by LDL-C reductions of 37% in the Lifestyle Heart Trial (278).

The AHA and the ATP III both recommend LDL-C concentrations of less than 2.6 mM (<100 mg/dl or 2.6 mM), which is substantially lower than that found in many high-risk patients (26,276). With novel drugs therapies, these targets can be reached in a majority of patients. Dietary phytosterols alone seem not to be enough to give the desired reduction for the individual to meet the lipid goals but offer a tool to be used in combination with drug treatment. Like ezetimibe, they may be successfully used in combination therapies with the benefit of a drug-sparing effect.

D.   Estimates of Coronary Heart Disease Risk Reduction by Phytosterols

The incorporation of phytosterols into food products may improve blood lipids in both high-risk subjects and the general population, and thus ultimately prevent CHD. This suggestion is based on an extrapolation of cohort and clinical trial data. Statins studies indicate that an LDL-C reduction of about 10% could decrease the risk of CHD by12–15% over 5 years (55,56). A meta-analysis of prospective cohort data involving 500,000 men and 18,000 CHD events showed that a reduction of LDL-C concentrations of 0.6 mmol/L, which corresponded to approximately 10% reduction, would decrease CHD risk by 27%, if the LDL-C reduction was sustained over 5 years (59). The calculated risk reductions were most pronounced in the younger population, with age-specific risk reductions of 54% at 40 years, falling to 20% at 70 years.

There are no data on the effect of phytosterol consumption on CHD development, which makes a direct claim about the effect on heart disease more difficult. However, the preventive effects of lowering plasma LDL-C concentrations are very consistent independent of type of treatment, be it drugs, diet, or through surgical procedures such as ileal bypass (induces bile acid malabsorption, which indirectly decreases the body pool of cholesterol) (59). This fact supports that even though phytosterols have a different mechanism of action than, say, statins, phytosterols also prevent CHD, by virtue of lowering LDL-C concentrations.

E. Who Consumes Phytosterol Ester–Enriched Foods?

Postmarketing surveys have been performed on the major European markets of phytosterol ester enriched–spreads, i.e., in the Netherlands, Germany, United Kingdom, France, and Belgium (281). The mentioned "postlaunch monitoring" was requested by the European Union following the introduction of a phytosterol-enriched spread on the European market and builds on market studies of real purchase data from approxmately 2000 households. It shows that the product reaches the target group of consumers, which are classified as "cholesterol concerned." The majority of consumers are older than 45 years and live in households of one or two people without children. The daily intake per household is less than the 20 g of spread per day within the range of 3–12 g/day (data from the Netherlands and Germany). Actual intake of sterols is thereby estimated to be less than 1.6 g/day. In a Finnish study, consumers of phytostanol ester enriched spreads were reported to have an age of 55 years or older, with better lifestyle, higher education, and higher income than nonusers. Moreover, 46% of them had CHD (282).

Generally, it seems as if the phytosterol/stanol-enriched foods reach one of their target groups, i.e., health-concerned individuals who may be at risk for or have established atherosclerosis. However, in order for functional foods to exert their suggested effect in disease prevention, they must reach younger individuals who are at risk. We need new means of educating society that prevention from a young age is the factor most likely to have an impact on morbidity and mortality.

VI. FUTURE CONCEPTS: COMBINATION THERAPY WITH OTHER INGREDIENTS

Besides a combination therapy of phytosterols with drugs, addition of other dietary supplements could also offer several benefits. There is increasing awareness that other food ingredients or bioactive components can affect LDL-C concentration further or even influence other CHD risk factors. One way to make phytosterols more effective for CHD risk reduction is to incorporate them into foods with other bioactive substances that work with different mechanisms for LDL-C lowering. A portfolio of dietary factors has been proposed by Jenkins and Kendall, which emphasises a reduction of saturated fat and dietary cholesterol in combination with weight loss of (5 kg), dietary intake of 5–10 g of soluble fiber, 25 g of soy protein, together with 1–3 g of phytosterols. A study of this diet showed that LDL-C was lowered by 29% (283), which is a reduction similar to that achieved by some older-type statins (31,258,284).

As pointed out by others, elevated levels of LDL-C are not the only risk factor in the development of atherosclerosis. Lowering of LDL-C concentrations alone is not the sole answer to reducing the risk of mortality and morbidity from CHD (285). A combination therapy with several dietary components addressing more than one CHD risk factor is therefore within the scope of future treatment. Examples for potential ingredients are n-3 fatty acids, which have been shown to lower the risk of sudden death in observational prospective studies (286,287) as well as in a secondary prevention trial (288). The trials suggest that the n-3 fatty acids from fish, namely, eicosapentaenoic acid and docosahexaenoic acid, are the active components, which work primarily through a decreased risk of arrhythmia (289). In a recent study with adult guinea pigs, phytosterols that had been esterified directly with n-3 fatty acids were tested. This study showed that phytosterols and n-3 in combination lowered TC and TG with 36% and 29%, respectively (290). Interesting was also that the analysis of the cardiac left ventricle indicated that generation of the proaggregatory arrhythmic eicosanoid thromboxane A(2) was more than 60% lower in the phytosterols plus (n-3)-supplemented guinea pigs than in controls (290). The results need to be verified in a human model but is an interesting example of a novel combination therapy.

Another example of an applied combination therapy is the addition of phytostanols to a functional oil, based on medium-chain triglycerides (MCTs). MCTs are not metabolized in a similar manner as long-chain fatty acids and have been shown to increase energy expenditure and induce weight loss in human subjects (0.5 kg in 3 weeks) (291). The phytostanols in this functional oil lowered LDL-C. Together with the other ingredient, MCT, this oil would offer two functionalities, weight loss and LDL-C lowering, which could prove to be an interesting combination for overweight people with elevated blood lipids.

VII. CONCLUSION

Elevated LDL-C concentration is an established major risk factor for CHD development, and phytosterol-rich foods entail simple novel means to lower these concentrations. Phytosterol consumption from natural food sources results in relatively modest effects on cholesterol absorption and lowering of LDL-C concentrations, though small changes of blood lipids also may have a meaningful impact on CHD risk reduction in a public health perspective.

Reduction of LDL-C concentrations by 10–15% requires a phytosterol intake of 1.5–3.0 g/day. This can only be achieved by consumption of phytosterol-enriched foods, i.e., functional food products. Doses higher than 3.0 g/day increase the effect, but the extra benefit is small. Current general estimates of CHD risk reduction with phytosterol consumption seem somehow exaggerated as they build on an assumption that a majority of the population will consume phytosterol-enriched products. This is currently not in line with different national nutrition guidelines for the general population. Still, phytosterol supplementation has great potential for cost-effective management of CHD if it reaches the high-risk individuals. Phytosterols may therefore have an important role in primary prevention. In secondary prevention their role seems less clear, but as phytosterols act in an additive way with drug treatment, they could provide an additional way to lower LDL-C concentrations and possibly also spare the need of higher drug doses.

Future scenarios involve the combination with other dietary ingredients that could either affect LDL-C concentrations further or modulate other CHD risk factors, such as endothelial function, TG concentrations, HDL concentrations, or blood pressure. Phytosterol supplementation is likely to get more and more common, with a larger public awareness of the health effects as a result.

ACKNOWLEDGMENTS

The authors thank the following: Dr. Richard Ostlund, Washington University, Dr. Peter Jones, McGill University, and Dr. William Connor, Oregon Health and Science University, for providing us with manuscripts and data in press. The authors also thank Dr. Mohammed Moghadasian and Dr. Min Li, University of British Columbia, as well as Dr. Guus Duchateau, Kevin Povey, Dr. Linda Lea, Ay Lin, and Dr. Hans Zevenbergen, Unilever Bestfoods, the Netherlands, for scientific advice and/or reviewing this paper.

ABBREVIATIONS

ABC	ATP-binding cassette protein
ACAT2	acyl CoA: cholesterol acyltransferase 2
AHA	American Heart Association
ATP III	Adult Treatment Panel III
Apo	apolipoprotein
CD36	cell cluster determinant antigen 36
CRP	C-reactive protein
CHD	coronary heart disease
CoA	coenzyme A
E%	energy percent
FABP2	fatty acid–binding protein-2
FXR	farnesoid X receptor (bile acid receptor)
HDL	high-density lipoprotein
HMG	3-hyroxy-3-methylglutaryl
IDL	intermediate-density lipoprotein
LDL-C	low-density lipoprotein cholesterol
LPL	lipoprotein lipase
LXR	liver X receptor (oxysterol receptor)
MCT	medium chain triglyceride
MI	myocardial infarction
MTP	microsomal triglyceride transfer protein
MUFA	monounsaturated fatty acid
NCEP	National Cholesterol Education Program
NHLBI	National Heart, Lung and Blood Institute
PUFA	polyunsaturated fatty acid
PS	Phytosterols
RXR	Retinoid X receptor
SR-B1	scavenger receptor class B type 1
TAD	typical American diet

TC	total cholesterol
TG	triacylglycerol
TLC	therapeutic lifestyle changes
VLDL	very-low-density lipoprotein

REFERENCES

1. The Lipid Research Clinics Coronary Primary Prevention Trial results. II. The relationship of reduction in incidence of coronary heart disease to cholesterol lowering. JAMA 251(3):365–374, 1984.
2. The Lipid Research Clinics Coronary Primary Prevention Trial results. I. Reduction in incidence of coronary heart disease. JAMA 251(3):351–364, 1984.
3. Anonymous. Randomised trial of cholesterol lowering in 4444 patients with coronary heart disease: the Scandinavian Simvastatin Survival Study (4S). Lancet 344(8934):1383–1389, 1994.
4. AW Caggiula, G Christakis, M Farrand, SB Hulley, R Johnson, NL Lasser, J Widdowson, G Widdowson. The multiple risk intervention trial (MRFIT). IV. Intervention on blood lipids. Prev Med 10(4):443–475, 1981.
5. MH Frick, O Elo, K Haapa, OP Heinonen, P Heinsalmi, P Helo, JK Huttunen, P Kaitaniemi, P Koskinen, V Manninen, et al. Helsinki Heart Study: primary-prevention trial with gemfibrozil in middle-aged men with dyslipidemia. Safety of treatment, changes in risk factors, and incidence of coronary heart disease. N Engl J Med 317(20):1237–1245, 1987.
6. AL Gould, JE Rossouw, NC Santanello, JF Heyse, CD Furberg. Cholesterol reduction yields clinical benefit. A new look at old data. Circulation 91(8):2274–2282, 1995.
7. MF Muldoon, SB Manuck, AB Mendelsohn, JR Kaplan, SH Belle. Cholesterol reduction and non-illness mortality: meta-analysis of randomised clinical trials. BMJ 322(7277):11–15, 2001.
8. SC Smith Jr, P Greenland, SM Grundy. AHA Conference Proceedings. Prevention conference V: Beyond secondary prevention: identifying the high-risk patient for primary prevention—executive summary. American Heart Association. Circulation 101(1):111–116, 2000.
9. S Sans, H Kesteloot, D Kromhout. The burden of cardiovascular diseases mortality in Europe. Task Force of the European Society of Cardiology on Cardiovascular Mortality and Morbidity Statistics in Europe. Eur Heart J 18(8):1231–1248, 1997.
10. S Tuljapurkar, N Li, C Boe. A universal pattern of mortality decline in the G7 countries. Nature 405(6788):789–792, 2000.
11. WHO. World Health Statistics Annual, 1997–1999.
12. 2002 Heart and stroke statistical update. Dallas: American Heart Association, 2001.
13. WD Rosamond, LE Chambless, AR Folsom, LS Cooper, DE Conwill, L Clegg, CH Wang, G Heiss. Trends in the incidence of myocardial infarction and in

mortality due to coronary heart disease, 1987 to 1994. N Engl J Med 339(13): 861–867, 1998.

14. H Kesteloot, S Sans, D Kromhout. Evolution of all-causes and cardiovascular mortality in the age-group 75–84 years in Europe during the period 1970–1996. A comparison with worldwide changes. Eur Heart J 23(5):384–398, 2002.

15. H Tunstall-Pedoe, K Kuulasmaa, M Mahonen, H Tolonen, E Ruokokoski, P Amouyel. Contribution of trends in survival and coronary-event rates to changes in coronary heart disease mortality: 10-year results from 37 WHO MONICA project populations. Monitoring trends and determinants in cardiovascular disease. Lancet 353(9164):1547–1557, 1999.

16. FB Hu, MJ Stampfer, JE Manson, F Grodstein, GA Colditz, FE Speizer, WC Willett. Trends in the incidence of coronary heart disease and changes in diet and lifestyle in women. N Engl J Med 343(8):530–537, 2000.

17. GL Khor. Cardiovascular epidemiology in the Asia–Pacific region. Asia Pac J Clin Nutr 10(2):76–80, 2001.

18. A Joseph, VR Kutty, CR Soman. High risk for coronary heart disease in Thiruvananthapuram city: a study of serum lipids and other risk factors. Indian Heart J 52(1):29–35, 2000.

19. R Gupta, H Prakash, S Majumdar, S Sharma, VP Gupta. Prevalence of coronary heart disease and coronary risk factors in an urban population of Rajasthan. Indian Heart J 47(4):331–338, 1995.

20. RJ Goldberg, EJ Gorak, J Yarzebski, DW Hosmer Jr, P Dalen, JM Gore, JS Dalen, JE Dalen. A communitywide perspective of sex differences and temporal trends in the incidence and survival rates after acute myocardial infarction and out-of-hospital deaths caused by coronary heart disease. Circulation 87(6):1947–1953, 1993.

21. PG McGovern, JS Pankow, E Shahar, KM Doliszny, AR Folsom, H Luepker, RV Luepker. Recent trends in acute coronary heart disease—mortality, morbidity, medical care, and risk factors. The Minnesota Heart Survey Investigators. N Engl J Med 334(14):884–890, 1996.

22. M de Onis, M Blossner. Prevalence and trends of overweight among preschool children in developing countries. Am J Clin Nutr 72(4):1032–1039, 2000.

23. MI Goran. Metabolic precursors and effects of obesity in children: a decade of progress, 1990–1999. Am J Clin Nutr 73(2):158–171, 2001.

24. AH Mokdad, MK Serdula, WH Dietz, BA Bowman, JS Marks, JP Koplan. The continuing epidemic of obesity in the United States. JAMA 284(13):1650–1651, 2000.

25. I Bengtsson, M Hagman, H Wedel. Age and angina as predictors of quality of life after myocardial infarction: a prospective comparative study. Scand Cardiovasc J 35(4):252–258, 2001.

26. ATPIII. Third report of the National Cholesterol Education Program (NCEP) Expert Panel on Detection, Evaluation, and Treatment of High Blood Cholesterol in Adults (Adult Treatment Panel III).: National Cholesterol Education Program. National Heart, Lung and Blood Institute. National Institutes of Health.; 2001. NIH Publication No. 01-3670.

27. D Tribble, R Krauss. Atherosclerotic Cardiovascular Disease. In: B Bowman, R Russel, eds. Present knowledge in nutrition. Washington, DC: International Life Sciences Institute, 2001, pp. 543–551.

28. Low-dose aspirin and vitamin E in people at cardiovascular risk: a randomised trial in general practice. Collaborative Group of the Primary Prevention Project. Lancet 357(9250):89–95, 2001.

29. R Collins, R Peto, J Armitage. The MRC/BHF Heart Protection Study: preliminary results. Int J Clin Pract 56(1):53–56, 2002.

30. J Muntwyler, CH Hennekens, JE Manson, JE Buring, JM Gaziano. Vitamin supplement use in a low-risk population of US male physicians and subsequent cardiovascular mortality. Arch Intern Med 162(13):1472–1476, 2002.

31. BG Brown, XQ Zhao, A Chait, LD Fisher, MC Cheung, JS Morse, AA Dowdy, EK Marino, EL Bolson, P Alaupovic, J Frohlich, JJ Albers. Simvastatin and niacin, antioxidant vitamins, or the combination for the prevention of coronary disease. N Engl J Med 345(22):1583–1592, 2001.

32. H Li, MI Cybulsky, MA Gimbrone Jr, P Libby. An atherogenic diet rapidly induces VCAM-1, a cytokine-regulatable mononuclear leukocyte adhesion molecule, in rabbit aortic endothelium. Arterioscler Thromb 13(2):197–204, 1993.

33. P Libby, PM Ridker, A Maseri. Inflammation and atherosclerosis. Circulation 105(9):1135–1143, 2002.

34. RJ Woodman, GF Watts, IB Puddey, V Burke, TA Mori, JM Hodgson, LJ Beilin. Leukocyte count and vascular function in Type 2 diabetic subjects with treated hypertension. Atherosclerosis 163(1):175–181, 2002.

35. HC Stary, AB Chandler, RE Dinsmore, V Fuster, S Glagov, W Insull Jr, ME Schwartz, CJ Schwartz, WD Wagner, RW Wissler. A definition of advanced types of atherosclerotic lesions and a histological classification of atherosclerosis. A report from the Committee on Vascular Lesions of the Council on Arteriosclerosis, American Heart Association. Arterioscler Thromb Vasc Biol 15(9):1512–1531, 1995.

36. HC Stary, AB Chandler, RE Dinsmore, V Fuster, S Glagov, W Insull Jr, ME Schwartz, CJ Schwartz, WD Wagner, RW Wissler. A definition of advanced types of atherosclerotic lesions and a histological classification of atherosclerosis. A report from the Committee on Vascular Lesions of the Council on Arteriosclerosis, American Heart Association. Circulation 92(5):1355–1374, 1995.

37. DS Celermajer, KE Sorensen, D Georgakopoulos, C Bull, O Thomas, J Deanfield, JE Deanfield. Cigarette smoking is associated with dose-related and potentially reversible impairment of endothelium-dependent dilation in healthy young adults. Circulation 88(5 Pt 1):2149–2155, 1993.

38. P Clarkson, DS Celermajer, AJ Powe, AE Donald, RM Henry, JE Deanfield. Endothelium-dependent dilatation is impaired in young healthy subjects with a family history of premature coronary disease. Circulation 96(10):3378–3383, 1997.

39. CF Ebenbichler, W Sturm, H Ganzer, J Bodner, B Mangweth, A Ritsch, A Lechleitner, M Lechleitner, B Foger, JR Patsch. Flow-mediated, endothelium-

dependent vasodilatation is impaired in male body builders taking anabolic-androgenic steroids. Atherosclerosis 158(2):483–490, 2001.

40. MA Sader, KA Griffiths, RJ McCredie, DJ Handelsman, DS Celermajer. Androgenic anabolic steroids and arterial structure and function in male body-builders. J Am Coll Cardiol 37(1):224–230, 2001.

41. LA Simons, D Sullivan, J Simons, DS Celermajer. Effects of atorvastatin monotherapy and simvastatin plus cholestyramine on arterial endothelial function in patients with severe primary hypercholesterolaemia. Atherosclerosis 137(1): 197–203, 1998.

42. RA Vogel, MC Corretti, GD Plotnick. Changes in flow-mediated brachial artery vasoactivity with lowering of desirable cholesterol levels in healthy middle-aged men. Am J Cardiol 77(1):37–40, 1996.

43. A Brouet, P Sonveaux, C Dessy, S Moniotte, JL Balligand, O Feron. Hsp90 and caveolin are key targets for the proangiogenic nitric oxide–mediated effects of statins. Circ Res 89(10):866–873, 2001.

44. O Feron, C Dessy, S Moniotte, JP Desager, JL Balligand. Hypercholesterolemia decreases nitric oxide production by promoting the interaction of caveolin and endothelial nitric oxide synthase. J Clin Invest 103(6):897–905, 1999.

45. AM Zeiher, V Schachlinger, SH Hohnloser, B Saurbier, H Just. Coronary atherosclerotic wall thickening and vascular reactivity in humans. Elevated high-density lipoprotein levels ameliorate abnormal vasoconstriction in early atherosclerosis. Circulation 89(6):2525–2532, 1994.

46. P.R.C.–U.S.A. Cardiovascular and Cardiopulmonary Epidemiology Research Group. An epidemiological study of cardiovascular and cardiopulmonary disease risk factors in four populations in the People's Republic of China. Circulation 85:1083–1096, 1992.

47. HC McGill Jr, JP Strong. The geographic pathology of atherosclerosis. Ann N Y Acad Sci 149(2):923–927, 1968.

48. A Keys, C Aravanis, H Blackburn. Seven countries—a multivariate analysis of death and coronary heart disease. Cambridge, MA: Harvard Universvity Press, 1980.

49. M Toor, A Katchalsky, J Agmon, D Allalouf. Atherosclerosis and related factors in immigrants. Circulation 22:265–279, 1960.

50. A Kagan, BR Harris, W Winkelstein Jr, KG Johnson, H Kato, SL Syme, GG Gay, ML Gay, MZ Nichaman, HB Hamilton, J Tillotson. Epidemiologic studies of coronary heart disease and stroke in Japanese men living in Japan, Hawaii and California: demographic, physical, dietary and biochemical characteristics. J Chron Dis 27(7–8):345–364, 1974.

51. MS Brown, JL Goldstein. A receptor-mediated pathway for cholesterol homeostasis. Science 232(4746):34–47, 1986.

52. J Stamler, D Wentworth, JD Neaton. Is relationship between serum cholesterol and risk of premature death from coronary heart disease continuous and graded? Findings in 356,222 primary screenees of the Multiple Risk Factor Intervention Trial (MRFIT). JAMA 256(20):2823–2828, 1986.

53. PW Wilson, RB D'Agostino, D Levy, AM Belanger, H Silbershatz, WB Kannel. Prediction of coronary heart disease using risk factor categories. Circulation 97(18):1837–1847, 1998.
54. Prevention of cardiovascular events and death with pravastatin in patients with coronary heart disease and a broad range of initial cholesterol levels. The Long-Term Intervention with Pravastatin in Ischaemic Disease (LIPID) Study Group. N Engl J Med 339(19):1349–1357, 1998.
55. JR Downs, M Clearfield, S Weis, E Whitney, DR Shapiro, PA Beere, A Stein, EA Stein, W Kruyer, Gotto, AM Jr. Primary prevention of acute coronary events with lovastatin in men and women with average cholesterol levels: results of AFCAPS/TexCAPS, Air Force/Texas Coronary Atherosclerosis Prevention Study. JAMA 279(20):1615–1622, 1998.
56. J Shepherd, SM Cobbe, I Ford, CG Isles, AR Lorimer, PW MacFarlane, JH Packard, CJ Packard. Prevention of coronary heart disease with pravastatin in men with hypercholesterolemia. West of Scotland Coronary Prevention Study Group. N Engl J Med 333(20):1301–1307, 1995.
57. FM Sacks, MA Pfeffer, LA Moye, JL Rouleau, JD Rutherford, TG Cole, L Warnica, JW Warnica, JM Arnold, CC Wun, BR Davis, E Braunwald. The effect of pravastatin on coronary events after myocardial infarction in patients with average cholesterol levels. Cholesterol and Recurrent Events Trial investigators. N Engl J Med 335(14):1001–1009, 1996.
58. JC LaRosa, J He, S Vupputuri. Effect of statins on risk of coronary disease: a meta-analysis of randomized controlled trials. JAMA 282(24):2340–2346, 1999.
59. MR Law, NJ Wald, SG Thompson. By how much and how quickly does reduction in serum cholesterol concentration lower risk of ischaemic heart disease? BMJ 308(6925):367–372, 1994.
60. DM Hegsted, RB McGandy, ML Myers, FJ Stare. Quantitative effects of dietary fat on serum cholesterol in man. Am J Clin Nutr 17(5):281–295, 1965.
61. A Keys, J Anderson, F Grande. Serum cholesterol response to changes in the diet. I. Iodine values of dietary fat versus 2S-P. Metabolism 14:747–758, 1965.
62. RP Mensink, MB Katan. Effect of dietary fatty acids on serum lipids and lipoproteins. A meta-analysis of 27 trials. Arterioscler Thromb 12(8):911–919, 1992.
63. HN Ginsberg, P Kris-Etherton, B Dennis, PJ Elmer, A Ershow, M Lefevre, T Roheim, P Roheim, R Ramakrishnan, R Reed, K Stewart, P Stewart, K Anderson, N Anderson. Effects of reducing dietary saturated fatty acids on plasma lipids and lipoproteins in healthy subjects: the DELTA Study, protocol 1. Arterioscler Thromb Vasc Biol 18(3):441–449, 1998.
64. RM Weggemans, PL Zock, MB Katan. Dietary cholesterol from eggs increases the ratio of total cholesterol to high-density lipoprotein cholesterol in humans: a meta-analysis. Am J Clin Nutr 73(5):885–891, 2001.
65. NM de Roos, ML Bots, MB Katan. Replacement of dietary saturated fatty acids by trans fatty acids lowers serum HDL cholesterol and impairs endothelial function in healthy men and women. Arterioscler Thromb Vasc Biol 21(7):1233–1237, 2001.

66. PM Kris-Etherton, TA Pearson, Y Wan, RL Hargrove, K Moriarty, V Fishell, TD Etherton. High-monounsaturated fatty acid diets lower both plasma cholesterol and triacylglycerol concentrations. Am J Clin Nutr 70(6):1009–1015, 1999.

67. S Yu-Poth, G Zhao, T Etherton, M Naglak, S Jonnalagadda, PM Kris-Etherton. Effects of the National Cholesterol Education Program's Step I and Step II dietary intervention programs on cardiovascular disease risk factors: a meta-analysis. Am J Clin Nutr 69(4):632–646, 1999.

68. JL Tang, JM Armitage, T Lancaster, CA Silagy, GH Fowler, HA Neil. Systematic review of dietary intervention trials to lower blood total cholesterol in free-living subjects. BMJ 316(7139):1213–1220, 1998.

69. AS Truswell. Review of dietary intervention studies: effect on coronary events and on total mortality. Aust N Z J Med 24(1):98–106, 1994.

70. WC Knowler, E Barrett-Connor, SE Fowler, RF Hamman, JM Lachin, EA Nathan, DM Nathan. Reduction in the incidence of type 2 diabetes with lifestyle intervention or metformin. N Engl J Med 346(6):393–403, 2002.

71. J Judd, D Baer, B Clevidence, R Muesing, C SC, G Meijer. Dietary fat and fatty acids and plasma modifying effects of sterol esters (Abstract 503.8). FASEB J 15:A639, 2001.

72. DW Peterson. Effect of soybean sterols in the diet on plasma and liver cholesterol in chicks. Proc Soc Exp Biol 78:143–147, 1951.

73. E Newsholme, A Leech. Biosyntesis and metabolism of cholesterol and the steroid hormones. Biochemistry for the Medical Sciences. Tiptree, Essex, UK: John Wiley and Sons Ltd, Courier International Ltd., 1993, pp. 710–771.

74. P Jones, A Papamandjaris. Lipids: cellular metabolism. In: B Bowman, R Russel, eds. Present Knowledge in Nutrition, 8th ed. Washington, DC: International Life Sciences Institute, 2001, pp. 104–114.

75. J Dietschy. Regulation of cholesterol metabolism in man and other species. Klin Wochenschr 62:338–345, 1984.

76. RP Troiano, RR Briefel, MD Carroll, K Bialostosky. Energy and fat intakes of children and adolescents in the united states: data from the national health and nutrition examination surveys. Am J Clin Nutr 72(5 Suppl):1343S–1353S, 2000.

77. LB Dixon, MA Winkleby, KL Radimer. Dietary intakes and serum nutrients differ between adults from food-insufficient and food-sufficient families: Third National Health and Nutrition Examination Survey, 1988–1994. J Nutr 131(4): 1232–1246, 2001.

78. A Lichtenstein, P Jones. Lipids: Absorption and transport. In: B Bowman, eds. Present knowledge in nutrition, 8th ed. Washington, DC: International Life Sciences Institute, 2001, pp. 92–103.

79. EJ Schaefer. Lipoproteins, nutrition, and heart disease. Am J Clin Nutr 75(2): 191–212, 2002.

80. MC Carey, DM Small, CM Bliss. Lipid digestion and absorption. Annu Rev Physiol 45:651–677, 1983.

81. MD Wilson, LL Rudel. Review of cholesterol absorption with emphasis on dietary and biliary cholesterol. J Lipid Res 35(6):943–955, 1994.

82. SM Grundy, AL Metzger. A physiological method for estimation of hepatic secretion of biliary lipids in man. Gastroenterology 62(6):1200–1217, 1972.

83. MS Bosner, LG Lange, WF Stenson, Ostlund, RE Jr. Percent cholesterol absorption in normal women and men quantified with dual stable isotopic tracers and negative ion mass spectrometry. J Lipid Res 40(2):302–308, 1999.

84. RE Ostlund Jr, MS Bosner, WF Stenson. Cholesterol absorption efficiency declines at moderate dietary doses in normal human subjects. J Lipid Res 40(8): 1453–1458, 1999.

85. SL Acton, PE Scherer, HF Lodish, M Krieger. Expression cloning of SR-BI, a CD36-related class B scavenger receptor. J Biol Chem 269(33):21003–21009, 1994.

86. S Acton, A Rigotti, KT Landschulz, S Xu, HH Hobbs, M Krieger. Identification of scavenger receptor SR-BI as a high density lipoprotein receptor. Science 271(5248):518–520, 1996.

87. H Stangl, G Cao, KL Wyne, HH Hobbs. Scavenger receptor, class B, type I-dependent stimulation of cholesterol esterification by high density lipoproteins, low density lipoproteins, and nonlipoprotein cholesterol. J Biol Chem 273(47): 31002–31008, 1998.

88. H Hauser, JH Dyer, A Nandy, MA Vega, M Werder, E Bieliauskaite, FE Compassi, S Compassi, A Gemperli, D Boffelli, E Wehrli, G Schulthess, MC Phillips. Identification of a receptor mediating absorption of dietary cholesterol in the intestine. Biochemistry 37(51):17843–17850, 1998.

89. SW Altmann, HR Davis Jr, X Yao, M Laverty, DS Compton, LJ Zhu, JH Caplen, MA Caplen, LM Hoos, G Tetzloff, T Priestley, DA Burnett, CD Graziano, MP Graziano. The identification of intestinal scavenger receptor class B, type I (SR-BI) by expression cloning and its role in cholesterol absorption. Biochim Biophys Acta 1580(1):77–93, 2002.

90. M Werder, CH Han, E Wehrli, D Bimmler, G Schulthess, H Hauser. Role of scavenger receptors SR-BI and CD36 in selective sterol uptake in the small intestine. Biochemistry 40(38):11643–11650, 2001.

91. MA Kawashiri, C Maugeais, DJ Rader. High-density lipoprotein metabolism: molecular targets for new therapies for atherosclerosis. Curr Atheroscler Rep 2(5):363–372, 2000.

92. JJ Repa, DJ Mangelsdorf. Nuclear receptor regulation of cholesterol and bile acid metabolism. Curr Opin Biotechnol 10(6):557–563, 1999.

93. JJ Repa, SD Turley, JA Lobaccaro, J Medina, L Li, K Lustig, B Shan, RA Dietschy, JM Dietschy, DJ Mangelsdorf. Regulation of absorption and ABC1-mediated efflux of cholesterol by RXR heterodimers. Science 289(5484):1524–1529, 2000.

94. J McNeish, RJ Aiello, D Guyot, T Turi, C Gabel, C Aldinger, KL Hoppe, ML Royer, LJ Royer, J de Wet, C Broccardo, G Chimini, OL Francone. High density lipoprotein deficiency and foam cell accumulation in mice with targeted disruption of ATP-binding cassette transporter-1. Proc Natl Acad Sci USA 97(8):4245–4250, 2000.

95. KE Berge, H Tian, GA Graf, L Yu, NV Grishin, J Schultz, P Kwiterovich, B Barnes, R Barnes, HH Hobbs. Accumulation of dietary cholesterol in sito-

sterolemia caused by mutations in adjacent ABC transporters. Science 290(5497):1771–1775, 2000.

96. JJ Repa, JM Dietschy, SD Turley. Inhibition of cholesterol absorption by SCH 58053 in the mouse is not mediated via changes in the expression of mRNA for ABCA1, ABCG5, or ABCG8 in the enterocyte. J Lipid Res 43(11):1864–1874, 2002.

97. RE Burrier, AA Smith, DG McGregor, LM Hoos, DL Zilli, Davis, HR Jr. The effect of acyl CoA: cholesterol acyltransferase inhibition on the uptake, ester-ification and secretion of cholesterol by the hamster small intestine. J Pharmacol Exp Ther 272(1):156–163, 1995.

98. G Salen, V Shore, GS Tint, T Forte, S Shefer, I Horak, E Horak, B Dayal, L Nguyen, AK Batta, et al. Increased sitosterol absorption, decreased removal, and expanded body pools compensate for reduced cholesterol synthesis in sitosterolemia with xanthomatosis. J Lipid Res 30(9):1319–1330, 1989.

99. LB Nguyen, G Salen, S Shefer, J Bullock, T Chen, GS Tint, IR Chowdhary, S Lerner. Deficient ileal 3-hydroxy-3-methylglutaryl coenzyme A reductase act-ivity in sitosterolemia: sitosterol is not a feedback inhibitor of intestinal cholesterol biosynthesis. Metabolism 43(7):855–859, 1994.

100. LB Nguyen, M Cobb, S Shefer, G Salen, GC Ness, GS Tint. Regulation of cholesterol biosynthesis in sitosterolemia: effects of lovastatin, cholestyramine, and dietary sterol restriction. J Lipid Res 32(12):1941–1948, 1991.

101. L Nguyen, G Salen, S Shefer, V Shore, GS Tint, G Ness. Unexpected failure of bile acid malabsorption to stimulate cholesterol synthesis in sitosterolemia with xanthomatosis. Comparison with lovastatin. Arteriosclerosis 10(2):289–297, 1990.

102. JA Hubacek, KE Berge, JC Cohen, HH Hobbs. Mutations in ATP-cassette binding proteins G5 (ABCG5) and G8 (ABCG8) causing sitosterolemia. Hum Mutat 18(4):359–360, 2001.

103. I Ikeda, K Tanaka, M Sugano, GV Vahouny, LL Gallo. Inhibition of cho-lesterol absorption in rats by plant sterols. J Lipid Res 29(12):1573–1582, 1988.

104. I Ikeda, Y Tanabe, M Sugano. Effects of sitosterol and sitostanol on micellar solubility of cholesterol. J Nutr Sci Vitaminol (Tokyo) 35(4):361–369, 1989.

105. J Plat, RP Mensink. Increased intestinal ABCA1 expression contributes to the decrease in cholesterol absorption after plant stanol consumption. Faseb J 16(10):1248–1253, 2002.

106. I Ikeda, M Sugano. Some aspects of mechanism of inhibition of cholesterol absorption by beta-sitosterol. Biochim Biophys Acta 732(3):651–658, 1983.

107. B Borgstrom. Partition of lipids between emulsified oil and micellar phases of glyceride-bile salt dispersions. J Lipid Res 8(6):598–608, 1967.

108. L Christiansen, M Karjalainen, R Serimaa, N Lönnroth, T Paakari, J Yliruusi. Phase behaviour of beta-sitosterol-cholesterol and beta-sitostanol precipitates. STP Pharma Sciences 11(2):167–173, 2001.

109. L Normén, P Dutta, A Lia, H Andersson. Soy sterol esters and beta-sitostanol ester as inhibitors of cholesterol absorption in human small bowel. Am J Clin Nutr 71(4):908–913, 2000.

110. PJ Jones, M Raeini-Sarjaz, FY Ntanios, CA Vanstone, JY Feng, WE Parsons. Modulation of plasma lipid levels and cholesterol kinetics by phytosterol versus phytostanol esters. J Lipid Res 41(5):697–705, 2000.
111. G Beaumier-Gallon, C Dubois, M Senft, MF Vergnes, AM Pauli, H Portugal, D Lairon. Dietary cholesterol is secreted in intestinally derived chylomicrons during several subsequent postprandial phases in healthy humans. Am J Clin Nutr 73(5):870–877, 2001.
112. FJ Field, E Born, SN Mathur. Effect of micellar beta-sitosterol on cholesterol metabolism in Caco-2 cells. J Lipid Res 38(2):348–360, 1997.
113. V Piironen, J Toivo, A-M Lampi. Natural sources of dietary plant sterols. J Food Comp Analys, 2000.
114. MT Cerqueira, MM Fry, WE Connor. The food and nutrient intakes of the Tarahumara Indians of Mexico. Am J Clin Nutr 32(4):905–915, 1979.
115. G Morton, S Lee, D Buss, P Lawrance. Intakes and major dietary sources of cholesterol and phytosterols in the British diet. J Hum Nut Diet 8:429–440, 1995.
116. AL Normen, HA Brants, LE Voorrips, HA Andersson, PA van den Brandt, RA Goldbohm. Plant sterol intakes and colorectal cancer risk in the Netherlands Cohort Study on Diet and Cancer. Am J Clin Nutr 74(1):141–148, 2001.
117. K Hirai, C Shimazu, R Takezoe, Y Ozeki. Cholesterol, phytosterol and poly-unsaturated fatty acid levels in 1982 and 1957 Japanese diets. J Nutr Sci Vitaminol (Tokyo) 32(4):363–372, 1986.
118. SS Strom, Y Yamamura, CM Duphorne, MR Spitz, RJ Babaian, PC Pillow, SD Hursting. Phytoestrogen intake and prostate cancer: a case-control study using a new database. Nutr Cancer 33(1):20–25, 1999.
119. PP Nair, N Turjman, G Kessie, B Calkins, GT Goodman, H Davidovitz, G Nimmagadda. Diet, nutrition intake, and metabolism in populations at high and low risk for colon cancer. Dietary cholesterol, beta-sitosterol, and stigmasterol. Am J Clin Nutr 40(4 Suppl):927–930, 1984.
120. L Ellegård, I Bosaeus, H Andersson. Will recommended changes in fat and fibre intake affect cholesterol absorption and sterol excretion? An ileostomy study. Eur J Clin Nutr 54:306–313, 2000.
121. RE Ostlund Jr, SB Racette, A Okeke, WF Stenson. Phytosterols that are naturally present in commercial corn oil significantly reduce cholesterol absorption in humans. Am J Clin Nutr 75(6):1000–1004, 2002.
122. JL Weihrauch, JM Gardner. Sterol content of foods of plant origin. J Am Diet Assoc 73(1):39–47, 1978.
123. TJ Howell, DE MacDougall, PJ Jones. Phytosterols partially explain differences in cholesterol metabolism caused by corn or olive oil feeding. J Lipid Res 39(4):892–900, 1998.
124. S Grundy, H Mok. Effects of low-dose phytosterols on cholesterol absorption in man. In: H Greten, ed. Lipoprotein Metabolism. Berlin: Springer-Verlag, 1976, pp 112–118.
125. AM Lees, HY Mok, RS Lees, MA McCluskey, SM Grundy. Plant sterols as cholesterol-lowering agents: clinical trials in patients with hypercholesterolemia and studies of sterol balance. Atherosclerosis 28(3):325–338, 1977.

126. FH Mattson, SM Grundy, JR Crouse. Optimizing the effect of plant sterols on cholesterol absorption in man. Am J Clin Nutr 35(4):697–700, 1982.
127. HT Vanhanen, J Kajander, H Lehtovirta, TA Miettinen. Serum levels, absorption efficiency, faecal elimination and synthesis of cholesterol during increasing doses of dietary sitostanol esters in hypercholesterolaemic subjects. Clin Sci (Lond) 87(1):61–67, 1994.
128. TA Miettinen, H Vanhanen. Dietary sitostanol related to absorption, synthesis and serum level of cholesterol in different apolipoprotein E phenotypes. Atherosclerosis 105(2):217–226, 1994.
129. RE Ostlund Jr, CA Spilburg, WF Stenson. Sitostanol administered in lecithin micelles potently reduces cholesterol absorption in humans. Am J Clin Nutr 70(5):826–831, 1999.
130. G Gremaud, E Dalan, C Piguet, M Baumgartner, P Ballabeni, B Decarli, ME Berger, A Berger, LB Fay. Effects of non-esterified stanols in a liquid emulsion on cholesterol absorption and synthesis in hypercholesterolemic men. Eur J Nutr 41(2):54–60, 2002.
131. C Spilburg, A Goldberg, J McGill, W Stenson, S Racette, J Bateman, T Ostlund, R Ostlund. Fat-free foods supplemented with soy stanol-lecithin powder reduce cholesterol absorption and LDL-cholesterol. J Am Diet Assoc 103:577–581, 2003.
132. H Gylling, TA Miettinen. Serum cholesterol and cholesterol and lipoprotein metabolism in hypercholesterolaemic NIDDM patients before and during sitostanol ester–margarine treatment. Diabetologia 37(8):773–780, 1994.
133. H Gylling, TA Miettinen. The effect of cholesterol absorption inhibition on low density lipoprotein cholesterol level. Atherosclerosis 117(2):305–308, 1995.
134. H Gylling, TA Miettinen. Effects of inhibiting cholesterol absorption and synthesis on cholesterol and lipoprotein metabolism in hypercholesterolemic non-insulin-dependent diabetic men. J Lipid Res 37(8):1776–1785, 1996.
135. H Gylling, R Radhakrishnan, TA Miettinen. Reduction of serum cholesterol in postmenopausal women with previous myocardial infarction and cholesterol malabsorption induced by dietary sitostanol ester margarine: women and dietary sitostanol. Circulation 96(12):4226–4231, 1997.
136. D Lutjohann, CO Meese, JR Crouse 3rd, K von Bergmann. Evaluation of deuterated cholesterol and deuterated sitostanol for measurement of cholesterol absorption in humans. J Lipid Res 34(6):1039–1046, 1993.
137. D Lutjohann, I Bjorkhem, UF Beil, K von Bergmann. Sterol absorption and sterol balance in phytosterolemia evaluated by deuterium-labeled sterols: effect of sitostanol treatment. J Lipid Res 36(8):1763–1773, 1995.
138. TA Miettinen, M Vuoristo, M Nissinen, HJ Jarvinen, H Gylling. Serum, biliary, and fecal cholesterol and plant sterols in colectomized patients before and during consumption of stanol ester margarine. Am J Clin Nutr 71(5):1095–1102, 2000.
139. CA Vanstone, M Raeini-Sarjaz, WE Parsons, PJ Jones. Unesterified plant sterols and stanols lower LDL-cholesterol concentrations equivalently in hypercholesterolemic persons. Am J Clin Nutr 76(6):1272–1278, 2002.
140. T Heinemann, O Leiss, K von Bergmann. Effect of low-dose sitostanol on serum

cholesterol in patients with hypercholesterolemia. Atherosclerosis 61(3):219–223, 1986.
141. T Heinemann, GA Kullak-Ublick, B Pietruck, K vonBergmann. Mechanisms of action of plant sterols on inhibition of cholesterol absorption. Comparison of sitosterol and sitostanol. Eur J Clin Pharmacol 40(Suppl 1):S59–63, 1991.
142. SM Grundy, HY Mok. Determination of cholesterol absorption in man by intestinal perfusion. J Lipid Res 18(2):263–271, 1977.
143. AM Langkilde, M Champ, H Andersson. Effects of high-resistant-starch banana flour [RS(2)] on in vitro fermentation and the small-bowel excretion of energy, nutrients, and sterols: an ileostomy study. Am J Clin Nutr 75(1):104–111, 2002.
144. LI Christiansen, PL Lahteenmaki, MR Mannelin, TE Seppanen-Laakso, RV Yliruusi, JK Yliruusi. Cholesterol-lowering effect of spreads enriched with microcrystalline plant sterols in hypercholesterolemic subjects. Eur J Nutr 40(2): 66–73, 2001.
145. R Schönheimer. New contributions in sterol metabolism. Science 74:579–584, 1931.
146. R Schönheimer. Über die Bedeutung der Pflanzen sterine für den tierischen Organismus. Hoppe-Seyler's Z Physiol Chem 180:1, 1929.
147. JA Weststrate, GW Meijer. Plant sterol-enriched margarines and reduction of plasma total- and LDL-cholesterol concentrations in normocholesterolaemic and mildly hypercholesterolaemic subjects. Eur J Clin Nutr 52(5):334–343, 1998.
148. T Slota, NA Kozlov, HV Ammon. Comparison of cholesterol and beta-sitosterol: effects on jejunal fluid secretion induced by oleate, and absorption from mixed micellar solutions. Gut 24(7):653–658, 1983.
149. T Heinemann, G Axtmann, K von Bergmann. Comparison of intestinal absorption of cholesterol with different plant sterols in man. Eur J Clin Invest 23(12):827–831, 1993.
150. G Salen, E Ahrens, S Grundy. Metabolism of beta-sitosterol in man. J Clin Invest 49:952–967, 1970.
151. G Salen, G Xu, GS Tint, AK Batta, S Shefer. Hyperabsorption and retention of campestanol in a sitosterolemic homozygote: comparison with her mother and three control subjects. J Lipid Res 41(11):1883–1889, 2000.
152. RE Ostlund Jr, JB McGill, CM Zeng, DF Covey, J Stearns, WF Stenson, CA Spilburg. Gastrointestinal absorption and plasma kinetics of soy Delta(5)-phytosterols and phytostanols in humans. Am J Physiol Endocrinol Metab 282(4):E911–E916, 2002.
153. KR Norum, AC Lilljeqvist, P Helgerud, ER Normann, A Mo, B Selbekk. Esterification of cholesterol in human small intestine: the importance of acyl-CoA:cholesterol acyltransferase. Eur J Clin Invest 9(1):55–62, 1979.
154. KR Norum, P Helgerud, LB Petersen, PH Groot, HR De Jonge. Influence of diets on acyl-CoA:cholesterol acyltransferase and on acyl-CoA:retinol acyltransferase in villous and crypt cells from rat small intestinal mucosa and in the liver. Biochim Biophys Acta 751(2):153–161, 1983.
155. FJ Field, SN Mathur. Beta-sitosterol: esterification by intestinal acylcoenzyme

A: cholesterol acyltransferase (ACAT) and its effect on cholesterol esterification. J Lipid Res 24(4):409–417, 1983.

156.   E Ros. Intestinal absorption of triglyceride and cholesterol. Dietary and pharmacological inhibition to reduce cardiovascular risk. Atherosclerosis 151(2): 357–379, 2000.

157.   F Karpe. Postprandial lipoprotein metabolism and atherosclerosis. J Intern Med 246(4):341–355, 1999.

158.   RS Tilvis, TA Miettinen. Serum plant sterols and their relation to cholesterol absorption. Am J Clin Nutr 43(1):92–97, 1986.

159.   SJ Robins, JM Fasulo, CR Pritzker, GM Patton. Hepatic transport and secretion of unesterified cholesterol in the rat is traced by the plant sterol, sitostanol. J Lipid Res 37(1):15–21, 1996.

160.   AK Bhattacharyya, WE Connor, DS Lin. The origin of plant sterols in the skin surface lipids in humans: from diet to plasma to skin. J Invest Dermatol 80(4):294–296, 1983.

161.   SJ Robins, JM Fasulo. High density lipoproteins, but not other lipoproteins, provide a vehicle for sterol transport to bile. J Clin Invest 99(3):380–384, 1997.

162.   KM Boberg, K Einarsson, I Bjorkhem. Apparent lack of conversion of sitosterol into C24-bile acids in humans. J Lipid Res 31(6):1083–1088, 1990.

163.   DS Goodman, RP Noble. Turnover of plasma cholesterol in man. J Clin Invest 47(2):231–241, 1968.

164.   FR Smith, RB Dell, RP Noble, DS Goodman. Parameters of the three-pool model of the turnover of plasma cholesterol in normal and hyperlipidemic humans. J Clin Invest 57(1):137–148, 1976.

165.   SM Grundy, EH Ahrens Jr, G Salen. Dietary beta-sitosterol as an internal standard to correct for cholesterol losses in sterol balance studies. J Lipid Res 9(3):374–387, 1968.

166.   DS Lin, WE Connor, BE Phillipson. Sterol composition of normal human bile. Effects of feeding shellfish (marine) sterols. Gastroenterology 86(4):611–617, 1984.

167.   MT Subbiah, A Kuksis. Differences in metabolism of cholesterol and sitosterol following intravenous injection in rats. Biochim Biophys Acta 306(1):95–105, 1973.

168.   I Ikeda, M Sugano. Comparison of absorption and metabolism of beta-sitosterol and beta-sitostanol in rats. Atherosclerosis 30(3):227–237, 1978.

169.   H Relas, H Gylling, TA Miettinen. Fate of intravenously administered squalene and plant sterols in human subjects. J Lipid Res 42(6):988–994, 2001.

170.   TA Miettinen, P Puska, H Gylling, H Vanhanen, E Vartiainen. Reduction of serum cholesterol with sitostanol-ester margarine in a mildly hypercholesterolemic population. N Engl J Med 333(20):1308–1312, 1995.

171.   J Plat, RP Mensink. Effects of plant stanol esters on LDL receptor protein expression and on LDL receptor and HMG-CoA reductase mRNA expression in mononuclear blood cells of healthy men and women. FASEB J 16(2):258–260, 2002.

172.   MA Hallikainen, ES Sarkkinen, MI Uusitupa. Plant stanol esters affect serum

cholesterol concentrations of hypercholesterolemic men and women in a dose-dependent manner. J Nutr 130(4):767–776, 2000.

173. HF Hendriks, JA Weststrate, T van Vliet, GW Meijer. Spreads enriched with three different levels of vegetable oil sterols and the degree of cholesterol lowering in normocholesterolaemic and mildly hypercholesterolaemic subjects. Eur J Clin Nutr 53(4):319–327, 1999.

174. S Blomqvist, M Jauhiainen, A von Tol, M Hyvonen, I Torstila, H Vanhanen, T Ehnholm, C Ehnholm. Effect of sitostanol ester on composition and size distribution of low- and high-density lipoprotein. Nutr Metab Cardiovasc Dis 3:158–164, 1993.

175. OA Matvienko, DS Lewis, M Swanson, B Arndt, DL Rainwater, J Stewart, DL Alekel. A single daily dose of soybean phytosterols in ground beef decreases serum total cholesterol and LDL cholesterol in young, mildly hypercholesterolemic men. Am J Clin Nutr 76(1):57–64, 2002.

176. H Gylling, P Puska, E Vartiainen, TA Miettinen. Serum sterols during stanol ester feeding in a mildly hypercholesterolemic population. J Lipid Res 40(4): 593–600, 1999.

177. H Gylling, TA Miettinen. Cholesterol reduction by different plant stanol mixtures and with variable fat intake. Metabolism 48(5):575–580, 1999.

178. HT Vanhanen, S Blomqvist, C Ehnholm, M Hyvonen, M Jauhiainen, I Torstila, TA Miettinen. Serum cholesterol, cholesterol precursors, and plant sterols in hypercholesterolemic subjects with different apoE phenotypes during dietary sitostanol ester treatment. J Lipid Res 34(9):1535–1544, 1993.

179. RP Mensink, S Ebbing, M Lindhout, J Plat, MM van Heugten. Effects of plant stanol esters supplied in low-fat yoghurt on serum lipids and lipoproteins, non-cholesterol sterols and fat soluble antioxidant concentrations. Atherosclerosis 160(1):205–213, 2002.

180. M Hallikainen. Role of plant stanol ester and sterol ester–enriched margarines in the treatment of hypercholesterolemia [PhD thesis]. Kuopio, Finland: University of Kuopio and Kuopio University Hospital; 2001.

181. TA Miettinen, TE Strandberg, H Gylling. Noncholesterol sterols and cholesterol lowering by long-term simvastatin treatment in coronary patients: relation to basal serum cholestanol. Arterioscler Thromb Vasc Biol 20(5):1340–1346, 2000.

182. H Vanhanen. Cholesterol malabsorption caused by sitostanol ester feeding and neomycin in pravastatin-treated hypercholesterolaemic patients. Eur J Clin Pharmacol 47(2):169–176, 1994.

183. LA Simons. Additive effect of plant sterol-ester margarine and cerivastatin in lowering low-density lipoprotein cholesterol in primary hypercholesterolemia. Am J Cardiol 90(7):737–740, 2002.

184. J Plat, RP Mensink. Vegetable oil based versus wood based stanol ester mixtures: effects on serum lipids and hemostatic factors in non-hypercholesterolemic subjects. Atherosclerosis 148(1):101–112, 2000.

185. F Nigon, C Serfaty-Lacrosniere, I Beucler, D Chauvois, C Neveu, P Giral, MJ Bruckert, E Bruckert. Plant sterol-enriched margarine lowers plasma LDL in

hyperlipidemic subjects with low cholesterol intake: effect of fibrate treatment. Clin Chem Lab Med 39(7):634–640, 2001.

186. C Bailey, P Bryla, A Malick. The use of the intestinal epithelial cell culture model, Caco-2, in pharmaceutical development. Adv Drug Deliv Rev 22(1–2): 85–103, 1996.

187. C Bailey, P Bryla, A Malick. The use of the intestinal epithelial cell culture model, Caco-2, in pharmaceutical development. Adv Drug Deliv Rev 22(1–2): 85–103, 1996.

188. H Gylling, P Puska, E Vartiainen, TA Miettinen. Retinol, vitamin D, carotenes and alpha-tocopherol in serum of a moderately hypercholesterolemic population consuming sitostanol ester margarine. Atherosclerosis 145(2):279–285, 1999.

189. TT Nguyen. The cholesterol-lowering action of plant stanol esters. J Nutr 129(12):2109–2112, 1999.

190. AF Vuorio, H Gylling, H Turtola, K Kontula, P Ketonen, TA Miettinen. Stanol ester margarine alone and with simvastatin lowers serum cholesterol in families with familial hypercholesterolemia caused by the FH-North Karelia mutation. Arterioscler Thromb Vasc Biol 20(2):500–506, 2000.

191. A Tammi, T Ronnemaa, H Gylling, L Rask-Nissila, J Viikari, J Tuominen, K Simell, O Simell. Plant stanol ester margarine lowers serum total and low-density lipoprotein cholesterol concentrations of healthy children: the STRIP project. Special Turku Coronary Risk Factors Intervention Project. J Pediatr 136(4):503–510, 2000.

192. J Plat, EN van Onselen, MM van Heugten, RP Mensink. Effects on serum lipids, lipoproteins and fat soluble antioxidant concentrations of consumption frequency of margarines and shortenings enriched with plant stanol esters. Eur J Clin Nutr 54(9):671–677, 2000.

193. J Plat, RP Mensink. Effects of diets enriched with two different plant stanol ester mixtures on plasma ubiquinol-10 and fat-soluble antioxidant concentrations. Metabolism 50(5):520–529, 2001.

194. P Borel, P Grolier, M Armand, A Partier, H Lafont, D Lairon, V Azais-Bra-esco. Carotenoids in biological emulsions: solubility, surface-to-core distribution, and release from lipid droplets. J Lipid Res 37(2):250–261, 1996.

195. L Normen, Y Pafumi, M Armand, D Lairon. Effects of free and esterified plant sterols on cholesterol solubility, distribution and release to the aqueous phase during emulsification and fat digestion. Manuscript 2002.

196. H Relas, H Gylling, TA Miettinen. Acute effect of dietary stanyl ester dose on postabsorptive alpha-tocopherol, beta-carotene, retinol and retinyl palmitate concentrations. Br J Nutr 85(2):141–147, 2001.

197. P Borel, P Grolier, Y Boirie, L Simonet, E Verdier, Y Rochette, MC Alexandre-Gouabau, B Beaufrere, D Lairon, V Azais-Braesco. Oxidative stress status and antioxidant status are apparently not related to carotenoid status in healthy subjects. J Lab Clin Med 132(1):61–66, 1998.

198. M Noakes, P Clifton, F Ntanios, W Shrapnel, I Record, J McInerney. An increase in dietary carotenoids when consuming plant sterols or stanols is ef-

fective in maintaining plasma carotenoid concentrations. Am J Clin Nutr 75(1): 79–86, 2002.
199. JD Barth. Which tools are in your cardiac workshop? Carotid ultrasound, endothelial function, and magnetic resonance imaging. Am J Cardiol 87(4A): 8A–14A, 2001.
200. MH Moghadasian, BM McManus, DV Godin, B Rodrigues, JJ Frohlich. Proatherogenic and antiatherogenic effects of probucol and phytosterols in apolipoprotein E–deficient mice: possible mechanisms of action. Circulation 99(13):1733–1739, 1999.
201. MH Moghadasian, BM McManus, PH Pritchard, JJ Frohlich. "Tall oil"– derived phytosterols reduce atherosclerosis in ApoE-deficient mice. Arterioscler Thromb Vasc Biol 17(1):119–126, 1997.
202. OL Volger, RP Mensink, J Plat, G Hornstra, LM Havekes, HM Princen. Dietary vegetable oil and wood derived plant stanol esters reduce atheroscler- otic lesion size and severity in apoE*3-Leiden transgenic mice. Atherosclerosis 157(2):375–381, 2001.
203. FY Ntanios, PJ Jones, JJ Frohlich. Dietary sitostanol reduces plaque formation but not lecithin cholesterol acyl transferase activity in rabbits. Atherosclerosis 138(1):101–110, 1998.
204. FY Ntanios, AJ van de Kooij, EA de Deckere, GS Duchateau, EA Trautwein. Effects of various amounts of dietary plant sterol esters on plasma and hepatic sterol concentration and aortic foam cell formation of cholesterol-fed hamsters. Atherosclerosis 169:41–50, 2003.
205. MH Moghadasian, DV Godin, BM McManus, JJ Frohlich. Lack of regression of atherosclerotic lesions in phytosterol-treated apoE-deficient mice. Life Sci 64(12):1029–1036, 1999.
206. T Ngyuen. Response to comments from Ntanios et al. regarding "The cho- lesterol lowering action of plant stanol esters." J Nutr 130(9):2391, 2000.
207. L Ellegård, I Bosaeus. Cholesterol absorption and excretion in ileostomy sub- jects on high- and low-dietary-cholesterol intakes. Am J Clin Nutr 59(1):48–52, 1994.
208. CJ Glueck, J Speirs, T Tracy, P Streicher, E Illig, J Vandegrift. Relationships of serum plant sterols (phytosterols) and cholesterol in 595 hypercholesterolemic subjects, and familial aggregation of phytosterols, cholesterol, and premature coronary heart disease in hyperphytosterolemic probands and their first-degree relatives. Metabolism 40(8):842–848, 1991.
209. T Sudhop, BM Gottwald, K von Bergmann. Serum plant sterols as a potential risk factor for coronary heart disease. Metabolism 51(12):1519–1521, 2002.
210. AB Awad, AJ Smith, CS Fink. Plant sterols regulate rat vascular smooth muscle cell growth and prostacyclin release in culture. Prostaglandins Leukot Essent Fatty Acids 64(6):323–330, 2001.
211. O Pollak, D Kritchevsky. Sitosterol. New York: Marcel Dekker, 1981.
212. OJ Pollak. Effect of plant sterols on serum lipids and atherosclerosis. Pharmacol Ther 31(3):177–208, 1985.
213. F Perez-Jimenez, A Espino, F Lopez-Segura, J Blanco, V Ruiz-Gutierrez, JL

Lopez-Miranda, J Lopez-Miranda, J Jimenez-Pereperez, JM Ordovas. Lipoprotein concentrations in normolipidemic males consuming oleic acid-rich diets from two different sources: olive oil and oleic acid-rich sunflower oil. Am J Clin Nutr 62(4):769–775, 1995.

214. A Pedersen, MW Baumstark, P Marckmann, H Gylling, B Sandstrom. An olive oil-rich diet results in higher concentrations of LDL cholesterol and a higher number of LDL subfraction particles than rapeseed oil and sunflower oil diets. J Lipid Res 41(12):1901–1911, 2000.

215. TE Strandberg, RS Tilvis, TA Miettinen. Metabolic variables of cholesterol during squalene feeding in humans: comparison with cholestyramine treatment. J Lipid Res 31(9):1637–1643, 1990.

216. TA Miettinen, H Vanhanen. Serum concentration and metabolism of cholesterol during rapeseed oil and squalene feeding. Am J Clin Nutr 59(2):356–363, 1994.

217. AF Stalenhoef, M Hectors, PN Demacker. Effect of plant sterol-enriched margarine on plasma lipids and sterols in subjects heterozygous for phytosterolaemia. J Intern Med 249(2):163–166, 2001.

218. MA Hallikainen, ES Sarkkinen, H Gylling, AT Erkkila, MI Uusitupa. Comparison of the effects of plant sterol ester and plant stanol ester-enriched margarines in lowering serum cholesterol concentrations in hypercholesterolaemic subjects on a low-fat diet. Eur J Clin Nutr 54(9):715–725, 2000.

219. P Nestel, M Cehun, S Pomeroy, M Abbey, G Weldon. Cholesterol-lowering effects of plant sterol esters and non-esterified stanols in margarine, butter and low-fat foods. Eur J Clin Nutr 55(12):1084–1090, 2001.

220. P Dutta, L Normen. Capillary column gas-liquid chromatographic separation of delta-5 unsaturated and saturated phytosterols. J Chromatog A 816:177–184, 1998.

221. S Kochar. Influence of processing on sterols of edible vegetable oils. Prog Lipid Res 22:161–188, 1983.

222. MA Hallikainen, MI Uusitupa. Effects of two low-fat stanol ester-containing margarines on serum cholesterol concentrations as part of a low-fat diet in hypercholesterolemic subjects. Am J Clin Nutr 69(3):403–410, 1999.

223. MN Vissers, PL Zock, GW Meijer, MB Katan. Effect of plant sterols from rice bran oil and triterpene alcohols from sheanut oil on serum lipoprotein concentrations in humans. Am J Clin Nutr 72(6):1510–1515, 2000.

224. EA Stein. Identification and treatment of individuals at high risk of coronary heart disease. Am J Med 112(Suppl 8A):3S–9S, 2002.

225. H Schulte, P Cullen, G Assmann. Obesity, mortality and cardiovascular disease in the Munster Heart Study (PROCAM). Atherosclerosis 144(1):199–209, 1999.

226. AD Sniderman, B Lamarche, J Tilley, D Seccombe, J Frohlich. Hypertriglyceridemic hyperapoB in type 2 diabetes. Diabetes Care 25(3):579–582, 2002.

227. H Gylling, TA Miettinen. Cholesterol absorption, synthesis, and LDL metabolism in NIDDM. Diabetes Care 20(1):90–95, 1997.

228. P Simonen, H Gylling, AN Howard, TA Miettinen. Introducing a new com-

ponent of the metabolic syndrome: low cholesterol absorption. Am J Clin Nutr 72(1):82–88, 2000.
229. PP Simonen, H Gylling, TA Miettinen. Body weight modulates cholesterol metabolism in non-insulin dependent type 2 diabetics. Obes Res 10(5):328–335, 2002.
230. H Kojima, H Hidaka, K Matsumura, Y Fujita, S Yamada, M Haneda, H Kikkawa, R Kikkawa, A Kashiwagi. Effect of glycemic control on plasma plant sterol levels and post-heparin diamine oxidase activity in type 1 diabetic patients. Atherosclerosis 145(2):389–397, 1999.
231. Y-M Lee, H Haastert, W Sherbaum, H Hauner. Ambulante Studie sur Wirkung von phytosterol-angerichter Margarine auf den Lipidstatus von Typ-2 Diabetikern. Proc Germ Nutr Soc 3(7 (abstract V9), 2001.
232. MA Austin, JL Breslow, CH Hennekens, JE Buring, WC Willett, RM Krauss. Low-density lipoprotein subclass patterns and risk of myocardial infarction. JAMA 260(13):1917–1921, 1988.
233. CL Williams, MC Bollella, BA Strobino, L Boccia, L Campanaro. Plant stanol ester and bran fiber in childhood: effects on lipids, stool weight and stool frequency in preschool children. J Am Coll Nutr 18(6):572–581, 1999.
234. Y Aggoun, D Bonnet, D Sidi, JP Girardet, E Brucker, M Polak, ME Safar, BI Levy. Arterial mechanical changes in children with familial hypercholesterolemia. Arterioscler Thromb Vasc Biol 20(9):2070–2075, 2000.
235. BW McCrindle, E Helden, G Cullen-Dean, WT Conner. A randomized crossover trial of combination pharmacologic therapy in children with familial hyperlipidemia. Pediatr Res 51(6):715–721, 2002.
236. S de Jongh, L Ose, T Szamosi, C Gagne, M Lambert, R Scott, P Perron, D Saborio, M Saborio, MB Tuohy, M Stepanavage, A Sapre, B Gumbiner, M van Trotsenburg, AS van Trotsenburg, HD Bakker, JJ Kastelein. Efficacy and safety of statin therapy in children with familial hypercholesterolemia: a randomized, double-blind, placebo-controlled trial with simvastatin. Circulation 106(17):2231–2237, 2002.
237. AL Amundsen, L Ose, MS Nenseter, FY Ntanios. Plant sterol ester–enriched spread lowers plasma total and LDL cholesterol in children with familial hypercholesterolemia. Am J Clin Nutr 76(2):338–344, 2002.
238. EA Stein. Treatment of familial hypercholesterolemia with drugs in children. Arteriosclerosis 9(1 Suppl):I145–I151, 1989.
239. RW Mahley. Apolipoprotein E: cholesterol transport protein with expanding role in cell biology. Science 240(4852):622–630, 1988.
240. F Kuipers, MC Jong, Y Lin, M Eck, R Havinga, V Bloks, HJ Verkade, MH Moshage, H Moshage, TJ Berkel, RJ Vonk, LM Havekes. Impaired secretion of very low density lipoprotein-triglycerides by apolipoprotein E-deficient mouse hepatocytes. J Clin Invest 100(11):2915–2922, 1997.
241. SH Quarfordt, B Oswald, B Landis, HS Xu, SH Zhang, N Maeda. In vivo cholesterol kinetics in apolipoprotein E-deficient and control mice. J Lipid Res 36(6):1227–1235, 1995.
242. SH Zhang, RL Reddick, JA Piedrahita, N Maeda. Spontaneous hypercho-

lesterolemia and arterial lesions in mice lacking apolipoprotein E. Science 258(5081):468–471, 1992.

243. OL Volger, H van der Boom, EC de Wit, W van Duyvenvoorde, G Hornstra, J Havekes, LM Havekes, RP Mensink, HM Princen. Dietary plant stanol esters reduce VLDL cholesterol secretion and bile saturation in apolipoprotein E*3-Leiden transgenic mice. Arterioscler Thromb Vasc Biol 21(6):1046–1052, 2001.

244. EJ Schaefer, S Lamon-Fava, S Johnson, JM Ordovas, MM Schaefer, WP Wilson, PW Wilson. Effects of gender and menopausal status on the association of apolipoprotein E phenotype with plasma lipoprotein levels. Results from the Framingham Offspring Study. Arterioscler Thromb 14(7):1105–1113, 1994.

245. JM Ordovas, L Litwack-Klein, PW Wilson, MM Schaefer, EJ Schaefer. Apolipoprotein E isoform phenotyping methodology and population frequency with identification of apoE1 and apoE5 isoforms. J Lipid Res 28(4):371–380, 1987.

246. F Schiele, D De Bacquer, M Vincent-Viry, U Beisiegel, C Ehnholm, A Evans, A Martins, MC Martins, S Sans, C Sass, S Visvikis, G De Backer, G Siest. Apolipoprotein E serum concentration and polymorphism in six European countries: the ApoEurope Project. Atherosclerosis 152(2):475–488, 2000.

247. YA Kesaniemi, C Ehnholm, TA Miettinen. Intestinal cholesterol absorption efficiency in man is related to apoprotein E phenotype. J Clin Invest 80(2):578–581, 1987.

248. K Ishiwata, Y Homma, T Ishikawa, H Nakamura, S Handa. Influence of apolipoprotein E phenotype on metabolism of lipids and apolipoproteins after plant stanol ester ingestion in Japanese subjects. Nutrition 18(7–8):561–565, 2002.

249. A Geelen, PL Zock, JH de Vries, MB Katan. Apolipoprotein E polymorphism and serum lipid response to plant sterols in humans. Eur J Clin Invest 32(10):738–742, 2002.

250. MA Hallikainen, ES Sarkkinen, MI Uusitupa. Effects of low-fat stanol ester enriched margarines on concentrations of serum carotenoids in subjects with elevated serum cholesterol concentrations. Eur J Clin Nutr 53(12):966–969, 1999.

251. KC Maki, MH Davidson, DM Umporowicz, EJ Schaefer, MR Dicklin, KA Chen, S Chen, JR McNamara, BW Gebhart, JD Ribaya-Mercado, G Perrone, SJ Robins, WC Franke. Lipid responses to plant-sterol-enriched reduced-fat spreads incorporated into a National Cholesterol Education Program Step I diet. Am J Clin Nutr 74(1):33–43, 2001.

252. MA Denke. Lack of efficacy of low-dose sitostanol therapy as an adjunct to a cholesterol-lowering diet in men with moderate hypercholesterolemia. Am J Clin Nutr 61(2):392–396, 1995.

253. F Pazzucconi, F Dorigotti, G Gianfranceschi, G Campagnoli, M Sirtori, G Sirtori, CR Sirtori. Therapy with HMG CoA reductase inhibitors: characteristics of the long-term permanence of hypocholesterolemic activity. Atherosclerosis 117(2):189–198, 1995.

254. SN Blair, DM Capuzzi, SO Gottlieb, T Nguyen, JM Morgan, NB Cater. In-

cremental reduction of serum total cholesterol and low-density lipoprotein cholesterol with the addition of plant stanol ester-containing spread to statin therapy. Am J Cardiol 86(1):46–52, 2000.

255. HA Neil, GW Meijer, LS Roe. Randomised controlled trial of use by hypercholesterolaemic patients of a vegetable oil sterol-enriched fat spread. Atherosclerosis 156(2):329–337, 2001.

256. M Becker, D Staab, K Von Bergman. Long-term treatment of severe familial hypercholesterolemia in children: effect of sitosterol and bezafibrate. Pediatrics 89(1):138–142, 1992.

257. H Gylling, T Miettinen. Serum cholesterol lowering by dietary sitostanol is associated with reduced absorption and synthesis of cholesterol and decreased transport of LDL Apoprotein B in men with type II diabetes. In: AJ Gotto, M Richter, W Richter, P Schwandt, eds. 4th International Symposium of the Treatment of Severe Dyslipidemia in the Prevention of Coronary Heart Disease.; 1992; Munich: Karger, 1992.

258. P Jones, S Kafonek, I Laurora, D Hunninghake. Comparative dose efficacy study of atorvastatin versus simvastatin, pravastatin, lovastatin, and fluvastatin in patients with hypercholesterolemia (the CURVES study). Am J Cardiol 81(5):582–587, 1998.

259. C Gagne, D Gaudet, E Bruckert. Efficacy and safety of ezetimibe coadministered with atorvastatin or simvastatin in patients with homozygous familial hypercholesterolemia. Circulation 105(21):2469–2475, 2002.

260. RC Pasternak, SC Smith Jr, CN Bairey-Merz, SM Grundy, JI Cleeman, C Lenfant. ACC/AHA/NHLBI clinical advisory on the use and safety of statins. Circulation 106(8):1024–1028, 2002.

261. M Evans, RA Anderson, J Graham, GR Ellis, K Morris, S Davies, SK Jackson, MJ Lewis, MP Frenneaux, A Rees. Ciprofibrate therapy improves endothelial function and reduces postprandial lipemia and oxidative stress in type 2 diabetes mellitus. Circulation 101(15):1773–1779, 2000.

262. B Staels, J Dallongeville, J Auwerx, K Schoonjans, E Leitersdorf, JC Fruchart. Mechanism of action of fibrates on lipid and lipoprotein metabolism. Circulation 98(19):2088–2093, 1998.

263. R Uauy-Dagach, E Hertrampf. Food-based dietary recommendations. In: B Russel, R Russel, eds. Present Knowledge in Nutrition. 8th ed. Washington, DC: International Life Sciences Institute, 2001, pp. 617–635.

264. Anonymous. Canada's Food Guide to Healthy Eating for People Four Years and Over. Canada: Minister of Public Works and Government Services, 1997.

265. USDA/USDHHS. Dietary Guidelines for Americans. USDA Center for Nutrition Policy and Prevention; 2000.

266. RM Krauss, RH Eckel, B Howard, LJ Appel, SR Daniels, RJ Deckelbaum, JW Erdman Jr, P Kris-Etherton, IJ Goldberg, TA Kotchen, AH Lichtenstein, WE Mitch, R Mullis, K Robinson, J Wylie-Rosett, S St Jeor, J Suttie, DL Bazzarre, TL Bazzarre. AHA Dietary Guidelines: revision 2000: A statement for healthcare professionals from the Nutrition Committee of the American Heart Association. Circulation 102(18):2284–2299, 2000.

267. A Bruce. Strategies to prevent the metabolic syndrome at the population level: role of authorities and non-governmental bodies. Br J Nutr 83(Suppl 1):S181–S186, 2000.

268. A FDA. Food Labeling: Health Claims; Plant Sterol/Stanol Esters and Coronary Heart Disease. Interim Final Rule. Washington, DC, 2000 September 8.

269. AM Garber, WS Browner, SB Hulley. Cholesterol screening in asymptomatic adults, revisited, Part 2. Ann Intern Med 124(5):518–531, 1996.

270. SB EUROASPIRE. Lifestyle and risk factor management and use of drug therapies in coronary patients from 15 countries; principal results from EUROASPIRE II Euro Heart Survey Programme. Eur Heart J 22(7):554–572, 2001.

271. K Nelson, K Norris, CM Mangione. Disparities in the diagnosis and pharmacologic treatment of high serum cholesterol by race and ethnicity: data from the Third National Health and Nutrition Examination Survey. Arch Intern Med 162(8):929–935, 2002.

272. Dietary Recommendations. Celia Villalobos, Spain: Spanish Cardiology Society/Spanish Health Minister; 2001 May.

273. Cholesterol lowering diet. Diet in the promotion of heart health. Association of Clinical and Public Health Nutritionists in Finland; 2002. Report No. 2.

274. Plant sterols and stanols. A position statement from the Heart Foundation's Nutrition and Metabolism Advisory Committee. National Heart Foundation of Australia; 1999 July.

275. DE Hilleman, JO Phillips, SM Mohiuddin, KL Ryschon, CA Pedersen. A population-based treat-to-target pharmacoeconomic analysis of HMG-CoA reductase inhibitors in hypercholesterolemia. Clin Ther 21(3):536–562, 1999.

276. SC Smith Jr, SN Blair, RO Bonow, LM Brass, MD Cerqueira, K Dracup, V Gotto, A Gotto, SM Grundy, NH Miller, A Jacobs, D Jones, RM Krauss, L Ockene, I Ockene, RC Pasternak, T Pearson, MA Pfeffer, RD Starke, KA Taubert. AHA/ACC Scientific Statement: AHA/ACC guidelines for preventing heart attack and death in patients with atherosclerotic cardiovascular disease: 2001 update: A statement for healthcare professionals from the American Heart Association and the American College of Cardiology. Circulation 104(13): 1577–1579, 2001.

277. I Hjermann, K Velve Byre, I Holme, P Leren. Effect of diet and smoking intervention on the incidence of coronary heart disease. Report from the Oslo Study Group of a randomised trial in healthy men. Lancet 2(8259):1303–1310, 1981.

278. D Ornish, LW Scherwitz, JH Billings, SE Brown, KL Gould, TA Merritt, S Armstrong, WT Armstrong, TA Ports, RL Kirkeeide, C Hogeboom, RJ Brand. Intensive lifestyle changes for reversal of coronary heart disease. JAMA 280(23): 2001–2007, 1998.

279. M de Lorgeril, P Salen, JL Martin, I Monjaud, J Delaye, N Mamelle. Mediterranean diet, traditional risk factors, and the rate of cardiovascular complications after myocardial infarction: final report of the Lyon Diet Heart Study. Circulation 99(6):779–785, 1999.

280. RB Singh, G Dubnov, MA Niaz, S Ghosh, R Singh, SS Rastogi, O Manor, D Berry, EM Berry. Effect of an Indo-Mediterranean diet on progression of coronary artery disease in high risk patients (Indo-Mediterranean Diet Heart Study): a randomised single-blind trial. Lancet 360(9344):1455–1461, 2002.
281. SCF. Opinion of the Scientific Committee on Food on a report on post launch monitoring of yellow-fat spreads with added phytosterol esters. Brussels:Health and Consumer Protection Directorate-General, European Union; October 4, 2002. Report No. SCF/CS/NF/DOS/21 ADD 2 Final.
282. M Anttolainen, R Luoto, A Uutela, J Boice, W Blot, J McLaughlin, et al. Characteristics of users and non-users of plant stanol ester margarine in Finland—an approach to study functional foods. J Am Diet Assoc 101:1365–1368, 2001.
283. DJ Jenkins, CW Kendall, D Faulkner, E Vidgen, EA Trautwein, TL Parker, A Koumbridis, G Koumbridis, KG Lapsley, RG Josse, LA Leiter, PW Connelly. A dietary portfolio approach to cholesterol reduction: combined effects of plant sterols, vegetable proteins, and viscous fibers in hypercholesterolemia. Metabolism 51(12):1596–1604, 2002.
284. Y Saito, Y Goto, N Nakaya, Y Hata, Y Homma, C Naito, H Hayashi, H Yamamoto, M Yamamoto, I Takeuchi, et al. Dose-dependent hypolipidemic effect of an inhibitor of HMG-CoA reductase, pravastatin (CS-514), in hypercholesterolemic subjects. A double blind test. Atherosclerosis 72(2–3):205–211, 1988.
285. R Vogel, E Schaefer. Should all patients with cardiovascular disease receive statin therapy? Am J Manag Care 7(5 Suppl):S117–S124, 2001.
286. CM Albert, H Campos, MJ Stampfer, PM Ridker, JE Manson, WC Willett, J Ma. Blood levels of long-chain n-3 fatty acids and the risk of sudden death. N Engl J Med 346(15):1113–1118, 2002.
287. T Rissanen, S Voutilainen, K Nyyssonen, TA Lakka, JT Salonen. Fish oil–derived fatty acids, docosahexaenoic acid and docosapentaenoic acid, and the risk of acute coronary events: the Kuopio ischaemic heart disease risk factor study. Circulation 102(22):2677–2679, 2000.
288. JT GISSI. Dietary supplementation with n-3 polyunsaturated fatty acids and vitamin E after myocardial infarction: results of the GISSI-Prevenzione trial. Gruppo Italiano per lo Studio della Sopravvivenza nell'Infarto miocardico. Lancet 354(9177):447–455, 1999.
289. IA Brouwer, PL Zock, LG van Amelsvoort, MB Katan, EG Schouten. Association between n-3 fatty acid status in blood and electrocardiographic predictors of arrhythmia risk in healthy volunteers. Am J Cardiol 89(5):629–631, 2002.
290. HS Ewart, LK Cole, J Kralovec, H Layton, JM Curtis, JL Wright, MG Murphy. Fish oil containing phytosterol esters alters blood lipid profiles and left ventricle generation of thromboxane a(2) in adult guinea pigs. J Nutr 132(6):1149–1152, 2002.
291. M-P St-Onge, R Ross, W Parsons, P Jones, Consumption of a functional oil containing medium chain triglycerides by overweight men increases energy ex-

pediture and decreases body adiposity compared to a diet rich in oilve oil. Obes Res (in press).

292. R Ayesh, JA Weststrate, PN Drewitt, PA Hepburn. Safety evaluation of phytosterol esters. Part 5. Faecal short-chain fatty acid and microflora content, faecal bacterial enzyme activity and serum female sex hormones in healthy normolipidaemic volunteers consuming a controlled diet either with or without a phytosterol ester-enriched margarine. Food Chem Toxicol 37(12):1127–1138, 1999.

293. MH Davidson, KC Maki, DM Umporowicz, KA Ingram, MR Dicklin, E Lane, RW Lane, JR McNamara, JD Ribaya-Mercado, G Perrone, SJ Robins, WC Franke. Safety and tolerability of esterified phytosterols administered in reduced-fat spread and salad dressing to healthy adult men and women. J Am Coll Nutr 20(4):307–319, 2001.

294. H Hendriks. One year follow-up study on the use of a low-fat spread enriched with plant sterol esters (abstract). In: 17th International Conference of Nutrition.; Vienna, Austria, 2001.

295. PJ Jones, T Howell, DE MacDougall, JY Feng, W Parsons. Short-term administration of tall oil phytosterols improves plasma lipid profiles in subjects with different cholesterol levels. Metabolism 47(6):751–756, 1998.

296. JT Judd, DJ Baer, SC Chen, BA Clevidence, RA Muesing, M Kramer, GW Meijer. Plant sterol esters lower plasma lipids and most carotenoids in mildly hypercholesterolemic adults. Lipids 37(1):33–42, 2002.

297. H Niinikoski, J Viikari, T Palmu. Cholesterol-lowering effect and sensory properties of sitostanol ester margarine in normocholesterolemic adults. Scand J Nutr 41(1):9–12, 1997.

298. X Pelletier, S Belbraouet, D Mirabel, F Mordret, JL Perrin, X Pages, G Debry. A diet moderately enriched in phytosterols lowers plasma cholesterol concentrations in normocholesterolemic humans. Ann Nutr Metab 39(5): 291–295, 1995.

299. A Sierksma, JA Weststrate, GW Meijer. Spreads enriched with plant sterols, either esterified 4,4-dimethylsterols or free 4-desmethylsterols, and plasma total- and LDL-cholesterol concentrations. Br J Nutr 82(4):273–282, 1999.

300. EH Temme, PG Van Hoydonck, EG Schouten, H Kesteloot. Effects of a plant sterol–enriched spread on serum lipids and lipoproteins in mildly hypercholesterolaemic subjects. Acta Cardiol 57(2):111–115, 2002.

301. A Andersson, B Karlstrom, R Mohsen, B Vessby. Cholesterol-lowering of a stanol ester–containing low-fat margarine in conjunction with a strict lipid-lowering diet. Eur Heart J 1(suppl S):S80–S90, 1999.

302. J De Graaf, PR De Sauvage Nolting, M Van Dam, EM Belsey, JJ Kastelein, P Haydn Pritchard, AF Stalenhoef. Consumption of tall oil-derived phytosterols in a chocolate matrix significantly decreases plasma total and low-density lipoprotein-cholesterol levels. Br J Nutr 88(5):479–488, 2002.

303. PJ Jones, FY Ntanios, M Raeini-Sarjaz, CA Vanstone. Cholesterol-lowering efficacy of a sitostanol-containing phytosterol mixture with a prudent diet in hyperlipidemic men. Am J Clin Nutr 69(6):1144–1150, 1999.

304. TT Nguyen, LC Dale, K von Bergmann, IT Croghan. Cholesterol-lowering effect of stanol ester in a US population of mildly hypercholesterolemic men and women: a randomized controlled trial. Mayo Clin Proc 74(12):1198–1206, 1999.

305. M Becker, D Staab, K Von Bergmann. Treatment of severe familial hypercholesterolemia in childhood with sitosterol and sitostanol. J Pediatr 122(2): 292–296, 1993.

306. H Gylling, MA Siimes, TA Miettinen. Sitostanol ester margarine in dietary treatment of children with familial hypercholesterolemia. J Lipid Res 36(8): 1807–1812, 1995.

7

Plant Sterols in Functional Foods

Robert A. Moreau
U.S. Department of Agriculture, Wyndmoor, Pennsylvania, U.S.A.

I. INTRODUCTION—DEFINITIONS OF NUTRACEUTICALS, FUNCTIONAL FOODS, AND RELATED PRODUCTS

The terms *nutraceutical* and *functional food* are often used interchangeably, with the latter currently being the more popular term.

In 1989, Dr. Stephen L. DeFelice of the Foundation for Innovation in Medicine coined the term nutraceutical and defined it as "any substance that is a food or part of a food that provides medical and/or health benefits, including the prevention and treatment of disease" (1). According to De-Felice, *nutraceuticals* include *foods, dietary supplements,* and *medical foods.* According to the Nutraceuticals Institute (Rutgers and St. Joseph Universities), "*nutraceuticals* (often referred to as *phytochemicals* or *functional foods*) are natural bioactive chemical compounds that have health promoting, disease preventing, and medicinal properties" (2). The Merriam–Webster Dictionary defines nutraceutical as "a foodstuff (as a *fortified food* or *dietary supplement*) that provides health benefits."

The concepts that led to the field of functional foods were developed in Japan in the 1980s (3). The term *physiological functional food* first appeared in a 1993 *Nature* article entitled "Japan Explores the Boundary Between Food and Medicine" (4).

A 1994 definition of functional foods, from the Washington, D.C.–based National Academy of Sciences Institute of Medicine, is "any food or

---

Mention of a brand or firm name does not constitute an endorsement by the U.S. Department of Agriculture above others of a similar nature not mentioned.

food ingredient which can offer health benefits in addition to those considered to be traditionally nutritional" (5). The definition used by the group Functional Foods for Health (University of Illinois) is "food that encompasses health benefit beyond the traditional nutrients it contains" (6).

In the United States there are no legal definitions for the terms nutraceutical and functional foods. The Food and Drug Administration (FDA) regulates functional foods no differently from other foods. Although the FDA does not have legal definitions for the terms nutraceutical and functional foods, it does define two related terms: *foods for special dietary use* (FSDU) and *medical foods*. (FSDUs) are defined as "foods that supply a dietary need that exists by reason of a physical, physiological, pathologic, or other condition including, but not limited to, diseases, convalescence, pregnancy, lactation, allergen hypersensitivity, underweight, or overweight" (7). Medical foods are defined as "foods that are formulated to be consumed or administered entirely under the supervision of a physician and are intended for the specific dietary management of a disease or condition for which distinctive nutritional requirements based on recognized scientific principles are established by medical evaluation" (7). Medical foods are distinguished from FDSU by being associated with specific disease conditions and requiring the supervision of a physician.

In 1994, the FDA approved a definition for *dietary supplement*: "a) a product (other than tobacco) that is intended to supplement the diet that bears or contains one or more of the following dietary ingredients: a vitamin, a mineral, an herb or other botanical, an amino acid, a dietary substance for use by man to supplement the diet by increasing the total daily intake, or a concentrate, metabolite, constituent, extract, or combinations of these ingredients. b) is intended for ingestion in pill, capsule, or liquid form. c) is not represented for use as a conventional food or as the sole item of a meal or diet. d) is labeled as a 'dietary supplement.' e) includes products such as an approved new drug, certified antibiotic, or licensed biologic that was marketed as a dietary supplement or food before approval, certification, or license (unless the Secretary of Health and Human Services waives this provision)" (8). According to the FDA: A *food* is defined as "a substance that must be ingested, at least in part, for taste, aroma, and nutritive value" (9), and a *drug* is defined as "any article intended for use in the diagnosis, cure, mitigation, treatment or prevention of disease in man or other animals" (9).

Until late 2001, the working definition for a nutraceutical in Canada was "a product that has been isolated or purified from foods and generally sold in medicinal forms not usually associated with food. Nutraceuticals have been shown to exhibit a physiological benefit or provide protection against chronic disease." Examples of Canadian products that were previously considered to be nutraceuticals include encapsulated and bottled oil products such as borage, evening primrose, flaxseed, hemp, fish oil, conjugated linoleic acid (CLA), pumpkin seed, and canola phytosterols. Health Canada recently decided that the product category of nutraceuticals will be eliminated and encompassed within proposed regulations for *natural health products* (*NHP*) (10). In September 2001, the NHP Directorate of Health Canada proposed *NHP* to include homeopathic preparations, substances used in traditional medicine, a mineral or trace element, a vitamin, an amino acid, an essential fatty acid, or other botanical-, animal-, or microorganism-derived substance. It is expected that this product category will not include foods. Furthermore, Health Canada has proposed NHP to include products manufactured, sold, or represented for use in (a) the diagnosis, treatment, mitigation or prevention or prevention of a disease, disorder, or abnormal physical state or its symptoms in humans; (b) resorting or correcting organic functions in humans, or (c) maintaining or promoting health or otherwise modifying organic functions in humans. Currently, functional foods are not legally defined in any global jurisdiction except Japan. In a Policy Options proposal released by Health Canada (1998), the working definition for a functional food included foods that are "generally accepted to be foods similar in appearance to conventional foods and consumed as part of the usual diet. These foods have demonstrated physiological benefits, and/or reduce the risk of chronic disease beyond basic nutritional functions" (10). In late 2001, Health Canada announced that a new regulatory definition for functional foods will not be required under the current Canadian Food and Drugs Act to permit health claims for foods. However, this term is used extensively in Canada to describe foods with demonstrated physiological benefits and/or that reduce the risk of chronic disease beyond basic nutritional functions. Examples of functional food ingredients that are currently available from Canadian companies include oat bran, wheat bran, high oleic sunflower oil, modified fatty acid canola oils, milled flaxseed, fenugreek, Saskatoon berry, and red clover (personal communication, K. Fitzpatrick).

In the European Union (EU) countries functional foods are considered to be foods that have health-promoting effects, and such foods are evaluated for possible health claims by individual countries. In contrast, novel foods have been regulated by the EU since adoption of the Novel Foods Act in 1997 (11). Novel foods include ones that contain new or modified ingredients (including food from genetically modified organisms), and a simple definition would include foods that were invented since 1997. In the United Kingdom, the Government's Food Advisory Committee defines functional foods as ordinary foods that have components or ingredients incorporated into them that confer a specific medical or physiological benefit, other than a purely nutritional effect (12).

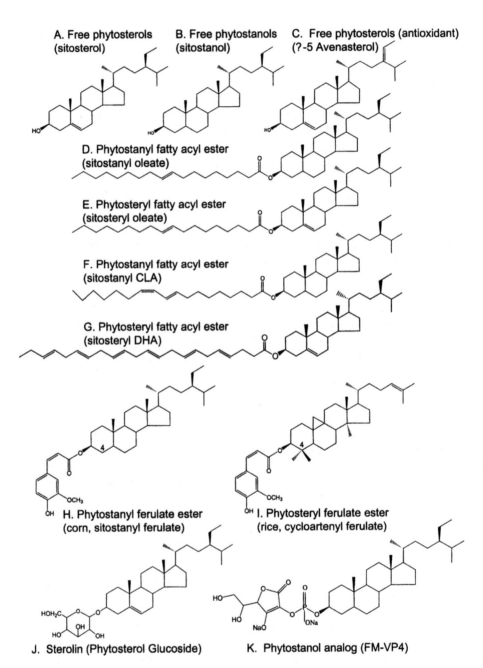

A. Free phytosterols (sitosterol)

B. Free phytostanols (sitostanol)

C. Free phytosterols (antioxidant) (?-5 Avenasterol)

D. Phytostanyl fatty acyl ester (sitostanyl oleate)

E. Phytosteryl fatty acyl ester (sitosteryl oleate)

F. Phytostanyl fatty acyl ester (sitostanyl CLA)

G. Phytosteryl fatty acyl ester (sitosteryl DHA)

H. Phytostanyl ferulate ester (corn, sitostanyl ferulate)

I. Phytosteryl ferulate ester (rice, cycloartenyl ferulate)

J. Sterolin (Phytosterol Glucoside)

K. Phytostanol analog (FM-VP4)

Figure 1 Examples of the chemical structures of phytosterols in current and future functional foods, dietary supplements, and pharmaceuticals.

Phytosterols (plant sterols) are natural components of all plant cells (see Chapter 3). Most natural phytosterols contain one or two carbon-carbon double bonds and are thus considered to be "unsaturated." Phytostanols are a subset of phytosterols that are completely saturated molecules (they contain no carbon-carbon double bonds). The first phytosterol product was a cholesterol-lowering pharmaceutical, Cytellin, that was marketed by Eli Lilly from 1957 to 1982 (13). The active ingredient in Cytellin was mainly free β-sitosterol (currently called sitosterol, as per the latest IUPAC rules of nomenclature) with lesser amounts of other phytosterols (Fig. 1A). The first phytosterol functional food was Benecol spread, which was launched in Finland in 1995. The active ingredients in Benecol are phytostanols esterified to natural fatty acids from vegetable oils (Fig. 1D). Since 1995 there has been a remarkable international interest in phytosterols by consumers and food scientists and a rapid development of new phytosterol functional food products. The purpose of this chapter is to highlight some of these developments.

II. LEGAL, REGULATORY, HEALTH CLAIM, AND HEALTH WARNING ISSUES

Since Benecol was first launched in Finland in 1995, phytosterol products have spread to more than 20 countries (Table 1).

A. Finland

When Benecol was launched in Finland in 1995, Raisio successfully petitioned and obtained approval from Finnish authorities. At that time there were no EU regulations for functional foods or novel foods. Currently, all commercial stanol ester–containing foods in Finland are registered as PARNUT (Foodstuffs for Particular Nutritional Purposes) products. Since 1997, Finland has adopted the EU guidelines, and novel foods must now meet EU criteria. Unilever's Becel Pro-Activ was launched in Finland and most other EU member countries in 2000, after a long process to obtain EU approval as a novel food.

B. Other EU Countries

The regulatory situation in the EU is quite complicated. The Novel Food Act was enacted in 1997. In the EU, new foods can be categorized in two separate systems: functional foods and novel foods. Currently, functional foods are regulated by individual EU member countries (an EU commission is being established to unify functional food regulations, but the date when it

Table 1 Countries Where Major Phytosterol Products Are Currently Being Marketed

Country	Year introduced	Benecol products	Unilever products	Phytosterol products other than margarine spreads and dietary supplements
Australia	2000		Y	
Austria, EU	2000		Y	
Belgium, EU	2000	Y	Y	
Brazil	2000		Y	
Czech Republic	2000		Y	
Denmark, EU	2000		Y	
Finland, EU	1995	Y	Y	Cream cheese spreads, milk, mayonnaise, cheese, pasta, yogurt, meat products, and snack bars
France, EU	2000		Y	
Germany, EU	2000		Y	
Greece, EU	2000		Y	
Ireland, EU	2000	Y	Y	
Japan	1999	Y	Y	Cooking oil, beverages
Korea	?		Y	Beverages
Netherlands, EU	2000		Y	
New Zealand	2000		Y	
Poland	2000	Y		
Portugal, EU	2000		Y	
South Africa	2000		Y	
Spain, EU	2000		Y	Yogurt
Switzerland	2000		Y	
Sweden, EU	2000	Y	Y	
UK, EU	1999	Y	Y	Snack bars, mayonnaise, and milk
US	1999	Y	Y	

might be ready and approved in member countries is unknown). This means that every member country can define its own rules, and health claims are approved on a variable basis in different countries. In 2000, EC Regulation No. 258/97 on Novel Foods and Food Ingredients was adopted for "yellow fat spreads" (various Unilever spreads) (14). The Novel Foods Act dictates that every new or modified component or a new processing technology must be approved before commercial use. This includes genetically modified organisms (GMOs). In that sense "novel" means something invented after May 1997. Sometimes this concept is confusing. Under Article 3 of this decision, companies are required to establish a surveillance program, referred to as post launch monitoring, to ensure that there is no evidence of adverse health effect. Novel foods are evaluated for their safety at the EU level, whereas functional foods are evaluated for their health-promoting effects at the national level. Also, a food item can be a functional food and not be a novel food, or a food item can be a novel food, but not a functional food.

In the UK, Flora Pro-Activ passed the necessary premarket review procedures and is required to carry a label advising that certain consumers may be placed at risk by consuming the product. In the United Kingdom, the Government's Food Advisory Committee defined functional foods as ordinary foods that have components or ingredients incorporated into them to give a specific medical or physiological benefit, other than a purely nutritional effect. The Leatherhead Food Research Association (LFRA) adds that to be classified as a functional food, a product should also carry a health claim on the packaging (12).

C. United States

In 1999, "no-question letters" were issued by the FDA for plant sterol esters (to Kleinfeld, Kaplan, and Becker, for Unilever) and for plant stanol esters (to McNeil Consumer Healthcare) in response to GRAS (Generally Recognized As Safe) notifications (9). In 2000, no-question letters for other phytosterol/phytostanol products were sent to Cargill, Novartis, ADM, and Proctor and Gamble (15) for additional GRAS notifications. In 2002, a no-question letter was sent to Teriaka (Helsinki, Finland) regarding GRAS notification of its microcrystalline phytosterol products (15).

Under provisions of the Nutrition Labeling and Education Act of 1990, FDA is authorized to regulate health claims on food labels to ensure that such claims are accurate and not misleading to consumers. The decision to allow a health claim must be substantiated by scientific evidence. In September 2000, the FDA issued an interim final rule allowing a health claim for reducing the risk of coronary heart disease for foods that contain plant stanol or sterol esters (9) and are low in saturated fat and cholesterol. This was only the 12th time the FDA has authorized a health claim. The rule included stanol esters in spreads, salad dressings, snack bars, and dietary supplements (soft gels). It included sterol esters in spreads and salad dressings only. To qualify for the health claim, a product had to contain no more than 13 g of total fat per serving or per 50 g. However, spreads and salad dressings were exempted from this requirement. Foods had to contain at least 0.65 g of plant sterol ester or 1.7 g of plant stanol ester per serving, and at least two servings had to be eaten at different times of the day for a total consumption of 1.3 and 3.4 g/day of sterol and stanol esters, respectively. The requirement for more than twice the amount of stanol esters than sterol esters seems contradictory to the preponderance of the scientific literature that indicates similar efficacy for both chemical species. However, the ruling was based on the data that were available at that time and the fact that many of the stanol ester studies were performed with higher dosages. The FDA allowed two separate comment periods with a final rule initially expected sometime in 2002. However, it is now expected that a final ruling will be made in 2003. Comments sent to FDA focused on issues surrounding dosage levels for sterol and stanol esters, whether free phytosterols and phytostanols should be included in the rule, additional food applications, and whether consumption of these ingredients may have a negative effect on absorption of dietary carotenoids. A review of these and other issues surrounding the health claim was recently published (16).

In addition to receiving approval by the FDA, phytosterol products have recently been evaluated by a coalition of major health organizations in the United States. In 2001, the National Cholesterol Education Program (NCEP), a multidisciplinary coalition of 40-plus major medical and health organizations (including the American Medical Association), voluntary health organizations, community programs, and government agencies (including the National Heart, Lung and Blood Institute at the National Institutes of Health), published a major report (the Adult Treatment Panel III, ATP-III), that recommended the use of phytosterols for the management of elevated cholesterol (17). The guidelines recommend plant sterols and stanols as "therapeutic dietary options to enhance lowering of LDL (low density lipoprotein) cholesterol"; 2 g of phytosterols or phytostanols per day, along with 10–25 g of soluble fiber, was recommended for significant cholesterol reduction. The American Heart Association recently published a statement on phytosterols that acknowledged their cholesterol-lowering efficacy, but the conclusion fell short of a recommendation, "Thus although foods containing plant sterols are a promising addition to dietary interventions aimed at improving cardiac risk profiles, more information is required before their routine ingestion is recommended in the general population as a step toward dietary prevention of coronary heart disease" (18).

D. Canada

For several months in 2001, Becel Pro-Activ was marketed in Canada. However, later in 2001, Health Canada advised consumers that Becel Pro-Activ was not in compliance with the Canadian Food and Drugs Act and Regulations. Health Canada stated that while phytosterols were being acknowledged for lowering cholesterol, they may pose health risks for certain groups, such as pregnant women, children, people predisposed to hemorrhagic strokes, and people on cholesterol-lowering medication (19). Consequently, no phytosterol functional foods are available in Canada and it is unclear whether Raisio or Unilever will continue to petition for approval to market their products. These Health Canada regulations do not cover supplements, and sale of phytosterol-containing supplements is permitted in Canada (K. Fitzpatrick, personal communication).

E. Japan

In 1984, the concept of functional foods was first introduced in Japan when the Foods for Specified Health Use (FOSHU) program was developed. In 1991, the Japanese Ministry of Health and Welfare had in place a policy that allows food manufacturers to declare that their product is a FOSHU (20). In 1999, Kao's Econa diacylglycerol oil with phytosterols was the first phytosterol product to be FOSHU approved (FOSHU no. 142). Recently, Unilever's Rama Pro-Activ (FOSHU no. 345) and Benecol margarine (personal communication, I. Wester) also received FOSHU approval (20).

F. Australia and New Zealand

Flora Pro-Activ is approved (A410) by the Australian and New Zealand Food Authority (ANZFA) (21). Flora Pro-Activ passed the necessary premarket review procedures and is required to carry a label advising that certain consumers may be placed at risk by consuming the product. Recently, ANZFA approval (A417) was granted for tall oil nonesterified phytosterols (21). Currently, Meadow Lea markets Logical spread, Coles markets Lo-Chol spread, and Redocol is approved for use in margarine (personal communication, J. Zawitstowski).

III. PHYTOSTEROL AND PHYTOSTEROL CONJUGATES AS FUNCTIONAL FOODS—SOURCES, CHEMISTRY, AND FORMULATIONS

A. Sources

The major sources of phytosterols for current functional foods and dietary supplements are tall oil and vegetable oil deodorizer distillate (Table 2). When Benecol was launched in Finland in 1995, the phytostanol fatty acyl esters (Fig. 1D) were obtained from tall oil. Tall oil is a by-product of the kraft pulping of wood to make paper. Tall oil contains a mixture of phytosterols and phytostanols. Partly due to the media attention that has focused on phytosterols and phytostanols, there have been several new patents on processes to obtain phytosterols from tall oil (22–27). For Benecol, the tall oil phy-

Table 2 Current and Future of Sources of Phytosterols for Functional Foods, Tall Oil, Deodorizer Distillate, Rice Bran Oil, and Corn Fiber Oil

Feedstock	Amount of phytosterols in fraction (wt %)	Types of phytosterols (see Fig. 1)	Ref.
Tall oil	10–20	A, B	22–28
Deodorizer distillate from soybean oil	15–30	A	29,30
Deodorizer distillate from other vegetable oils (palm, rape, peanut, etc.)	15–30	A	30
Corn fiber oil	10–15	A, D, E, H	32–35
Rice bran oil	1–2	A, I	36,37,40

tosterol fraction is converted to phytostanols via catalytic hydrogenation and the resulting free stanols are esterified to rapeseed oil fatty acids (Table 2).

Tall oil is also the feedstock for free phytosterols in Forbes Medi-Tech's Reducol. Forbes Medi-Tech has carried out research at the University of British Columbia (UBC) focused on using forest industry by-products as base materials for the production of pharmaceutical products. From this research, a unique process has been developed for extracting plant sterols from tall oil soap, a residue from the paper pulping process, as well as fermentation technology capable of converting plant sterols to pharmaceutical steroid intermediates. Forbes Medi-Tech, under a license agreement with UBC, acquired exclusive worldwide commercialization rights to these patented technologies (28). (In this article, many U.S. patents are cited and most are filed as part of a "family" of international patents that can be accessed at www.uspto.gov.) Forbes has refined the original licensed technology as well as developing improved technologies in the area of wood sterols whereby phytosterols are obtained from tall oil pitch. Obtaining phytosterols from tall oil pitch rather than tall oil soap has been made more cost efficient by utilization of improved extraction technology. Through these research initiatives, the company has developed a diversified line of health care products targeted to three global markets: nutraceuticals; over-the-counter dietary supplements, and pharmaceuticals.

Another major source of phytosterols is the deodorizer distillate fraction from vegetable oil refining. A process for the production of phytosterols and tocopherols from deodorizer distillates was developed by Eastman and patented in 1995 (29). Although deodorizer distillates are produced during the refining of most vegetable oils, soybean oil is the most popular oil in the United States, and soybean oil deodorizer distillates contain 15–30% phytosterols and have been used as the major feedstock for phytosterol production in the United States (Table 2). Vegetable oil processing involves refining (to remove the free fatty acids), bleaching (to remove off colors), and deodorization (to remove off flavors), often abbreviated as RBD. During deodorization the lower molecular weight compounds are removed from the crude oil and sequestered in the "deodorizer distillate" by-product fraction. Free phytosterols are a major component (15–30%) of the contents of the deodorizer distillate fraction. Tocopherols are also abundant in the deodorizer distillate (at concentrations of 9–11%), and this is a major starting material to obtain tocopherols for natural vitamin E supplements (30).

Although soybean oil deodorizer distillate is a major source for phytosterols in the United States, deodorizer distillates from other vegetable oils could potentially be used as a source of phytosterols (Table 2). Since deodorizer distillate from each oilseed species would have a unique phytosterol composition, the approval specifications in other countries may include phytosterols derived from deodorizer distillates from most common seed oils.

Stanols derived from tall oil contain primarily sitostanol and campestanol in a ratio of about 92:8. Stanols derived from soy oil have a sitostanol/campestanol ratio of about 68:32. A recent clinical study indicated that stanol ester margarines from both sources had equal cholesterol-lowering efficacy (31).

Corn fiber oil is obtained by extracting corn fiber, which is a pericarp-rich fraction obtained from the wet milling of corn. The wet milling process was developed to facilitate the economical removal of starch from corn kernels. Unlike commercial corn oil (which is obtained by extracting corn germ and contains 99% triacylglcerols and about 1% phytosterols), corn fiber oil is obtained by extracting corn fiber, and the oil contains about 10–15% phytosterols (32–35). Corn fiber oil contains three different types of phytosterol lipid classes, which will be discussed in detail in Sec. V.

Rice bran oil also contains about 1–2% phytosterols, most of which are composed of oryzanol, which is a mixture of ferulate phytosterol esters, with cycloartenyl ferulate being the most abundant ester (Fig. 1I). Oryzanol is a ferulate ester, similar to that in corn fiber oil, but instead of a common "desmethyl" phytosterol (no methyls on the number 4 ring carbon), it has a dimethyl (note the two methyls on the number 4 ring carbon) phytosterol. Oryzanol has been marketed as a dietary supplement for a number of years. There is some evidence of its cholesterol-lowering properties (36,37). However, in two well-controlled clinical studies, rice bran oil did not have a cholesterol-lowering effect (38,39). Body builders have also claimed that oryzanol is useful in building muscle mass. Gamma oryzanol has been approved in Japan for several conditions, including menopausal symptoms, mild anxiety, stomach upset, and high cholesterol. Each year Japan manufactures 7500 tons of gamma oryzanol from rice bran (40).

Although deodorization removes some phytosterols from crude vegetable oils, commercial vegetable oils still contain phytosterols (in the range of about 0.1–1.0%). Theoretically, vegetable oils could be used to obtain phytosterols, but at these low levels they are not an economically feasible starting material. Among the common commercial vegetable oils, corn oil (from germ) and avocado oil contain the highest levels of phytosterols (1–2%, Table 2), but at these low levels processes to obtain phytosterols from these oils are costly and therefore have not been commercially developed.

B. Chemistry

Phytosterols and phytostanols can occur in many forms in plant tissues (see Chapter 5). Eleven types of natural, synthetic, and "semisynthetic" phytosterols and/or phytosterol conjugates have been incorporated edible products (Fig. 1).

Standard processes for solvent extraction of vegetable oils utilize nonpolar solvents and mainly extract the two nonpolar phytosterols (free and phytosterol fatty acyl esters). The sterols in the common phytostanol ester (Fig. 1D) and phytosterol ester (Fig. 1E) products are produced by chemically esterifying the sterol to fatty acids derived from common vegetable oils (oleic acid is shown in the example for each because high-oleic acid rape/canola is a commonly used vegetable oil for this purpose). In corn, rice, and other grains, a major portion of the phytosterols also occur as phytosterol phenolic acid esters. In corn fiber oil, sitostanol ferulate (Fig. 1H) is the predominant phytosterol phenolic ester. In rice bran oil, cycloartenyl ferulate is the most abundant phytosterol phenolic ester (Fig. 1I). Cycloartenol is an unusual type of phytosterol because it is a 4,4-dimethyl phytosterol (it has two methyl groups on its number 4 ring carbon), whereas most phytosterols are desmethyl (meaning that they have no methyl groups on their number 4 ring carbon). Tall oil (and products such as Reducol that are made from tall oil phytosterols) contains mostly free phytoterols but may also contain a significant proportion of free phytostenols (31). As mentioned previously, phytostanols can also be produced by reducing phytosterols via catalytic hydrogenation. Benecol is produced by first hydrogenating phytosterols (from tall oil or soy bean oil deodorizer distillate) and then esterifying the phytostanols to rapeseed oil fatty acids (Fig. 1D). A newly developed phytostanol ester product is made by esterifying phytostanols to CLA (Fig. 1F) (41). It has been suggested that this product will combine the health benefits of phytostanols and CLA. Another newly developed product is a phytosterol ester made by esterifying the phytosterols to docosahexaenoic acid (DHA) (Fig. 1G), an ω-3 fatty acid (42). It has been suggested that this product will also combine the health-promoting properties of phytosterols and DHA.

In addition to the cholesterol-lowering (see Chapter 6) and anticancer (see Chapter 5) properties of phytosterols, there is also evidence of antioxidant activity in some common phytosterols such as Δ5-avenasterol (Fig. 1C), Δ7-avenasterol, fucosterol, and other minor phytosterols that contain an ethylidene group. White and Armstrong (43) demonstrated that Δ5-avenasterol, which occurs in high levels in oats and other plant tissues, may have valuable antioxidant activity. Additional studies have confirmed that vegetable oils that contain Δ5-avenasterol exhibit antioxidant and antipolymerization properties, which is especially valuable during frying (44–46). Although the antioxidant properties of these types of phytosterols have been demonstrated under frying conditions, their potential antioxidant health benefits have not been evaluated.

In recent years, the term "sterolin" (or "phytosterolin") has been used to refer to the sterol glucoside fraction of phytosterols (Fig. 1J). There is evidence that a mixture of free phytosterols and sterolins can enhance the performance of the immune system (47–49), and several supplement companies are marketing various sterol/sterolin supplements.

As with other types of rational drug design, efforts have been devoted to start with phytosterols or phytostanols and design a pharmaceutical that has increased efficacy compared to the natural phytosterols. Forbes Medi-Tech has developed a synthetic phytostanol analogue, FM-VP4 (Fig. 1K), which appears to have more potent cholesterol-lowering properties in comparison to natural phytosterols (50–52).

C. Formulations

In the last 5 years much effort has been devoted to developing various types of formulations/dispersions for phytosterols and phytostanols (Table 3). Although the "first generation" phytosterol products focused on sterol/stanol esters formulated in high-fat foods, the "second generation" sterol/stanol products are mainly low-fat formulations. Early phytosterol products, such as Cytellin, contained free phytosterols such as sitosterol (34). Large dosages (25 + g/day) of Cytellin were recommended and the cholesterol-lowering efficacy was not reliable, probably because the free phytosterols were not adequately dispersed. Scientists at Proctor and Gamble first suggested esterifying phytosterols to fatty acids to make them soluble in fat matrices such as vegetable oils (13), but they failed to produce a commercial product.

Table 3 Dispersion/Formulation Methods for Phytosterols

Chemical structure formulation method	Chemical structure in Fig. 1	Ref.
Esters		
a Phytostanol fatty acyl esters, in full-fat and low-fat spreads, salad dressings, mayonnaise, etc.	B	53–66
b Phytosterol fatty acyl esters, in full-fat and low-fat spreads, salad dressing, cooking oils, etc.	A	67–72
c Phytostanol CLA ester in various matrices	F	41
d Phytosterol DHA ester in various matrices	G	42
e Phytostanol ferulate ester in triacylglycerols	H	32–35
Free sterols		
f Free phytosterols—not dispersed (Cytellin)	A	
g Free tall oil phytosterols and phytosterols—dispersed, proprietary	A, B	74
h Free phytosterols—microcrystalline matrix	A	75–78
i Free phytosterols—nanoparticles	A	79
j Free phytosterols—lecithin dispersion	A	80–82
k Free phytosterols—dispersed in diacylglycerol	A	83–85
l Free phytosterols—dispersed in protein matrix	A	86
m Free phytosterols—dispersed with emulsifiers and mesophase-stabilized compositions	A	87,88
n Free phytosterols—dispersed in ground beef	A	89
o Free phytosterols and minerals—dispersed in bread, sausage, and yogurt	A	90,91
p Free phytosterols—dispersed in yogurt	A	92,93
q Free phytostanols and phytosterol fatty acyl esters dispersed in bread and breakfast cereal	A, E	94
r Designer oil containting free and esterified Δ5-avenasterol	A	95,96
s Free phytosterols in chocolate	A	97
Sterol conjugates (natural)		
t Sterolins (sterol glycosides) blended with free phytosterols	J	47–49
Sterol analogues (synthetic)		
u Phytostanol analogue (FM-VP4) as a pharmaceutical	K	50–52

Raisio was the first company to develop and market a phytosterol ester product. Raisio first chose to focus on a phytostanol fatty acyl ester delivered in a fat matrix such as margarine or salad dressing. Raisio and McNeil (its U.S. licensee) have sponsored many clinical studies of various forms of their products (see Chapter 6) and hold many patents (53–66). In addition to its spreads and salad dressings, Raisio and partners now market (in Finland) phytostanol ester products dispersed in a variety of food matrices: cream cheese spreads, semihard ripened cheese, pasta, milk, mayonnaise, yogurt, meat products, and snack bars (Table 1).

Unilever's (with its Take Control, Becel Pro-Activ, Flora Pro-Activ, and Rama Pro-Activ products) approach has been to focus on phytosterol fatty acyl esters initially delivered in high-fat spreads and salad dressings but now also marketed in low-fat formulations. Unilever and Unilever Best-foods North America (its U.S. affiliate) have also sponsored many clinical studies of various forms of products (see Chapter 6) and hold many patents (67–72).

As mentioned previously, two recent phytosterol fatty acyl ester products have been developed based on unique fatty acids, one a phytostanol ester of CLA (41) and the other a phytosterol ester of DHA (42), but both concepts are in the developmental stage and the types of formulations that will be used for each have not been announced (Table 4).

Corn fiber oil is a natural extract that contains phytosterol fatty acyl esters (5–9%), free phytosterols (1–2%), and phytostanol ferulate esters (4–6%) in a high-fat triacylglycerol matrix (32–35). Rice bran oil also contains 1–2% of a ferulate phytosterol ester called oryzanol (mainly cycloartenyl ferulate) in a triacylglycerol matrix. These natural ferulate phytosterol esters can be found in corn fiber oil and rice bran oil and in several other grains. Condo et al. (73) recently published a procedure to synthesize ferulate phytosterol esters at high yield.

Much recent effort has been devoted to developing dispersion methods for free phytosterols. Forbes Medi-Tech's approach for their Reducol formulation involves a proprietary process to disperse free phytosterols (74). Forbes Medi-Tech is actively developing many products (breakfast cereal, chocolate, beverages, etc.) with Reducol formulated into various types of food matrices.

Christiansen et al. (75–77) have developed a process to prepare "microcrystalline" phytosterols that can be formulated into many types of low-fat foods. This process is being commercialized by Teriaka (Helsinki). Interestingly, a second Finnish company, Suomen Sokeri Oy, has a U.S. patent on another process to make microcrystalline phytosterol for food use (78). A seemingly similar technology from scientists at Cognis involves a process to

Table 4 Major Phytosterol/Phytostanol Products Marketed in 2003

Product trade names ® = registered in the USA	Phytosterol active component(s)	Structure Fig. 1	Formulation Table 3	Type of food	Manufacturer	Website	Marketing status and refs.
Functional Foods							
Benecol	Phytostanol esters	D	a	Many; see Table 1	Raisio/McNeil	*www.benecol.com*	Presently available US, EU, and Japan
Take Control (US), Flora, Becel, Rama Pro-activ™ (Int.)	Phytosterol esters	E	b	Spread, salad dressing	Unilever Bestfoods	*www.takecontrol.com*	Presently available US, EU, FOSHU, and ANZFA approved
Logicol®	Phytosterols	A	Unknown	Spread	Meadow Lea	*www.logical.com.au*	Spreads now available Australia
Lo-Chol™	Phytosterols	A	Unknown	Spread	Coles	www.coles.com.au	Spreads now available Australia
CholZero™	Ucole brand of phytosterols	A	Unknown	Beverages and cooking oil	Eugene Sciences	*www.eugene21.com*	Available in Korea
Healthy Econa® oil for lower cholesterol	Phytosterols in DAG oil	A	k	Cooking oil and mayonnaise	Kao	*www.kao.com*	Available, FOSHU approved

Product	Composition			Form	Company	Website	Availability
"Kenkou Sarara" oil	Phytosterols in TAG oil	A	Unknown	Cooking oil	Ajinomoto Inc.	In Japanese	Available, FOSHU approved
Functional Food Ingredients							
Reducol™	Free phytosterol, phytostanol mixture	A, B	g	Ingredient	Forbes Medi-Tech	www.forbesmedi.com/s/TechnologyAndProducts.asp	Presently available as ingredients
Dietary Supplements							
Cholesterol Success®	Reducol™	A	g	Tablet	Twinlabs Co.	www.twinlab.com	Available
Cholest-Off®	Reducol™	A	g	Tablet	Nature Made	www.naturemade.com	Available
ChoLESStolife®	Phytosterols and Phytostanols	A	j	Tablet	Lifeline Technologies	www.lifelinetechnologies.com	Available
Natur-Leaf®	Phytosterols and sterolins (glycosides) from sprouts	A, J	f, t	Capsule	Nature Leaf	www.naturleaf.com	Available
Cholestatin®	Free phytosterols	A	Unknown	Tablet	Degussa Bioactives	www.bioactives.de	Presently available
Prosterol®	Free phytosterols and Policosanols			Tablet			In development
Kolestop™ and Sitosan™	Phytosterols	A	Unknown	Tablet	Hankintatukku Oy	www.hankinatukku.com	Available

make "nanoscale" dispersions of phytosterols (nanoparticles of sterols and/ or sterol esters with particle diameters of 10–300 nm) (79).

In addition to dispersing phytosterols, lecithin also appears to play a valuable role in increasing the bioavailability of free phytosterols (80,81). A recent study (82) showed that sitostanol powder (1 g) reduced cholesterol absorption in human subjects by about 11%. In contrast, only 300 mg of sitostanol administered in lecithin micelles reduced cholesterol absorption by 34%. The authors concluded that free sitostanol was not effective due to its slow dissolution in artificial bile.

In Japan, Kao has pioneered a diacylglycerol oil and it has also developed a form of the oil that is enriched in phytosterols (83–84). A recent study (85) demonstrated that the effect of only 500 mg/day of phytosterols in this product reduced serum low-density lipoprotein cholesterol (LDL-C) levels by about 8% compared to the same amount of phytosterols in a triacylglycerol base, which caused no decrease in LDL-C levels. Kao and ADM have recently formed a partnership (and received a GRAS no-question letter; see Ref. 15) to launch a diacylglycerol oil in the United States under the name Enova, but no plans have been revealed to also market the phytosterol-enriched diacylglycerol oil in the United States.

Monsanto recently received a patent on a "phytosterol protein complex" comprising phytosterols, proteins, and edible oil (86). The complex is said to "increase the bioavailability of phytosterols" and "is most preferred to extract the phytosterols from corn fiber oil." Kraft has developed a phytosterol dispersion that encompasses emulsifiers and mesophase-stabilized compositions (87,88).

A recent clinical study indicated that phytosterols (two-thirds esterified and one-third free) could be successfully formulated into ground beef (89).

Finnish scientists developed a product called Multi-Bene, a phytosterol formulation that combines phytosterols with calcium and other minerals (90). The ingredients in this product are intended to reduce serum cholesterol levels and blood pressure (91).

Two clinical studies indicated that free phytosterols and stanol esters could be formulated into a low-fat yogurt (92,93).

Another clinical study demonstrated that free phytostanols and phytosterol fatty acyl esters could be formulated in bread and breakfast cereals (94).

It has been suggested that the presence of Δ5-avenasterol in virgin olive oil may contribute to its high oxidative stability and possible health benefits (95). Some have suggested that "antioxidant phytosterols" (Δ5-avenasterol and other phytosterols that possess an ethylidene group) could be formulated into "antioxidant oils" (96).

To prove that functional foods can have a very appealing flavor, Forbes Medi-Tech recently reported promising cholesterol-lowering results with a phytosterol-enriched chocolate product. In December 2002 Forbes Medi-Tech announced that the results of its study on the cholesterol-lowering functional food chocolate had been published (97). The results indicated that participants eating the phytosterol Reducol-enriched chocolate reduced their LDL cholesterol by 10.3%.

Sterolins (sterol glycosides) have been formulated with free phytosterols and sold as dietary supplements to enhance immune function (47–49).

IV. A COMPARISON OF PHYTOSTEROL PRODUCTS CURRENTLY AVAILABLE IN WESTERN AND JAPANESE MARKETS

Phytosterol products are currently available in more than 20 countries (Table 1), including 13 EU countries.

The development of phytosterol products has been influenced by the appearance of other types of cholesterol-lowering products (Fig. 2). Phytosterol products first appeared in the 1950s, at about the same time as the bile

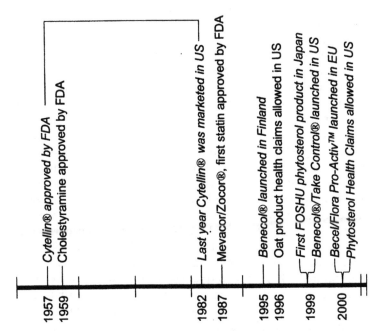

Figure 2 Timeline of important milestones in phytosterol products (*italics*) and competing products for lowering serum cholesterol.

salt chelating drug cholestyramine (Fig. 2). During the last 7 years substantial R&D effort has been devoted to developing phytosterol functional foods. The drug Cytellin (primarily sitosterol) was prescribed for more than 20 years and had an excellent safety record. Concern had been expressed (13) that ingestion of large amounts of phytosterols might lead to high serum levels of phytosterols, which might possibly elicit estrogenic or atherogenic effects. There is no evidence that phytosterols or phytostanols cause estrogenic or atherogenic effects in normal subject. This is one reason that much work on commercial products in the early 1990s was focused on the largely unabsorbed stanols and stanol esters.

Currently, only a few major international companies market functional food phytosterol products and ingredients (Table 4). A number of companies market phytosterol dietary supplements and five of the major companies are listed in Table 4. As described in Sec. II, the regulatory process for phytosterol functional foods has slowed their introduction and has limited the number of products. Stanol ester products entered the market in Finland in 1995, followed by both stanol and sterol ester products in other European countries (Table 4). Stanol and sterol ester products entered the U.S. market in 1999 after considerable regulatory discussions and delay. McNeil Consumer Healthcare (with North American marketing rights to Raisio's Benecol) intended to launch Benecol in the fall of 1998 as a dietary supplement (which, under the DSHEA act, is exempt from many FDA regulations). However, the FDA considered Benecol to be a conventional food and threatened to seize any products McNeil planned to launch in a test market in Portland, Oregon. Eventually McNeil agreed to market Benecol as a conventional food (a spread), but that decision required either a new food additive petition or recognition of GRAS status. In 1999, a panel of independent experts in the United States concluded that plant sterol esters were GRAS for use as an ingredient in vegetable oil–based spreads in amounts not to exceed 20%. Based on this GRAS recognition, the FDA had no questions about the safety of spreads containing up to 20% of plant sterol and stanol esters. In May 1999 Lipton (now called Unilever Bestfoods North America), and its parent company, Unilever, launched Take Control (which contains vegetable oil-derived sterol esters) and McNeil launched Benecol (which contains stanol esters) in the United States. Because Take Control's active ingredients, sterol esters, were simpler and less expensive to prepare (no hydrogenation required), Lipton's product could be sold at a lower price than Benecol. This initially did not seem to be a major advantage since sterol esters were considered by many to be inferior in efficacy to stanol esters. However, with the present evidence that sterol and stanol esters are equivalent in reducing serum cholesterol levels, the lower cost of Take Control may give it a small market advantage over stanol ester products (13).

In October 2000, Forbes Medi-Tech announced the first consumer product test market of its phytosterol-based cholesterol-lowering ingredient, Phytrol (unesterified tall oil phytosterols) in Australia (subsequently, ANSFA decided to only grant approval for spreads) and the United States. In 1999, Phytrol was exclusively licensed worldwide to Novartis Consumer Health, but in 2001 Novartis returned rights (and the rights to market and use the trademark Reducol) to Forbes Medi-tech. Forbes is in the process of incorporating Reducol into a variety of food products.

Even though the era of phytosterol functional foods only began in 1995, there have already been several products whose commercial "life" was short. During 2000 and 2001, Procter and Gamble test marketed a new line of phytosterol-containing cooking oils under the brand name CookSmart. These oils contained soy phytosterol esters. Procter and Gamble was the first company to market a phytosterol-containing cooking oil which conceivably could add cholesterol-lowering phytosterols to fried foods such as French fries. In 2001, Altus Food Company (a joint venture of Quaker Oats Company and Novartis) test marketed a Take Heart line of phytosterol-enriched breakfast cereals, snack bars, and fruit juice beverages, but the alliance was abandoned and these products were removed from the market place.

V. A COMPARISON OF PHYTOSTEROL PRODUCTS IN DEVELOPMENT AND FUTURE PROSPECTS

Many of the phytosterol products that are currently in the development "pipeline" are centered around the new formulation/dispersion technologies listed in Table 3. Some of these will likely be commercialized in the near future (Table 5).

Forbes Medi-Tech announced in December 2000 its development of a "designer oil" that reduces LDL-C and increases an individual's energy expenditure, which may prevent people from gaining weight. The research was conducted at McGill University between October 1999 and May 2000. The oil contains Forbe's phytosterol-based ingredient, Reducol, which is incorporated into the oil by a proprietary process that preserves the clarity of the oil (the oil also contains medium-chain triacylglycerols and ω-3 fatty acids). Clinical studies demonstrated that consumption of this designer oil by overweight men resulted in significantly decreased body weight (98).

A unique oil was discovered in corn fiber (a fiber-rich by-product from corn wet milling) by a team of scientists with USDA's Agricultural Research Service (ARS) (32–35) and proposed as a natural cholesterol-lowering oil. Corn fiber oil, called Amaizing Oil, was shown to lower serum cholesterol

Table 5 Major Types of Phytosterol/Phytostanol Products in Development in 2003

Product trade names ® = registered in the USA	Phytosterol active component(s)	Structure Fig. 1	Formulation Table 3	Type of food	Manufacturer	Website	Marketing status and refs.
Functional Foods							
"Designer oil"	Reducol®, medium-chain TAGs and omega-3 fatty acids.	A	g	Cooking oil	Forbes Medi-Tech	www.forbesmedi.com	In development
Corn fiber oil	Phytosterol fatty acyl esters, free phytosterols, phytosterol ferulate esters	A, D, E, H	e	Cooking oil	MBI International	www.mbi.org	In development
CLA One®	Phytosterol CLA esters in TAG oil	F	c	Cooking oil	PharmaNutrients Inc.	www.pharmanutrients.com/pres/cla_one/cla_pr3.html	In development
Nisshin Balance Oil™	Phytosterols in TAG oil		Unknown	Cooking oil	Nisshin	In Japanese	In development
Multi-Bene®	Phytosterols and minerals	A	o	Various foods	Multi-Bene	None	In development
Diminicol®	Free phytosterols	A	h	Various foods	Teriaka	www.teriaka.com	In development
Natucol™	Phytosterols	A, B	g	Spread	GFA Brands Inc.	www.earthbalance.net	In development in the USA
Sweet Life™	Phytosterol esters	E	Unknown	Chocolate beverage	Cargill	www.cargillnutraceuticals.com/news_pr_7.html	Concept beverage
Nutra-Chocolate™	Microcrystaline phytosterols	A	s	Chocolates	Nutra-Chocolate	www.nutrachocolate.com	Unknown
Functional Food Ingredients							
CardioAid® CardioAid-L	Phytosterols; Phytosterols and lecithin	A	j	Ingredient	ADM	www.admworld.com	In development
CardioAid-P	Phytosterols and soy proteins						
CoroWise® (PS, FP, and SE)	Various free and esterified phytosterols	A, E	Various	Ingredient	Cargill	www.cargillhft.com/products_se.html	Presently available as ingredients
Pharmaceutical							
FM-VP3™	Phytostanol analogue	K	u	Tablet	Forbes Medi-Tech	www.forbesmedi.com	In development

levels in an animal model at the University of Massachusetts (UMass) and was then patented (32) as a joint invention of ARS/UMass. Corn fiber oil has both phytosterol and phytostanol esters and phytosterol ferulate esters (Fig. 1H), and the latter may impart functional properties not found in any current commercial phytosterol product. Unlike phytosterols in soy or tall oil, most of the phytosterols in corn fiber oil are naturally esterified with fatty acids or phenolic acids, such as ferulic acid, a powerful antioxidant (13). Furthermore, corn fiber oil contains a high level of sitostanol in the ferulic acid ester fraction. In fact, corn fiber oil appears to be the richest source of natural stanols (and stanol esters) ever reported. Corn fiber oil also contains γ-tocopherol and various carotenoids, with important antioxidant properties. The levels of total phytosterols in corn fiber oil range from about 15% to more than 50%, depending on extraction and fiber pretreatment conditions (32–35). Corn fiber oil's combination of natural cholesterol-lowering components and antioxidants, which could potentially prevent oxidation of LDL-cholesterol, could give a unique combination in the fight against heart disease. An exclusive license for manufacturing and use of the oil for non-cooking oil applications has been granted to MBI International (Table 5).

The Belgian company Nutra-Chocolate recently announced plans to commercialize a phytosterol-containing chocolate product. According to the company's website, the phytosterols will be in the microcrystalline form (76). Interestingly, the clinical study describing the cholesterol lowering of phytosterol-enriched (with Forbes Medi-Tech's Reducol phytosterols) chocolates was reported at almost the same time, and most of the clinical studies were conducted in a neighboring Benelux country, the Netherlands (97).

The preliminary results with Forbes Medi-Tech's phytosterol analogue (FM-VP4) indicate that its approach to rational drug design has resulted in a very potent inhibitor of cholesterol absorption. Although it has not yet been tested in humans, preliminary testing in animal models indicates an impressive level of cholesterol reduction (50–52).

VI. CONCLUSION

During the last 10 years there has been an unprecedented escalation of interest in phytosterols. Most of this interest has focused on the cholesterol-lowering properties of phytosterols and phytostanols. Evidence of this phenomenon includes more than 40 patents on phytosterol products and more than 10 commercial phytosterol products currently being marketed in many parts of the world.

REFERENCES

1. SL DeFelice. FIM Rationale and proposed guidelines for the Nutraceutical Research and Education Act, 2002, *www.fimdefelice.org/archives/arc.researchact.html*.

2. Anonymous. What are functional foods, 2002, *www.foodsci.rutgers.edu/nci/#what*.

3. S Arai, Y Morinaga, T Yoshikawa, E Ichiisi, Y Kiso, M Yamazaki, M Morotomi, M Shimizu, T Kuwata, S Kamingawa. Recent trends in functional food science and the industry in Japan. Biosci Biotechnol Biochem 66:2017–2029, 2002.

4. D Swinbanks, J O'Brien. Japan explores the boundary between food and medicine. Nature 364:180, 1993.

5. Anonymous. Opportunities in the nutrition and food sciences. In: PR, Thomas, R, Earl, eds. Institute of Medicine/National Academy of Sciences. Washington DC, 1994, p. 109.

6. Anonymous. FDA approves new health claim for plant sterol and stanol esters and reduced risk of coronary heart disease, 2000, *www.ag.uiuc.edu/~ffh*.

7. S Ross. Functional foods: The Food and Drug Administration perspective. Am J Clin Nutr Supple 71:1735S–1738S, 2000.

8. Anonymous. Dietary Supplement Health and Education Act of 1994, *http://vm.cfsan.fda.gov/~dms/dietsupp.html*.

9. Anonymous. Plant sterol/stanol esters and coronary heart disease; health claims. Federal Register 65:54685–54739, 2000, *www.access.gpo.gov/su docs/fedreg/a000908c.html*.

10. Anonymous. Health Canada, Policy Paper on nutraceuticals/functional foods and health claims on foods, 2002, *http://www.hc-sc.gc.ca/food-aliment/ns-sc/ne-en/health claims-allegations sante/e nutra-funct foods.html*.

11. Anonymous. Regulation (EC) No 258/97 of the European Parliament and of the Council of 27 January 1997 concerning novel foods and novel food ingredients. Official Journal L 043, 14/02/1997 P 0001–0006, http://europa.eu.int/comm/food/fs/novel_food/nf_index_en.html.

12. Anonymous. United Kingdom—Chaos Reigns Supreme—Regulatory Requirements, 1998, UK, www.cspinet.org/reports.

13. RA Moreau, BD Whitaker, KB Hicks. Phytosterols, phytostanols, and their conjugates in foods: structural diversity, quantitative analysis, and health-promoting uses. Prog Lipid Res 41:457–500, 2002.

14. Anonymous. Commission Decision (2000/500/EC) of 24 July 2000 on authorizing the placing on the market of "yellow fat spreads with added phytosterol esters" as a novel food or novel food ingredient under Regulation (EC) No 258/97 of the European Parliament and of the Council (OJ L200, 8.8.2000, page 59) http://europa.eu.int/smartapi/cgi/sga_doc?smartapi! celexapi!prod! CELEXnumdoc&lg = EN&numdoc = 32000D0500&model = guichett.

15. Anonymous. Summary of all GRAS notices, FDA, *http://www.cfsan.fda.gov/~rdb/opagras.html*.

16. Anonymous. As more sterol/stanol foods reach market, FDA asked to modify the rules. INFORM 12:904–907, 2001.
17. JI Cleeman. Executive summary of the third report of the National Cholesterol Education Program (NCEP) Expert panel on detection, evaluation, and treatment of high blood cholesterol in adults (Adult Treatment Plan III). JAMA 285:2486–2497, 2001, www.nglbi.nih.gov/guidelines/cholesterol/atp3_rpt.htm.
18. AH Lichtenstein, RJ Deckelbaum. Stanol/sterol ester–containing foods and blood cholesterol levels—a statement for healthcare professionals from the Nutrition Committee of the Council on Nutrition, Physical Activity, and Metabolism of the American Heart Association. Circulation 103:1177–1179, 2001.
19. Anonymous. Health Canada Advisory, 2001, *http://www.hcsc.gc.ca/english/protection/warnings/2001/2001_106e.htm.*
20. Anonymous. Announcing the completion of FOSHU 2002 report, 2002, *www.npicenter.com/NBPrintDocuments.asp?DocumentID = 2604.*
21. Anonymous. Application A410-Phytosterol esters derived from vegetable oils, 2000, *www.foodstandards.gov.au.*
22. PL Robinsion, TJ Cuff, JE Parker III. Isolation and purification of sterols from neutrals fraction of tall oil pitch by single decantation crystallization. US Patent 6,057,462, 2000.
23. MA Fuenzalida Diaz, A Markovits Rojas, RB Leiva Hinojosa, E Markovits Schersl. Process for obtaining unsaponifiable compounds from black-liquor soaps, tall oil, and their by-products. US Patent 6,297,353, 2001.
24. MAF Diaz, AM Rojas. Fractionation process for the unsaponifiable material derived from black-liquor soaps. US Patent 6,462,210, 2002.
25. EM Schersl, High efficiency process for the recovery of the high pure sterols. US Patent 6,465,665, 2002.
26. DTA Huibers, AM Robbins, DH Sullivan. Method for separating sterols from tall oil. US Patent 6,107,456, 2000.
27. DTA Huibers, AM Robbins, DH Sullivan. Method for separating sterols from tall oil. US Patent 6,414,111, 2002.
28. JP Kutney, E Novak, PJ Jones. Process of isolating a phystosterol composition from pulping soap. US Patent 5,770,749, 1998.
29. CE Sumner, Jr., SD Barnicki, MD Dolfi, Process for the production of sterol and tocopherol concentrates. US Patent 5,424,457, 1995.
30. T Wang. Soybean oil. In: FD Gunstone, ed. Vegetable Oils in Food Technology: Composition, Properties, and Uses. Oxford: Blackwell, 2002, pp 1–58.
31. J Plat, RP Mensink. Vegetable oil based versus wood based stanol ester mixtures: effects on serum lipids and hemostatic factors in non-hypercholesterolemic subjects. Atoscerolsis 148:101–112, 2000.
32. RA Moreau, KB Hicks, RJ Nicolosi, RA Norton. Corn fiber oil, its preparation and use. US Patent 5,843,499, 1998.
33. RA Moreau. Corn Oil. In: FD Gunstone, ed. Vegetable Oils in Food Technology: Composition, Properties, and Uses. Oxford: Blackwell, 2002, pp 278–296.
34. KB Hicks, RA Moreau. Phytosterols and phytostanols: functional food cholesterol busters. Food Technol 55:63–67, 2001.

35. RA Moreau, MJ Powell, KB Hicks. The extraction and quantitative analysis of oil from commercial corn fiber. J Agric Food Chem 44:2149–2154, 1996.

36. CJ Berry, ML Bierenbaum. Anticholesterolemic edible oil. US Patent 6,277,431, 2001.

37. DW Hoffpauer, SL Wright III Fortified rice bran food product and method for promoting cardiovascular health. US Patent 6,436,431, 2002.

38. JA Westrate, GW Meijer. Plant sterol-enriched margarines and reduction of plasma total- and LDL-cholesterol concentrations in normocholesterolaemic and mildly cholesterolaemic subjects. Eur J Clin Nutrit 52:334–343, 1998.

39. MN Vissers, PL Zock, GW Meijer, MB Katan. Effect of plant sterols from rice bran oil and triterpene alcohols from sheanut oil on serum lipoprotein concentration in humans. Am J Clin Nutr 72:1510–1515, 2000.

40. Anonymous. Gamma oryzanol, 2000, http://www.principalhealthnews.com/topic/topic100587786.

41. L Liu. Sterol esters of conjugated linoleic acids and process for their production. US Patent 6,413,571, 2002.

42. HS Ewart, LK Cole, J Kralovec, H Layton, JM Curtis, JLC Wright, MG Murphy. Fish oil containing phytosterol esters alters blood lipid profiles and left ventricle generation of thromboxane A_2 in adult guinea pigs. J Nutr 132:1149–1152, 2002.

43. PJ White, LS Armstrong. Effect of selected oat sterols on deterioration of heated soybean oil. J Am Oil Chem Soc 63:525–529, 1986.

44. LL Tian, PJ White. Antipolymerization activity of oat extract in soybean and cottonseed oils under frying conditions. J Am Oil Chem Soc 71:1087–1094, 1994.

45. SP Kochhar. Stable and healthful frying oil for the 21st century. INFORM 11:642–647, 2000.

46. SP Kochhar. Stabilization of frying oils with natural antioxidative compounds. Eur J Lip Sci Technol 102:552–559, 2000.

47. PJD Bouic, A Clark, W Brittle, JH Lamprecht, M Freestone, RW Liebenberg. Plant sterol/sterolin supplement use in a cohort of South African HIV-infected patients—effects on immunological and virological surrogate markers. S Afr Med J 91:848–850, 2001.

48. PJD Bouic. The role of phytosterols and phytosterolins in immune modulation: a review of the past 10 years. Curr Opin Clin Nutr Metab Care 4:471–475, 2001.

49. PJD Bouic. Sterols and sterolins: new drugs for the immune system? Drug Discov Today 7:775–778, 2002.

50. KM Wasan, S Najafi, KD Peteherych, PH Pritchard. Effects of a novel hydrophobic phytostanol analog on plasma lipid concentrations in gerbils. J Pharm Sci 90:1795–1799, 2001.

51. KM Wasan, S Najafi, J Wong, M Kwong, PH Pritchard. Assessing plasma lipid levels, body weight, hepatic and renal toxicity following chronic oral administration of a water-soluble phytstanol compound, FM-VP4, to gerbils. J Pharm Pharmaceut Sci 4:228–234, 2001.

52. M Ramaswamy, E Yau, KM Wasan, KD Boulanger, M Li, PH Pritchard. Influence of phytostanol phosphoryl ascorbate, FM-VP4, on pancreatic lipase activity and cholesterol accumulation within Caco-2 cells. J Pharm Pharmaceut Sci 5:29–38, 2002.
53. T Miettenen, H Vanhanen, I Wester, Use of a stanol fatty acid ester for reducing serum cholesterol level. US Patent 5,502,045, 1996.
54. T Miettenen, H Vanhanen, I Wester, Substance for lowering high cholesterol level in serum and methods for preparing and using the same. US Patent 5,958,913, 1999.
55. I Wester. Fat compositions for use in foods. US Patent 6,162,483, 2000.
56. T Miettenen, I Wester, H Vanhanen. Substance for lowering cholesterol level in serum and methods for preparing and using the same. US Patent 6,174,560, 2001.
57. JD Higgins III. Preparation of sterol and stanol esters. US Patent 5,892,068, 1999.
58. JD Higgins, III. Preparation of sterol and stanol esters. US Patent 6,147,236, 2000.
59. A Roden, JL Williams, R Bruce, F Detrano, MH Boyer, JD Higgins, III. Preparation of sterol and stanol esters. US Patent 6,184,397, 2001.
60. B Burruano, RD Bruce, MR Hoy, JD Higgins III. Method for producing water dispersible sterol formulations. US Patent 6,054,144, 2000.
61. B Burruano, RD Bruce, MR Hoy. Method for producing water dispersible sterol formulations. US Patent 6,110,502, 2000.
62. CK Dartey, JD Higgins III, RD Bruce, BT Burruano. Stable salad dressings. US Patent 6,123,978, 2000.
63. RD Bruce, B Burruano, MR Hoy, NR Paquette. Method for producing dispersible sterol and stanol compounds. US Patent 6,242,001, 2001.
64. RD Bruce, JD Higgins, SA Martellucci. Sterol esters in tableted solid dosage forms. US Patent 6,376,481, 2002.
65. RD Bruce, B Burruaro, MR Hoy, NR Paquette. Method for producing dispersible sterol and stanol compounds. US Patent 6,387,411, 2002.
66. CK Dartey, JD Higgins III, RD Bruce, BT Burruano. Stable salad dressings. US Patent 6,399,137, 2002.
67. MP van Amerongen, LC Lievense. Stanol ester composition and production thereof. US Patent 6,031,118, 2000.
68. MP van Amerongen, LC Lievense. Process for the production of stanol esters, and use thereof. US Patent 6,106,886, 2000.
69. MP van Amerongen, LC Lievense. Fat based food products comprising sterols. US Patent 6,117,475, 2000.
70. MP van Amerongen, LC Lievense. Stanol comprising compositions. US Patent 6,231,915, 2001.
71. MP van Amerongen, LC Lienvense, CW Van Oosten. Method of manufacturing sterol ester mixture. US Patent 6,492,538, 2002.
72. AW Traska, M Patrick. Aqueous dispersion. US Patent 6,423,363, 2002.
73. AM Condo, DC Baker, RA Moreau, KB Hicks. Improved method for the

synthesis of transferuloyl-beta-sitosantol. J Agre Food Chem 49:4961–4964, 2001.

74. DJ Stewart, R Milanova, J Zawistowski, SH Wallis. Phytosterol compositions and use thereof in foods, beverages, pharmaceuticals, nutraceuticals and the like. US Patent 6,087,353, 2000.

75. LI Christiansen, JT Rantanen, AK von Bonsdorff, MA Karjalainen, YK Yliruusi. A novel method of producing a microcrystalline beta-sitosterol suspension in oil. Eur J Pharm Sci 15:261–269, 2002.

76. LI Christiansen, PLA Lähteenmäki, MR Mannelin, TE Seppänen-Laakso, RVK Hiltunen, YK Yliruusi. Cholesterol-lowering effect of spreads enriched with microcrystalline plant sterols in hypercholesterolemic subjects. Eur J Nutr 40:66–73, 2001.

77. LI Christiansen, M Karjalainen, R Serimaa, N Lönnroth, T Paakkari, YK Yliruusi. Phase behavior of β-sitosterol-cholesterol and β-sitosterol-cholesterol coprecipitates. STP Pharma Sci 11:167–173, 2001.

78. I Tiainen, J Nurmi, L Klasi. Product, a method for its production, and its use. US Patent 6,129,944, 2000.

79. C Kropf, T Foerster, B Fabry, M Hollenbrock. Use of nanoscale sterols and sterol esters. US Patent 6,316,030, 2001.

80. RE Ostlund Jr. Sitostanol formulation to reduce cholesterol absorption and method for preparing and use of same. US Patent 5,932,562, 1999.

81. RE Ostlund Jr. Sitostanol formulation with emulsifier to reduce cholesterol absorption and method for preparing and use of same. US Patent 6,063,776, 2000.

82. RE Ostlund, CA Spilburg, WF Stenson. Sitostanol administered in lecithin micelles potently reduces cholesterol absorption in humans. Am J Clin Nutr 70:826–831, 1999.

83. N Goto, T Nishide, Y Tanaka, T Yasukawa, K Masui. Oil or fat composition containing phytosterol. US Patent 6,139,897, 2000.

84. N Goto, T Nishide, Y Tanaka, T Yasukawa, K Masui. Oil or fat composition containing phytosterol. US Patent 6,326,050, 2001.

85. S Meguro, K Higashi, T Hase, Y Honda, A Otsuka, I Tokimitsu, H Itakura. Solubilization of phytosterols in diacylglycerol versus triacylglycerol improves the serum cholesterol-lowering effect. Eur J Clin Nutr 55:513–517, 2001.

86. G Corliss, JW Finley, HN Basu, F Kincs, L Howard. Phytosterol protein complex. US Patent 6,113,972, 2000.

87. A Akashe, M Miller. Plant sterol-emulsifier complexes. US Patent 6,267,963, 2001.

88. A Akashe, M Miller. Use of mesophase-stabilized compositions for delivery of cholesterol-reducing sterols and stanols in food products. US Patent 6,274,574, 2001.

89. OA Matvienko, DS Lewis, M Swanson, B Arndt, DL Rainwater, J Stewart, DL Alekel. A single dose of soybean phytosterols in ground beef decreases serum total cholesterol and LDL cholesterol in young, mildly hypercholesterolemic men. Am J Clin Nutr 76:57–64, 2002.

90. H Karppanen, PH Karppanen, PKML Karppanen, MLS Nevalainen, and T Vaskonen. Food seasoning, food ingredients and food item compositions and methods for their preparation. US Patent 6,136,349, 2000.
91. MJ Tikkanen, P Högström, J Tuomilehto, S Keinänen-Kiukaanniemi, J Sundvall, H Karppanen. Effect of a diet based on low-fat foods enriched with non-esterified plant sterols and mineral nutrients on serum cholesterol. Am J Cardiol 88:1157–1162, 2001.
92. R Volpe, L Niittynen, R Korpela, C Sirtori, A Bucci, N Fraone, F Pazzucconi. Effect of yoghurt enriched with plant sterols on serum lipids in patients with moderate hypercholesterolaemia. B J Nutr 86:233–239, 2001.
93. RP Mensick, S Ebbing, M Lindhout, J Plat, MMA van Heugten. Effects of plant stanol esters supplied in low-fat yoghurt on serum lipids and lipoproteins, non-cholesterol sterols and fat soluble antioxidant concentrations. Atherosclerosis 160:205–213, 2002.
94. P Nestel, M Cehum, S Pomeroy, M Abbey, G Weldon. Cholesterol-lowering effects of plant sterol esters and non-esterified stanols in margarine, butter and low-fat foods. Eur J Clin Nutr 55:1084–1090, 2001.
95. F Gutierrez, B Jimenez, A Ruiz, MA Albi. Effect of olive ripeness on the oxidative stability of virgin olive oil extracted from the varieties Picual and Hojiblanca and on the different components involved. J Agric Food Chem 47:121–127, 1999.
96. A Silkeberg, SP Kochhar. Refining of edible oil retaining maximum antioxidative potency. US Patent 6,033,706, 2000.
97. J de Graaf, PRWD Nolting, M van Dam, EM Belsey, JJP Kastelein, PH Pritchard, AFH Stalenhoef. Consumption of tall oil-derived phytosterols in a chocolate matrix significantly decreases plasma total and low-density lipoprotein-cholesterol levels. Br J Nutra 88:479–488, 2002.
98. M-P St-Onge, R Ross, WD Parsons, PJH Jones. Medium-chain triglycerides increase energy expenditure and decrease adiposity in overweight men. Obes Res 11:395–402, 2003.

8
Safety of Phytosterols and Phytosterol Esters as Functional Food Components

David Kritchevsky
The Wistar Institute, Philadelphia, Pennsylvania, U.S.A.

I. EARLY HISTORY

In 1911 Ellis and Gardner (1) demonstrated the unabsorbability of phytosterols. In the late 1920s Schönheimer and his colleagues found that sitosterol, stigmasterol, ergosterol, and brassicasterol were not absorbed by rabbits or rats (2–6). Peterson (7,8) provided the first evidence that plant sterols could interfere with cholesterol absorption in chickens. Peterson also found that addition of 1.3% soy sterol to a cholesterol (1%)–containing diet inhibited atherogenesis (9); his findings were confirmed in studies with ^{14}C-labeled cholesterol (10,11).

Pollak (12) was the first to show that addition of a plant sterol to the diet could reduce cholesterol levels in humans. Pollak fed 26 healthy men plant sterols (average dose 8.1 g/day) for 2 weeks and observed a 28% reduction in plasma cholesterol levels. Pollak's report was followed by a flood of papers almost all of which showed that phytosterols exerted a hypocholesterolemic effect in a broad spectrum of human subjects ranging from normal to severely hypercholesterolemic. The early history of phytosterol effects has been reviewed by Pollak and Kritchevsky (13). In the late 1970s, Sugano and his colleagues demonstrated the hypocholesterolemic effects of sitostanol (14–16). Thus, the cholesterol-lowering effects of both sitosterol and sitostanol had been demonstrated by the early 1980s.

In early studies of the hypocholesterolemic effects of phytosterols the compounds were regarded as pharmaceutical agents and administered as such. A resurgence in interest in phytosterol effects came after it was demonstrated that phytostanols (17) or phytosterols (18) were active when incorporated into the diet as spreads. It is now evident that they can be effective in humans when administered as lecithin micelles (19) or when solubilized in a diglyceride (20). Subcutaneous administration of phytosterols or stanols (in an olive oil-ethanol 6:1 vehicle) causes hypocholesterolemia in hamsters (21).

II. TOXICITY STUDIES

In the late 1950s the Eli Lilly Company introduced a cholesterol lowering preparation that was a suspension of phytosterols. Shipley et al. (22) carried out extensive safety studies using mixed soy sterols or β-sitosterols. Rats, rabbits, and dogs were fed large amounts of phytosterols and the authors found no detectable effects on growth, serum proteins, blood urea nitrogen, or gross or microscopic appearance of any tissue or organ. The doses of phytosterol fed were 4% (rabbits) or 5% (rats) or 0.5–1.0 g/kg/day (dogs). The studies were carried out over periods of 18–22 months.

There are now available exhaustive studies of the safety of phytosterol esters. Hepburn et al. (23) fed various preparations of mixed phytosterol esters to male and female rats. The major phytosterols present in the diet were β-sitosterol (48.7%), campesterol (25.8%), and stigmasterol (26.1%). Diets containing 0.1, 1.0, 2.0, or 5.0% (w/w) phytosterol esters were fed for 90 days. Each group (one control and four treatment groups) was composed of 20 male and 20 female rats. At the end of the feeding period, blood was taken for an large range of hematology tests; 21 blood clinical parameters were measured, and a number of tissues were weighed and the weights adjusted for body weight. The authors observed no treatment-related changes that were considered to be of toxicological significance.

Another study (24) compared absorption and tissue distribution of several ^{14}C-or tritium-labeled phytosterols in rats. Adrenals, ovaries, and intestinal epithelia showed the highest levels and longest retention of radioactivity; with one exception (4-^{14}C cholesterol in males), total recovery of radioactivity was greater than 90% (Table 1). In the single exception recovery was 81 ± 13%. Malini and Vanithakumari (25) previously found no differences in a number of blood parameters in rats given graded doses of sitosterol.

A two-generation reproduction study in rats found no adverse effects of orally administered phytosterol esters in two successive generations of

Table 1 Recovery of ^{14}C and ^{3}H from Rats Gavaged with Labeled Sterols or Stanols

Substance	No. of rats	Gender[a]	% Recovery Feces	% Recovery Total
[4-^{14}C]-β-Sitosterol	4	M	96.54±7.85	100.00±7.67
[4-^{14}C]-β-Sitosterol	4	F	85.23±3.84	97.85±6.20
[4-^{14}C]-β-Sitostanol	4	M	87.93±4.00	95.96±3.43
[4-^{14}C]-β-Sitostanol	4	F	88.09±3.61	99.47±2.98
[5,6-^{3}H]-β-Sitostanol	4	F	75.07±15.86	95.03±2.24
[3α-^{3}H]- Campesterol	3	F	76.45±2.88	94.17±3.14
[5,6-^{3}H]- Campestanol	3	F	90.39±1.05	97.97±2.46
[3-^{3}H]- Stigmasterol	3	F	84.65±2.64	95.03±0.33
[4-^{14}C]- Cholesterol	3	M	51.32±17.02	81.23±12.85
[4-^{14}C]- Cholesterol	3	F	58.70±5.59	91.55±1.15

[a] M, male; F, female.
Source: Ref. 24.

Wistar rats (26). There were no effects on fertility, fecundity, gestation time, live birth index, or ratio of male to female pups. The phytosterol esters were fed at concentrations as high as 8.1% (Table 2). Another study showed no indication of estrogenicity of the mixed phytosterol esters (27). Awad et al. (28) have reported a 33% reduction in serum testosterone in rats fed a diet containing 2% phytosterols and 0.2% cholic acid. A 55% reduction in aromatase activity was the only effect observed in the testes. The authors suggest that phytosterol feeding may reduce the risk of prostate cancer.

Studies were also conducted in healthy, normolipidemic human volunteers (12 male, 12 female) ingesting a designated diet containing a phytosterol ester–enriched margarine. In the first study (29), fecal neutral sterol concentrations were increased by 375% and fecal neutral sterol metabolite concentrations were increased by 67%. Fecal secondary bile acid concentration was reduced by 21%, suggesting more rapid transit time. In another study (30) in normiolipidemic volunteers, dietary phytosterol esters had no significant effects of fecal short-chain fatty acids or fecal bacterial enzymes. The dietary treatment did not affect levels of reproductive hormones (follicle-stimulating hormone, luteinizing hormone, progesterone, estradiol, sex hormone–binding globulin, estrone) in the female volunteers.

Findings in toxicity experiments with phytostanol esters reflect those obtained with phytosterol esters. They were shown not to be genotoxic in in vitro gene mutation assays using bacterial or mammalian cells (31). Similar

Table 2 Reproductive Performance of Female Rats Fed Phytosterol Esters

		Phytosterol esters (% of diet)			
Parameters	Generation	0	1.6	3.2	8.1
Mating index (%)[a]	F_o	100	100	100	100
	F_1	100	100	100	96
Fertility index(%)[b]	F_o	100	96	86	93
	F_1	96	93	96	89
Fecundity index (%)[c]	F_o	100	96	86	93
	F_1	96	93	96	93
Gestation index[d]	F_o	100	100	96	96
	F_1	100	100	96	100

[a] (No. females mated/no. females placed with males) × 100.
[b] (No. females pregnant/no. females placed with males) × 100.
[c] (No. females pregnant/no. females mated) × 100.
[d] (No. females with live pups/no. females pregnant) × 100.
Source: Ref. 26.

results have been obtained with phytosterol esters (32). Fed to Wistar rats at levels of 0.5 g/kg body weight per day, phytostanol esters conferred no adverse effects (33). Tests (in vitro and in vivo) for estrogenic activity of phytostanol esters proved negative (34). Tested in rats phytostanol esters had no adverse effects on reproduction or development (35,36) (Table 3).

While the toxicological data indicate that phytosterols and phytostanols are safe, questions have arisen about possible physiological effects.

Table 3 Reproductive Performance of Female Rats Fed Phytostanol Esters

		Phytostanol esters (% of diet)			
Parameter[a]	Generation	0	1.75	4.38	8.76
Mating index (%)[a]	F_o	100	100	100	96
	F_1	100	96	100	96
Fertility index (%)[b]	F_o	75	96	89	89
	F_1	96	93	89	100
Fecundity index (%)[c]	F_o	75	96	100	93
	F_1	96	96	89	100
Gestation index[d]	F_o	95	100	89	100
	F_1	100	96	92	100

[a] See footnote Table 2.
Source: Ref. 36.

Table 4 Phytosterol Absorption and Turnover in Humans (10 Subjects)[a]

	Absorption (%)	Turnover ($t_{1/2}$, days)
Sitostanol	0.0441±0.0041	1.84±0.24
Sitosterol	0.512±0.038	2.94±0.15
Campestanol	0.155±0.017	1.69±0.11
Campesterol	1.89±0.27	4.06±0.32

[a] Study used intravenously and orally administered deuterium-labeled phytosterols and phytostanols.
Source: Ref. 40.

Although early studies suggested that phytosterols were totally unabsorbed, isotopic studies indicate that 3–6% of β-sitosterol is absorbed (13,37,38). Heinemann et al. (39) compared the absorption of cholesterol and several phytosterols in 10 human subjects. Mean hourly absorption (%) was cholesterol, 31.2 ± 9.1; campestanol, 12.5 ± 4.8; campesterol, 9.6 ± 13.8; stigmasterol, 4.8 ± 6.5; and sitosterol 4.2 ± 4.2. Ostlund et al. (40) have studied absorption of phytosterols in humans using intravenous dideuterated phytosterols or phytostanols followed 2 weeks later by orally administered pentadeuterated phytosterols. The oral-to-intravenous tracer ratio provided a measure of absorption. Absorption of sitosterol and campesterol was 0.512 ± 0.038% and 1.89 ± 0.27%, respectively. Absorption of the corresponding stanols was considerably lower being 0.0441 ± 0.004% for sitostanol and 0.155 ± 0.017% for campestanol (Table 4). Conversion of sitosterol to acidic metabolites extracted from feces has been reported in rats (41) and vervet monkeys (42).

III. CONCERNS RELATIVE TO PHYTOSTEROL EFFECTS

Lees and Lees (43) state that "the ideal drug for the treatment of hypercholesterolemia should be effective, free of subjective side effects and of objective toxicity. The cholesterol analog, beta sitosterol, comes close to that ideal." However, after analyzing sera of subjects treated with two different plant sterol preparations and finding relatively high levels of campesterol (mean 16 mg/dl) in five patients, they raised the possibility of iatrogenic atherosclerosis caused by plant sterols (44). Sterols other than cholesterol have indeed been isolated from aortic tissues. Hardegger et al. (45) chemically analyzed a large mass of human aortas and found cholesterol plus small amounts of oxidized derivatives of cholesterol. Their study was carried out before the availability of paper or gas chromatography. Brooks et al.

(46) and Hodis et al. (47) have recovered cholesterol oxidation products, 7α-
and 7β-hydroxycholesterol and 24- and 26-hydroxycholesterol, from human
aortas. In analyzing human atherosclerotic lesions, Vaya et al. (48) found 7-
keto- and β-epoxycholesterol as well as the oxidation products reported by
Brooks and Hodis and their coworkers. Presence of phytosterols was not
reported.

Experimentally, Cook et al. (49) fed rabbits (2–3 per group) 1%
cholesterol, cholestanol, 7-dehydrocholesterol, or Δ^7-cholestenol (lathosterol) for 13–25 days. Cholesterol or cholestanol feeding resulted in well-
defined atherosclerotic plaques; the other two sterols gave barely visible
lesions. We (50) fed rabbits normal or atherogenic diets with or without 0.2%
triparanol. Triparanol inhibits cholesterogenesis late in the cycle and leads to
accumulation of desmosterol (24-dehydrocholesterol). Both normal and
atherosclerotic aortas contained desmosterol, and we also recovered small
amounts of cholestanol and coprostanol but no hydroxylated derivatives
of cholesterol.

We (51) recently fed rabbits semipurified diets containing sitosterol or
sitostanol esters. Serum cholesterol was present in milligram (per deciliter)
quantities and serum phytosterols in microgram (per deciliter) quantities
(Table 5). Aortic cholesterol was present in microgram quantities and aortic
phytosterols in nanogram quantities. No aortic lesions were seen in any of
the animals. Apparently, an unsubstituted eight-carbon side chain is required
for incorporation of sterols into the arterial wall.

Table 5 Serum and Aorta Cholesterol and Phytosterol Levels in Rabbits Fed
0.266–0.295 g[a]

Group	Cholesterol	Sitosterol	Campesterol	Sitostanol
		Serum		
	mg/ml	µg/ml	µg/ml	µg/ml
Control	105	9.49	15.56	0.11
Sitosterol-fed	76	16.17	44.82	15.07
Sitostanol-fed	76	0.41	0.29	1.13
		Aorta		
	µg/ml	ng/ml	ng/ml	ng/ml
Control	75	1055	2470	50
Sitosterol-fed	55	1115	3489	40
Sitostanol-fed	57	507	1479	169

[a] Phytosterol daily for 60 days.
Source: Ref. 51.

Sitosterolemia is a rare lipid storage disease that results in atherosclerosis and xanthomatosis (52,53). The condition is precipitated by mutations in the ATP-binding cassette proteins ABCG5 and ABCG8, which are found in the liver and intestines (54). Sitosterolemia is a rare disease; in 1998, Berger et al. (55) estimated that there were 34 cases worldwide. The genetics of sitosterolemia has been reviewed by Lee et al. (56). The possibility that excess dietary phytosterol may lead to this condition, though slight, nevertheless exists. Glueck et al. (57,58) found that subjects with premature coronary disease and their first-degree relatives exhibited elevated plasma levels of both cholesterol and phytosterol. The authors implied that phytosterolemia may have contributed to premature coronary disease. Salen et al. (59) found that in homzygotes cholesterol accounted for over 80% of plasma, tissue, and atheroma sterol. In seven subjects in the Glueck study (58), serum cholesterol levels were 7.09 mmol/L whereas the concentration of campesterol, stigmasterol, and sitosterol combined was only 43.86 µmol/L. The implication is that a slight excess of serum phytosterols may increase sterol deposition in the aorta. Sudhop et al. (60) reported on serum sterol levels in subjects with or without a family history of coronary heart disease. There were no differences in levels of serum lipids, lipoproteins, or cholesterol precursors. The average serum cholesterol levels of the two groups were identical: 242 ± 46 mg/dl in 27 controls and 242 ± 31 mg/dl in 26 test subjects. The serum campesterol level in the control group was 0.38 mg/dl compared to 0.50 mg/dl in the group with a family history of heart disease; sitosterol levels in the two groups were 0.31 mg/dl and 0.40 mg/dl, respectively. Levels of phytosterol in the sera of the test groups were small, representing 0.21% (campesterol) or 0.17% (sitosterol) of total serum sterol. In the control group, the percentage of serum sterol represented by campesterol is 0.16% and by sitosterol is 0.13%. If the effect is real, it is very potent indeed.

IV. CONCERNS ABOUT CAROTENOIDS

The effect of dietary phytosterols on circulating carotenoids has emerged as an issue of concern, although there are no data regarding optimal plasma levels of these compounds nor of effects of relatively small changes in their concentration. There is little consistency in the published data. All of the compounds of interest are fat soluble and are transported with the plasma lipoproteins: hence, attempts have been made to normalize the data to all or some of the plasma lipids, but no consistency emerges even then.

Hallikainen et al. studied the effects of stanol and sterol esters in hypercholesterolemic subjects. In one study (61), margarines prepared with stanols isolated from wood or from vegetable oil were compared over a period of 8 weeks. Serum beta-carotene and α-tocopherol levels were reduced significantly on both margarine preparations. When corrected for total cholesterol, beta-carotene levels in the test population were not different from controls and α-tocopherol levels rose. In a second study (62), increasing doses of plant sterol esters—0.0, 0.8, 1.6, 2.4, or 3.2 g/day—were fed for 4 weeks. The α-tocopherol levels were found to be reduced significantly in all subjects and lycopene levels to be reduced significantly only in women. When the data were recalculated as millimoles of carotenoid or tocopherol per mole of cholesterol, the differences disappeared. In a study in which the effects of margarines containing sterol and stanol esters were compared, both margarines significantly reduced serum beta-carotene and α-tocopherol levels; data were not corrected for serum cholesterol or other lipids (63). Maki et al. (64) fed subjects the NCEP Step 1 diet, 92 subjects received the same diet including a low level (1.1 g/day) of margarine containing phytosterol esters, and 40 subjects were fed the NCEP Step 1 diet plus a high (2.2 g/day) level of the same margarine. After 5 weeks plasma lipids had fallen in both groups but serum concentrations of carotenoids and fat-soluble vitamins were within reference ranges at baseline and at the study's end. Nestel et al. (65) included plant sterol esters or plant stanols in the diets of 37 subjects. They found no reduction in either plasma carotenoids or tocopherols.

Weststrate and Meijer (18) compared effects of margarine containing phytosterol esters derived from wood, soybean oil, rice bran oil, or sheanut oil to 100 subjects over a 2.5- to 3.5-week period. They found 7–16% reductions in serum α-tocopherol plus beta-carotene and serum lycopene levels in all but the subjects fed rice bran oil sterols. Correcting the findings by dividing them by the total of serum cholesterol plus glycerol did not change the observed differences. A similar finding was observed in a study in which graded levels of plant sterols were fed in the form of margarine (66). Judd et al. (67) fed a group of 26 men and 27 women diets in which plant sterol esters were included in the form of salad dressing. Significant reductions in plasma total and LDL-cholesterol and triglycerides were observed. Plasma total carotenoids were reduced significantly. Ntanios et al. (68) report that a plant sterol–enriched margarine lowered plasma cholesterol and lipoproteins but did not affect levels of vitamins A and E in Japanese subjects. They observed a 21% reduction in beta-carotene levels. When Amundsen et al. (69) included a plant sterol–enriched spread in the diet of children with familial hypercholesterolemia, plasma total and LDL-cholesterol were reduced significantly. Plasma α-tocopherol levels were unaffected, but lycopene and beta-carotene levels were reduced. Dividing the plasma levels by total cholesterol plus total triglycerides significantly raised the retinol and α tocopherol values. Lycopene and beta-carotene levels remained significantly depressed. Noakes et al. (70) have reported that addition of carotenoids to the diet of subjects consuming plant sterols or stanols is effective in maintaining plasma carotenoid levels. The findings are summarized in Table 6.

The foregoing suggests that studying the effects of phytosterols on plasma carotenoids and tocopherols is an interesting exercise with no rigorous physiological end point. The report that increasing levels of dietary carotenoids can vitiate the phytosterol effect suggests that very careful examination of the diets used in the phytosterol studies may yield clues to the disparate findings. The medical literature is replete with studies of effects of hypolipidemic agents. The studies have involved many more subjects than considered here and were usually carried out over much longer periods. If lowering of LDL levels results in reduced levels of carotenoids and tocopherols, the pharmacological studies would probably show greater reductions than those reported in the phytosterol studies. The mechanism

Table 6 Summary of Human Studies Involving Sterol or Stanol Ester

No. of subjects	Dose[a]	Duration (wk)	Δ Total chol. (%)	Δ LDL-chol. (%)	Δ Carotenoids[b] (%)	Ref.
18	2.31 NW	8	−18.4	−23.3	−12.2	61
14	2.6 NV	8	−16.0	−18.8	−8.4	
22	0.8 N	4	−2.6	−1.6	−5.0	62
	1.6 N	4	−6.9	−6.1	−7.3	
	2.4 N	4	−10.6	−10.6	−8.5	
	3.2 N	4	−11.7	−11.5	−11.4	
75	1.1 R	5	−2.6	−4.9	−6.5	64
35	2.2 R	5	−4.6	−5.4	−0.7	
22	2.4 N	4	−8.6	−13.6	NC	65
	2.4 R	4	−3.6	−8.4	NC	
95	1.5–3.3 R	3.5	−7.2	−12.0	−22	18
79	0.83 R	3.5	−4.9	−6.7	−11.7	66
79	1.61 R	3.5	−5.9	−8.5	−11.0	
79	3.24 R	3.5	−6.8	−9.9	−19.2	
53	3.6 R	6	−13.8	−12.5	−11.2	67
53	1.8 R	6	−5.8	−9.1	−20.8	68
38[c]	1.6 R	8	−7.4	−10.2	−4.3	69

[a] NW, wood-derived stanol; NV, vegetable-derived stanol; N, stanol; R, sterol.
[b] Different authors measured different carotenoids, so that the results are not totally comparable.
[c] Children writh familial hypercholesterolemia.

by which cholesterol lowering is achieved, i.e., inhibition of cholesterol absorption vs. inhibition of cholesterol synthesis, for instance, might affect carotenoid transport in LDL. The end point of the hypocholesterolemia drug studies is changes in plasma or serum lipids and lipoproteins, and since the investigators conducting those studies are generally not nutritionally oriented, few if any data are available relative to plasma carotenoids and tocopherols. The fact that none of those studies have reported side effects attributable to reduced levels of carotenoids and tocopherols suggests that the findings discussed above, while interesting, may not be immediately relevant to health. The Lipid Research Clinics Coronary Primary Prevention Trial (71) entailed treatment of 1906 asymptomatic, hypercholesterolemic men with a bile acid–binding resin (cholestyramine) for an average of 7.4 years. There were 1906 controls. After 1 year plasma cholesterol levels had fallen by 14.9% and after 7 years by 8.4%. Serum carotenoids in the test group had fallen by 25.5% after 1 year and by 11.4% after 7 years, whereas no change in serum carotenoids was observed in the control group.

It may be instructive to point out that a panel of the Food and Nutrition Board of the Institute of Medicine, National Academy of Sciences (USA) has deliberated on dietary reference intakes for carotenoids. In their summary (72) they state, "This evidence, although consistent, cannot be used to establish a requirement for β carotene or carotenoid intake.... While there is evidence that β-carotene is an antioxidant in vitro, its importance to health is not known."

V. SIDE EFFECTS

The only published report of side effects in humans is an uncited reference to diarrhea that appeared in a review by Pollak (73). However, possible toxic effects of phytosterols have been studied in animal models that may provide clues to human disease. Ratnayake et al. (74) have studied effects of phytosterols in stroke-prone spontaneously hypertensive (SHRSP) rats that may be a model for human essential hypertension and hemorrhagic stroke (75). Ratnayake et al. found that vegetable oils containing high levels of phytosterols increase rigidity of erythrocytes and shorten the life span of SHRSP rats. They studied the effects of canola and soybean oils alone or augmented with phytosterols, and they also studied corn and olive oils. The phytosterol content of the diets ranged from 27.4 mg/100 g (olive oil) to 206.7 mg/100 g (canola oil). Mean survival times (days) were: canola, 88.4 ± 10.4; canola plus phytosterol, 87.6 ± 5.2; soybean oil, 102.1 ± 15.4; and soybean oil plus phytosterol, 88.3 ± 16.5. Survival times for corn oil– and olive oil–fed rats were 85.8 ± 10.5 and 80.8 ± 6.6 days, respectively. The phytosterol content of the two oils was 114.0 mg/100 g diet (corn oil) and 27.4 mg/100 g diet (olive oil). If the observed effect is due to dietary phytosterol alone, the seemingly anomalous behavior of the olive oil deserves further study. The total phytosterol/cholesterol ratio of red blood cells, liver, and kidney of SHRSP rats fed canola or soybean oils was three- to four-fold higher than in controls fed a fat designed to mimic Canadian fat intake. Human subjects with a history of hemorrhagic stroke exhibit relatively low concentrations of cholesterol in cell membranes (75,76) and blood (77–79). Ikeda et al. (80) compared the effects of phytosterol feeding in SHRSP, WKY, Wistar, and WKA rats for 2 weeks. Their test diet contained 10% safflower oil and 0.5% of a plant sterol mixture. Plant sterol content of the serum, liver aorta, adrenals, and adipose tissue of the SHRSP and WKY rats was significantly elevated. Generally the tissues of the SHRSP and WKY rats contained three to four times as much plant sterol as did those of the Wistar or WKA rats. Ikeda et al. (80) concluded that enhanced deposition of plant sterol in the tissues of rats of the three HRSP and WKY strains is due to a combination of increased absorption and decreased excretion.

Nieminen et al. (81) studied the effects of several dietary levels (1, 5, or 50 mg/kg/day) of a purified β-sitosterol preparation (88.7% β-sitosterol) on a number of biological parameters in male and female European polecats (*Mustela putorius*). There was no effect on growth. Plasma testosterone and estradiol levels increase in both males and females. There was no effect on triiodothyronine (T_3) levels, and thyroxine levels were elevated only in males on 5 or 50 mg sitosterol/kg/day. Cortisol and leptin levels were unchanged but ghrelin levels were reduced significantly in polecats fed the highest dose. Liver glycogen levels were increased at high levels of sitosterol intake and liver lipase esterase activity decreased significantly in the animals fed 5 or 50 mg/kg. Analysis of the serum showed total cholesterol to increase with higher doses of sitosterol, being 24% higher than the control in the 50 mg/kg/day group. Serum LDL levels were 54% higher than control levels in the group fed 50 mg/kg/day of sitosterol. HDL and triglyceride levels were unchanged. The plasma sterols were analyzed spectrophotometrically, so there was no chance to discriminate between cholesterol and sitosterol, which have very similar absorption spectra (82). One might speculate that the European polecat fed 50 mg sitosterol/kg/day may have absorbed some of the sterol. The authors state that "none of the observed effects of phytosterols on the various parameters measured is deleterious, harmful, or, on the other hand, useful by itself." Laraki et al. (83) found that in rats fed ratios of phytosterol to cholesterol of 1:1, 2:1, or 4:1 liver acetyl-CoA carboxylase and malic enzymes were reduced significantly.

Moghadasian (84) reviewed the pharmacological properties of plant sterols. Most of the adverse effects were seen in animals bearing high concentrations of phytosterol in their plasma or in animals in which phytosterols were applied topically or injected intraperitoneally. The data that have been reviewed suggest that, in general, dietary plant sterols are a safe and effective means for lowering blood cholesterol levels. Earlier reviews (85,86) have arrived at similar conclusions. With the expanding use of dietary plant sterols must come increased vigilance with respect to those subjects who may be susceptible to possible deleterious effects. This admonition applies to all dietary and pharmacologic treatments.

ACKNOWLEDGMENT

Supported in part by a Research Career Award (HL00734) from the National Institutes of Health.

REFERENCES

1. GW Ellis, JA Gardner. The origin and destiny of cholesterol in the organism. VIII. On the cholesterol content of the liver of rabbits under various diets and during inanition. Proc Roy Soc (London) B84:461–470, 1912.
2. R Schöheimer. Versuch einer bilanz am kaninchen bei futterring mit sitosterin. Hoppe-Seylers Z Physiol Chem 180:24–32, 1929.
3. R Schöheimer. Über die sterine des kaninchenkotes. Hoppe Seylers Z Physiol Chem 180:32–37, 1929.
4. R Schöheimer, H vonBehring, R Hummel. Über die spezifität der resorption von sterinen abhängig von ihrer konstitution. Hoppe-Seylers Z Physiol Chem 192:117–124, 1930.
5. R Schöheimer. Die spezifität der cholesterin resorption und ihre biologische bedeutung. Klin Wochschr 11:1793–1796, 1932.
6. R Schönheimer. New contributions in sterol metabolism. Science 74:79–84, 1931.
7. DW Peterson. Effect of soybean sterols in the diet on plasma and liver cholesterol in chicks. Proc Soc Exp Biol Med 78:143–147, 1951.
8. DW Peterson, EA Shneour, NF Peck, HW Gaffey. Dietary constituents affecting plasma and liver cholesterol in cholesterol-fed chicks. J Nutr 50:191–201, 1953.
9. DW Peterson, CW Nichols Jr, EA Shneour. Some relationships among dietary sterols, plasma and liver cholesterol levels, and atherosclerosis in chicks. J Nutr 47:25–57, 1952.
10. HH Hernandez, DW Peterson, IL Chaikoff, WG Dauben. Absorption of cholesterol-4-C^{14} in rats fed mixed soybean sterols and sitosterol. Proc Soc Exp Biol Med 83:498–499, 1953.
11. HH Hernandez, IL Chaikoff. Do soy sterols interfere with absorption of cholesterol? Proc Soc Exp Biol Med 84:541–544, 1954.

12. OJ Pollak. Reduction of blood cholesterol in man. Circulation 7:702–706, 1953.
13. OJ Pollak, D Kritchevsky. Sitosterol. Basel, S Karger, 1981.
14. M Sugano, F Kamo, I Ikeda, H Morioka. Lipid-lowering activity of phytostanols in rats. Atherosclerosis 24:301–309, 1976.
15. M Sugano, H Morioka, I Ikeda. A comparison of hypocholesterolemic activity of β sitosterol and β sitostanol in rats. J Nutr 107:2011–2019, 1977.
16. I Ikeda, M Sugano. Comparison of absorption and metabolism of β sitosterol and β sitostanol in rats. Atherosclerosis 30:227–237, 1978.
17. TA Miettinen, P Puska, H Gylling, H Vanhaven, E Vartainen. Reduction of serum cholesterol with sitostanol-ester margarine in a mildly hypercholesterolemic population. N Engl J Med 333:1308–1312, 1995.
18. JA Weststrate, GW Meijer. Plant sterol-enriched margarines and reduction of plasma total-and LDL-cholesterol concentrations in normocholesterolaemic and mildly hypercholesterolaemic subjects. Eur J Clin Nutr 52:334–343, 1998.
19. RE Ostlund Jr, CA Spilburg, WF Stenson. Sitosterol administration in lecithin micelles potently reduces cholesterol absorption in human subjects. Am J Clin Nutr 70:826–831, 1999.
20. S Meguro, K Higashi, T Hase, Y Honda, A Otsuka, I Tokimitsu, H Itakura. Solubilization of phytosterols in diacylglycerol versus triacylglycerol improves the serum cholesterol-lowering effect. Eur J Clin Nutr 55:513–517, 2001.
21. CA Vanstone, M Raeini-Sarjaz, PJH Jones. Injected phytosterols/stanols suppress plasma cholesterol levels in hamsters. J Nutr Biochem 12:565–574, 2001.
22. RE Shipley, RR Pfieffer, MM Marsh, RC Anderson. Sitosterol feeding: chronic animal and clinical toxicology and tissue analysis. Circulation 6:373–382, 1958.
23. PA Hepburn, SA Horner, M Smith. Safety evaluation of phytosterol esters. Part 2. Subchronic 90-day oral toxicity study on phytosterol esters—a novel functional food. Food Chem Toxicol 37:521–532, 1999.
24. DJ Sanders, HJ Minter, D Howes, PA Hepburn. The safety evaluation of phytosterol esters. Part 6. The comparative absorption and tissue distribution of phytosterols in the rat. Food Chem Toxicol 38:485–491, 2000.
25. T Malini, G Vanithakumari. Rat toxicity studies with β sitosterol. J Ethnopharacol 28:221–234, 1990.
26. DH Waalkens-Berendsen, APM Wolterbeek, MVW Wijnands, M Richold, PA Hepburn. Safety evaluation of phytosterol esters. Part 3. Two generation reproduction study in rats with phytosterol esters—a novel functional food. Food Chem Toxicol 37:683–696, 1999.
27. VA Baker, PA Hepburn, SJ Kennedy, PA Jones, LJ Lea, JP Sumpter, J Ashby. Safety evaluation of phytosterol esters. Part 1. Assessment of oestrogenicity using a combination of in vivo and in vitro assays. Food Chem Toxicol 37:13–22, 1999.
28. AB Awad, MS Hartate, CS Fink. Phytosterol feeding induces alteration in testosterone metabolism in rat tissue. J Nutr Biochem 9:712–717, 1998.
29. JA Weststrate, R Ayesh, C Bauer-Plank, PN Drewitt. Safety evaluation of phytosterol esters. Part 4. Faecal concentrations of bile acids and neutral sterols in healthy normolipidaemic volunteers consuming a controlled diet with or

without a phytosterol ester-enriched margarine. Food Chem Toxicol 37:1063–1071, 1999.

30. R Ayesh, JA Weststrate, PN Drewitt, PA Hepburn. Safety evaluation of phytosterol esters. Part 5. Faecal short-chain fatty acid and microflora content, faecal bacterial enzyme activity and serum female sex hormones in healthy normolipidaemic volunteers consuming a controlled diet either with or without a phytosterol ester-enriched margarine. Food Chem Toxicol 37:1127–1138, 1999.

31. D Turnbull, VH Frankos, JHM van Delft, N DeVogel. Genotoxicity evaluation of wood-derived and vegetable oil-derived stanol esters. Reg Toxicol Pharmacol 29:205–210, 1999.

32. AM Wolfreys, PA Hepburn. Safety evaluation of phytosterol esters. Part 7. Assessment of mutagenic activity of phytosterols, phytosterol esters, and the cholesterol derivative, 4-cholestene-3-onc. Food Chem Toxicol 40:461–470, 2002.

33. D Turnbull, MH Whittaker, VH Frankos, D Jonker. Thirteen-week oral toxicity study with stanol esters in rats. Reg Toxicol Pharmacol 29:216–226, 1999.

34. D Turnbull, VH Frankos, WR Leeman, D Jonker. Short term tests of estrogenic potential of plant stanols and plant stanol esters. Reg Toxicol Pharmacol 29:211–215, 1999.

35. RS Slesinski, D Turnbull, VH Frankos, APM Wolterbeek, DH Waalkens-Berendsen. Development toxicology study of vegetable oil-derived stanol fatty acid esters. Reg Toxicol Pharmacol 29:227–233, 1999.

36. MH Whittaker, VH Frankos, APM Wolterbeek, DH Waalkens-Berendsen. Two generation reproductive toxicity study of plant stanol esters in rats. Reg Toxicol Pharmacol 29:196–204, 1999.

37. RG Gould. Absorbability of β-sitosterol. Trans NY Acad Sci 18:129–134, 1955.

38. G Salen, EH Ahrens Jr, SN Grundy. Metabolism of β sitosterol in man. J Clin Invest 49:952–967, 1979.

39. T Heinemann, G Axtmann, K von Bergman. Comparison of intestinal absorption of cholesterol with different plant sterols in man. Eur J Clin Invest 23:827–831, 1993.

40. RE Ostlund Jr, JA McGill, C-M Zeng, DF Covey, J Stearns, WF Stenson, CA Spilburg. Gastrointestinal absorption and plasma kinetics of soy Δ^5-phytosterols and phytostanols in humans. Am J Physiol Endocrinol Metab 282:911–916, 2002.

41. KM Boberg, E Lund, J Olund, I Bjorkheim. Formation of C21 bile acids from plant sterols in rats. J Biol Chem 265:7967–7975, 1990.

42. D Kritchevsky, LM Davidson, EM Mosbach, BL Cohen. Identification of acidic steroids in feces of monkeys fed β-sitosterol. Lipids 16:77–78, 1981.

43. RS Lees, AM Lees. Effect of sitosterol therapy on plasma lipid and lipoprotein concentration. In: H Greten ed. Lipoprotein Metabolism. Berlin: Springer-Verlag, 1976, pp. 119–124.

44. AM Lees, HYI Mok, RS Lees, MA McCluskey, SM Grundy. Plant sterols as cholesterol lowering agents: clinical trials in patients with hypercholesterolemia and studies of sterol balance. Atherosclerosis 28:325–338, 1977.

45. E Hardegger, L Ruzicka, E Tagmann. Untersuchungen über organe extracte.

Zur kenntnis der unseifbaren lipoide aus arteriosklerotischen aorten. Helv Chim Acta 30:2205–2221, 1943.

46. CJW Brooks, G Steel, JD Gilbert, WA Starland. Lipids of human atheroma. Part 4. Characterization of a new group of polar sterol esters from human atherosclerotic plaques. Atherosclerosis 13:223–237, 1971.

47. HN Hodes, DW Crawford, A Sevanian. Cholesterol feeding increases plasma and aortic tissue cholesterol oxide levels in parallel: further evidence for the role of cholesterol oxidation in atherosclerosis. Atherosclerosis 89:117–126, 1991.

48. J Vaya, M Aviram, S Mahmood, T Hayek, E Grenader, H Hoffman, S Milo. Selective distribution of oxysterols in atherosclerotic lesions and human plasma lipoproteins. Free Rad Res 34:485–497, 2001.

49. RP Cook, A Kliman, LF Fieser. The absorption and metabolism of cholesterol and its main companions in the rabbit—with observations on the atherogenic nature of the sterols. Arch Biochem 52:439–450, 1954.

50. D Kritchevsky, R Fumagalli, F Cattabeni, SA Tepper. Effect of triparanol on sterol composition in normal and atherosclerotic rabbit aorta. Riv Farmacol Terapia 1:455–463, 1970.

51. D Kritchevsky, SA Tepper, SK Czarnecki, B Wolfe, KDR Setchell. Serum and aortic levels of phytosterols in rabbits fed sitosterol or sitostanol ester preparations. Submitted.

52. AK Bhattacharyya, WE Connor. β-Sitosterolemia xanthamatosis: a newly described lipid storage disease in two sisters. J Clin Invest 53:1033–1043, 1974.

53. G Salen, V Shore, GS Tint, T Forte, S Shefer, I Horak, E Horak, B Dayal, L Nguyen, AK Batta, T Lindgren, PO Kwiterovich. Sitosterolemia. J Lipid Res 30:1319–1330, 1989.

54. KE Berge, H Tian, GA Graf, L Yu, NV Grishn, J Schultz, P Kwiterovich, B Barnes, R Barnes, HH Hobbs. Accumulation of dietary cholesterol in sitosterolemia caused by mutations in adjacent ABC transporters. Science 290: 1771–1775, 2000.

55. GMB Berger, RJ Pegoraro, SA Patel, P Naidee, L Rom, H Hidaka, AD Marais, A Jadhav, RP Naoumova, GR Thompson. HMG-CoA reductase is not the site of the primary defect in phytosterolemia. J Lipid Res 39:1046–1054, 1998.

56. M-H Lee, K Lu, SB Patel. Genetic basis of sitosterolemia. Curr Opin Lipidol 12: 141–149, 2001.

57. CJ Glueck, J Speirs, T Tracy, P Streicher, E Illeg, J Vandergrift. Relationships of serum plant sterols (phytosterol) and cholesterol in 595 hypercholesterolemic subjects, and familial aggregion of phytosterols, cholesterol, and premature coronary heart disease in hyperphytosterolemic probands and their first degree relatives. Metabolism 40:842–848, 1991.

58. CJ Glueck, P Streicher, E Illeg. Serum and dietary phytosterols, cholesterol and coronary heart disease in hypercholesterolemic probands. Clin Biochem 25: 331–334, 1992.

59. G Salen, I Horak, M Rothkopf, JL Cohen, J Speck, GS Tint, V Shorg, B Dayal, T Chen, S Shefer. Lethal atherosclerosis associated with abnormal plasma and

tissue sterol composition in sitosterolemia with xanthomatosis. J Lipid Res 26:1126–1133, 1985.

60. T Sudhop, BM Gottwald, K vonBergmann. Serum plant sterols as a potential risk factor for coronary heart disease. Metabolism 51:1519–1521, 2002.

61. MA Hallikainen, MIJ Uusitupa. Effects of 2 low-fat stanol-ester containing margarines on serum cholesterol concentrations as part of a low-fat diet in hypercholesterolemic subjects. Am J Clin Nutr 69:403–410, 1999.

62. MA Hallikainen, ES Sarkkinen, MIJ Uusitupa. Plant stanol esters affect serum cholesterol concentrations of hypercholesterolemia men and women in a dose-dependent manner. J Nutr 130:767–776, 2000.

63. MA Halllikainen, ES Sarkkinen, H Gylling, AT Erkkkila, MIJ Uusitupa. Comparison of the effects of plant sterol esters and plant stanol ester–enriched margarines in lowering serum cholesterol concentrations in hypercholester-olaemic subjects on a low fat diet. Eur J Clin Nutr 54:715–725, 2000.

64. KC Maki, MH Davidson, DM Umporowicz, EJ Schaefer, MR Cicklin, KA Chen, S Chen, JR McNamara, BW Gebhart, JD Ribaya-Mercado, G Perrone, SJ Robins, WC Franke. Lipid responses to plant-sterol-enriched reduced-fat spreads incorporated into a national cholesterol education program Step 1 diet. Am J Clin Nutr 74:33–43, 2001.

65. P Nestel, M Cehun, S Pomeroy, M Abbey, G Weldon. Cholesterol-lowering effects of plant sterol esters and non-esterified stanols in margarine, butter, and low fat foods. Eur J Clin Nutr 55:1084–1090, 2001.

66. HFJ Hendriks, JA Weststrate, T van Vliet, GW Meijer. Spreads enriched with three different levels of vegetable oil sterols and the degree of cholesterol lowering in normocholesterolaemic and mildly hypercholesterolaemic subjects. Eur J Clin Nutr 53:319–327, 1999.

67. JT Judd, DJ Baer, SC Chen, BA Clevidence, RA Muesing, M Kramer, GW Meijer. Plant sterol esters lower plasma lipids and most carotenoids in mildly hypercholesterolemic adults. Lipids 37:33–42, 2002.

68. FY Ntanios, Y Homma, S Ushiro. A spread enriched with plant sterol-esters lowers blood cholesterol and lipoproteins without affecting vitamins A and E in normal and hypercholesterolemic Japanese men and women. J Nutr 132: 3650–3655, 2002.

69. AL Amundsen, L Ose, MS Nenseter, FY Ntanios. Plant sterol ester–enriched spread-lowers plasma total and LDL cholesterol in children with familial hypercholesterolemia. Am J Clin Nutr 76:338–344, 2002.

70. M Noakes, P Clifton, F Ntanios, W Shrapnel, I Record, J McInerney. An increased in dietary carotenoids when consuming plant sterols or stanols is effective in maintaining plasma carotenoid concentrations. Am J Clin Nutr 75:19–86, 2002.

71. Lipid Research Clinics Program. The Lipid Research Clinics Coronary Primary Prevention Trial Results. 1. Reduction in incidence of coronary heart disease. JAMA 251:351–364, 1984.

72. Panel on Dietary Antioxidants and Related Compounds. Food and Nutrition Board. Institute of Medicine, Dietary Reference Intakes for Vitamin C, Vitamin

E, Selenium and Carotenoids p. 325. Washington DC: National Academy Press, 2000.

73. OJ Pollak. Effect of plant sterols on serum lipids and atherosclerosis. Pharmacol Ther 31:177–208, 1985.

74. WMN Ratnayake, MR L'Abbe', R Mueller, S Hayward, L Plouffe, R Trick, K Trick. Vegetable oils high in phytosterols make erythrocytes less deformable and shorten the lifespan of stroke-prone spontaneously hypertensive rats. J Nutr 130:1166–1178, 2000.

75. Y Yamori, R Horie, H Tanase, K Fujiwara, Y Nara, W Lovenberg. Possible role of nutritional of cerebral lesions in stroke-prone spontaneously hypertensive rats. Hypertension 6:49–53, 1984.

76. K Tsuda, Y Veno, I Nishio, Y Masuyama. Membrane fluidity as a genetic marker of hypertension. Clin Exp Pharmacol Physiol 19:11–16, 1992.

77. A Kagan, JS Popper, GG Rhodes. Factors related to stroke incidence in Hawaiian Japanese men; the Honolulu Heart Study. Stroke 11:14–21, 1980.

78. H Tanaka, Y Veda, M Hayashi, C Date, T Baba, H Yamashita, H Shoji, Y Owada, D Owada, R Detels. Risk factors for cerebral hemorrhage and cerebral infarction in a Japanese rural community. Stroke 13:62–73, 1982.

79. JD Neaton, H Blackburn, D Jacobs, L Kuller, DJ Lee, R Sherwin, J Shih, J Wentworth, D Wentworth. Serum cholesterol level and mortality findings for men screened in the multiple risk factor intervention trial. Arch Intern Med 152:1490–1500, 1992.

80. I Ikeda, H Nakagiri, M Sugano, S Ohara, T Hamada, M Nonaka, K Imaizumi. Mechanisms of phytosterolemia in stroke-prone spontaneously hypertensive and WKY rats. Metabolism 50:1361–1368, 2001.

81. P Nieminen, AP Mustonen, P Lindström-Seppä, J Asikainen, H Mussalo-Raubamaa, JVK Kukkonen. Phytosterols act as endocrine and metabolic disruptors in the European polecat (*Mustela putorius*). Toxicol Appl Pharmacol 178:22–28, 2002.

82. D Kritchevsky, SA Tepper. Assay of plant sterols by use of cholesterol oxidase. Clin Chem 25:1464–1465, 1979.

83. L Laraki, X Pelletier, J Mourot, G Debry. Effects of dietary phytosterols on liver lipids and lipid metabolism enzymes. Ann Nutr Metab 37:129–133, 1993.

84. MH Moghadasian. Pharmacological properties of plant sterols. In vivo and in vitro observations. Life Sci 67:605–615, 2000.

85. WH Ling, PJH Jones. Dietary phytosterols: a review of metabolism, benefits and side effects. Life Sci 57:195–206, 1995.

86. DJA Jenkins, CWC Kendall. Plant sterols, health claims and strategies to reduce cardiovascular risk. J Am Coll Nutr 18:559–562, 1999.

9

Potential Health Risks Associated with Large Intakes of Plant Sterols

W. M. Nimal Ratnayake and Elizabeth J. Vavasour
Health Canada, Ottawa, Ontario, Canada

I. INTRODUCTION

Plant sterols, which are structurally related to cholesterol, have long been known to reduce serum total and low-density lipoprotein (LDL) cholesterol levels when consumed in the diet by lowering the intestinal absorption of dietary and biliary cholesterol (1–4). Plant sterols interfere with cholesterol absorption by decreasing the micellar solubility of cholesterol. Plant sterols are not synthesized by animals or humans but are obtained in the diet primarily from vegetable oils, margarines, and other vegetable oil–based food products. The majority of vegetable oils contain 0.1–0.5 g total plant sterols per 100 g oil, while some oils such as canola and corn may contain up to 1% (w/w). The major plant sterols in the diet are the unsaturated plant sterols, i.e., β-sitosterol, campesterol, and stigmasterol. Saturated plant sterols (i.e., stanols), such as sitostanol and campestanol, are present in vegetable oils only in trace amounts. Individuals consuming typical Western diets take in 160–360 mg/day unsaturated plant sterols and 20–50 mg/day stanols (5). At these levels of consumption, plant sterols and stanols have little effect on cholesterol absorption. However, it has been recognized over the last four decades that very large intakes (ranging from 0.8 to 4 g/day) can interfere with intestinal absorption of cholesterol and significantly reduce (10–15%) serum cholesterol levels (1–4). Increasing daily intakes above 4 g/day is not expected to lead to any further decreases in serum total or LDL cholesterol levels.

There are now for sale in many countries special margarines, cereals, and other commercial foods fortified with about 8–12% by weight of plant sterols (isolated from vegetable oil deodorizer distillate, a by-product of the vegetable oil refining process) or stanols (derived from hydrogenation of sterols obtained from wood pulp). These products are aimed at helping to reduce elevated serum total and LDL cholesterol levels (6). In order to achieve effective (10–15%) serum cholesterol reduction, the level of intake of plant sterols or stanols must be about 2–3 g/day (7,8). This corresponds to an approximately 10-fold increase in comparison to the normal intake of plant sterols from various food sources.

II. ABSORPTION OF PLANT STEROLS AND STANOLS

The safety of large intakes of plant sterols and stanols on a daily basis by the general population needs to be evaluated (9) even though some review articles about the beneficial effects of plant sterols have generally assumed that there are no adverse effects associated with them (3,6,10,11). The safe consumption of plant sterols is to a large extent related to their lack of absorption in the upper small intestine. Most experimental evidence at present indicates that the absorption of plant sterols and stanols is generally poor. The estimated overall absorption of plant sterols by disease-free humans is 5% and laboratory animals when typical amounts of plant sterols are consumed (2). The actual absorption rate varies with the particular plant sterol. Campesterol is more easily absorbed than sitosterol in rats (12), dogs (13), and rabbits (14). A study investigating the fate of radiolabeled phytosterols and phytostanols administered by gavage to rats (15) demonstrated that the sterols are absorbed to a much greater extent than the stanols (campesterol, 13%; β-sitosterol, 4%; stigmasterol 4%; campestanol, β-sitostanol, each 1–2%), although both plant sterols and stanols are absorbed at lower levels than cholesterol (27%). In humans also, campesterol has the highest rate of absorption at 9.6%, followed by stigmasterol at 4.8% and β-sitosterol at 4.2% (16). However, there are large individual differences among subjects in percentage absorption and rates of absorption for each of these sterols (Table 1). The absorption for brassicasterol, Δ5-avenasterol, and other minor, naturally occurring plant sterols has not been measured so far, but from tissue and blood levels, it may be assumed that their absorption would be less than 1% of the dietary amount. Among the 5α saturated derivatives, sitostanol appears to be nonabsorbable. However, the absorption of campestanol (the saturated derivative of campesterol) is considerably higher than other saturated derivatives and estimated as 12.5%, when infused into the small intestine at a lower concentration than campesterol, sitosterol, and stigmasterol (16). Cholesterol absorption is much more efficient than that of plant sterols and stanols; between 20% and 42% of dietary cholesterol is absorbed (Table 1).

Table 1 Mean Hourly and Percentage Absorption of Different Sterols in Man[a]

Sterol	Hourly absorption (mg/h)			Percentage absorption		
	Minimum	Maximum	Mean ± SD	Minimum	Maximum	Mean ± SD
Cholesterol	8.1	23.7	15.7 ± 5.7	20.5	42.4	31.2 ± 9.1
Campesterol	0.24	0.62	0.46 ± 0.06	5.8	20.9	9.6 ± 13.8
Stigmasterol	0	0.29	0.08 ± 0.11	0	17	4.8 ± 6.5
Sitosterol	0.03	2.43	0.91 ± 0.67	0.2	10.9	4.2 ± 4.2
Campestanol	0.01	0.11	0.07 ± 0.03	1.7	17.7	12.5 ± 4.8

[a] Values are measured for 10 healthy subjects by an intestinal perfusion technique over a 50-cm segment of the upper jejunum using sitostanol as non-absorbable marker.
Source: Ref. 16.

The underlying mechanism for the relatively lower absorption of plant sterols and stanols compared to cholesterol is not yet known (see Chapter 6). However, recent molecular studies, and in particular investigations of subjects with phytosterolemia (discussed below), have increased the understanding of intestinal sterol absorption and the key roles played by ATP-binding cassette (ABC) transporters (proteins) in transporting sterols into epithelial or surface cells in the lining of the intestine (17,18). In particular, ABCG5 and ABCG8 have been identified as key transporters providing an effective barrier against the absorption of plant sterols. Furthermore, plant sterols and stanols are poorly esterified by intestinal acyl-coenzyme A: cholesterol acyltransferase (ACAT) (19). After absorption, only about 12–25% of sitosterol in the thoracic lymph is esterified, as compared with 70–90% esterification of cholesterol (20,21). Since sterols need to be esterified for incorporation into chylomicrons, this inefficient esterification may explain poor absorption of plant sterols into the lymph. Because of their low rate of absorption, very little plant sterol and stanol is incorporated into tissues. Only small amounts of plant sterols are normally found in plasma of normal, healthy humans. The total plasma plant sterols in healthy humans following a normal diet may range from 0.3 to 1.0 mg/dl (8,21). The average value is approximately 0.5 mg/dl, thus accounting for less than 0.5% of the total plasma sterols. Sitostanol and campestanol are virtually undetectable (8).

III. HYPERABSORBERS OF PLANT STEROLS AND STANOLS

A. Phytosterolemia (Sitosterolemia)

In contrast to healthy humans, some individuals can absorb higher amounts of plant sterols than others and consequently have elevated levels of serum

Ratnayake and Vavasour

plant sterols. A well-documented group of high absorbers of plant sterols and stanols are the patients with the rare autosomal recessive trait phytosterolemia (also known as sitosterolemia) (21–30), which was first described in 1973 by Bhattacharyya and Connor (21). The true prevalence of this condition is not known, but it appears that phytosterolemia has been diagnosed worldwide in 50 families (31). A characteristic biochemical feature of phytosterolemia is the high plasma levels of sitosterol, campesterol, and stigmasterol, and accumulation of these non-cholesterol sterols in tendon xanthomas, adipose tissue, skin, aorta, and virtually all other tissues, except the brain. Various reports have shown that the serum plant sterol concentrations in phytosterolemic patients range from 18 to 72 mg/dl (Table 2) and account for about 7–30% of their total plasma sterols (21–30). The absolute amount is 50–60 times higher than the concentrations in healthy subjects, whose total plant sterol levels normally range from 0.3 to 1.0 mg/dl. In addition to the common unsaturated forms of sterols, the saturated sterols sitostanol, campestanol, and cholestanol are also present in increased amounts

Table 2 Plasma Sterol and Stanol Concentrations (mg/dl) in Some Reported Instances of Familial Phytosterolemia

Patient's age and sex		Cholesterol	Cholestanol	Campesterol	Campestanol	Sitosterol	Sitostanol
—	F[a]	203	NA	10	NA	37	NA
—	F[a]	206	NA	8	NA	17	NA
24	F[b]	245 ± 39	6.7 ± 1.1	12 ± 1.1	2.3 ± 0.4	20 ± 2.3	4.2 ± 1.1
18	F[b]	202 ± 25	4.7 ± 1.0	8 ± 3.1	1.4 ± 0.2	14 ± 4.1	2.2 ± 0.7
22	F[b]	233 ± 12	3.8 ± 1.4	10 ± 5.1	1.9 ± 1.0	21 ± 8.3	5.4 ± 2.5
16	M[b]	249 ± 39	7.5 ± 2.4	13 ± 1.5	2.6 ± 0.9	20 ± 5.5	3.9 ± 1.1
52	M[b]	134	1.6	13	1.5	27	3
7	M[b]	202	1.2	12	1.6	26	3.1
32	F[b]	207	2.5	10	1.9	28	2
30	F[b]	368	3.6	29	9	65	8
41	F[b]	169	1.2	18	7	42	4.0
22	M[b]	324	3.9	27	3	60	6
20	M[b]	271	3.8	19	2	42	4
32	F[b]	256	3.1	15	1	29	3
24	F[b]	482	11	24	4	56	6
38	F[b]	336	4.9	20	1.2	45	3.2
41	F[c]	118 ± 10	NA	6.67 ± 0.74	0.72 ± 0.06	18.5 ± 2.0	5.45 ± 0.55
32	M[d]	261	NA	NA	NA	43	NA

NA, values not available.
[a] Ref. 21.
[b] Ref. 26.
[c] Ref. 46.
[d] Ref. 41.

in plasma (Table 2). In a group of 14 patients the mean concentrations of sitostanol, campestanol, and cholestanol in plasma were 4.1, 2.9, and 4.2 mg/dl, respectively (26). These dihydro derivatives are not normally detected in the plasma of healthy subjects. Since the diet contains only small amounts of sitostanol, campestanol, and cholestanol, these dihydro derivatives are probably produced endogenously by phytosterolemic patients (26).

The plant sterols are deposited in all tissues of these individuals in approximately the same increased ratio as is present in plasma. In patients studied by Salen et al. (27) and Bhattacharyya et al. (32), the total exchangeable pool of sitosterol is expanded to a size of 3500–6200 mg as compared with healthy control levels of 120–290 mg. In the xanthomas, the plant sterols account for 15–20% of the total sterols (21,33). In a patient who died of phytosterolemia, the aorta was extensively atherosclerotic and contained more than twice the quantity of plant sterols (33).

The large accumulation of plant sterols in blood and tissues in phytosterolemia is most likely related to three abnormal mechanisms: hyperabsorption and impaired biliary secretion of plant sterols, and reduced cholesterol synthesis. Healthy, normal humans generally absorb 5% or less of ingested sitosterol, whereas phytosterolemic patients absorb extremely high levels; absorption rates ranging from 19% to 63% have been found in studies with phytosterolemic subjects (21,24,27,28). On the other hand, cholesterol absorption of phytosterolemic subjects seems normal (24). Recent molecular studies have identified the genetic defect in phytosterolemia and found that hyperabsorption of plant sterols is caused by mutations in either ABCG5 or ABCG8 genes located on chromosome 2p21, which normally control processes that act as an intestinal barrier for absorption of plant sterols and mediate their excretion from the liver (17,18,34). The higher absorption is accompanied by a defect in elimination of sterols in bile (24,32). Miettinen found that the biliary excretion of sitosterol in his patient was less than 20% of normal (24). There was also a low rate of biliary cholesterol excretion, which would decrease the intestinal dilution of plant sterol. Moreover, phytosterolemic patients have a reduced cholesterol 7α-hydroxylase activity, which is the rate-limiting step in bile acid synthesis (35). In two patients with phytosterolemia, the cholesterol 7α-hydroxylase activity was about 70% of the controls (35). The reduced activity is primarily due to inhibition of cholesterol 7α-hydroxylase activity by the elevated levels of sitosterol and cholestanol as demonstrated by Shefer et al. in both rat and human liver microsomes (36). Furthermore, whole-body cholesterol synthesis is reduced in phytosterolemic subjects (24,27,32), and it has been suggested that high intracellular sterol pools (cholesterol + plant sterols) are detected by sterol sensors regulating cellular cholesterol homeostasis (37). Average cholesterol synthesis is about 50% lower in sitosterolemic subjects than in healthy controls (24,38,32). 3-Hydroxy-3-methylglutaryl–coenzyme A (HMG-CoA) reductase activity was also reduced in liver cell microsomes of patients with sitosterolemia as compared with controls (15 vs. 98 pmol/mg protein/min) (38). Honda et al. (39) demonstrated that the reduced cholesterol synthesis in phytosterolemic subjects results from inhibition of all the enzymes mediating the cholesterol biosynthetic pathway, including HMG-CoA reductase, the rate-controlling enzyme for cholesterol biosynthesis in the liver and the mononuclear cells. Clearly, the increased absorption of sterols seen in phytosterolemia is primarily due to a functional mutation in a transporter that normally actively excludes plant sterols from the intestinal brush border and secretes into the bile. The lowering of HMG-CoA reductase activity might be due to down-regulation of the enzyme.

The major clinical manifestations of sitosterolemia include development of tendon and tuberous xanthomas at an early age and predisposition for early development of atherosclerosis (28,29). Early development of atherosclerosis has been present in several cases of sitosterolemia particularly in young males. The increased risk of atherosclerosis was observed in three young men who died at age 13, 18, and 32, respectively, because of acute myocardial infarctions associated with extensive coronary and aortic arteriosclerosis (33,40,41). Coronary bypass surgery has been performed on a 29-year-old man for extensive coronary atherosclerosis (33). Although premature atherosclerosis is very common, most of the phytosterolemic subjects are only mildly or moderately hypercholesterolemic (Table 2). Only in one case was plasma cholesterol as high as 750 mg/dl. Since the cholesterol absorption of phytosterolemic subjects seems normal (24) and cholesterol synthesis is reduced, the hypercholesterolemia seen in some subjects might be a consequence of poor clearance of cholesterol as bile acids. Sitosterol impairs the conversion of cholesterol to bile acids by blocking cholesterol 7α-hydroxylase; therefore, the high circulating levels of sitosterol may be partly responsible for the rise in plasma cholesterol and early development of atherosclerosis in phytosterolemia. Other important clinical manifestations of phytosterolemia include episodes of hemolysis, splenomegaly, platelet abnormalities (25,42,43), and painful arthritis of the knee and ankle joints (22).

Since high systemic concentrations of plant sterols produce harmful effects, it has been recommended that patients with phytosterolemia consume a diet with the lowest possible amount of plant sterols (28). Thus, the diet should not contain vegetable oils, shortenings, margarines, or other products containing measurable amounts of plant sterols whether added or not. Fruits and vegetables are allowed because they contain very little fat and sterols. Since absorption of cholesterol is not altered in phytosterolemia, food derived from animal sources with cholesterol as the dominant sterol is allowed. Such a regimen has been shown to decrease elevated levels of plasma plant sterols rapidly (32,44,45), but full normalization is difficult to achieve (25,40). In addition to plant sterols, it would also be advisable to avoid foods fortified with plant stanols because, as with the unsaturated plant sterols, phytosterolemic patients hyperabsorb and retain dietary campestanol (46).

B. Stroke-Prone Spontaneously Hypertensive Rats

Stroke-prone spontaneously hypertensive (SHRSP) rats, initially developed in Japan, are considered to be a model for human essential hypertension and hemorrhagic stroke (47). These animals have been widely used to study the effects of environmental factors on hypertension and stroke (47–50). A characteristic feature of SHRSP rats is that they develop severe hypertension (systolic blood pressure >200 mm Hg) at a very young age (4–5 weeks) and maintain this high blood pressure until they die of stroke. Hemorrhagic stroke is the primary cause of death, but some SHRSP rats develop both hemorrhagic and thrombotic strokes. Various dietary factors influence the development of stroke. As with humans with a positive family history of hypertension, SHRSP rats are also sensitive to dietary sodium; excess salt intake causes a sharp rise in blood pressure. Salt-induced hypertension and its related complications can be controlled and prevented in SHRSP rats by modifying dietary factors. Sodium reduction coupled with increased intakes of potassium, calcium, protein, dietary fiber, fish oil, or cholesterol prevented development of hypertension-induced stroke (47). The applicability of some of these dietary factors has been confirmed by clinical and epidemiological studies in humans (50).

Recent studies conducted in Canada and Japan have demonstrated that the development of stroke in SHRSP rats is also influenced by the dietary amount of plant sterols and cholesterol (51–55). Cholesterol is a beneficial dietary component for SHRSP rats. A diet containing cholesterol (1 g per 100 g diet) significantly increased plasma cholesterol concentration, and this consequently delayed the onset of stroke and prolonged the life span of SHRSP rats by approximately 40% in both NaCl-loaded and unloaded conditions (51). In comparison, diets with no added cholesterol greatly shortened post stroke survival. A primary defect of SHRSP rats is that they have defective, less deformable and fragile red blood cell membranes than in other rat strains due to the low amount of cholesterol in cell membranes. The cell membrane abnormalities are of pathogenic importance in hypertensive vascular lesions because the cerebral hemorrhage and infarction noted in SHRSP rats are commonly caused by arterionecrosis (47). Therefore, the beneficial effects of dietary cholesterol on stroke prevention and longevity on SHRSP rats are most likely due to incorporation of exogenous cholesterol into cell membranes, which leads to an improvement of cell membrane physical characteristics.

In contrast to dietary cholesterol, plant sterols have adverse effects on SHRSP rats (52–54). Ratnayake et al. (52,53) have shown that dietary fats high in plant sterols as well as addition of plant sterols to vegetable oils naturally low in plant sterols exacerbates the development of stroke and consequently shortens the life span of SHRSP rats. Plant sterol–fortified soybean oil, as opposed to unfortified soybean oil, reduced the survival rate considerably. In general, the survival times were 10–12% lower in groups fed diets enriched in plant sterols (100–200 mg/100 g diet) compared to those groups fed lower amounts of plant sterols (27–36 mg/100 g diet). On the other hand, Canadian and Japanese fat mimic diets that contained some cholesterol and lower amounts of plant sterols apparently do not shorten the average life span (52,54). The life-shortening effects were linked to excessive absorption and incorporation of plant sterols into plasma, red blood cells, and various other tissues (52) (Table 3). The proportion of plant sterols in total sterols in tissues of SHRSP rats fed diets enriched plant sterols was about 30% (52) (Table 3). These values were similar to those observed for phytosterolemic patients.

Recent studies by Ikeda et al. (54) have confirmed the increased accumulation of plant sterols by SHRSP rats. They further demonstrated that tissue levels of plant sterols in SHRSP rats are three to four times higher than those for healthy Wistar and WKA strains (Table 4). In addition, they observed that not only SHRSP rats but also WKY rats and spontaneously

Table 3 Cholesterol and Total Plant Sterol Concentration of Plasma, Red Blood Cells, and Liver of Stroke-Prone Spontaneously Hypertensive Rats Fed Various Fats with Plant Sterols[a]

Test fat	PS content in diet mg/100 g diet	Plasma (mmol/L) CH	Plasma (mmol/L) PS	Red cells (μmol/100g) CH	Red cells (μmol/100g) PS	Liver (μmol/100g) CH	Liver (μmol/100g) PS
Canola oil	98	1.39	0.5	194	46	473	122
Canola oil + PS	207	1.63	0.92	181	83	487	226
Soybean oil	37	1.14	0.25	236	30	720	84
Soybean oil + PS	202	1.44	0.9	173	86	453	216
Corn oil	114	1.37	0.61	173	52	644	180
Olive oil	27	0.26	0.15	171	15	787	49
Canadian Fat Mimic	27	1.28	0.13	275	24	794	59

[a]Rats were fed fats with differing amounts of plant sterols for 30–32 days.
CH, cholesterol; PS, total plant sterols.
Source: Ref. 52.

Table 4 Sterol Compositions in Several Tissues of Wistar, WKA, WKY, and SHRSP Rats Fed a 0.5% Plant Sterol Diet for 2 Weeks

Tissue	Total sterol (mg/dl or mg/g)	Composition (wt %)				
		Cholesterol	Campesterol	Sitosterol	Sitostanol	Total plant sterol
Serum						
Wistar	88.4 ± 3.4	88.5 ± 0.9	6.6 ± 0.4	4.7 ± 0.4	0.2 ± 0.0	11.5 ± 0.9
WKA	69.1 ± 3.4	90.4 ± 0.4	5.4 ± 0.3	4.1 ± 0.2	0.1 ± 0.0	9.6 ± 0.4
WKY	114 ± 3	69.1 ± 0.8	15.8 ± 0.4	14.5 ± 0.4	0.6 ± 0.0	30.9 ± 0.8
SHRSP	93.8 ± 2.0	65.7 ± 1.1	18.1 ± 0.8	15.5 ± 0.7	0.7 ± 0.2	34.3 ± 1.1
Liver						
Wistar	2.8 ± 0.1	90.8 ± 0.6	6.2 ± 0.4	2.9 ± 0.2	0.1 ± 0.0	9.2 ± 0.6
WKA	3.5 ± 0.2	93.8 ± 0.3	4.3 ± 0.2	1.8 ± 0.1	0.1 ± 0.0	6.2 ± 0.3
WKY	2.4 ± 0.0	70.8 ± 1.0	16.6 ± 0.5	12.1 ± 0.4	0.6 ± 0.0	29.2 ± 0.9
SHRSP	2.4 ± 0.1	70.2 ± 0.9	18.0 ± 0.7	11.4 ± 0.3	0.5 ± 0.0	29.8 ± 0.9
Aorta						
Wistar	1.5 ± 0.0	93.9 ± 0.6	3.9 ± 0.4	2.0 ± 0.3	0.2 ± 0.0	6.1 ± 0.6
WKA	1.8 ± 0.1	95.4 ± 0.2	3.0 ± 0.1	1.4 ± 0.1	0.2 ± 0.0	4.6 ± 0.2
WKY	1.7 ± 0.1	85.7 ± 1.1	8.3 ± 0.6	5.4 ± 0.5	0.5 ± 0.1	14.3 ± 1.1
SHRSP	1.6 ± 0.2	86.1 ± 0.9	8.4 ± 0.5	5.0 ± 0.1	0.6 ± 0.1	13.9 ± 0.9
Adrenals						
Wistar	14.7 ± 1.3	89.9 ± 0.7	6.8 ± 0.5	3.3 ± 0.3	0.1 ± 0.0	10.2 ± 0.8
WKA	18.1 ± 1.4	91.7 ± 0.3	5.6 ± 0.2	2.6 ± 0.2	0.1 ± 0.0	8.3 ± 0.4
WKY	27.0 ± 2.3	71.9 ± 0.7	16.6 ± 0.4	10.9 ± 0.3	0.6 ± 0.1	28.1 ± 0.7
SHRSP	11.4 ± 1.0	68.2 ± 0.8	19.2 ± 0.7	12.1 ± 0.2	0.5 ± 0.0	31.8 ± 0.8
Adipose						
Wistar	0.3 ± 0.0	90.4 ± 0.7	5.7 ± 0.4	3.2 ± 0.3	0.7 ± 0.1	9.6 ± 0.9
WKA	0.2 ± 0.0	90.9 ± 0.3	5.3 ± 0.2	3.0 ± 0.1	0.9 ± 0.1	9.1 ± 0.3
WKY	0.4 ± 0.0	75.2 ± 0.6	13.4 ± 0.3	10.2 ± 0.3	1.3 ± 0.1	24.8 ± 0.6
SHRSP	0.4 ± 0.4	72.6 ± 1.1	15.7 ± 0.7	10.6 ± 0.4	1.2 ± 0.0	27.5 ± 1.1

Source: Ref. 54.

hypertensive rats (SHR) can deposit plant sterols in tissues to a significant extent. The SHRSP and SHR rats were initially derived from WKY rats (49). Therefore, it is apparent that these genetically related rat strains have an ability to deposit plant sterols efficiently. In all tissues, the deposition of campesterol was higher than that of sitosterol, although the proportion of campesterol in dietary plant sterols was much lower than that of sitosterol (52–54) (Table 4).

Dietary plant stanols (campestanol and sitostanol) were also found to affect the survival of SHRSP rats (unpublished work of W. M. N. Ratnayake and L. J. Plouffe). The survival rates of SHRSP rats fed commercial margarines fortified with either 11% plant stanols or 12% plant sterols were significantly lower than those of SHRSP rats fed soybean oil (Fig. 1). Between

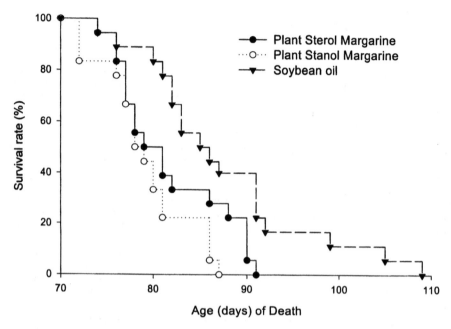

Figure 1 Survival curves of stroke-prone spontaneously hypertensive (SHRSP) rats fed diets containing soybean oil, plant sterol–fortified margarine, and plant stanol–fortified margarine. The mean survival time (mean ± SD, n = 18) of each dietary group was 87.6 ± 9.3, 81.7 ± 5.8, and 78.8 ± 4.3 days, respectively based on unpublished data of W. M. N. Ratnayake and J. L. Plouffe.

the two margarines, the plant sterol margarine reduced the survival rates to a slightly greater extent than did the plant stanol margarine. Levels of campestanol and sitostanol in tissues analyzed (plasma, red blood cells, liver, and kidney) of SHRSP rats fed the plant stanol margarine were about 100 times greater than the levels for SHRSP rats fed plant sterol margarine or soybean oil and achieved levels slightly exceeding those of campesterol and sitosterol. Similarly high levels of plant stanols have been reported in a clinical study of a phytosterolemic patient (30).

The increased accumulation of plant sterols and stanols in SHRSP rats might be due to the combined effects of their enhanced intestinal absorption and poor excretion in bile (54). The biliary excretion of cholesterol was also lower in this rat strain (55). This may be attributed to the inhibition of the activity of cholesterol 7α-hydroxylase by high circulating levels of plant sterols (36). All of these biochemical characteristics are similar to those described previously for phytosterolemic patients. These similarities, therefore, suggest that the hyperabsorption of plant sterols in SHRSP rats and related rat strains might be due to a mutation in a transporter protein expressed in intestinal and hepatic cells.

The large accumulation of plant sterols into cell membranes was associated with a poorer red cell membrane integrity and impaired deformability (52). Impaired red cell deformability can induce hemorrhagic episodes because inextensible membranes are easily ruptured during physical stress (56). Like SHRSP rats, humans with a history of hemorrhagic stroke also have rigid, stiff, and fragile red cells (57–59). Yamori et al. (57) noted that these red cell abnormalities were partially dependent on cholesterol content of cell membranes, which was significantly decreased in men with predisposition to stroke. Unfortunately, the dietary or tissue levels of plant sterol in the subjects of Yamori et al. were not reported; therefore, it is not known whether plant sterols had any influence on the observed abnormalities of red cells in men with a history of stroke. Furthermore, in prospective studies, hemorrhagic stroke has been found to occur at higher rates in persons with low levels of total blood cholesterol rather than in persons with higher levels (60–64). In the Multiple Risk Factor Intervention Trial, for intracanial hemorrhage, plasma cholesterol levels less than 4.14 mmol/L (160 mg/dl) were associated with a twofold increase in risk (64). Although the reasons behind these associations are not known, it is quite plausible that, similar to SHRSP rats, the weakening of cell membranes that result from low circulating cholesterol may also promote arterionecrosis in humans. It will be important for future studies to consider whether large intakes of plant sterols exacerbate the development of hemorrhagic stroke in humans in a fashion similar to that seen in SHRSP rats.

In addition to the detrimental effects on life span and cell membranes, increased intakes of plant sterols substantially lowered the platelet number of SHRSP rats (52,53). The physiological effect of the reduction in platelets is not known, but the strong positive correlation observed between platelet number and mean survival time in Ratnayake's studies (52,53) might indicate that the lowering of platelet number by dietary plant sterols may also be contributing to the development of hemorrhagic stroke.

IV. HYPERCHOLESTEROLEMIC-HYPERPHYTOSTEROLEMIC SUBJECTS

Several lines of evidence suggest that even a slight elevation in serum plant sterols could also provoke premature coronary heart disease (CHD) in some hypercholesterolemic subjects. In a study in the United States, the relationship between serum plant sterols and CHD was studied in 595 hypercholesterolemic subjects (65). These subjects composed the top serum cholesterol quintile in screening of 3472 self-referred subjects. Twenty-one subjects (or 3.5% of the full cohort) had elevated levels of serum campesterol and stigmasterol (median total serum total plant sterol value was 14.65 vs. 5.92 µg/ml for the remainder of the cohort). Glueck et al. (65) referred to these 21 subjects as hypercholesterolemic-hyperphytosterolemic patients. The median serum cholesterol level of these 21 subjects was 262 mg/dl and this value was not different from that of the remainder of the cohort (260 mg/dl). The high serum campesterol (median value 5.94 µg/ml) and stigmasterol levels (median value 3.74 µg/ml) in the 21 hypercholesterolemic-hyperphytosterolemic subjects were associated with a personal or family history of CHD in subjects aged 55 years or younger (premature CHD). Premature CHD was nearly twice as common in subjects in the top serum stigmasterol decile than in the bottom stigmasterol decile. This difference in premature CHD was independent of serum cholesterol levels. Furthermore, 34 first-degree relatives of the 21 hypercholesterolemic-hyperphytosterolemic patients also had high serum concentrations of plant sterols that were closely correlated with increased premature CHD. A history of premature CHD was twice as common in the relatives as it was in those in the full cohort. In a follow-up study it was shown that the serum levels of plant sterols in hyperphytosterolemic patients could be normalized by eating diets low in plant sterols (66). This mild hyperphytosterolemia associated with hypercholesterolemia may be a heritable atherogenic trait separate from the rare recessive familial phytosterolemia sitosterolemia. Therapeutic interventions for mildly hyperphytosterolemic-hypercholesterolemic subjects may also include diets relatively low in enriched sources of plant sterols.

Two other studies (67,68) have also found that CHD is more common in subjects with hypercholesterolemia associated with hyperphytosterolemia than in hypercholesterolemic subjects without the condition of hyperphytosterolemia. In the study by Sutherland et al. (67) in New Zealand, angiographically determined severity of coronary artery disease (CAD) and plasma lipids, lipoproteins, and sterols were measured in 44 patients (21 men and 23 women) with ischemic heart disease. Women had significantly higher plasma concentrations of sitosterol than men (10.14 ± 11.08 vs. 6.10 ± 3.23 µmol/L for men). Furthermore, in women, high plasma sitosterol concentrations and high plasma sitosterol/cholesterol ratio were associated with increased extent and severity of CAD. These associations were weaker for men. Age, body mass index, plasma total-, LDL-, and HDL-cholesterol were not correlated with CAD. Women with a greater extent and severity of CAD also had a lower ratio of plasma lathosterol to sitosterol. A low ratio of lathosterol to sitosterol is considered to be an indicator of impaired cholesterol synthesis and increased absorption of cholesterol (69). Since plasma cholesterol levels of the female CAD patients were not correlated with severity of CAD, a lower plasma lathosterol/sitosterol ratio would indicate that the absorption of cholesterol might not have been sufficient to compensate for a lower endogenous synthesis of cholesterol or that higher absorption of cholesterol was compensated for by reduced endogenous synthesis. In addition, a lower ratio of lathosterol to sitosterol might have been partly due to the higher plasma plant sterol content resulting from increased absorption of sitosterol. Therefore, the results of the Sutherland et al. study (67) suggest that CAD patients, and particularly women with slightly elevated levels of sitosterol, and with a lower endogenous cholesterol synthesis are at increased risk of cardiac events.

Results similar to those of Sutherland et al. (67) were reported by Rajaratnam et al. (68) for a cohort of angiographically verified CAD patients in Finland. In this study, in addition to conventional coronary risk factors, serum sterols and squalene were measured in 44 postmenopausal women with angiographically verified CAD and in 61 age-matched healthy controls. The coronary patients had elevated levels of plasma campesterol (610.9 ± 44.3 µg/dl), sitosterol (332.3 ± 21.7 µg/dl), squalene (91.3 ± 2.4 µg/dl), desmosterol (212.4 ± 17.9 µg/dl), and lower lathosterol value (373.8 ± 20.0 µg/dl) compared to controls. The corresponding values for controls were 493.4 ± 24.8 µg/dl, 271.0 ± 13.2 µg/dl, 60.8 ± 2.1 µg/dl, 163.3 ± 4.8 µg/dl, and 425.5 ± 16.5 µg/dl. However, plasma cholesterol values were not significantly different between the two groups (5.94 for cases vs. 5.71 mmol/L for controls). After adjusting for age, body mass index, family history of CAD, smoking, hypertension, serum triglycerides, LDL and HDL cholesterol level, the ratios of squalene, desmosterol, campesterol, and sitosterol to cholesterol were significantly associated with risk of CAD. In addition, serum lathosterol was inversely related to the presence of CAD. The independent positive associations of campesterol and sitosterol and the inverse association of lathosterol with CAD suggest that low synthesis of cholesterol and high absorption of sterols (both cholesterol and plant sterols) may be related to atherosclerosis in women independently of other lipid and nonlipid risk factors.

The Scandinavian Simvastatin Survey Study (4S study) also identified a subpopulation of CHD patients with low endogenous synthesis of cholesterol and higher absorption of cholesterol and plant sterols (70–72). The subjects of this subpopulation had the highest levels of circulating plant sterols and the highest risk of recurrent coronary events despite lower levels of serum cholesterol due to simvastatin ingestion. All these studies (67–72) imply that individuals with low endogenous cholesterol synthesis are at increased risk for CHD when consuming diets enriched with plant sterols. It remains to be shown what the lowest plant sterol levels are that contribute to enhanced CAD risk and the significance of the ratio of plant sterols, or of sitosterol, to cholesterol. There have been a large number of clinical studies examining the cholesterol lowering efficacy of plant sterols and stanols (6), but unfortunately most of the studies were of short duration and have not been long enough to observe whether dietary plant sterols have prevented CHD. Future studies should be aimed at determining the increased risk of CHD associated with increased intakes of plant sterols particularly in some subpopulation of hypercholesterolemic subjects with difficulty synthesizing endogenous cholesterol for their tissue needs.

V. OTHER DELETERIOUS EFFECTS OF PLANT STEROLS

Further deleterious effects of the tissue accumulation of plant sterols can be drawn from studies done in children on long-term parenteral nutrition (73–75). Total parenteral nutrition (TPN) provides life-saving treatment for children who have had extensive gut resections or who suffer from severe gut failure. However, long-term TPN in infants is complicated in 30–50% of cases by progressive cholestatic liver disease (76). The incidence may be as high as 50% in infants who have received TPN for 2 months. Cholestatic liver disease may progress to cirrhosis and to liver failure. Commercial lipid emulsions (e.g., Intralipid, Vitlipid) used for parenteral nutrition are made from soybean oil and therefore contain considerable amounts of plant sterols (1 mmol/L) (73). Clayton et al. (73) recently observed that infants receiving daily infusion of TPN had elevated plasma concentrations of campesterol, stigmasterol, and sitosterol. The concentrations of plant sterols in the red cells were similar to the plasma concentrations. All the plant sterols detected in plasma were also present in commercial Intralipid. The plasma concentrations of plant sterols in the children with severe liver disease (1.3–1.8 mmol/L) were as high as those seen in patients with phytosterolemia. This large accumulation of plant sterols has been identified as a possible cause of parenteral nutrition-associated cholestasis (73–75). In addition to cholestasis, this accumulation of plant sterols contributed to several other disorders, namely, thrombocytopenia due to excessive peripheral destruction of platelets, large and giant platelets, abnormal red cell morphology and increased red cell breakdown, hemolytic anemia, increases in plasma bilirubin and aspartate aminotransferase levels, and decreased liver function (74). A reduction in intake of lipid emulsion was associated with a decrease in plasma plant sterol concentration and improvement in liver function tests (74). A link between cholestasis and the lipid used for TPN has also been noted in adults (77).

The mechanism of plant sterol–induced cholestasis has not yet been elucidated. In vitro, plant sterols inhibit cholesterol 7α-hydroxylase, the rate-limiting step in bile acid synthesis (28). This could lead to reduced bile acid–dependent bile flow. However, basal bile flow was not affected in piglets daily injected with plant sterols in amounts similar to those given to infants who receive TPN (75). Iyer et al. (75) explained this observation by suggesting that plant sterols may bind to sterol carrier proteins, impairing movement of cholesterol, bile acid precursors, and other lipids across the cell.

VI. CONSIDERATIONS IN THE SAFETY ASSESSMENT OF PLANT STEROL AND STANOL ESTERS

Extraction of sterols from various plant sources enables their addition to processed foods at levels that could not normally be obtained through the diet. To improve their solubility in food matrices, formulations of plant sterols and stanols intended for addition to foods are esterified to dietary fatty acids with the intention of marketing the resulting foods as a means for consumers to lower their cholesterol levels through the diet. Because these substances are physiologically active, it is necessary to establish whether their addition to an unlimited number of food products would introduce the potential for induction of adverse effects. The end point of this investigation would be to define an upper limit of consumption for dietary plant sterols and stanols.

Studies addressing the safety for humans of a plant sterol ester formulation derived from vegetable oil distillates and plant stanol ester formulations derived from wood and vegetable oil have been conducted in rats and in vitro systems in support of applications for GRAS (Generally Recognized As Safe) status of these formulations in the United States, and, as part of the requirements for GRAS status, are available in the published literature. For both the sterol and stanol ester formulations, the databases include a 90-day repeat dose feeding study in rats (78,79), a two-generation reproduction study in rats (80,81), genotoxicity assays (82,83), short-term studies of estrogenic potential in rats and in vitro systems (84,85), and, for the phytostanol ester formulation, a developmental toxicity study in rats (86). In addition, clinical studies of the safety and efficacy of several plant sterol and plant stanol ester formulations in humans have been published (reviewed in 87; 88–93).

The sterol and stanol esters were incorporated into the diet of rats at concentrations equivalent to 0.1–5.0% of the free phytosterols or phytostanols. In the subchronic studies, the resulting dose range was equivalent to 80–5000 mg/kg body weight/day sterols/stanols. This is well above the dose found to provide an optimal cholesterol-lowering effect in humans (1.6–2.4 g, about 25–35 mg/kg body weight/day) (92,94) or from naturally occurring dietary sources (250 g/day; 3–4 mg/kg body weight/day). Except for two high-dose clinical studies conducted with plant sterol esters (88,89,91), the doses administered in the clinical studies (i.e., 0.8–3.2 g phytosterols/day) were within the consumption range recommended for a cholesterol-lowering effect.

Because dietary consumption of both sterol or stanol esters results in increases in serum and tissue concentrations of the unesterified sterols or stanols in experimental animals and humans (8,79,92,93), it is necessary to consider the consequences of increased intake of sterols and stanols in terms of the potential for adverse systemic effects. The biological end points for potential adverse effects of elevated systemic concentrations of plant sterols or stanols can be predicted based on their chemical similarity to cholesterol and would relate to the role of cholesterol in cell membranes or as the precursor for bile acids, steroid hormones, and vitamin D. The effects observed in individuals with high systemic concentrations of plant sterols as described in the previous sections of this chapter (reduced bile acid synthesis, atherosclerosis, xanthomatoma, cholestatic liver disease, abnormal red cell morphology and increased rigidity of membranes, hemolytic anemia, and thrombocytopenia) are consistent with the interference of plant sterols in the metabolic roles and cell membrane functions of cholesterol. In rats, the greatest accumulation of plant sterols following oral or parenteral administration, was observed in the ovaries, testes, and adrenal glands (15; cited in 28), all sites of steroid hormone synthesis from cholesterol.

Plant sterols are absorbed from the intestinal tract at a greater rate than are plant stanols, as reflected in the serum concentrations of human volunteers consuming similar doses ($\sim$2 g/person/day) of plant sterol or stanol esters in the diet (92). Even at the highest dose of plant stanol esters given to humans (3.0 g stanols/person/day) in the absence of supplementation with plant sterols, their concentration in serum, both as the absolute amount, and relative to cholesterol, was about 10-fold less than the concentration of plant sterols (93). In fact, consumption of a plant stanol ester formulation inhibited the absorption of plant sterols present in the usual diet of both humans and rats, resulting in a 30–45% reduction of serum plant sterols in humans consuming 3.0 g stanols/day (93), and an 80–90% reduction in rats consuming 5% stanols in the diet (79). Thus, the overall effect of consumption of stanol esters in rats and humans was the reduction of serum concentrations of total plant sterols and stanols compared with controls.

A. Preclinical Studies

In the 90-day feeding study with plant sterol esters (composition: campesterol, 25.8%; β-sitosterol, 48.7%; stigmasterol, 21.6%; β-sitostanol, 1.8%) in Wistar rats (78), small but statistically significant dose-related increases in serum concentrations of membrane-bound enzymes aspartate aminotransferase (AST) and alamine aminotransferase (ALT) in both sexes, and decreases in platelet counts in females were observed (Table 5). Except for the platelet counts in the high-dose male rats, a statistically significant dose-related

Table 5 Selected Hematology and Clinical Chemistry Parameters from the 90-day Feeding
Study with Plant Sterol Esters in Wistar Rats

	Males					Females				
Dietary conc.	0 (control)	0.1%	1.0%	2.0%	5.0%	0 (control)	0.1%	1.0%	2.0%	5.0%
Cholesterol (mmol/L)	3.13	3.15	3.29	3.50[b]	3.29	2.39	2.5	2.68[a]	2.71[a]	2.46
Platelets ($\times 10^9$/L)	787	759	704[b]	698[b]	740	796	726[b]	710[b]	707[b]	706[b]
Prothrombin time (s)	16.5	16.5	16.3	16.1[b]	16.2	17.6	17.3[a]	17.2[a]	17.2[a]	17.0[b]
APTT (s)	20.4	21.3	21.3	22.3[a]	22.4[a]	19.7	20.5	21.7[a]	19.1	19.4
AP (IU/L)	234	217	259	255	337[b]	130	135	166[b]	196[b]	172[b]
ALT (IU/L)	40.7	47.8	51.2[a]	50.2[a]	56.9[b]	30	31.4	33.9	39.9[a]	46.4[b]
AST (IU/L)	103.3	111.7	104.9	108.9	109.6	85.4	83.9	90.7	91.6	106.4[b]

APTT, activated partial prothrombin time; AP, alkaline phosphatase; ALT, alanine aminotransferase; AST, asparatate
aminotransferase.
[a] $p < 0.05$
[b] $p < 0.01$
Source: Ref. 78.

decrease of platelet count was also noted in the male rats. The decrease in
platelet count was the most sensitive indicator of treatment-related effects and
was observed in females down to the lowest dose tested (0.1% plant sterols in
the diet, about 80 mg plant sterols/kg body weight/day). Because of the much
longer turnover time of red blood cells compared with platelets, effects on red
blood cell membranes would be more difficult to detect in a 90-day study. The
effects on serum enzymes and platelet count were not observed in the 90-day
feeding study in the Wistar rat with a vegetable oil–derived and a wood-
derived stanol formulation using the same dose range (79). Interestingly, no
effects on platelet count and only slightly increased serum enzymes AST and
ALT (achieving statistical significance only for ALT in male, but not female,
rats) were observed up to a dose of 9000 mg/kg body weight/day in a recently
published 90-day feeding study in Sprague–Dawley rats (95) receiving
extremely high doses of a plant sterol ester formulation (campesterol,
27.9%; β-sitosterol, 49.4%; stigmasterol, 18.5%) administered by gavage
(1000, 3000, and 9000 mg/kg body weight/day). The lack of effects in rats
administered higher doses of plant sterol esters (95) may be explained by
differences in plant sterol absorption following gavage, compared with dietary administration of the plant sterol ester formulation. Serum concentrations of sterols other than cholesterol that would allow comparison of the
systemic dose of sterols were not available for the studies using the plant sterol
ester formulations.

Neither type of ester formulation had an effect on reproductive capacity, was genotoxic, or exerted estrogenic activity as assessed by the assays used.

One study has documented the effects of feeding a plant sterol/stanol mixture in the diet at doses of 1, 5, and 50 mg/kg body weight/day for 2 weeks on endocrine function in prebreeding European polecats (*Mustela putorius*) (96). The mixture in question had a β-sitosterol/β-sitostanol content of 89% and a campesterol/campestanol content of 9%. A nonsignificant increase in serum testosterone concentrations in males and in serum estradiol concentrations in females was noted at the two higher dose levels. These results are not in agreement with the predicted effects of competitive inhibition of cholesterol metabolism to steroid hormones, which would have resulted a decrease, rather than an increase, in testosterone and estradiol. For example, studies conducted in goldfish (*Carassius auratus*) injected intraperitoneally with β-sitosterol demonstrated a decrease in gonadal steroid biosynthetic capacity (97). Effects of treatment noted in the polecats that achieved statistical significance and dose relationship were a reduction in serum ghrelin, a hormone that regulates food intake and body mass in response to food intake; increased liver glycogen content; increased kidney, but not liver, glucose-6-phosphatase activity; and increased serum LDL cholesterol at the highest dose tested, 50 mg/kg body weight/day. No effects of treatment on body weight, length, body mass index, liver or testicle weight were noted. These results do not appear to be consistent with the predicted effects of elevated dietary and serum concentrations of plant sterols. The short duration of the study should be noted.

In two studies (35,98), large doses of β-sitosterol administered by injection to male Sprague–Dawley rats that produced plasma concentrations of 25% and 7% total plasma sterols, respectively, resulted in competitive inhibition of cholesterol 7α-hydroxylase activity (one of the enzymes mediating bile acid synthesis) and concomitant increases in plasma cholesterol concentrations.

It is notable that administration of plant sterol or stanol esters to Wistar rats in the diet or by gavage at doses that ranged from 80 to 9000 mg/kg body weight/day had no effect in altering serum cholesterol concentrations. In comparison, doses of 0.8–3.2 g/day sterols or stanols (about 10–45 mg/kg body weight/day), as esters, were effective in lowering serum total and LDL cholesterol in normal and hypercholesterolemic adults as by as much as 8–13% after 3.5–8 weeks (8,90,92–94,99). In most of these studies, the effect on serum LDL cholesterol concentrations was somewhat greater than the effect on total serum cholesterol concentrations. Hallikainen et al. (93) showed that no further cholesterol-lowering effect was noted when the dose of stanols was increased from 2.3 g/day to 3.0 g/day in hypercholesterolemic adults. A marked cholesterol-lowering effect was noted in male Sprague–Dawley rats fed 2% free plant sterols and 0.2% cholic acid (to enhance plant sterol absorption) in the diet (100) and in male Mongolian gerbils (*Meriones unquiculatus*) fed 0.75% free phytosterols, sterol esters, or stanol esters with 0.15% added cholesterol in the diet (101).

B. Clinical Studies

The effects of plant sterol ester consumption on fecal bile acid excretion in both men and women an on sex steroid hormone levels in women were assessed in a short-term clinical study (88,89) conducted in normal subjects with a dose of plant sterol esters in margarine (sterol composition of 46% sitosterol, 26% campesterol, and 20% stigmasterol) equivalent to 8.6 g plant sterols/day. This dose exceeded the range usually suggested for cholesterol-lowering effects (0.8 to 3.2 g/day). The control group and, during the run-in period, the treatment group received margarine with 1% added cholesterol, equivalent to an additional 460 mg/day. The dose of plant sterols (120 mg/kg body weight/day for a 70-kg person) exceeded the dietary dose in female rats at which effects on platelet count were observed after 90 days (80 mg/kg body weight/day). However, no data on the serum enzymes, AST or ALT, or on hematology parameters were provided in the published results for comparison with the rat data. Consumption of the plant sterol-containing margarine for 3 or 4 weeks (men and women, respectively) resulted in a statistically significant decrease in fecal concentration of total and secondary bile acids in both sexes (fecal weight was not affected by treatment), and a decrease in serum progesterone and estrogens, estradiol and estrone, that was statistically significant for progesterone. In addition, there was also a small increase, statistically significant in women, in fecal concentration of 4-cholesten-3-one, a precursor of cholestanol, a normal metabolite of cholesterol that is accumulated in inborn errors of bile acid biosynthesis and in sitosterolemia (28). While an effort was made to synchronize weekly hormone measurements with the women's estrous cycles over the 4-week study period, the number of subjects was small and the means of the aggregated data for the weekly measurements showed large variations that reduced the ability to determine statistical significance for the steroid hormones.

Since no compensatory increases in fecal primary bile acids were noted, as might have been expected if bacterial conversion of primary to secondary bile acids had been suppressed by treatment, the results could be interpreted to indicate that the nearly 25% decrease in fecal total bile acid concentration was due to an inhibitory effect on bile acid synthesis. It was not possible to determine whether the observed effects of treatment on bile acids and steroid hormones could be explained by competitive inhibition of the enzymes mediating their synthesis from cholesterol as a result of increases in absorbed plant sterols, or whether cholesterol availability for these reactions was affected by the 18% decrease in serum total cholesterol.

The results of a study that assessed the effects of plant stanol ester consumption on fecal bile acid excretion (102) suggest that the effect on cholesterol availability is the correct interpretation. In this study, fecal bile acid excretion was reduced by 17% after 7 days consumption of the test product by colectomized patients consuming 2 g stanols/day in margarine containing phytostanol esters. This effect was associated with a 44% decrease in the efficiency of cholesterol absorption and a reduction of serum cholesterol concentration of 12% from baseline. Since plant stanol consumption has been shown to result in a net reduction of serum total plant sterol concentration, the comparable effects on fecal bile acid excretion noted with consumption of sterols and stanols suggest that this effect was due to reduced cholesterol availability rather than an effect of increased serum phytosterol levels on activity of enzymes involved in bile acid synthesis. It is likely that this explanation would also apply to the effects of phytosterol intake on steroid hormones.

Another high-dose clinical study, conducted with a similar plant sterol ester formulation using a range of doses (3, 6, and 9 g plant sterols/day) for 8 weeks (91), addressed most of the design problems of the first study (i.e., duration of study, number of treated groups, number of subjects per group, control margarine without added cholesterol, assessment of clinical chemistry parameters and serum concentrations of plant sterols). However, most of the data, including clinical chemistry end points and serum plant sterol concentrations, were statistically analyzed using nonparametric tests appropriate for nonnormally distributed data, and data were reported as medians and upper and lower values rather than means and standard deviations. Using these statistical techniques, the plant sterol formulation appeared to be ineffective in lowering serum total and LDL cholesterol, even at the highest dose of 9 g plant sterols/day. Previously cited studies (8,90,92–94,99) had determined the optimal dose range for a cholesterol-lowering effect of dietary sterols as 0.8–3.2 g/day. Increases in median serum concentrations of campesterol (60%, 70%) and β-sitosterol (20%, 30%) were noted both as the actual measurement and relative to serum total cholesterol, indicating both compliance with study product consumption and plant sterol absorption. Some evidence of treatment-related increases in serum ALT and creatine kinase activities was observed; a nonsignificant increase in alkaline phosphatase and AST activities was also noted in the high-dose group compared with controls. Full reporting of hematology data, including platelet counts, was not provided. A four- and fivefold increase in reported adverse effects of a nonspecific nature (e.g., hayfever, exhaustion) was observed in the group consuming 6 and 9 g/

day of plant sterols in food products containing plant sterol esters compared with the 3 g/day group and controls (5%, 0%, 21%, and 26% for the 0, 3, 6, and 9 g/day groups, respectively). Because of their general nature, the effects were deemed by the authors to be unrelated to the study product. Gastro-intestinal complaints were also twice as common in the high-dose group compared with controls. These results suggest that the limit of tolerability for the product might be less than 9 g/day.

Clinical studies pertaining to the therapeutic use of a pharmaceutical preparation containing β-sitosterol as its major component are available in the published literature (103–108). The doses used were high (up to 25 g/day) in comparison with those anticipated from addition of sterol/stanol esters to food products (1.6–3.4 g/day free sterols/stanols). While no significant adverse effects were reported in these studies, only a limited number of parameters, relating mostly to the efficacy of the preparation to reduce serum cholesterol, were assessed, and the maximal observation period was 4 years. It was not possible to conclude from these studies whether therapeutic use of β-sitosterol was associated with an increase in severity of coronary heart disease or in incidence of hemorrhagic stroke in the study population as suggested in the observational epidemiological studies described previously.

C. Derivation of an Upper Limit of Consumption

Because plant stanol ester consumption results in a net decrease in serum plant sterol/stanol concentrations at high doses administered to the Wistar rat (79) as well as at the more modest doses consumed by humans (92,93), and no effects of plant stanol ester consumption were observed at high doses in animal toxicity studies, the potential for systemic effects of plant stanols consumed in the diet appears to be very low. Consequently, the focus of the safety assessment of phytostanol esters for the general population would be on effects of reduced absorption of fat-soluble vitamins and any effects associated with reduced cholesterol availability.

On the other hand, the effects on platelet count and serum activity of membrane-bound enzymes AST and ALT, observed in the 90-day feeding study with plant sterol esters in rats, suggest that an upper limit of consumption for dietary plant sterols could be established for systemic effects of absorbed plant sterols on the basis that these represent preliminary effects of increased incorporation of plant sterols into cell membranes (74,109).

While the studies in animal models are useful for identifying the effects of high dietary doses of plant sterol esters, extrapolation of such doses from the Wistar rat to humans is problematic for a number of reasons. One of these is the markedly different serum and tissue sterol content of the Wistar rat compared with humans and with Sprague–Dawley rats, which complicates the interpretation of observed effects in Wistar rats attributed to increased systemic plant sterol concentrations. Available data for untreated Wistar rats indicated that the plant sterol content of serum and red blood cells was about 11% and 8% of total sterol content, respectively (79; W. M. N. Ratnayake, unpublished data), compared with plasma plant sterol content of total sterols of 1.5% in Sprague–Dawley rats (100) and about 0.5% in humans (8,92,93,99). Most of the difference between Wistar rats and humans or Sprague–Dawley rats was due to the higher concentration of plant sterols rather than lower cholesterol concentration. Additional factors complicating the use of this rat strain for quantitative extrapolation are differences in dietary factors influencing the rate of phytosterol uptake (since there is no cholesterol in the diet of laboratory rats to compete for uptake in the intestinal tract); differences in cholesterol metabolism, including normal serum concentrations, in the rat compared with humans; and the lack of change in serum cholesterol in the Wistar rat in response to large doses of plant sterol concentrations consumed in the diet or administered by gavage.

In any case, it was not possible to assess the effect of dietary administration of plant sterol esters on serum plant sterol concentrations in the Wistar rat as these data were not available from the 90-day study (78). While serum plant sterol concentration data were available in a 2-week study for similarly treated Wistar rats (dietary administration of 0.5% of a plant sterol mixture (55% sitosterol, 38% campesterol, 4% stigmasterol) equal to 400 mg/kg body weight/day) (54), in the absence of control values for serum sterol content, the effects of treatment on this parameter could not be determined. The requirement for concurrent control data to support any conclusions about the effects of feeding plant sterols on serum concentrations is illustrated by the observation that the values for plant sterol and cholesterol concentrations in treated rats from this study (10.1/78.2 mg/dl) were nearly identical to those for untreated Wistar rats (9.4/80.9 mg/dl) in a separate study (79).

The data from the 8-week high dose clinical study with plant sterol esters (91) suggest that a dose of 9 g/day might approach an upper limit of intake based on effects on serum enzymes, although unconventional methods used for data analysis made this difficult to assess. No other clinical study that assessed platelet count and serum enzymes, and was of sufficient duration to observe effects on these parameters, was available in the published literature.

The data from the presently available rat and human studies are not sufficient for a characterization of the dose–response relationship that would assist in interpretation of the small increases in plant sterol levels in epidemiology studies that were associated with increased severity or incidence of coronary artery disease. Further investigation of the factors contributing to the increased uptake of plant sterols from the diet, such as apolipoprotein E polymorphisms, might contribute to an understanding of the causative factors for increased severity or incidence of coronary artery disease.

VII. CONCLUSIONS

Because they are poorly absorbed, it has been generally considered that addition of plant sterols to the diet at moderate levels could be considered safe for the majority of the population. While the effects of increased consumption of plant sterols as fatty acid esters on absorption of fat-soluble vitamins can be used as an end point to establish limits for their addition to foods, attention should also be directed to a quantitative assessment of the adverse health outcomes related to increases of systemically absorbed phytosterols. In extreme cases, hyperabsorption of plant sterols combined with poor excretion has been shown to lead to a variety of deleterious effects, including xanthomata, atherosclerosis, hemorrhagic stroke, thrombocytopenia, hemolytic anemia, reduced bile acid synthesis, abnormal red cell morphology, cell membrane rigidity, decreased liver function, and cholestatic liver disease. The minimal serum concentrations of plant sterols that could contribute to adverse health outcomes, and their relation to dietary intakes, have yet to be identified.

In addition to those individuals with the rare genetic condition of phytosterolemia, increasing the dietary intake of plant sterols might present a risk for individuals in whom biological propensity or exogenous factors such as statin therapy contribute to a higher rate of intestinal sterol absorption. Abnormalities of red blood cell membranes related to lowered cholesterol content have also been identified in individuals prone to hemorrhagic stroke; large intakes of plant sterols would also be of concern in terms of exacerbation of this condition. Since hyperabsorbers of plant sterols can only be identified through analysis of the blood sterol profile, unrestricted fortification of foods with plant sterols for sale to the general population may pose a problem in terms of alerting those individuals for whom the product would be inappropriate. Thus, large amounts of plant sterols should be taken only under the supervision of a physician who will monitor carefully for accumulation of plant sterols in blood and tissues.

REFERENCES

1. OJ Pollak, D Kritchevsky. Sitosterol. In: TB Clarkson, D Kritchevsky, OJ Pollak, eds. Monographs on Atherosclerosis, Vol. 10, New York: Basel, 1981.

2. OJ Pollak. Effect of plant sterols on serum lipids and atherosclerosis. Pharmacol Ther 31:177–208, 1985.
3. WH Ling, PJH Jones. Dietary phytosterols: a review of metabolism, benefits and side effects. Life Sci 57:196–206, 1995.
4. NB Cater, SM Grundy. Safety aspects of dietary plant sterols and stanols. Managing high cholesterol. In: WO Roberts, ed. A Postgraduate Medicine Special Report. Minneapolis, MN: McGraw-Hill, 1998, pp. 6–14.
5. TA Miettinen, RS Tilvis, YA Kesäniemi. Serum plant sterols and cholesterol precursors reflect cholesterol absorption and synthesis in volunteers of a randomly selected male population. Am J Epidemiol 31:20–31, 1990.
6. M Law. Plant sterol and stanol margarines and health. Br Med J 320:861–864, 2000.
7. I Wester. Dose responsiveness to plant stanol esters. Eur Heart J (suppl 1), 104–108, 1999.
8. JA Weststrate, GW Meijer. Plant sterol-enriched margarines and reduction of plasma total- and LDL-cholesterol concentrations in normocholesterolaemic and mildly hypercholesterolaemic subjects. Eur J Clin Nutr 52:334–343, 1998.
9. AH Lichtenstein, RJ Deckelbaum, for the American Heart Association Nutrition Committee. Stanol/sterol ester-containing foods and blood cholesterol levels. Circulation 103:1177–1179, 2001.
10. J Plat, RP Mensink. Effects of plant sterols and stanols on lipid metabolism and cardiovascular risk. Nutr Metab Cardiovasc Dis 11:31–40, 2001.
11. NCW Wong. The beneficial effects of plant sterols on serum cholesterol. Can J Cardiol 17:715–721, 2001.
12. M Sugano, H Morioka, Y Kida, I Ikeda. The distribution of dietary plant sterols in serum lipoproteins and liver subcellular fraction of rats. Lipids 13:427–432, 1978.
13. A Kuksis, TC Huang. Differential absorption of plant sterols in the dog. Can J Biochem Physiol 40:1493–1504, 1962.
14. AK Bhattacharyya, LA Lopez. Absorbability of plant sterols and their distribution in rabbit tissues. Biochim Biophys Acta 574:146–153, 1979.
15. DJ Sanders, HJ Minter, D Howes, PA Hepburn. The safety evaluation of phytosterol esters. Part 6. The comparative absorption and tissue distribution of phytosterols in the rat. Food Chem Toxicol 38:485–491, 2000.
16. T Heinemann, G Axtmann, K von Bergmann. Comparison of intestinal absorption of cholesterol with different plant sterols in man. Eur J Clin Invest 23:827–831, 1993.
17. KE Berge, H Tian, GA Graf, L Yu, NV Grishin, J Schultz, P Kwiterovich, B Shan, R Barnes, HH Hobbs. Accumulation of dietary cholesterol in sitosterolemia is caused by mutations in adjacent ABC transporters. Science 290:1771–1775, 2000.
18. MH Lee, K Lu, S Hazard, H Yu, S Shulenin, H Hidaka, H Kojima, R Allikmets, N Sakuma, R Pegoraro, AK Srivastra, G Salen, M Dean, SB Patel. Identification of a gene, ABCG5, important in the regulation of dietary cholesterol absorption. Nature Genet 27:79–83, 2001.

19. FJ Field, SN Mathur. β-Sitosterol esterification by intestine acylcoenzyme A:cholesterol acyltransferase (ACAT) and its effect on cholesterol esterification. J Lipid Res 24:409–417, 1983.
20. I Ikeda, K Tanaka, M Sugano, GV Vahouny, LL Gallo. Discrimination between cholesterol and sitosterol for absorption in rats. J Lipid Res 29:1583–1591, 1988.
21. AK Battacharyya, WE Connor. Beta-sitosterolemia and xanthomatosis. A newly described lipid storage disease in two sisters. J Clin Invest 68:1033–1043, 1974.
22. RS Schulman, AK Bhattacharyya, WE Connor, DS Fredrickson. Beta-sitosterolemia and xanthomatosis. N Engl J Med 294:482–483, 1976.
23. RS Lees, AM Lees. Effects of sitosterol therapy on plasma lipid and lipoprotein concentrations. In: H Greten, ed. Lipoprotein Metabolism. pp. 119–130. Berlin: Springer-Verlag, 1976.
24. TA Miettinen. Phytosterolaemia, xanthomatosis and premature atherosclerotic arterial disease: a case with high plant sterol absorption, impaired sterol elimination and low cholesterol synthesis. Eur J Clin Invest 10:27–35, 1980.
25. C Wang, HJ Lin, TK Chan, G Salen, WC Chan, TF Tse. A unique patient with coexisting cerebrotendinous xanthomatosis and β-sitosterolemia. Am J Med 71:313–319, 1981.
26. G Salen, PO Kwiterovich Jr, S Shefer, GS Tint, I Horak, V Shore, B Dayal, E Horak. Increased plasma cholesterol and 5α-saturated plant sterol derivatives in subjects with sitosterolemia and xanthomatosis. J Lipid Res 26:203–209, 1985.
27. G Salen, V Shore, GS Tint, T Forte, S Shefer, I Horak, F Horak, B Dayal, L Nguyen, AK Batta, FT Lindgren, PO Kwiterovich Jr. Increased sitosterol absorption, decreased removal, and expanded body pools compensate for reduced cholesterol synthesis in sitosterolemia with xanthomatosis. J Lipid Res 30:1319–1330, 1989.
28. I Björkhem, KM Boberg, E Leitersdorf. Inborn errors in bile acid biosynthesis and storage of sterols other than cholesterol. In: CR Scriver, AL Beaudet, WS Sly, D Valle, eds. The Metabolic and Molecular Bases of Inherited Disease. New York: McGraw-Hill, 2001, pp. 2961–2988.
29. G Salen, S Shefer, LB Nguyen, GC Ness, GS Tint. Sitosterolemia. In: D Betteridge, DR Illingworth, J Shepard, eds. Lipoproteins in Health and Disease. London: Oxford University Press, 1999, pp. 815–827.
30. G Salen, G Xu, GS Tint, AK Batta, S Shefer. Hyperabsorption and retention of campestanol in a sitosterolemic homozygote: comparison with her mother and three control subjects. J Lipid Res 41:1883–1889, 2000.
31. S Patel. Sitosterolemia. Dietary cholesterol absorption. Lancet 358 (Supplement 1):63, 2001.
32. AK Bhattacharyya, WE Connor, DS Lin, MM McMurray, RS Shulman. Sluggish sitosterol turnover and hepatic failure to excrete sitosterol into bile cause expansion of body pool sitosterol in patients with sitosterolemia with xanthomatosis. Atheriosclerol Thromb 11:1287–1294, 1991.

33. G Salen, I Horak, M Rothkopf, JL Cohen, J Speck, GS Tint, V Shore, B Dayal, T Chen, S Shefer. Lethal atherosclerosis associated with abnormal plasma and tissue concentration in sitosterolemia with xanthomatosis. J Lipid Res 26:1126–1133, 1985.

34. S Heimer, T Langmann, C Moehle, R Mauerer, M Dean, FU Beil, K Von Bergmann, G Schmitz. Mutations in the human ATP-binding cassette transporters ABCG5 and ABCG8 in sitosterolemia. Hum Mutat 20:151, 2002.

35. S Shefer, G Salen, J Bullock, LB Nguyen, GC Ness, Z Vhao, PF Belamarich, I Chowdhary, S Lerner, AK Batta, GS Tint. The effect of increased hepatic sitosterol on the regulation of 3-hydroxy-3-methylglutaryl-coenzyme A reductase and cholesterol 7α-hydroxylase in the rat and sitosterolemic homozygotes. Hepatology 20:213–229, 1994.

36. S Shefer, G Salen, L Nguyen, AK Batta, V Packlin, GS Tint, S Hauser. Competitive inhibition of bile acid synthesis by endogenous cholestanol and sitosterol in sitosterolemia with xanthomatosis; effect of cholesterol 7α-hydroxylase. J Clin Invest 82:1833–1839, 1988.

37. SB Patel, G Salen, H Hidaka, PO Kwiterovich, AFH Stalenhoef, TA Miettinen, SM Grundy, M Lee, JS Rubenstein, MH Polymeroloulos, MJ Brownstein. Mapping a gene involved in regulating dietary cholesterol absorption. The sitosterolemia locus is found at chromosome 2p21. J Clin Invest 102:1041–1044, 1998.

38. L Nguyen, G Salen, S Shefer, GS Tint, V Shore, G Ness. Decreased cholesterol biosynthesis in sitosterolemia with xanthomatosis: diminished mononuclear leukocyte 3-hydroxy-3-methylglutaryl coenzyme A reductase activity and enzyme protein associated with increased low-density lipoprotein receptor function. Metab Clin Exp 39:426–443, 1990.

39. A Honda, G Salen, LB Nguyen, GS Tint, AK Batta, S Shefer. Down-regulation of cholesterol biosynthesis in sitosterolemia: diminished activities of acetoacetyl-CoA thiolase, 3-hydroxy-3-methylglutaryl-CoA synthase, reductase, squalene synthase, and 7-dehydrocholesterol Δ^7 reductase in liver and mononuclear leukocytes. J Lipid Res 39:44–50, 1998.

40. PO Kwiterovich Jr, PS Bachorik, HH Smith, VA McKusick, WE Connor, B Teng, AD Sniderman. Hyperapobetaliproteinemia in two families with xanthomatosis and phytosterolemia. Lancet 1:466–469, 1979.

41. BK Grahlke. Xanthome der Achillessehnen als Leitsymptom der Sitosterinämie (Xanthoma of the Achilles tendon as the cardinal sign of sitosterolemia). Dtsch Med Wochenschr 116:335, 1991.

42. B Skrede, I Björkhem, O Bergesen, HJ Kayden, S Skrede. The presence of 5α-sitostanol in the serum of a patient with phytosterolemia, and its biosynthesis from plant steroids in rats with bile fistula. Biochim Biophys Acta 836:368–375, 1985.

43. C Wang, HJ Ling, TK Chan, G Salen, WC Chan, TF Tse. A unique patient with coexisting cerebrotendinous xanthomatosis and β-sitosterolemia. Am J Med 71:313–319, 1981.

44. AK Bhattacharyya, WE Connor. Familial diseases with storage of sterols

other than cholesterol: cerebrotendinous xanthomatosis, and β-sitosterolemia and xanthomatosis. In: JB Stanbury, JB Wyngaarden, DS Fredrickson, eds. The Metabolic Basis of Inherited Disease. 3rd ed. New York: McGraw-Hill, 1978, pp. 656–669.

45. A Kuksis, JJ Myher, L Marai, JA Little, RG McArthur, DAK Roncari. Fatty acid composition of individual plasma steryl esters in phytosterolemia and xanthomatosis. Lipids 21:371–377, 1986.

46. G Salen, G Xu, GS Tint, AK Batta, S Shefer. Hyperabsorption and retention of campestanol in a sitosterolemic homozygote: comparison with her mother and three control subjects. J Lipid Res 41:1883–1889, 2000.

47. Y Yamori. Predictive and preventive pathology of cardiovascular diseases. Acta Pathol Japanica 39:683–705, 1989.

48. Y Yamori. Physiology of the various strains of spontaneously hypertensive rats. In: J Genest, O Kuchel, P Hamet, M Cantin, eds. Hypertension. Montreal: McGraw-Hill, 1983, pp. 656–669.

49. Y Yamori. The stroke-prone spontaneously hypertensive rat: contribution to risk factor analysis and prevention of diseases. In: W de Jong, ed. Handbook of Hypertension, Vol. 4, Experimental and Genetic Models of Hypertension. Amsterdam: Elsevier Science, 1984, pp. 240–255.

50. Y Yamori, Y Nara, S Mizushima, S Murakami, K Ikeda, M Saimiri, T Nabika, R Horie. Gene-environment interaction in hypertension, stroke and atherosclerosis in experiment models and supportive findings from a worldwide cross-sectional epidemiological survey: A WHO-Cardiac study. Clin Exp Pharmacol Physiol 19:43–52, 1992.

51. M Hamano, S Mashiko, T Onda, I Tomita, T Tomita. Effects of cholesterol-diet on the incidence of stroke and life-span in malignant stroke prone spontaneously hypertensive rats. Jpn Heart J 36:511, 1995.

52. WMN Ratnayake, MR L'Abbé, R Mueller, S Hayward, L Plouffe, R Hollywood, K Trick. Vegetable oils high in phytosterols make erythrocytes less deformable and shorten the life span of stroke prone spontaneously hypertensive rats. J Nutr 130:1166–1178, 2000.

53. WMN Ratnayake, L Plouffe, R Hollywood, MR L'Abbé, N Hidiroglou, G Sarwar, R Mueller. Influence of sources of dietary oils on the life span of stroke-prone spontaneously hypertensive rats. Lipids 35:409–420, 2000.

54. I Ikeda, H Nakagiri, M Sugano, S Ohara, T Hamada, M Nonaka, K Imaizumi. Mechanisms of phytosterolemia in stroke-prone spontaneously hypertensive and WKY rats. Metabolism 60:1361–1368, 2001.

55. N Watanabe, Y Endo, K Fujimoto. Effects of fat mixtures similar to Japanese diet on the life span of stroke-prone spontaneously hypertensive rats (SHRSP). J Oleo Sci 51:183–190, 2002.

56. RI Weed. The importance of erythrocyte deformability. Am J Med 49:147–150, 1970.

57. Y Yamori, Y Nara, R Horie, A Ooshima. Abnormal membrane characteristics of erythrocytes in rat models and men with predisposition to stroke. Clin Exp Hypertension 2:1009–1021, 1980.

58. K Tsuda, Y Ueno, I Nishio, Y Masuyama. Membrane fluidity as a genetic marker of hypertension. Clin Exp Pharmacol Physiol 19:11–16, 1992.
59. M Konishi, A Terao, M Dol. Osmotic resistance and cholesterol content in of the erythrocyte membrane in cerebral hemorrhage. Igaku Ayumi 120:30–32, 1982.
60. K Ueda, Y Hasuo, J Kiyohara, J Wada, H Kawano, I Kato, T Fujii, T Yanai, T Omae, M Fujishima. Intracerebral haemorrhage in Japanese Community, Hysayma: incidence, changing pattern during follow-up, and related factors. Stroke 19:52–58, 1988.
61. H Iso, DR Jacobs, D Wentworth, JD Neaton, JD Cohen. Serum cholesterol levels and six-year mortality from stroke in 350, 977 men screened for the multiple risk factor intervention trial. N Encl J Med 320:904–910, 1989.
62. C Iribarren, DM Reed, CM Burchfiel, JH Dwyer. Serum total cholesterol and mortality: confounding and risk modification in Japanese-American men. JAMA 273:1926–1932, 1995.
63. C Iribarren, DR Jacobs, M Sadler, AJ Claton, S Sidney. Low total serum cholesterol and intracerebral haemorrhagic stroke: is the association confined to elderly men? Stroke 27:1993–1998, 1996.
64. JD Neaton, H Blackburn, D Jacob, L Kuller, DJ Lee, R Sherwin, J Stamler, D Wentworth. Serum cholesterol level and mortality findings for men screened in the Multiple Risk Factor Intervention Trial. Arch Intern Med 152:1490–1500, 1992.
65. CJ Glueck, J Spiers, T Tracy, P Streicher, E Illig, J Vandegrift. Relationships of serum plant sterols (phytosterols) and cholesterol in 595 hypercholesterolemic subjects, and familial aggregation of phytosterols, cholesterol, and premature coronary heart disease in hyperphytosterolemic probands and their first degree relatives. Metabolism 40:842–848, 1991.
66. CJ Glueck, P Streicher, E Illig. Serum and dietary phytosterols, cholesterol, and coronary heart disease in hyperphytosterolemic probands. Clin Biochem 25:331–334, 1992.
67. WHF Sutherland, MJA Williams, ER Nye, NJ Restieaux. SA de Jong, HL Walker, Associations of plasma noncholesterol sterol levels with severity of coronary artery disease. Nutr Metab Cardiovas Dis 8:386–391, 1998.
68. RA Rajaratnam, H Gylling, TA Miettinen. Independent association of serum squalene and noncholesterol sterols with coronary artery disease in postmenopausal women. J Am Coll Cardiol 35:1185–1191, 2000.
69. TA Miettinen, RS Tilvis, YA Kesäniemi. Serum plant sterols and cholesterol precursors reflect cholesterol absorption and synthesis in volunteers of a randomly selected male population. Am J Epidemiol 131:20–31, 1990.
70. TA Miettinen, H Gylling, T Strandberg, S Sarna. for the Finnish 4S Investigators of the Scandinavian Simvastatin Survival Study Group. Noncholesterol sterols and cholesterol lowering by long-term simvastatin treatment in coronary patients. Relation to basal serum cholestanol. Arterioscler Thromb Vascular Biol 20:1340–1346, 2000.
71. TA Miettinen, RS Tilvis, YA Kesäniemi. Serum cholestanol and plant sterol

levels in relation to cholesterol metabolism in middle-aged men. Metabolism 38:136–140, 1989.

72. TA Miettinen, H Gylling, T Strandberg, S Sarna, for the Finnish 4S investigators. Baseline serum cholestanol as predictor of recurrent coronary events in subgroup of Scandinavian simvastatin survival study. B Med J 316:1127–1130, 1998.

73. PT Clayton, A Bowron, KA Mills, A Massoud, M Casteels, PJ Milla. Phytosterolemia in children with parental nutrition-associated cholestatic liver disease. Gastroenterology 105:1806–1813, 1993.

74. PT Clayton, P Whitefield, K Iyer. The role of phytosterols in the pathogenesis of liver complications of pediatric parenteral nutrition. Nutrition 14:158–164, 1998.

75. KR Iyer, L Spitz, P Clayton. New insight into mechanisms of parenteral nutrition-associated cholestasis: Role of plant sterols. J Pediatr Surg 33:1–6, 1998.

76. EMM Quigley, MN Narsh, JL Shafer, RS Markin. Hepatobiliary complications of total parenteral nutrition. Gastroenterology 104:286–301, 1993.

77. M Gerard-Bonconpain, JP Claudel, P Gaussorgues, F Salord, M Sirdot, M Chevallier, D Robert. Hepatic cytolytic and cholestatic changes related to a change of lipid emulsions in four long-term parenteral nutrition patients with short bowel. J Parenter Enteral Nutr 16:78–83, 1992.

78. PA Hepburn, SA Horner, M Smith. Safety evaluation of phytosterol esters. Part 2. Subchronic 90-day oral toxicity study on phytosterol esters—a novel functional food. Food Chem Toxicol 37:521–532, 1999.

79. D Turnbull, MH Whittaker, VH Frankos, D Jonker. 13-Week oral toxicity study with stanol esters in rats. Regul Toxicol Pharmacol 29:216–226, 1999.

80. DH Waalkens-Berendsen, APM Wolterbeek, MVW Wijnands, M Richold, PA Hepburn. Safety Evaluation of phytosterol esters. Part 3. Two-generation reproduction study in rats with phytosterol esters–a novel functional food. Food Chem Toxicol 37:683–696, 1999.

81. MH Whittaker, VH Frankos, APM Wolterbeek, DH Waalkens-Berendsen. Two-generation reproductive toxicity study of plant stanol esters in rats. Regul Toxicol Pharmacol 29:196–204, 1999.

82. D Turnbull, VH Frankos, JHM van Delft, N DeVogel. Genotoxicity evaluation of wood-derived and vegetable oil-derived stanol esters. Regul Toxicol Pharmacol 29:205–210, 1999.

83. AM Wolfreys, PA Hepburn. Safety Evaluation of phytosterol esters. Part 7. Assessment of mutagenic activity of phytosterols, phytosterol esters and the cholesterol derivative, 4 cholesten-3-one. Food Chem Toxicol 40:461–470, 2002.

84. VA Baker, PA Hepburn, SJ Kennedy, PA Jones, LJ Lea, JP Sumpter, J Ashby. Safety evaluation of phytosterol esters. Part 1. Assessment of oestrogenicity using a combination of in vivo and in vitro assays. Food Chem Toxicol 37:13–22, 1999.

85. D Turnbull, VH Frankos, WR Leeman, D Jonker. Short-term tests of estro-

genic potential of plant stanols and plant stanol esters. Regul Toxicol Pharmacol 29:211–215, 1999.

86. RS Slesinski, D Turnbull, VH Frankos, APM Wolterbeek, DH Waalkens-Berendsen. Developmental toxicity study of vegetable oil-derived stanol fatty acid esters. Regul Toxicol Pharmacol 29:227–233, 1999.

87. MH Moghadasian, JJ Frohlich. Effects of dietary phytosterols on cholesterol metabolism and atherosclerosis: clinical and experimental evidence. Am J Med 107:588–594, 1999.

88. R Ayesh, JA Weststrate, PN Drewitt, PA Hepburn. Safety evaluation of phytosterol esters. Part 5. Faecal short-chain fatty acid and microflora content, faecal bacterial enzyme activity and serum female sex hormones in healthy normolipidaemic volunteers. Food Chem Toxicol 37:1127–1138, 1999.

89. JA Weststrate, R Ayesh, C Bauer-Plank, PN Drewitt. Safety evaluation of phytosterol esters. Part 4. Faecal concentrations of bile acids and neutral sterols in healthy normolipidaemic volunteers consuming a controlled diet either with or without a phytosterol ester–enriched margarine. Food Chem Toxicol 37: 1063–1071, 1999.

90. MA Hallikainen, IJ Uusitupa. Effects of 2 low-fat stanol ester–containing margarines on serum cholesterol concentrations as part of a low-fat diet in hypercholesterolemic subjects. Am J Clin Nutr 69:403–410, 1999.

91. MH Davidson, KC Maki, DM Umporowicz, KA Ingram, MR Dicklin, E Schaefer, RW Lane, JR McNamara, JD Ribaya-Mercado, G Perrone, SJ Robins, WC Franke. Safety and tolerability of esterified phytosterols administered in reduced-fat spread and salad dressing to healthy adult men and women. J Am Coll Nutr 20:307–319, 2001.

92. MA Hallikainen, ES Sarkkinen, H Gylling, AT Erkkilä, MIJ Uusitupa. Comparison of the effects of plant sterol ester and plant stanol ester–enriched margarines in lowering serum cholesterol concentrations in hypercholesterolaemic subjects on a low-fat diet. Eur J Clin Nutr 554:715–725, 2000.

93. MA Hallikainen, ES Sarkkinen, MIJ Uusitupa. Plant stanol esters affect serum cholesterol concentrations of hypercholesterolemic men and women in a dose-dependent manner. J Nutr 130:767–776, 2000.

94. HFJ Hendriks, JA Weststrate, T van Vliet, GW Meijer. Spreads enriched with three different levels of vegetable oil sterols and the degree of cholesterol lowering in normocholesterolaemic and mildly hypercholesterolaemic subjects. Eur J Clin Nutr 53:319–327, 1999.

95. J-C Kim, B-H Kang, C-C Shin, Y-B Kim, H-S Lee, C-Y Kim, J Han, K-S Kim, D-W Chung, M-K Chung. Subchronic toxicity of plant sterol esters administered by gavage to Sprague–Dawley rats. Food Chem Toxicol 40: 1569–1580, 2002.

96. P Nieminen, A-M Mustonen, P Linström-Seppä, J Asikainen, H Mussalo-Rauhamaa, JVK Kukkonen. Phytosterols act as endocrine and metabolic disruptors in the European polecat (*Mustela putorius*). Toxicol Appl Pharmacol 178:22–28, 2002.

97. DL Mcatchy, GJ van den Kraak. The phytoestrogen β-sitosterol alters the

reproductive endocrine status of goldfish. Toxicol Appl Pharmacol 134:305–312, 1995.

98. KM Boberg, J-E Åkerlund, I Björkhem. Effect of sitosterol on the rate-limiting enzymes in cholesterol synthesis and degradation. Lipids 24:9–12, 1989.
99. KC Maki, MH Davidson, DM Umporowicz, EJ Schaefer, MR Dicklin, KA Ingram, S Chem, JR McNamara, BW Gebhart, JD Ribaya-Mercado, G Perrone, SJ Robins, WC Franke. Lipid responses to plant-sterol-enriched reduced-fat spreads incorporated into a National Cholesterol Education Program Step I diet. Am J Clin Nutr 74:33–43, 2001.
100. AB Awad, MD Garcia, CS Fink. Effect of dietary phytosterols on rat tissue lipids. Nutr Cancer 29:212–216, 1997.
101. KC Hayes, A Pronczuk, V Wijendran, M Beer. Free phytosterols effectively reduce plasma and liver cholesterol in gerbils fed cholesterol. J Nutr 132:1983–1988, 2002.
102. TA Miettinen, M Vuoristo, M Nissinen, HJ Järvinen, H Gylling. Serum, biliary, and fecal cholesterol and plant sterols in colectomized patients before and during consumption of stanol ester margarine. Am J Clin Nutr 71:1095–1102, 2000.
103. JMR Beveridge, HL Haust, WF Connell. Magnitude of the hypocholesterolemic effect of dietary sitosterol in man. J Nutr 83:119–122, 1964.
104. CH Duncan, MM Best. Effects of sitosterol on serum lipids of hypercholesterolemic subjects. J Clin Invest 34:930, 1955.
105. JW Farquhar, RE Smith, ME Dempsey. The effect of beta-sitosterol on the serum lipids of young men with arteriosclerotic heart disease. Circulation 14:77–82, 1956.
106. CR Joyner, PT Kuo. The effect of sitosterol administration upon the serum cholesterol level and lipoprotein pattern. Am J Med Sci 230:636–647, 1955.
107. AM Lees, HYI Mok, RS Lees, MA McCluskey, SM Grundy. Plant sterol as cholesterol-lowering agents: clinical trials in patients with hypercholesterolemia and studies of sterol balance. Atherosclerosis 28:325–338, 1977.
108. JM Lesesne, CW Castor, SW Hoobler. Prolonged reduction in human blood cholesterol levels induced by plant sterols. Univ Mich Med Bull 21:13–17, 1955.
109. AA Spector, MA Yorek. Membrane lipid composition and cellular function. J Lipid Res 26:1015–1035, 1985.

10

Chemistry, Analysis, and Occurrence of Phytosterol Oxidation Products in Foods

Paresh C. Dutta
Swedish University of Agricultural Sciences, SLU, Uppsala, Sweden

I. INTRODUCTION

The most common phytosterols in plant lipids are sitosterol, campesterol, and stigmasterol, belonging to the group known as 4-desmethylsterols (1). Brassicasterol, another 4-desmethylsterol, is present in considerable amounts in the lipids of Cruciferae (1). In addition, some cereal lipids contain a considerable proportion of the saturated counterpart of plant sterols (phytosterols), generally called stanols (2). The formation, analysis, occurrence, and biological effects of the oxidation products of cholesterol (COPs), the main sterol in animal lipids, have been extensively studied (3). However, research on these aspects with the corresponding phytosterol oxidation products (POPs) has been very limited (4–6). Based on limited number of studies on phytosterol oxidation it has been concluded that the main oxidation route is generally the same as that in cholesterol oxidation (7). Similarly, the analysis of POPs also follows the same procedures as in the case of COPs (5). In this chapter, pathways of phytosterol oxidation, results from the recent research in the area of analysis, and their occurrence of phytosterols in foods are reviewd. Analysis of POPs in biological samples is highlighted in Chapter 4 of this book.

II. MECHANISMS OF AUTOXIDATION OF FREE AND ESTERIFIED PHYTOSTEROLS

Cholesterol and phytosterols are unsaturated steroid alcohols that are susceptible to oxidation under various conditions. The two main pathways elucidated for cholesterol oxidation are enzymic oxidations and nonenzymic oxidations. Detailed mechanisms of these pathways of cholesterol oxidation have been reviewed (7–9). Chemical structures of cholesterol, the most abundant phytosterol (sitosterol), the saturated counterpart of sitosterol (sitostanol), and some POPs that were detected in foods, similar to those from cholesterol, are presented in Fig. 1. Listed in Table 1 are the systematic, trivial, and short names of POPs generated from some common phytosterols (4).

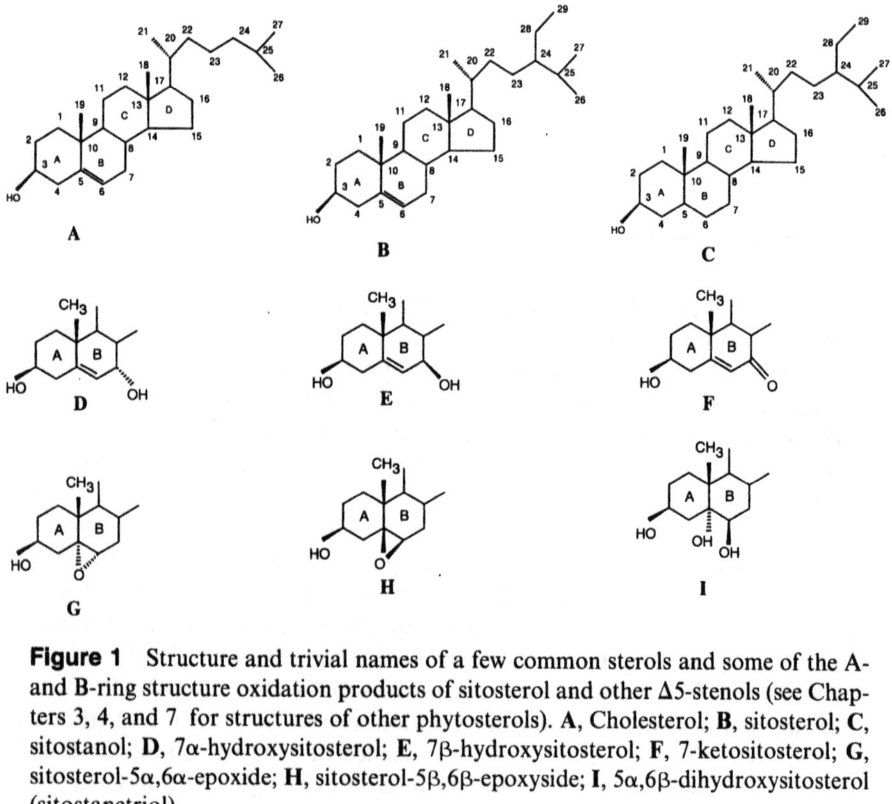

Figure 1   Structure and trivial names of a few common sterols and some of the A- and B-ring structure oxidation products of sitosterol and other Δ5-stenols (see Chapters 3, 4, and 7 for structures of other phytosterols). **A**, Cholesterol; **B**, sitosterol; **C**, sitostanol; **D**, 7α-hydroxysitosterol; **E**, 7β-hydroxysitosterol; **F**, 7-ketositosterol; **G**, sitosterol-5α,6α-epoxide; **H**, sitosterol-5β,6β-epoxyside; **I**, 5α,6β-dihydroxysitosterol (sitostanetriol).

Table 1 Nomenclature of Some Common Oxidation Products of Sitosterol, Stigmasterol, and Campesterol

Short name	Trivial name	Systematic name
Sitosterol	Sitosterol	(24R)-Ethylcholest-5-en-3β-ol
7α-OH-sitosterol	7α-Hydroxysitosterol	(24R)-Ethylcholest-5-en-3β,7α-diol
7β-OH-sitosterol	7β-Hydroxysitosterol	(24R)-Ethylcholest-5-en-3β,7β-diol
7-Ketositosterol	7-Ketositosterol	(24R)-Ethylcholest-5-en-3β-ol-7-one
α-Epoxysitosterol	Sitosterol-5α,6α-epoxide	(24R)-5α,6α-Epoxy-24-ethylcholestan-3β-ol
β-Epoxysitosterol	Sitosterol-5β,6β-epoxide	(24R)-5β,6β-Epoxy-24-ethylcholestan-3β-ol
Sitostanetriol	Sitostanetriol	(24R)-Ethylcholestan-3β,5α,6β-triol
Stigmasterol	Stigmasterol	(24S)-Ethylcholest-5,22-dien-3β-ol
7α-OH-stigmasterol	7α-Hydroxystigmasterol	(24S)-Ethylcholest-5,22-dien-3β,7α-diol
7β-OH-stigmasterol	7β-Hydroxystigmasterol	(24S)-Ethylcholest-5,22-dien-3β,7β-diol
7-Ketostigmasterol	7-Ketostigmasterol	(24S)-Ethylcholest-5,22-dien-3β-ol-7-one
α-Epoxystigmasterol	Stigmasterol-5α,6α-epoxide	(24S)-5α,6α-Epoxy-24-ethylcholest-22-en-3β-ol
β-Epoxystigmasterol	Stigmasterol-5β,6β-epoxide	(24S)-5β,6β-Epoxy-24-ethylcholest22-en-3β-ol
Stigmastentriol	Stigmastentriol	(24S)-Ethylcholest-22-en-3β,5α,6β-triol
Campesterol	Campesterol	(24R)-Methylcholest-5-en-3β-ol
7α-OH-campesterol	7α-Hydroxycampesterol	(24R)-Methylcholest-5-en-3β,7α-diol
7β-OH-campesterol	7β-Hydroxycampesterol	(24R)-Methylcholest-5-en-3β,7β-diol
7-Ketocampesterol	7-Ketocampesterol	(24R)-Methylcholest-5-en-3β-ol-7-one
α-Epoxycampesterol	Campesterol-5α,6α-epoxide	(24R)-5α,6α-Epoxy-24-methylcholestan-3β-ol
β-Epoxycampesterol	Campesterol-5β,6β-epoxide	(24R)-5β,6β-Epoxy-24-methylcholestan-3β-ol
Campestanetriol	Campestanetriol	(24R)-Methylcholestan-3β,5α,6β-triol

Source: Adapted from Ref. 4.

A. Oxidation in the Ring Structure

The initial reactions in the cholesterol autoxidation process involve predominantly abstraction of a reactive allylic 7-hydrogen atom forming a free radical followed by reaction with molecular oxygen ($^3O_2$), yielding 3β-hydroxycholest-5-en-7-peroxyl radicals. These products are stabilized in turn by hydrogen abstraction producing more stable products, 7-hydroperoxides. These compounds further decomposed to the 7α- and 7β-hydroxycholesterol. The epimeric 7-hydroperoxides and 7-hydroxycholesterol can be further thermally degraded to 7-ketocholesterol and subsequently can be converted to cholesta-3,5-dien-7-one (7–9).

Autoxidation of cholesterol may occur in solid state, in solution, or in aqueous dispersions in the presence of molecular oxygen forming isomeric 5,6-epoxides and 3β,5α,6β-triol (7–9). The isomeric epoxycholestanols are formed from the interaction of cholesterol as secondary oxidation products with sterol hydroperoxides and other oxidizing agents. The ratio of formation of the 5α,6α-epoxycholestanol and 5β,6β-epoxycholestanol in solvent and dispersion is determined by the oxidizing agent and pH. However, formation of 5β,6β-epoxycholestanol is more favorable than formation of 5α,6α-epoxycholestanol during the epoxidation of the B ring with organic hydroperoxides. It may be noted that the ratio of the formation of the epoxides can be different depending on other oxidizing agents as well. The epimers of epoxycholestanol can be further hydrolyzed to cholestane-3β,5α,6β-triol. The formation of cholestane-3β,5α,6β-triol is also dependent in the reaction environment, e.g., at acidic pH 5β,6β-epoxycholestanol is less stable than 5α,6α-epoxycholestanol (7–9).

Yanishlieva and Schiller (10) studied the oxidation of sitosterol in a model lipid system. In this study, 5% sitosterol in tristearin was oxidized at 120°C for 7 h. In addition to epimeric 7-hydroxysitosterols, sitosterol-5,6-epoxides, sitostanetriol, and other oxidation products were also identified, as was reported by the authors (11). It was demonstrated from that study that, in contrast to pure sitosterol oxidation, the sitosterol oxidation in tristearin medium produced epimeric sitosterol-5,6-epoxides and stigmasta-4,6-dien-3-one. In addition, the formation of 6-hydroxystigmast-4-en-3-one and sitostanetriol was considerably lower than in pure sitosterol oxidation.

The effect of lipid media unsaturation on the oxidation of sitosterol was studied by the same researchers (12). It was demonstrated that when sitosterol was added at the 5% level in different triacylglycerols having variable degrees of unsaturation, the percentage of changed sterol increased with increasing unsaturation of the lipid medium. The formation of sitistanetriol was only observed in sunflower oil, whereas no formation of this triol was observed in tristearin and lard with 5% sitosterol; the epimeric sitosterol-5,6-epoxides were present in both cases (12).

Although studies on the oxidation of sterol esters are not as extensive as those on free sterols, it is generally known that sterol esters are less sensitive to air oxidation than the corresponding free sterol (7). Comparative oxidizability of free sitosterol and its esterified form with stearic acid was conducted (13). It was demonstrated from the kinetics of autoxidation of pure sitosterol and sitosteryl stearate that during the initial stage sitostryl stearate was oxidized more quickly than sitosterol. However, during the later stage when higher amounts of peroxides and 7-hydroxy and 7-keto derivatives were formed, the difference in the rate of oxidation beteween these two compounds diminished considerably. The authors explained that this might be due to stronger initiating action of the free sitosterol peroxides and of the other initially formed sitosterol oxidation products in comparison with the action of the corresponding oxidation products from sitosteryl stearate (13).

During oxidation of sitosteryl stearate at 150°C for 1 h, some other basic differences in oxidation products from sitosterol and sitosteryl stearate are seen (14). It was demonstrated that in contrast to oxidation of free sitosterol under similar conditions in which 3β-hydroxy group was oxidized to a keto group and the Δ5 double bond was relocated to Δ4 position. However, no such oxidation products or 6-hydroperoxide, an important primary oxidation product of sitosterol, were formed as a result of oxidation of sitosteryl stearate. In addition to the formation of the other common oxidation products in the ring structures, the formation of 25-hydroperoxisitosteryl stearate was confirmed in that study. It is worth mentioning that these studies were conducted with a saturated fatty acid, stearic acid, esterified at the 3β position in the ring structure of sitosterol and the reactivity of the unsaturated fatty acids may be different (14).

Some oxidized products of sitosterol were characterized after heat treatment at 100°C of pure sitosterol (15). These authors characterized 7α-hydroxysitosterol, 7β-hydroxysitosterol, and 7-ketositosterol by thin-layer chromatography (TLC) and gas chromatography–mass spectrometry (GC-MS) along with the epimeric sitosterol-5,6-epoxides, Δ4-sitosterol-3-6-dione, Δ4-sitosterol-3-one, and Δ5-sitosterol-3-one.

Gordon and Magos (16) isolated some oxidized products of Δ5-avenasterol produced by heat treatment at 180°C in trioleylglycerol containing 0.1% Δ5-avenasterol. After isolation of the unsaponifiables and enrichment by preparative TLC, epimeric 7-hydroxy, 5,6-epoxy, and 7-keto derivatives generated from Δ5-avenasterol were identified by comparison with standard cholesterol oxidation products.

In a study on oxidation of stigmasterol in triacylglycerols heated at 180°C, eight oxidation products of stigmasterol were identified and characterized (17). The identified products were stigmasta-3,5,22-triene, stigmasta-3,5,22-trien-7-one, stigmasta-4,22-dien-3-one, stigmasta-4,6,22-trien-3-one, stigmasterol-5,6-epoxides, 7α-hydroxystigmasterol, 7β-hydroxystigmasterol, and stigmastentriol. Of these products the first four are considered as unpolar compounds. It is worth noting here that some of these nonpolar compounds are postulated to be at least partially generated via some of the polar oxidation products, during refining processes (18,19).

B. Oxidation in the Side Chain

The pathway for the generation of side chain autoxidation products from cholesterol, e.g., 20-hydroperoxide, 24-hydroperoxide, 25-hydroperoxide, and 26-hydroperoxide derivatives, and their thermal decomposition products have been clearly demonstrated (8). It has also been reported that the side chain autoxidation products of cholesterol are quantitatively much less than the autoxidation products at the allylic C-7 position (8). With an additional methyl or ethyl group at C-24, sitosterol and campesterol presumably can form similar side chain oxidation products as cholesterol. Despite the unambiguous possibility of generating side chain oxidation products of phyto-sterols, very limited research has been conducted regarding their identifi-cation in foods (11,20,21). However, some side chain oxidation products of sitosterol and campesterol were synthesized enzymatically or chemically, and were characterized by TLC, GC, and GC-MS (22).

Sitosterol was subjected to oxidation either by heating at 150°C for 1 hr or by leaving at room temperature in air for 6 months, and the oxida-tion products were characterized (11). In addition to common ring structure oxidation products, three possible side chain oxidation products were pre-dicted in that paper, i.e., 24- and 25-hydroxysitosterol and 24-dehydrositos-terol; however, only the last one was isolated. That compound was postu-lated to be generated through dehydration of 24- or 25-hydroxysitosterol.

III. PHOTOOXIDATION AND OTHER PATHWAYS

In this pathway, singlet oxygen ($^1O_2$) instead of triplet oxygen ($^3O_2$) started the initial reaction. The generation of singlet oxygen can be initiated by chemical reactions involving metal ions or the so called photosensitizers cholorophylls and some heme proteins in the presence of light. The singlet oxygen is very reactive and can attach to any side of the double bonds of the sterol molecule by the cyclic ene mechanism. The products formed following this pathway differ qualitatively and quantitatively from those produced by autoxidation. 5α-Hydroperoxide is considered as the major product of photo-oxidation of cholesterol, and traces of epimeric 3β-hydroxycholes-4-ene-6α- and 6β-hydroperoxides are also formed due to the allylic shift of the Δ5 double bond (8). Recently, this mechanism of photooxidation was demon-strated with stigmasterol induced by the sensitizer methylene blue under light during 0–90 min (23).

In addition to singlet oxygen, there are several active oxygen species that can generate specific oxidation products from cholesterol. Furthermore, several COPs are intermediates in bile acids biosynthesis in animals; they can be generated by oxidative enzymes, and their formation can be induced by xenobiotics. Detailed mechanisms involving these pathways for choles-terol oxidation have been reviewed; however, similar studies with phytoster-ols are not available (8,9).

IV. OXIDATION OF SATURATED STEROLS (STANOLS)

Some comments on the possibility of oxidation of stanols, the saturated counterpart of sterols, are of interest. Stanols do not contain a double bond at the Δ5 position (see Fig. 1), and are considered to be more stable than their counterparts containing an olefinic unsaturation (7). Since cholesterol is autoxidized in the side chain producing a varity of hydroperoxides, the saturated counterpart, the stanol, might also generate similar products. In addition, the free stanols may also generate 3-keto compound as in the case of free unsaturated sterols. Up to now, no systematic studies have been conducted on the autoxidation of stanols.

V. ANALYSIS OF PHYTOSTEROL OXIDATION PRODUCTS

Several major phytosterols are present in plant lipid, and these can be oxidized to a large number of POPs (4,5). The basic principles of analysis of POPs are similar to these for the analysis of COPs, with the following steps: extraction of lipids, saponification, enrichment of POPs, and qualitative and quantitative determination by GC and GC-MS, or high-performance liquid chromatography (HPLC) and HPLC-MS (5). Publications are limited in the area of POPs. Special attention may also be needed to extract lipids from certain plant-based food products because of the difficulties in completing lipid extraction.

Commercial wheat flour samples were analyzed for POPs after lipids were extracted in water-saturated *n*-butanol (24). Crude lipids were purified by partitioning in chloroform-methanol-water, and the lipids were saponified. The unsaponifiables were extracted with diethyl ether, and qualitative and quantitative analyses were performed by color development on TLC plates. Prior to further analyses, the unsaponifiables were subjected to further enrichment for POPs by preparative TLC. For qualitative examination, the TLC (silica) plates were developed twice, first in the solvent system ethyl acetate-heptane (1:1, v/v), and then in the solvent system acetone-heptane (1:1, v/v). For preparative analysis, the TLC plates were developed three times in the solvent system ethyl acetate-heptane (1:1, v/v), and the POPs were extracted from silica gel using chloroform. Further purification of the POPs was accomplished by preparative TLC using various solvent systems, and by utilizing a reversed-phase HPLC system prior to identification by MS. Quantitative determination of POPs was performed using TLC. For this purpose, *para*-toluenesulfonic acid was used for developing color, and the POPs were quantified by planimetry. Preparative TLC has also been used to enrich POP from the unsaponifiable fraction in vegetable oils and subsequent spectrophotometric quantification by reaction of POPs with cholesterol oxidase (25).

Among the few studies on the levels of POPs in foods, potato chips was one of the products investigated (26). Lipids from potato chips, fried exclusively in cottonseed oil, was extracted with chloroform-methanol, following the method of Bligh and Dyer (27). After saponification, the POPs were further separated from fatty acids and pigments by using normal and argented column chromatography. For the quantitative analysis, normal-phase HPLC column was used for separation by using the solvent system hexane-isopropanol (10:1, v/v). Detection of POPs was accomplished using a differential refractometer. Although results on sitosterol-5β,6β-epoxide and epimers of 7-hydroxysitosterol were shown, it is not known if the HPLC system was able to separate 5,6-epoxides or hydroxysterols generated from campesterol and stigmasterol.

POPs were determined in some vegetable oils and wheat flour (28). Lipids from wheat flour and vegetable oils were further separated to different fractions by using Lipidex-5000 and solid-phase extraction (SPE) chromatography using an NH_2 column. Further quantification of trimethylsilyl (TMS) ether derivatives of POPs was performed by GC. Only isomeric sitosterol-5,6-epoxides and epimeric 7-hydroxysitosterols were analyzed. Since analysis of POPs in this method was done only on the free POPs, the result may not reflect the total POP levels in the samples because sterol esters are also oxidized and the rate of oxidation may be even more rapid than for free sterols (13).

Various coffee samples were examined for content of POPs (29). The total lipids were extracted by chloroform and further fractionated to polar fraction containing POP, by using a 500-mg SPE silica column. This fraction was further analyzed by reversed-phase HPLC, and detection was carried out by UV at 245 nm. Although the separation efficiency of the HPLC system was not shown, the authors observed that interferences were induced during HPLC analysis.

The levels of 7-ketcholesterol and 7-ketositosterol in infant milk formulas and milk cereals were reported recently (30). Lipids were extracted from samples by hexane and isopropanol (31), and cold saponification of the lipid extract and extraction of unsaponifiables were done according to Park and Addis (32). Subsequent enrichment of mixed sterol oxidation products (SOPs) was done by SPE by using a 500-mg silica column. Separation of 7-ketosterols was done on reversed-phase HPLC, and detection and quantification were done by diode array detector at 240 nm. Separation between 7-ketocholesterol and 7-ketositosterol was very good, but the presence of other 7-ketosterols that could be originated from other phytosterols was not investigated.

Potato products, prepared in several vegetable oils in industrial conditions, were investigated (33,34). Quantification of POPs was achieved by GC; a considerable number of POPs were synthesized in the author's laboratory for identification. Further confirmation of POPs was accomplished by GC-MS. It was reported that any unoxidized phytosterols left in the enriched fraction after SPE would overlap with some POPs in the capillary column used in those studies. It was therefore emphasized that complete removal of unoxidized phytosterols was necessary for subsequent GC analysis, and the method developed for this purpose is described below.

Unoxidized phytosterols from the unsaponifiable fraction of the lipids extracted from potato chips were removed completely by double SPE chromatography. For this purpose, a 0.5-g silica cartridge was solvated by 3 ml hexane. The unsaponifiable fraction from about 0.5 g of oil, which was dissolved in 1 ml hexane-diethyl ether (75:25, v/v) in a glass tube, was charged to the solvated SPE column. The tube was washed with an additional 2 ml hexane-diethyl ether (75:25, v/v) and was eluted through the column at a rate of about 4 ml/min. Thereafter, the column was eluted with 6 ml hexane-diethyl ether (60:40, v/v) and the elutes were discarded. The sterol oxidation products and the remaining sterols were eluted with 5 ml acetone. The acetone was dried under nitrogen and the residue was again dissolved in 1 ml hexane-diethyl ether (75:25). Another 0.5-g silica column was prepared as before and the sample was charged again and was eluted with 3 ml hexane-diethyl ether (60:40, v/v) and the eluate was discarded. Finally, the column was eluted with 5 ml acetone; the acetone was dried under nitrogen and the POPs were derivatized to TMS ether for subsequent analysis by GC and GC-MS. Prior to derivatization of POPs to TMS ether, analytical TLC was used to check the separation of unoxidized phytosterols and POPs. It was demonstrated using this double SPE system that the POP fraction was completely free from unoxidized phytosterols, which made it possible to avoid any overlapping of unoxidized phytosterol peaks and POP peaks during GC and GC-MS analyses (33,34).

Stability of some common phytosterols was studied in canola, coconut, peanut, and soybean oils spiked with sitosterol (0.9 mg/g oil) at various temperatures for 20 h (35). The main steps were somewhat similar as those described before, i.e., cold saponification, enrichment of POPs by SPE, and subsequent analysis by GC and GC-MS. In this study a medium polar column was used in the analysis of POPs. However, separation efficiencies of POPs by this column was not shown, but the authors reported that during GC-MS analysis at 290°C the background noise due to column bleeding was high (35).

Thermo oxidation products of plant sterols were studied after developing a new method of enrichment by SPE and subsequent analysis by GC capillary column (20). The enrichment of POPs after complete separation of unoxidized phytosterol from the cold saponified samples was accomplished by single-silica-column SPE (500 mg). It is well known by all analysts that separation of the unoxidized phytosterol is necessary to analyze POPs by GC flame ionization detection (FID) because of the possibility of overlapping with unoxidized phytosterols. The column used in this study was a 60 m × 0.32 mm i.d. × 0.1 μm of the type 5% diphenyl-95% dimethyl polysiloxane. The authors reported that separation of a few POPs was not achieved, and a few POP peaks were contaminated with interfering substances. Variable recoveries were reported by using standard samples of COPs. The lowest recovery was of 7-ketocholesterol (54.1%), cholestanetriol (72.4%), 5α,6α-epoxycholesterol (94.5%), and 7β-OH-cholesterol (108.6%).

Some primary oxidation products of stigmasterol after induced photo-oxidation were successfully determined by HPLC (23). The epimers of 6-OOH-stigmasterol, 5α-OOH-stigmasterol, and 7α-OOH-stigmasterol were determined by normal-phase HPLC using a fluorescence detector after post-column diphenyl-1-pyrenylphosphine reagent without prior derivatization and sample pretreatment. Further confirmation of the specific photooxidation products of stigmasterol was done by GC-MS.

Oxidation products of phytosterols in some crude vegetable oils, and during refining processes of these oils, were determined (36). The 7α- and 7β-hydroxy derivatives of sitosterol, stigmasterol, and campesterol, and the 7-ketositosterol as TMS derivatives were analyzed by GC-MS of the cold saponified oils and the enriched POP fraction by SPE. The authors reported that 7β-hydroxycampesterol could not be separated from sitosterol by the column used in this study. However, the problem of quantification was overcome by selective ion mode of the GC-MS.

VI. CONTENT OF PHYTOSTEROL OXIDATION PRODUCTS IN FOODS

The presence of small amounts of 7-hydroxysitosterols in crude oils, neutralized oils, and soapstocks was reported in the late 1960s (18,37). In one study (18), the author explained that oxidation of sterols in plant oils occurs very slowly and never attains a high level, at least in oils fit for human consumption. It was observed in that study that oxidation was accelerated during the neutralization process of oils, when the oil comes in contact with air for a long time. The author reported that mainly 7-hydroxy- and 7-ketosterols appeared during deacidification of the oil. In contrast, the neutralized oil containing oxidized sterols when subjected to bleaching at temperatures of 90–95°C favored the dehydration of sterols.

In another investigation (37), the authors demonstrated that 7-hydroxysterols are formed when soybean oil was heated in the presence of oxygen. Further treatment of 7-hydroxysterols with bleaching earth led to the formation of sterol hydrocarbons and ketosteroids. No quantitative data on POPs were presented in either study (18,37).

Recently, some additional nonpolar steroidal hydrocarbons were determined (19). The authors supported that these compounds were dehydrated from several side chain hydroxyphytosterols during bleaching of olive oils under laboratory conditions.

In a study by Nourooz-Zadeh and Appelqvist (28), no POPs could be detected in crude soybean oil, freshly refined soybean oil, refined soybean oils stored at 4°C for a year, and freshly opened olive oil samples. However, olive oil stored at room temperature for 30 months had trace amounts of POPs. In this study wheat flour was also included for studying the formation of some POPs in flour during storage. It was shown that total POPs in 2-month-old wheat flour was up to 35 µg/g and the levels increased to 328 µg/g total POPs after 36 months of storage. The POPs analyzed were epimeric sitosterol-5,6-epoxides and the epimeric 7-hydroxysitosterols (28).

Commercial wheat flours, processed in various ways, were analyzed by a combination of TLC and HPLC for the content of POPs (24). It was reported that there was a significant increase in the level of POPs in the bleached flour, compared with the unbleached flours. The mean value of sterol-5,6-epoxides were 367 µg/g, 133 µg/g, 200 µg/g, and 150 µg/g in the lipids extracted from bleached, unbleached, bromated, and whole wheat flour, respectively. The levels of total 7-hydroxysterols in these samples were 160 µg/g, 116 µg/g, 120 µg/g, and 280 µg/g, respectively. It was observed in this study that there was a significant increase in the amount of oxidized sterols in the bleached flours compared with the unbleached flours. The author also pointed out that significant differences in the amounts of POPs were observed in the samples of bleached and unbleached brands. The cause of these differences among same type of samples, as speculated by the author, was the use of different types of bleaching agents. Randomly chosen samples could also have been subjected to different storage conditions and time. The author concluded that storage and processing conditions affect sterol oxidation in food products.

The effect of processing on the generation of POPs was studied in 20 samples of different types of coffees by analyzing 7-ketositosterol using HPLC (29). It was reported that the level of 7-ketositosterol was negligible, less than 10 µg/kg in the lipids extracted from coffee samples investigated. The authors explained that the limited formation of POPs could be attributed to the fact that oxidation conditions occur only in the heating step of the initial stages of roasting because, after the formation of the initial fumes caused by pyrolysis of the more heat-susceptible components, a reducing atmosphere is established inside the roasting autoclave. Then, only the surface of the coffee bean is exposed to oxidation, and that part of the product contributes very little to the total amount of sterol-containing lipids. It was also demonstrated that the decaffeination process did not have any effect on the generation of POPs.

Published reports from the author's laboratory demonstrated that refined vegetable oils contain a considerable amount of some oxidation products generated from sitosterol, campesterol, and stigmasterol. Partial results are presented in Table 2 (34). The level of total POPs in the refined oils was 41, 40, and 46 μg/g in palm/rapeseed oil blend, sunflower, and high-oleic sunflower oil, respectively.

The major POPs in palm/rapeseed oil blends were epoxy derivatives of both campesterol and sitosterol. The other components were 7α-and 7β-hydroxysitosterol, sitostanetriol, and 7-ketositosterol. The major difference in the sterol oxidation products in sunflower oil compared with the rapeseed oil/palm oil blend was marked in the higher contents of 7α-and 7β- hydroxysitosterol, sitostanetriol, and 7-ketositosterol, and lower content of sterol-5,6-epoxides. A similar pattern was observed in the high-oleic sunflower oil, except that this oil contained more 7-ketositosterol and sterol-5,6-epoxides than sunflower oil. Small amounts of 7-hydroxystigmasterol were also detected, ranging from 0.4 to 0.8 μg/g in most of the oils (Table 2).

Two papers have been published on POPs in potato chips fried exclusively in vegetable oil, during storage (26,33). The contents of POPs in potato chips, fried in cottonseed oil and packaged in foil during storage at 23°C and 40°C for 150 and 95 days, respectively, were determined (26). It was demonstrated that the oxidation products were a result of elevated temper-

Table 2 Content of Phytoterol Oxidation Products (POPs) in Oil Samples of Rapeseed Oil/Palm Oil Blend, Sunflower Oil, and High-Oleic Sunflower Oil

Oil sample	7α-OH	7β-OH	7-Keto	Epoxy[a]	Triol	Total
RP						
Sitosterol	4.4	1.3	1.6	17.2	2.9	41.0
Campesterol	1.1	nd	nd	10.8	1.7	
Stigmasterol	nd[b]	nd	nd	nd	nd	
SO						
Sitosterol	5.8	6.6	12.9	5.3	4.9	39.9
Campesterol	1.7	0.5	nd	0.9	nd	
Stigmasterol	0.7	0.6	nd	nd	nd	
HOSO						
Sitosterol	7.7	5.4	14.1	7.8	5.7	46.7
Campesterol	2.8	1.0	nd	1.4	nd	
Stigmasterol	0.4	0.4	nd	nd	nd	

POP(μg/g)

[a] Includes both 5α,6α-epoxy- and 5β,6β-epoxysterols.
[b] Not detected.
RP, blend of hydrogenated rapeseed/palm oil; SO, sunflower oil; HOSO, high-oleic sunflower oil.
Source: Ref. 34.

ature, since in samples stored at 23°C for 150 days, oxidation products were not detected; in contrast, chips stored at 40°C for 95 days had 28 µg/g of the sum of the three sitosterol oxidation products analyzed. Although the samples stored at 40°C had high peroxide values and a rancid taste after 24 days of storage, no sterol oxidation products were detected (26).

The level of POPs in chips fried in different vegetable oils prepared under industrial conditions and vacuum packed was studied during storage of up to 25 weeks at room temperature (33). The levels of total POPs in the lipids of chips fried in palm oil were increased from 6 µg/g at 10 weeks to 9 µg/g at 25 weeks of storage (Table 3). The contribution from different sitosterol oxidation products was almost at equal levels from 0.6 to 0.9 µg/g, except for the sitostanetriol, which was at a considerably lower level, i.e., 0.2 µg/g in the lipids.

The content of POPs in chips fried in sunflower oil was virtually unchanged during storage of up to 25 weeks (Table 4). 7-Ketositosterol was present in the highest amount, about 16 µg/g, and remained almost unchanged during storage.

The content of POPs in the chips fried in high-oleic sunflower oil was 55 µg/g after 10 weeks of storage. This amount increased to about 59 µg/g after 25 weeks of storage (Table 5).

The proportions of different oxidation products of campesterol and sitosterol in the chips prepared in high-oleic sunflower oil were similar, and were also dominated by 7-ketocampesterol and 7-ketositosterol. Results from that study show that storage at 25°C up to 25 weeks did not increase the amounts of POPs to any great extent, except for chips prepared in high-

Table 3 Content of Phytosterol Oxidation Products (POPs) at 0, 10, and 25 Weeks of Storage in Lipids of Potato Chips Fried in Palm Oil

Sterol	Weeks	POP (µg/g)					
		7α-OH	7β-OH	7-Keto	Epoxy[a]	Triol	Total
Sitosterol	0	0.6	0.8	0.9	0.9	0.2	5.0
Campesterol		0.3	0.2	0.7	0.3	0.1	
Sitosterol	10	0.5	1.0	1.3	0.8	0.2	6.1
Campesterol		0.3	0.3	1.0	0.6	0.1	
Sitosterol	25	1.2	1.4	2.0	1.9	0.9	8.6
Campesterol		0.1	0.2	0.7	0.2	nd[b]	

[a] Includes both 5α,6α-epoxy- and 5β, 6β-epoxy-sterols.
[b] Not detected.
Source: Ref. 33.

Table 4 Content of Phytosterol Oxidation Products (POPs) at 0, 10, and 25 Weeks of Storage in Lipids of Potato Chips Fried in Sunflower Oil

Sterol	Weeks	7α-OH	7β-OH	7-Keto	Epoxy[a]	Triol	Total
Sitosterol	0	4.4	9.9	16.1	3.5	1.2	5.8
Campesterol		0.9	1.5	4.7	2.4	1.2	
Sitosterol	10	4.9	10.4	16.5	4.4	1.2	49.4
Campesterol		1.1	1.9	6.0	2.1	0.9	
Sitosterol	25	6.6	8.8	15.7	4.6	0.8	47.1
Campesterol		0.6	1.5	5.3	2.6	0.6	

[a] Includes both 5α,6α-epoxy- and 5β,6β-epoxy sterols.
Source: Ref. 33.

oleic sunflower oil, where the content of sterol oxidation products tended to increase after 10 weeks of storage.

French fries, fried in various vegetable oils at 200°C, were investigated for the content of POPs (34). The levels of total POP products in the lipids of French fries fried at 200°C in rapeseed oil/palm oil blend, sunflower oil, and high-oleic sunflower oil were 32, 37, and 54 μg/g, respectively. 7-Keto-sitosterol and 7-ketocampesterol were the dominating oxidation products in all the samples, followed by epimers of sterol-5,6-epoxides (Table 6).

In addition to epimers of 7-hydroxysterols originating from both campesterol and sitosterol, both campestanetriol and sitostanetriol were present

Table 5 Content of Phytosterol Oxidation Products (POPs) at 0, 10, and 25 Weeks of Storage in Lipids of Potato Chips Fried in High-Oleic Sunflower Oil

Sterol	Weeks	7α-OH	7β-OH	7-Keto	Epoxy[a]	Triol	Total
Sitosterol	0	2.8	7.8	10.9	3.1	1.7	35.1
Campesterol		1.0	1.7	3.3	2.0	0.8	
Sitosterol	10	2.7	11.3	14.2	4.2	2.3	54.8
Campesterol		1.3	2.4	12.2	2.9	1.3	
Sitosterol	25	5.3	11.5	18.1	6.3	3.6	58.5
Campesterol		0.7	1.5	6.2	3.5	1.8	

[a] Includes both 5α,6α-epoxy- and 5β,6β-epoxy sterols.
Source: Ref. 33.

Table 6 Content of Phytosterol Oxidation Products (POPs) in the Lipids of French Fries Samples Fried in Rapeseed Oil/Palm Oil Blend, Sunflower Oil, and High-Oleic Sunflower Oil, Prepared at 200°C for 15 min

Sample	POP (µg/g)					
	7α-OH	7β-OH	7-Keto	Epoxy[a]	Triol	Total
RP						
Sitosterol	2.9	3.7	4.1	7.6	0.5	32.0
Campesterol	1.4	2.6	3.3	5.4	0.5	
SO						
Sitosterol	3.8	7.3	13.1	2.2	1.1	36.9
Campesterol	0.3	1.3	5.9	1.3	0.6	
HOSO						
Sitosterol	4.7	9.7	13.5	5.4	2.8	53.7
Campesterol	1.4	1.8	9.2	3.6	1.6	

[a] Includes both 5α,6α-epoxy- and 5β,6β-epoxy sterols.
RP, lipids from French fries fried in a blend of palm/rapeseed oil; SO, lipids from French fries fried in sunflower oil; HOSO, lipids from French fries fried in high-oleic sunflower oil.
Source: Ref. 34.

in relatively low levels at 1 µg/g in palm oil/rapeseed oil blend, 1.7 µg/g in regular sunflower oil, and 4.4 µg/g in high-oleic sunflower oil. In general, the content of POPs in French fries prepared in different vegetable oils reflects those of oils used for the frying operations (Table 2).

The interest in fortifying food products with phytosterols has increased during the last decade because of their cholesterol lowering property in humans (38–40). The author (41) studied some polar oxidation products of phytosterols in raw materials (wood sterols), and in a number supplement tablet preparations based on phytosterols commercially available in Finland. A sample of a pure phytosterol mixture, which was subjected to oxidation by treatment with high temperature, was analyzed against the unheated raw materials (Table 7).

It is to be noted that even the recrystallized sterol sample contained oxidized POPs, though in small amounts (35 mg/100 g), whereas this level increased dramatically in the heat treated sterol mixture in which the level was 893 mg/100 g (41). It might be very important to carefully control the temperature treatment process during industrial production of new foods enriched with phytosterols.

Mixed sterol oxidation products (SOPs) in infant formulas and milk cereals were determined (30). The levels of 7-ketocholesterol and 7-ketositosterol were analyzed by HPLC. The samples generally contained milk fat

Table 7 Content of Phytosterol Oxidation Products (POPs) in Three Samples of Pure Phytosterol Mixtures and in Three Tablet Supplements Manufactured in Finland

	POP (µg/g)						
	7α-OH	7β-OH	7-Keto	α-Epoxy	β-Epoxy	Triol	Total
Raw materials							
Wood sterols							
Sitosterol	7.3	4.3	12.3	8.0	22.6	0.4	60.0
Campesterol	0.8	0.3	1.3	0.5	2.2	nd[a]	
Heat-treated sterols							
Sitosterol	35.7	40.5	408.7	111.0	194.5	12.0	892.8
Campesterol	5.9	8.7	43.7	11.1	21.0	nd	
Recrystrallized sterols							
Sitosterol	3.7	2.3	5.0	4.1	13.9	0.4	34.6
Campesterol	0.5	0.4	0.6	1.4	2.3	nd	
Tablet preparations							
Anti K-steroli							
Sitosterol	0.8	1.8	2.7	1.6	2.8	0.2	10.7
Campesterol	0.2	0.2	0.2	tr[b]	0.2	nd	
Tri Tolosen Kasvisteroli							
Sitosterol	1.8	0.8	4.0	1.7	9.1	0.3	21.5
Campesterol	0.2	0.2	0.3	0.6	2.5	nd	
Kolestop							
Sitosterol	2.0	3.8	5.0	3.1	9.9	0.2	26.0
Campesterol	0.3	0.2	0.4	0.3	0.8	nd	

[a] Less than 0.05 mg/100 g.
[b] Not detected.
Source: Ref. 41.

and unspecified vegetable oils or unspecified animal fats. The level of 7-ketositosterol was usually less than 4 µg/g lipids, except for a few samples. The authors concluded that industrial production technologies might have a significant influence on lipid oxidation.

Very recently, POPs were quantified in several crude vegetable oils. Among the vegetable oils analyzed, sunflower oil and maize oil contained the highest amounts of POPs, ranging from 4 to 68 ppm (36). The main POPs were 7α-,7β-hydroxy derivatives of sitosterol, campesterol, and stigmasterol, and 7-ketositosterol in all the samples. The authors suggested that the large differences in POP contents in various vegetable oils might be influenced by fatty acid composition, i.e., oils containing more unsaturated fatty acids were more prone to oxidation. In addition, different conditions and times of storage of the raw materials and the crude oils before the refining process were also influenced the formation of POPs in the oils. The POPs content in some of the crude oils analyzed in this study is shown in Table 8.

The authors also studied the formation and degradation of POPs during refining processes using these crude vegetable oils. During bleaching of sunflower oil at 80°C for 1 h with 1% and 2% of both acidic and neutral earths, it was observed that the dihydroxyphytosterols were dehydrated, producing steroidal hydrocarbons with three double bonds in the ring structures; results from similar studies were reviewed (42). It was interesting to note that during deodorization of the same sunflower oil at 180°C for 1 h under vacuum, no hehydration products of the dihydroxyphytosterols occurred. The authors suggested that dehydration of dihydroxyphytosterols would require higher temperatures as reported previously (36).

Oxidation of phytosterols in test food systems using various vegetable oils spiked with 0.9 mg/g sitosterol was studied (35). It was demonstrated in that study that oils containing higher levels of polyunsaturated fatty acids facilitated phytosterol oxidation. The authors also observed that a greater variety of POPs was identified at 100°C than at 150°C or 180°C. It was further explained that a highly unsaturated oil could utilize more oxygen before the total decomposition of the phytosterols through an oxidative pathway, resulting in recovery of more unoxidized phytosterols and more POP than

Table 8 Content of Phytosterol Oxidation Products (POPs) in Some Samples of Crude Peanut Oils, Sunflower Oils, Corn Oils, and Lampante Olive Oils (µg/g)

POP	Peanut oils		Sunflower oils		Corn oils		Lampante olive oils	
	PE1	PE4	SF1	SF5	CO1	CO5	LOO2	LOO6
7α-OH-sitosterol	1.0	2.7	14.8	1.9	2.2	15.3	1.0	1.0
7β-OH-sitosterol	0.7	1.9	10.6	1.1	1.5	10.4	0.4	0.5
7α-OH-campesterol	0.3	0.7	1.8	0.6	0.6	1.9	0.1	<0.1
7β-OH-campesterol	0.2	0.5	1.4	0.4	0.5	1.6	<0.1	<0.1
7α-OH-stigmasterol	0.3	0.6	2.4	0.3	0.3	3.0	<0.1	<0.1
7β-OH-stigmasterol	0.2	0.5	1.7	0.2	0.2	2.0	<0.1	<0.1
7-Ketositosterol	<0.1	2.7	34.8	<0.1	3.0	22.3	<0.1	1.0
Total POPs	2.7	9.6	67.5	4.5	8.3	60.1	1.5	2.5

PE1, peanut oil sample 1; PE4, peanut oil sample 4; SF1, sunflower oil sample 1; SF5, sunflower oil sample 5; CO1, corn oil sample 1; CO5, corn oil sample 5; LOO2, lampante virgin olive oil sample 2; LOO6, lampante virgin olive oil sample 6.
Source: Adapted from Ref. 36.

with coconut oil (a very saturated oil). In all the samples containing different levels of phytosterols and unsaturation, the main POPs were epimers of 5,6-epoxy, 7α-hydroxy, and 7-keto derivatives of campesteol, stigmasterol, and sitosterol. It was demonstrated in this study that most POPs were further degraded and could not be detected after heating at 180°C for 20 h.

Pure stigmasterol and rapeseed oil were subjected to thermal oxidation at 180°C during various time periods (20). During heat treatment from 15 min to 1 h, the levels of all POP analyzed increased linearly. In addition to the common polar POPs, 25-hydroxystigmasterol was also verified and quantified in that study. In contrast, the levels of individual and total POPs generated from campesterol and sitosterol were shown to increase even after 24 h of heating rapeseed oil at the same temperature as with pure stigmasterol. There were no qualitative differences in the generation of POPs in pure stigmasterol or phytosterols in the rapeseed oil subjected to similar conditions of thermal oxidation. Similar to 25-hydroxystigmasterol during oxidation of pure stigmasterol, 25-hydroxysitosterol and 25-hydroxycampesterol were also detected and quantified in oxidized rapeseed oil. The quantitative results on various POP in oxidized rapeseed oil from this study are presented in Table 9.

From the results of that study it was concluded that thermal oxidation and autoxidation of phytosterols are dependent on the sample area exposed to air and the area-to-volume ratio of the samples (20).

Table 9 Content of Phytosterol Oxidation Products (POPs) in a Sample of Rapeseed Oil After Heating at 180°C

Heating time (h)	POP (μg/g)						
	7α-OH	7β-OH	7-Keto	5α,6α-Epoxy	5β,6β-Epoxy	25-OH	Total
Sitosterol							
0	nd[a]	nd	nd	nd	nd	nd	nd
6	31	52	12	27	38	2	162
12	71	105	42	58	92	4	372
24	109	154	102	107	172	2	646
Campesterol							
0	nd	nd	nd	nd	nd	nd	nd
6	21	27	9	13	31	3	104
12	52	59	31	33	72	6	253
24	83	87	72	69	131	11	452

[a] Not detected.
Source: Ref. 20.

VII. CONCLUSIONS

Based on the limited published reports, it can be concluded that oxidation of phytosterols follows similar pathways to that of cholesterol. At present there are only a few data available on the side chain POPs and their presence in foods; additional studies are needed. The kinetics of oxidation of different phytosterols is also of interest for future research. Although GC-MS or HPLC-MS can be used for detection and quantification, it is still necessary to correctly identify a particular peak because of the possible large numbers of POPs. For reliable quantification, separation of POPs by GC or HPLC columns is necessary, and this challenge should be overcome in the future. Development of harmonized methods is also emphasized for reliable quantitative analysis of POPs in foods and biological samples.

REFERENCES

1. SP Kochhar. Influence of processing on sterols of edible vegetable oils. Prog Lipid Res 22:161–188, 1983.
2. V Piironen, DG Lindsay, TA Miettinen, J Toivo, A-M Lampi. Plant sterols: biosynthesis, biological function and their importance to human nutrition. J Sci Food Agric 80:939–966, 2000.
3. F, Guardiola, PC Dutta, R Codony, GP Savage, eds. Cholesterol and Phytosterol Oxidation Products: Analysis, Occurrence, and Biological Effects. Champaign, IL: AOCS Press, 2002, pp. 1–394.
4. PC Dutta, GP Savage. Formation and content of phytosterol oxidation products in foods. In: F Guardiola, PC Dutta, R Codony, GP Savage, eds. Cholesterol and Phytosterol Oxidation Products: Analysis, Occurrence, and Biological Effects. Champaign, IL: AOCS Press, 2002, pp. 319–334.
5. PC Dutta. Determination of phytosterol oxidation products in foods and biological samples. In: F Guardiola, PC Dutta, R Codony, GP Savage, eds. Cholesterol and Phytosterol Oxidation Products: Analysis, Occurrence, and Biological Effects. Champaign, IL: AOCS Press, 2002, pp. 335–374.
6. A Grandgirard. Biological effects of phytosterol oxidation products, future research areas and concluding remarks. In: F Guardiola, PC Dutta, R Codony, GP Savage, eds. Cholesterol and Phytosterol Oxidation Products: Analysis, Occurrence, and Biological Effects. Champaign, IL: AOCS Press, 2002, pp. 375–382.
7. LL Smith. Cholesterol Autoxidation. Plenum Press, New York, 1980, pp. 1–249.
8. LL Smith. The oxidation of cholesterol. In: S-K Peng, RJ Morin, eds. Biological Effects of Cholesterol Oxides. Boca Raton: CRC Press, 1992, pp. 7–31.
9. G Lercker, MT Rodriguez-Estrada. Cholesterol oxidations mechanisms. In: F Guardiola, PC Dutta, R Codony, GP Savage, eds. Cholesterol and Phytosterol

Oxidation Products: Analysis, Occurrence, and Biological Effects. Champaign, IL: AOCS Press, 2002, pp. 1–25.

10. NVI Yanishlieva, H Schiller. Effect of sitosterol on autoxidation rate and product composition in a model lipid system. J Sci Food Agric 35:219–224, 1983.

11. N Yanishlieva, H Schiller, E Marinova. Autoxidation of sitosterol. II: Main products formed at ambient and high temperature treatment with oxygen. La Rivista Italiana Delle Sostanze Grasse LVII:572–576, 1980.

12. NVI Yanishlieva-Maslarova, EM Marinova-Tasheva. Effect of the unsaturation of lipid media on the autoxidation of sitosterol. Grasas y Aceites 37:343–347, 1986.

13. N Yanishlieva, E Marinova. Autoxidation of sitosterol I: kinetic studies on free and esterified sitosterol. La Rivista Italiana Delle Sostanze Grasse LVII:477–480, 1980.

14. N von Yanishlieva-Maslarova, H Schiller, A Seher. Die autoxidation von sitosterin III: sitosterylstearat. Fette Seifen Anstrichmittel 84:308–310, 1982.

15. GG Daly, ET Finocchiaro, T Richardson. Characterization of some oxidation products of β-sitosterol. J Agric Food Chem 31:46–50, 1983.

16. MH Gordon, P Magos. Products from the autoxidation of Δ5-avenasterol. Food Chem 14:295–301, 1984.

17. G Blekas, D Boskou. Oxidation of stigmasterol in heated triacylglycerols. Food Chem 33:301–310, 1989.

18. H Niewiadomski. The sterol hydrocarbons in edible oils. Die Nahrung 19:525–536, 1975.

19. R Bortolomeazzi, MDe Zan, L Pizzale, LS Conte. Identification of new steroidal hydrocarbons in refined oils and the role of hydroxy sterols as possible precursors. J Agric Food Chem 48:1101–1105, 2000.

20. A-M Lampi, L Juntunen, J Toivo, V Piironen. Determination of thermooxidation products of plant sterols. J Chromatogr B 777:83–92, 2002.

21. L Johnsson, PC Dutta. Chromatographic and mass spectral characterisation of some side chain oxidation products of phytosterols, the 21st Nordic Lipid Symposium, June 5–8, 2001, Bergen, Norway. Poster.

22. L Aringer, L Nordstrm. Chromatographic properties and mass spectrometric fragmentation of dioxygenated C27-. C28-, C29-steroids. Biomed Mass Spectrom 8:183–203, 1981.

23. S Säynäjoki, S Sundberg, L Soupas, A-M Lampi, V Piironen. Determination of stigmasterol primary oxidation products by high-performance liquid chromatography. Food Chem 80:415–421, 2003.

24. ET Finocchiaro. Sterol oxides in grated cheese, wheat flour and in model systems, PhD dissertation, University of Wisconsin, Madison, 1983.

25. VK Lebovics, M Antal, Ö Gaál. Enzymatic determination of cholesterol oxides. J Sci Food Agric 71:22–26, 1996.

26. K Lee, AM Herian, NA Highly. Sterol oxidation products in French fries and in stored potato chips. J Food Prot 48:158–161, 1985.

27. EG Bligh, WJ Dyer. A rapid method of total lipid extraction and purification. Can J Biochem Physiol 37:911–917, 1959.

28. J Nourooz-Zadeh, L-Å Appelqvist. Isolation and quantitative determination of sterol oxides in plant-based foods: soybean oil and wheat flour. J Am Oil Chem Soc 69:288–293, 1992.

29. E Turchetto, G Lercker, R Bortolomeazzi. Oxisterol determination in selected coffees. Toxicol Ind Health 9:519–527, 1993.

30. P Zunin, C Calcagno, F Evangelisti. Sterol oxidation in infant milk formulas and milk cereals. J Dairy Res 65:591–598, 1998.

31. A Hara, NS Radin. Lipid extraction of tissues with a low-toxicity solvent. Anal Biochem 90:420–426, 1978.

32. SW Park, PB Addis. Capillary-column gas-liquid chromatographic resolution of oxidized cholesterol derivatives. Anal Biochem 149:275–283, 1985.

33. PC Dutta, L-Å Appelqvist. Studies on phytosterol oxides I: effect of storage on the content in potato chips prepared in different vegetable oils. J Am Oil Chem Soc 74:647–657, 1997.

34. PC Dutta. Studies on phytosterol oxides II: content in some vegetable oils and in French fries prepared in these oils. J Am Oil Chem Soc 74:659–666, 1997.

35. LL Oehrl, AP Hansen, CA Rohrer, GP Fenner, LC Boyd. Oxidation of phytosterols in a test food system. J Am Oil Chem Soc. 78:1073–1078, 2001.

36. R Bortolomeazzi, F Cordaro, l Pizzale, LS Conte. Presence of phytosterol oxides in crude vegetable oils and their fate during refining. J Agric Food Chem 51:2394–2401, 2003.

37. HP Kaufmann, E Vennekel, Y Hamza. Über die veränderung der sterine in fetten und ölen bei der industriellen bearbeitung derselben I. Fette Seifen Anstrichmittel 72:242–246, 1970.

38. T Heinemann, G Aztmann, Kvon Bergmann. Comparison of intenstinal absorption of cholesterol with different plant sterols in man. Eur J Clin Invest 23:827–831, 1993.

39. TA Miettinen, P Puska, H Gylling, H Vanhanen, E Vartiainen. Reduction of serum cholesterol with sitostanol-ester margarine in a mildly hypercholesterolemic population. N Engl J Med 333:1308–1312, 1995.

40. PJ Jones, DE MacDougall, F Ntanios, CA Vanstone. Dietary phytosterols as cholesterol-lowering agents in humans. Can J Physiol Pharmacol 75:217–227, 1997.

41. PC Dutta. Phytosterol oxides in some samples of pure phytosterols mixture and in a few tablet supplement preparations in Finland. In: JT Kumpulainen, JT Salonen, eds. Proceedings of the Second International Conference on Natural Antioxidants and Anticarcinogens in Nutrition, Health and Disease. Helsinki, 1998. Royal Society of Chemistry, Cambridge, 1999, pp. 316–319.

42. PC Dutta, R Przybylski, L-.Å Appelqvist, NAM Eskin. Formation and analysis of oxidized sterols in frying fat. In: EG Parkins, MD Erickson, eds. Deep Frying—Chemistry, Nutrition, and Practical Applications. Champaign: AOCS Press, 1996, pp. 112–150.

11

Biological Effects and Safety Aspects of Phytosterol Oxides

Lisa Oehrl Dean and Leon C. Boyd
North Carolina State University, Raleigh, North Carolina, U.S.A.

I. INTRODUCTION

As minor components of plant lipids, phytosterols function as structural components in membranes and as precursors to steroid hormones. They are similar in structure to, and thus analogous in functionality to, cholesterol in animal lipids. Due to their structural similarities to cholesterol, phytosterols are expected to undergo similar chemical reactions including oxidation (1). Many types of oxidation products have been found to result from cholesterol. This oxidation is thought to proceed by a free-radical process resulting in phytosterol oxides and phytosterol hydroperoxides. The refining of plant oils and food processing conditions commonly expose lipid components in foodstuffs to contact with heat, light, air, and metal surfaces that induce this free-radical oxidation. The many types of cholesterol oxide products (COPs) found in foods and their biological properties have been reviewed (2). The biological effects of such compounds are overwhelmingly negative. COPs have been implicated in the onset of arteriosclerosis and coronary heart disease in humans, with the epoxides having been found to be mutagenic (3,4). More recently, a very comprehensive review of the many COPs and their biological activity has appeared (5). The oxidation of phytosterols compared to cholesterol seems to be much more difficult, and the biological and safety aspects of phytosterol oxidation products (POPs) remain unclear (6). Much of the confusion with regard to the effects of POPs has resulted from the lack of research studies on the safety of phytosterols, as well as from the levels of consumption and thus concern by the general population.

Problems with the analysis and isolation of resulting products have slowed research concerning their effects on biological systems (7).

II. IMPACT OF DIETARY LIPID CONSUMPTION PATTERNS AND ABSORPTION LEVELS

Dietary lipid sources and patterns have changed dramatically over the last decade. In many Western populations, the heavy reliance on the use of animal fat for home and commercial frying operations has shifted from animal fats to plant lipids due to health concerns related to overconsumption of cholesterol and saturated fats. These shifts in fat sources have resulted in a general trend toward increased consumption of plant lipids and thus phytosterols (7). Absorption levels of dietary phytosterols by humans are much lower than that of cholesterol. It is thought that only 5% of the phytosterol intake is absorbed compared to 30% of cholesterol (8). This may change as diets increase in plant lipid content as opposed to animal lipids according to current recommendations. Vegetarians and formula-fed infants have been found to have higher blood levels of phytosterols than adults consuming a mixed diet (9). COPs have been proven to be absorbed from the diet by human subjects, and it remains questionable whether this will be the case for phytosterol oxides as consumption of foods containing them increases, especially as more infants are being formula fed (10,11).

Rat studies have shown that there is lymphatic absorption of POPs (12); thus, their effect on human health may become a concern. In rats, phytosterol oxides have been found to cross the intestinal barrier and be absorbed. Lymphatic recoveries were on the order of 2.5–7%. Campesterol oxides were better absorbed than sitosterol oxides, indicating that the size of the side chain is important in moving through the intestinal barrier. There is some indication that the epoxides absorbed are transformed in part to their corresponding triols in the stomach or that the cells of the intestines produce hydrolases that create the triols. More recently, feeding studies with hamsters have shown that incorporation of oxidized phytosterols into the tissue of test animals is dose dependent (13). When hamsters were fed 100 µg/g POPs in their diet, no POPs were recovered from the liver, heart, kidneys, plasma, or aorta tissues. When the POP level was increased above 500 µg/g, significant amounts of various POPs were found. The oxycampesterol compounds were in the majority. The lowest levels were the α-epoxy and 7α-hydroxy compounds. These were thought to be better metabolized in vivo. Based on their potential to be absorbed, the bioactivity of phytosterol oxidation products should be of interest.

III. BIOACTIVITY OF ISOLATED LIPID FRACTIONS CONTAINING PHYTOSTEROL OXIDES

The medicinal healing powers of plants and their extracts have long been exploited without full knowledge of the components responsible for their bioactivity. This is especially true in Eastern cultures. Consumption of juices from unripe fruits and leaves or powders made from dried roots has been used in the practice of traditional medicine since ancient times. More recently, scientific methods have been applied to these materials in an effort to identify the source of the effects. The results indicate that many of these bioactive products contain small quantities of compounds related to phytosterols.

Alcoholic extracts from the root of the stinging nettle (*Urtica dioica* L.) have been used for the treatment of prostate adenoma in folk medicine (14). Isolation and identification of the compounds present in stinging nettle revealed six steroid compounds of which four were the oxides of sitosterol. They were 7β-hydroxysitosterol, 7α-hydroxysitosterol, and their corresponding glucosides. This indicates that one or more of the steryl derivatives may be responsible for the bioactivity or pharmacological effects shown by these extracts.

Euphorbia fischeriana is a perennial herbaceous plant that is native to Mongolia and Siberia. The polar fractions isolated from the dried roots were found to have cytotoxic activity against tumor cells (15). Upon isolation, the compounds present were found to be the 7-oxo, 7α-hydroxy-, and 7β-hydroxy derivatives of campesterol, stigmasterol, and sitosterol. The authors theorized that these compounds are derived from the nonselective autoxidation of the parent sterols or from some nonselective biological oxidation and that these types of compounds may occur more commonly in nature than currently thought.

The Chinese drug *Bombyx cum Botryte* has been used for the traditional treatment of cancerous-type diseases. From this compound, several 7-oxo and 7-hydroxysterols were isolated and their effects on hepatoma cells compared to the same compounds synthesized from commercial sterols (16). The most active compound was found to be 7β-hydroxycampesterol, with the least activity shown by 7α- and 7β-hydroxystigmasterol. Synergism with other components was considered, as crude extracts may have better bioavailability than pure steroids in cultures.

The bioactivity of several plant extracts containing sterol derivatives was tested in an albino rat model to detect any antizygotic, blastocystotoxic, antiplantation activity, and early abortifacient activity (17). Dried alcoholic extracts obtained from the flowers, seeds, leaves, and bark of *Aloe barbadensis*

Mill Syn., *Ananas comosus* Merr Syn., *Apium graveolens* Linn., *Butea Mono-sperma* Lam. Kuntz, and *Gossypium herbaceum* Linn. were screened for bioactivity. The choice of these plants was based on traditional claims of their antifertility effects. Of the several plants tested, only the juice of unripened *Ananas comosus* (pineapple) showed antifertility activity. Solvent isolation of the juice from the leaves and the unripe fruit produced fractions containing a variety of steroids and their derivatives. Diols of stigmastene were tentatively identified (18). Antifertility effects were seen in mice exposed to these fractions. Further work with pure materials confirmed that phytosterol oxides have abortifacient effects on mice (19). Through a series of studies, the authors identified some of the compounds present in the leaves to be ergosterol peroxide, β-sitosterol, 5-stigmastene-3β,7α-diol, 5-stigmastene-3β,7β-diol, 7-oxo-5-stigmasten-3β-ol, and 5α-stigmastane-3β,5,6β-triol.

In a subsequent study designed to compare the effects of pure compounds to crude compounds isolated from the leaves of *Ananas comosus*, oral doses of the compounds were fed to mice. All compounds were abortifacient when administered on day 1 of the experiment, but β-sitosterol and 7-oxo-5-stigmasten-3β-ol showed no activity when given 7 days into the pregnancy. The most effective was ergosterol peroxide, which showed activity at both stages of pregnancy. Other side effects observed in the treated animals include loss of weight, lethargy, and anemia. Animal treated with 5-stigmastene-3β,7α-diol and 5α-stigmastane-3β,5,6β-triol before and after implantation showed the most consistent effect. The mechanism of the effect was absence or disruption of the implantation sites.

IV. BIOACTIVITY OF ISOLATED PHYTOSTEROL OXIDES

Additional models of specific compounds have more recently been studied using purified phytosterol derivatives (20). The 5,6α-, 5,6β-epoxides and the corresponding 3,5,6-trihydroxysteranes of sitosterol and stigmasterol were tested by direct injection into mealworms (*Tenebrio molitor* L.). After elimination of the effects of the injection method and solvent carrier, comparisons of the mortalities of the test animals when injected with the phytosterol oxides, cholesterol epoxides, and linoleic acid hydroperoxides were made. As both COPs and linoleic acid hydroperoxides have been established to be cytotoxic, comparisons were made with POPs using equal concentrations of the three types of oxides. The more highly oxidized compounds of cholesterol and plant sterols were found to be the most lethal. Mealworms treated with 3,5,6-trihydroxycholestane or the corresponding 3,5,6-trihydroxysteranes (Fig. 1) showed the highest death rates. The 5,6-epoxides (Fig. 2) and 5,6-chlorohydrins of cholesterol and phytosterols produced half the activity

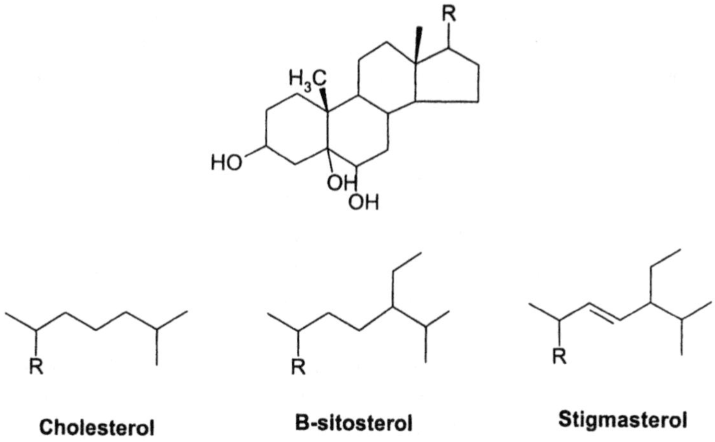

Figure 1   General structure of 3,5,6-trihydroxysteranes.

of the trihydroxy derivatives. Phytosterol oxides were found to be five times less active than the cholesterol oxides of the same type, although the effects were comparable. No differences were found between the cytotoxicity of the derivatives of stigmasterol and sitosterol. When rats were injected intravenously with 7-keto- and 7-hydroxy-β-sitosterol, no effects on cholesterolgenesis were seen (21), whereas the same experiment using the corresponding oxides of cholesterol caused inhibition of cholesterol synthesis. The authors attributed this to the inability of the liver enzymes metabolizing cholesterol to metabolize the phytosterols possessing additional functional groups on the side chain.

The liver is the site of cholesterol synthesis in mammals. Enzymes are produced there that act on a variety of compounds derived from sterols. As discussed above, sterol oxides have been shown to interfere with cholesterolgenesis. Hydroxylation of both cholesterol and plant sterols has been shown to occur in cell-free liver extracts (22). Once formed, they were subsequently

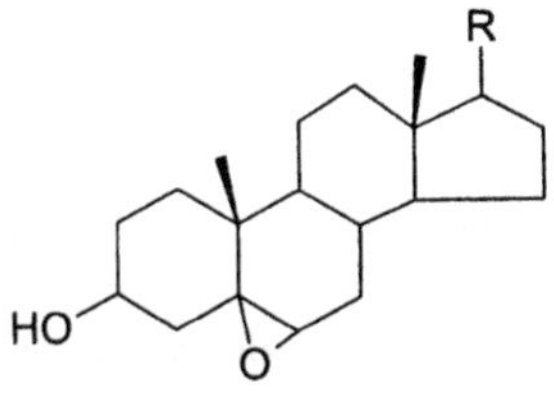

Figure 2   General structure of 5,6-epoxides. (See Figure 1 for R configurations.)

converted by microsomal hydrolases to their corresponding triols (23). Cholesterol again was more efficiently converted than the β-sitosterol. The cholesterol oxides were found to inhibit the hydrolases, but the β-sitosterol oxides did not. Such compounds have also been shown to be dehydroxylated in these types of experiments with 7α sterols being converted three times faster than 7β isomers (24). The transformations of these compounds from cholesterol and plant sterols are believed to utilize the same mechanism. Cholesterol or β-sitosterol is believed to serve as a precursor to bile acids, and hydroxylation of the sterol is considered to be the first step. These transformations have been studied using rat liver homogenates on cholesterol and the plant sterols campesterol and β-sitosterol (25,26).

Mouse microphage cells were used to compare effects of cholesterol oxides with their analogs from phytosterols (27). Macrophage cells are found to accumulate cholesterol and triglycerides in cytoplasmic droplets, which

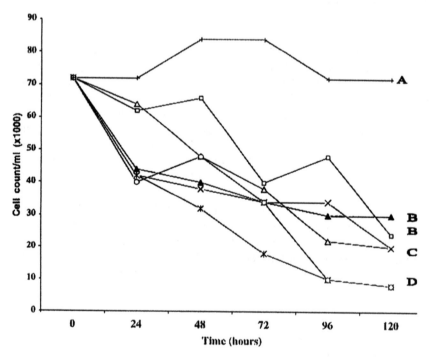

Figure 3   Cell viability expressed as the number of live cells per milliliter ($\times 1000$) over a 120-h observation period: control (+); cholesterol (□); cholesterol oxide (∗); 7-ketocholesterol (△); 5α,6α-epoxide (▲); β-sitosterol (x); β-sitosterol oxides (○). Lines followed by A–D were significantly different at $p < 0.05$. (From Ref. 27. Figure reproduced with permission of the *Journal of Agricultural and Food Chemistry*).

leads to the formation of foam cells. Foam cells isolated from the atherosclerotic plaque in human arteries have been found to contain various oxidation products of cholesterol (28). An oxidized mixture of campesterol and β-sitosterol was compared to cholesterol oxidized by the same oven heating method by adding to media used to culture the cells. Cell damage was monitored by lactate dehydrogenase leakage, cell viability, and mitochondrial dehydrogenase activity. Figure 3 shows the decrease in cell viability with time after the addition of sterol oxides. The damage caused by cholesterol oxides was greater than that caused by phytosterol oxides or by controls of untreated sterols. Both COPs and POPs caused similar types of cell damage.

V. BIOACTIVITY OF DEGRADATION PRODUCTS OF PHYTOSTEROL OXIDES

The degradation products of phytosterol oxides can also have negative biological effects. Female mosquito fish (Gambusia affinis) were found to develop anal fins that were elongated like those of male fish when inhabiting a stream polluted by paper mill waste (29). Although they displayed typical male reproductive behavior, the female fish still possessed reproductive capabilities slightly less than those of normal female fish. The abnormality was apparently only external. The males in the stream were normal. Paper mill effluents from facilities pulping pinewood chips contained tall oil, which is known to contain up to 3% phytosterols, of which 85% is sitosterol and campesterol (30). Degradation of these compounds by certain species of Mycobacterium results in C_{19} steroids, which were thought to have induced the masculinization of the female fish (31). Figure 4 shows the structure of one such compound, androstenedione, that was identified in river water containing paper mill effluent where deformed fish were also found (32). This compound shows the characteristic four-ring structure of sterols and is believed to be a bacterial metabolite of plant sterols. Such compounds have been reported to be produced by Mycobacterium strains through an oxidation mechanism in culture (33).

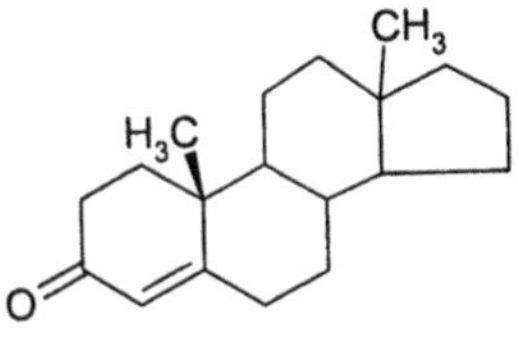

Figure 4   Androstenedione.

VI. THEORIES ON THE MECHANISMS FOR THE BIOACTIVITY OF PHYTOSTEROL OXIDATION PRODUCTS

The question could be proposed as to why the oxidation products of sterols have toxicity when their production is part of the normal biosynthesis of steroid hormones (34). Oxidation of the C-6 position of campesterol has been identified as the starting point for the formation of brassinosteroids (35). These compounds are plant growth regulators with high growth-promoting activity and are ubiquitous in nature. The key may be the site of the oxidation illustrated in Figure 5. Spatial positioning of the oxygen atoms and the type of oxygen functionality involved, for e.g., hydroxyl, ketone, or lactone, is thought to explain the growth-promoting activity, whereas damaging COPs are more often oxidized at the C-7 position, on a side chain, or in the epoxide form.

Another theory of toxicity may not involve oxidation products per se but rather the absence of functioning phytosterols. The major phytosterols campesterol and β-sitosterol are considered to regulate membrane fluidity. When this function is interfered with, cell death may result. Oxidation of phytosterols would result in the loss of intact sterols so that this function cannot be maintained. The level at which this occurs has yet to be defined (36). The toxicity of oxidation could theoretically be assigned to loss of intact sterols rather than effects of the oxides. Severe abnormalities in the development and fertility of tobacco plants were seen when transgenetic plants containing the gene for cholesterol oxidase were produced. Inclusion of the gene was targeted at expressing the protein in hopes that feeding boll weevils would be subjected to death or arrested development upon digestion. Although the total levels of phytosterols remained relatively the same between experimental plant lines and wild-type controls, the levels of free sterols were reduced, resulting in reduced growth. It has been suggested that altered sterol ratios also influence cell division (37).

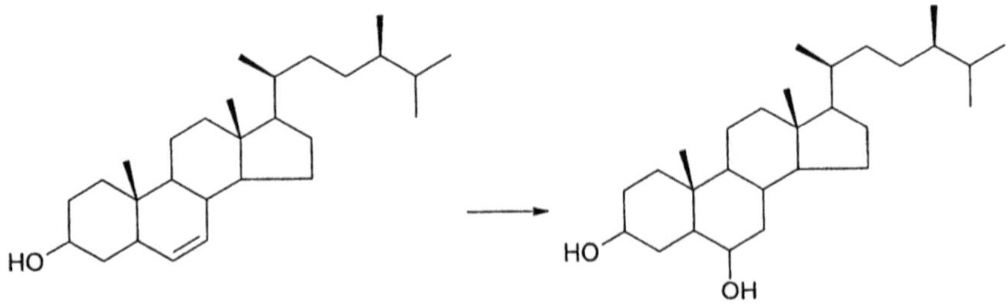

Figure 5   Oxidation at the C-6 position of campesterol.

Epoxides in nature often have detrimental effects on metabolism of living creatures. As an example, certain trees have been found to produce reactive epoxides called preconcenes, which are ring structures similar to phytosterol oxides (38). These compounds thus act as natural insectides in that they interfere with the biosynthesis, transport, or secretion of hormones needed for the sexual development of insects. The proposed mechanism of action is that the oxidized compound reacts with macromolecular cellular elements to condense the cytoplasm, resulting in cell death. This makes it impossible for the insect to biosynthesize or secrete the hormones necessary for regulating metamorphosis and for adult tissue development.

Cell aging has been found to increase the amounts of phytosterol oxides present in tissue cell cultures (39). The free-radical theory of lipid peroxidation would support that aging increases oxidative stress. The presence of lipid hydroperoxides results in an increase in phytosterol oxides (40). Epoxides can be formed enzymatically or via the hydroperoxides of unsaturated fatty acids. Simple epoxides, such as ethylene and propylene epoxides, are known to react with peptides by the formation of hydroxyethyl derivatives from the opening of the epoxide ring (41). One such peptide, glutathione, is an SH-containing nucleophile that is known to react with epoxides formed from cholesterol and unsaturated fatty acids. The loss of glutathione to the γ-glutamyl cycle will interfere with the transport of amino acids across cell membranes and with cell maintenance. Glutathione is also a major part of the detoxification enzymes termed glutathione conjugates (42). These enzymes protect cell membranes from peroxidizing compounds. It is suggested that the damage to aging cells from sterol oxides consists of deactivation of these enzymes (39).

VII. FUTURE DIRECTIONS

Compared to the amount of research already completed on COPs, phytosterol oxidation products have scarcely been studied. Scattered references go back more than 30 years, but with advances in separation science and identification techniques, these findings need to be confirmed. In depth studies have been done which describe the effects of COPs on membrane packing, functionality, and on the inhibition of sterol synthesis (43,44). Alterations in cell function, morphology, and viability have been described. Extrapolation of these results to analogous products from plant sterols cannot be taken for granted. The much lower rates of production of POPs in food products and the difficulties of oxidizing phytosterols in model systems in comparison with COPs indicate that there may be some basic differences in the mechanism of production that are not apparent. The available studies are very limited to analyses or model systems or are extremely dated. This field of study is just beginning to open and may gain in importance as human diets increase in plant lipids.

REFERENCES

1. LL Oehrl, AP Hansen, CA Rohrer, GP Fenner, LC Boyd. Oxidation of phytosterols in a test food system. J Am Oil Chem Soc 78:1073–1078, 2001.
2. R Paniangvait, AJ King, AD Jones, BG German. Cholesterol oxides in foods of animal origin. J Food Sci 60:1159–1174, 1995.
3. N Kumar, OP Singhal. Cholesterol oxides and atherosclerosis: a review. J Sci Food Agric 5:497–510, 1991.
4. A Sevanian, AR Peterson. Cholesterol epoxide is a direct-acting mutagen. Proc Nat Acad Sci USA 81:4198–4202, 1984.
5. GJ Schroepfer. Oxysterols: modulators of cholesterol metabolism and other processes. Phys. Rev. 80:361–554, 2000.
6. JC Knight, MG Wovcha. Microbial degradation of the phytosterol side-chain to 24-oxo products. Steroids 36:723–730, 1980.
7. PC Dutta, R Przbylsi, L-Å Appelqvist. formation and analysis of oxidized sterols in frying fat. In: EG Perkins, MD Erickson, eds. Deep Frying: Chemistry, Nutrition, and, Practical Applications. Champaign, IL: AOCS Press, 1996, pp. 112–150.
8. K Lee, AM Herian, NA Higley. Sterol oxidation products in french fries and stored potato chips. J Food Prot 48:158–161, 1985.
9. M Vuoristo, TA Miettinen. Absorption, metabolism, and serum concentrations of cholesterol in vegetarians: effects of cholesterol feeding. Am. J. Clin. Nutr. 59:1325–1331, 1994.
10. HA Emanuel, CA Hassel, PB Addis, SD Bermann, JH Zavoral. Plasma cholesterol oxidation products (oxysterols) in human subjects fed a meal rich in oxysterols. J Food Sci 56:843–847, 1991.
11. J Linseisen, G Wolfram. Absorption of cholesterol oxidation products from ordinary foodstuff in humans. Ann. Nutr. Metab. 42:221–230, 1998.
12. A Grandgirard, J-P Sergiel, M Nour, J Demaison-Meloche, C Giniès. Lymphatic absorption of phytosterol oxides in rats. Lipids 34(6):563–570, 1999.
13. A Grandgirard. Biological effects of phytosterol oxidation products: future research areas and concluding remarks. In: F Guardiola, PC Dutta, R Codony, GP Savage, eds. Cholesterol and Phytosterol Oxidation Products. Champaign, IL: AOCS Press, 2002, pp. 375–382.
14. N Chaurasia, M Wichtl. Sterols and steryl glycosides from *Urtica dioica*. J Nat Prod 50(5):881–885, 1987.
15. G Schroeder, M Rohmer, JP Beck, R Anton. 7-Oxo, 7α-hydroxy- and 7β-hydroxysterols from *Euphorbia fischeriana*. Phytochemistry 19:2213–2215, 1980.
16. H Nagano, JP Poyser, KP Cheng, L Bang, G Ourisson. Chemistry and

biochemistry of Chinese drugs: Part II. Hydroxylated sterols, cytotoxic toward cancerous cell: synthesis and testing. J Chem Res (M):2522–2571, 1977.

17. SK Garg, SK Saksena, RR Chaudhury. Antifertility screening of plants. Part VI. Effect of five indigenous plants on early pregnancy in albino rats. Indian J Med Res 58(9):1285–1289, 1970.

18. SC Pakrashi, B Achari, PC Majumdar. Studies on Indian medicinal plants: Part XXXII. Constituents of *Ananas comosus* (Linn.) Merr. leaves. Indian J Chem 13:755–756, 1975.

19. SC Pakrashi, B Basak. Abortifacient effect of steroids from *Ananas comosus* and their analogues on mice. J Reprod Fertil 46:461–462, 1976.

20. W Meyer, H Jungnickel, M Jandke, K Dettner, G Spiteller. On the cytotoxicity of oxidized phytosterols isolated from photoautotrophic cell cultures of *Chenopodium rubrum* tested on meal-worms *Tenebrio molitor*. Phytochemistry 47:789–797, 1998.

21. G Kakis, A Kuksis, JJ Myher. Injected 7-oxo cholesterol and plant sterol derivatives and hepatic cholesterogenesis. Atherosclerosis. Adv Exp Med Biol 82:297–299, 1977.

22. L Aringer, P Eneroth. Studies on the formation of C_7-oxygenated cholesterol and β-sitosterol metabolites in cell-free preparations of rat liver. J. Lipid Res 14:563–572, 1973.

23. L Aringer, P Eneroth. Formation and metabolism in vitro of 5,6-epoxides of cholesterol and β-sitosterol. J Lipid Res 15:389–398, 1974.

24. KM Boberg, A Stabursvik, I Björkhem, O Stokke. Dehydroxylation of a 7β-hydroxy-C_{27} plant sterol in rat liver. Biochim Biophys Acta 1004:321–326, 1989.

25. L Aringer. Conversion of 7α-hydroxycholesterol and 7α-hydroxy-β-sitosterol to 3α,7α-dihydroxy- and 3α,7α,12α-trihydroxy-5β-steroids in vitro. J Lipid Res 16:426–433, 1975.

26. L Aringer, P Eneroth, L Nordström. Side chain hydroxylation of cholesterol, campesterol, and β-sitosterol in rat liver mitochondria. J Lipid Res 17:263–272, 1976.

27. C Adcox, L Boyd, L Oehrl, J Allen, G Fenner. Comparative effects of phytosterol oxides and cholesterol oxides in cultured macrophage-derived cell lines. J Agric Food Chem 49:2090–2095, 2001.

28. DM van Reyk, W Jessup. The macrophage in atherosclerosis: modulation of cell function by sterols. J Leukocyte Biol 66:557–561, 1999.

29. WM Howell, TE Denton. Gonopodial morphogenesis in female mosquitofish, *Gambusia affinis affinis*, masculinized by exposure to degradation products from plant sterols. Envir Biol Fish 24:43–51, 1989.

30. AH Conner, JW Rowe. Neutrals in southern pine tall oil. J Am Oil Chem Soc 52:334–338, 1975.

31. TE Denton, WM Howell, JJ Allison, J McCollum, B Marks. Masculinization of female mosquitofish by exposure to plant sterols and *Mycobacterium smegmatis*. Bull Envir Contam Toxicol 35:627–632, 1985.

32. R Jenkins, RA Angus, H McNatt, WM Howell, JA Kemppainen, M Kirk, EM Wilson. Identification of androstenedione in a river containing paper mill effluent. Environ Toxic Chem 20:1325–1331, 2001.

33. W Marshek, JS Kraychy, RD Muir. Microbial degradation of sterols. Appl Microbiol 23:72–77, 1972.
34. C Grunwald. Plant Sterols. Annu Rev Plant Physiol 26:209–236, 1975.
35. C Brosa. Biological effects of brassinosteroids. In: EJ Parish, WD Nes, eds. Biochemistry and Function of Sterols. Boca Raton, FL: CRC Press, 1997, pp. 201–220.
36. DR Corbin, RJ Grebenok, TE Ohnmeiss, JT Greenplate, JP Purcell. Expression and chloroplast targeting of cholesterol oxidase in transgenic tobacco plants. Plant Phys 126:1116–1128, 2001.
37. H Shaller, P Bouvier-Naveí, P Benveniste. Overexpression of an *Arabidopsis* cDNA encoding a sterol-C24-methyltransferase in tobacco modifies the ratio of 24-methylcholesterol to sitosterol and is associated with growth reduction. Plant Phys 118:461–469, 1998.
38. WS Bowers. How anti-juvenile hormones work. Amer Zool 21:737–742, 1981.
39. W Meyer, G Spiteller. Oxidized phytosterols increase by ageing in photoautotrophic cell cultures of *Chenopodium rubrum*. Phytochemistry 45:297–302, 1997.
40. F Ursini, M Maiorino, A Sevanian. Membrane hydroperoxides. In: H Sies, ed. Oxidative Stress: Oxidants and Antioxidants. New York: Academic Press, 1991, pp. 319–336.
41. M Tornqvist. Formation of reactive species that lead to hemoglobin adducts during storage of blood samples. Carcinogenesis 11:51–54, 1990.
42. D Tsikas, G Brunner. High-performance liquid chromatography of glutathione conjugates. II. Ion-exchange chromatography and fluorescence detection. Fresenius J Anal Chem 343:330–334, 1992.
43. M Rooney, W Tamura-Lis, S Yachnin, O Kucuk, JW Kauffman. The influence of oxygenated sterol compounds on dipalmitoylphosphatidylcholine bilayer structure and packing. Chem Phys Lipids 41:81–92, 1986.
44. S Yachnin, RA Streuli, LI Gordon, RC Hsu. Alteration of peripheral blood cell membrane function and morphology by oxygenated sterols; a membrane insertion hypothesis. Curr Top Hematol 2:245–271, 1979.

12

Prospects of Increasing Nutritional Phytosterol Levels in Plants

Tatu A. Miettinen
Biomedicum Helsinki, University of Helsinki, Helsinki, Finland

Helena Gylling
University of Kuopio, and Kuopio University Hospital, Kuopio, Finland

I. INTRODUCTION

Recent clinical intervention studies have shown that long-term reduction of serum cholesterol by statins markedly reduces clinical manifestations of atheromatous arterial diseases, such as myocardial infarction and stroke (1,2). However, the studies have shown that before the curves of the event rates start to differ from each other in the placebo and statin treatment groups, the treatment should continue for up to a year or longer. Since dietary measures, including consumption of large amounts of phytosterols since the early 1950s (3), are being increasingly used for serum cholesterol lowering, their effects should also be long lasting, i.e., at least 1 year. However, despite the fact that many different phytosterol preparations lower serum cholesterol in humans in short-term studies (cf. 4,5), only plant stanol (mixture of campestanol and sitostanol) ester margarine has been shown to be effective for at least 1 year in both men and women as compared with the baseline or control values (6). In contrast, consumption of vegetable sterol (mixture of mainly sitosterol, campesterol, and stigmasterol) ester spread showed some cholesterol reduction for up to 1 year as compared with controls (7), but it was ineffective in women and, despite initial reduction, the 1-year cholesterol values were not significantly different from the baseline values (8). Plant sterol ester spread occasionally had no effect on serum cholesterol (9), but in general the plant sterol and stanol spreads are identical in their lowering of serum cholesterol levels in short-term studies (cf. 4,5).

Stanol esters also lower serum plant sterol concentrations (10), which are increased during plant sterol consumption (11,12). Since the isolation and saturation of plant sterols, and the subsequent esterification of stanols, or isolation and esterification of plant sterols, increases the production price of the respective functional foods, the increase of phytosterols in vegetables or in vegetable oils by plant breeding has aroused interest (13). Solubilization of free phytosterols, suggested to be important for their effect in inhibiting intestinal cholesterol absorption, may be obtained without esterification (14). This chapter deals briefly with the chemical structure of food plant sterols, their synthesis, and prospects for changing and increasing the synthesis of specific phytosterols by plant breeding.

II. DIETARY PLANT STEROLS

A normal human diet contains 200–400 mg of plant sterols/day. The consumption of plant sterols is higher in vegetarians, increasing to as high as 1 g/day (15–17). Plant sterols in normal human food are positively related to the variables of cholesterol synthesis, and negatively related to those of cholesterol absorption and serum cholesterol (18), indicating that even relatively small amounts of plant sterols interfere detectably with cholesterol absorption and lower serum cholesterol. The significance of dietary plant sterols on cholesterol absorption has recently been reviewed by Ostlund (19,20).

Sitosterol is the major dietary plant sterol, followed by campesterol, stigmasterol, and avenasterol. The amount of saturated plant stanols, sitostanol and campestanol, is usually only about 10% of the total plant sterol intake, and usually less than 50 mg/day (17,21). Dietary vegetable material contains plant sterols, and their highest contents occur in vegetable oils. Thus, the latter ones and cereal products, nuts, vegetables, fruits, and berries form the major sources of our daily dietary phytosterols. Crude corn oil and rice bran oil are the richest plant sterol nutrients, while olive oil contains usually only small amounts of phytosterols, but its squalene content can be high. Rapeseed oil, soybean oil, sesame seed oil, coffee oil, and chestnut oil are also rich in the usual plant sterols. Phytostanols (5α-saturated plant sterols) are obtained especially from rye, wheat, corn, and rice (22). Tall oil sterols are rich in sitosterol, about 90%, and they contain up to one-third of plant stanols, mainly sitostanol, of the total sterol contents (23). It should be borne in mind that sterol contents and composition vary greatly in different tissues of plants during growing (e.g., germs are rich in sterols), in genetically modified plants, and during storage conditions (cf. 24).

III. CHEMICAL STRUCTURES OF PLANT STEROLS

Chemically all the usual plant sterols resemble cholesterol, except that the side chain of sterol nucleus is acylated at C-24. Thus, campesterol has a methyl group at C-24 in the α position, and sitosterol an ethyl group at C-24, also in α position, while stigmasterol has, in addition to the ethyl group at C-24, a double bond at C22–23, and avenasterol an ethylidene group at C-24. Accordingly, the steroid nucleus of plant sterols contains 3β-hydroxy group and Δ5 double bond, which is saturated to 5α form in plant stanols (sitostanol and campestanol). Cholesterol is, correspondingly, saturated to cholestanol, except in intestinal contents (mainly in colon), where dietary and biliary Δ5 unsaturated sterols, cholesterol and plant sterols, are converted by bacterias to corresponding 3-keto-5β-coprostanones and corresponding coprostanols.

Five different forms of plant sterols are found in plant materials. One of them is the free (unesterified) sterol fraction, and four others are conjugated derivatives with the 3β-hydroxy group (24,25). Conjugated fraction includes fatty acid esters, glycosidic linkage (usually with glucose) with 3β-hydroxy, esterified glycoside form, and hydroxycinnamic acid (ferulic or p-cumaric acid) esters. Free cholesterol functions in mammals as an important membrane structure, whereas the esterified cholesterol is primarily a storage form. In plants unesterified phytosterols, to a lesser extent glucosides and glycoside esters, also play an important role in cellular structure and function, whereas the excess of phytosterols is mainly stored as esterified conjugates. A reason why seed oils are rich in plant sterols, especially in esterified ones, is that they form a storage needed for growing seedling and developing plant. Also in egg yolk or roe the high cholesterol content is intended for the developing embryo because it is not able to synthesize enough cholesterol for rapidly growing tissues. Plant sterol mixtures contain usually only small amounts of cholesterol, but the amount can increase occasionally up to several percentages of the total sterols.

IV. OUTLINES OF PHYTOSTEROL SYNTHESIS

According to the chemical structures, biosynthesis of phytosterols can be expected to follow that of cholesterol. In fact, synthesis of squalene seems to be roughly similar (24), even though regulation of its rate of production can be regulated differently by hydroxymethylglutaryl–coenzyme A reductase (HMG-CoA-R). In olive oil the squalene content can be very high, suggesting that its reduced cyclization and further conversion could explain the low plant sterol content. In contrast to conversion of squalene to lanosterol in mammalian and fungus organisms, squalene is converted to cycloartenol in plant cells. Subsequently, two side chain acylation steps are needed for the further plant sterol synthesis. These are mediated by cycloartenol C24 methyltransferase (SMT1) and SMT2 (SAM-24-methylene lophenol-C24 methyltransferase). SMT1 converts cycloartenol to C24-methylene cyclo-artenol, which is then converted to an unsaturated (Δ7) C-4 monomethyl C24-methylene lophenol, and further through C4 demethylation and double bond rearrangement to C24-methylene cholesterol, and finally to campesterol. SMT2, on the other hand, converts C24-methylene lophenol to C24-ethyl-idene lophenol, a parent compound for the C24-ethyl phytosterols. Analo-gously to the synthesis of campesterol, this intermediate is converted to sitosterol, and further through the side chain dehydration to stigmasterol. Activity of SMT1 and SMT2 determines the ratio of campesterol to sitosterol, and also effectively regulates overall plant sterol synthesis. This may contrib-ute to the high squalene/plant sterol ratio often found in olives. High activity of the two enzymes apparently favors plant sterol production, and a high SMT2 activity in relation to that of SMT1 increases the sitosterol/campesterol ratio. Current clinical studies favor the high sitosterol/campesterol ratio because the absorption of sitosterol and its subsequent increase in serum is less than that of campesterol during lowering of serum cholesterol by dietary consumption of plant sterols.

V. OUTLINES OF PHYTOSTANOL SYNTHESIS

As indicated in the introduction, clinical studies have shown that consump-tion of phytostanols (mixture of campestanol and sitostanol) in sufficiently soluble form not only lowers serum cholesterol but also lowers serum plant sterols. In addition, absorption of plant stanols, especially sitostanol, is markedly lower than that of the respective sterols in humans, and their elimination from the body seems more rapid than that of plant sterols. Fur-thermore, more attention should be paid to plant stanols because the con-sumption of plant stanol esters, as compared with that of plant sterol esters, lowers serum cholesterol concentrations more consistently in long-term con-sumption. Dietary plant stanols are known to be obtained especially from wheat and rye products, but not from those of oat and barley (22), raising a question of whether the latter two cereal sources were not able to saturate the Δ5 double bond. Accordingly, it might be worth looking at how nature synthesizes plant stanols from the respective C24 alkylated Δ 5-sterols. In fact, this seems to take place in plants similarly to the conversion of cholesterol to cholestanol in mammalian organism (26). For instance, campesterol has been shown to be transformed to 4-en-3β-ol, further to 4-en-3-one, to 3-one, and finally to campestanol (27). It is interesting that cholesterol and campestanol can be converted to cholestanol, 6-oxocholestanol and 6-oxocampes-tanol (28). It can be expected that plant breeding will allow us to develop, e.g., from cereal plants or oil-forming vegetables, varieties with high sitosterol contents and ratios to campesterol, which would then be converted to the corresponding stanols, and hopefully even esterified with relatively unsaturated fatty acids.

VI. PROSPECT OF INCREASING PHYTOSTEROLS BY PLANT BREEDING

A. Comparison of Plant Sterols and Stanols in Clinical Studies

Earlier free sterol/stanol studies suggested that the stanols inhibited cholesterol absorption more effectively than the sterols in experimental animals (29–31) and in humans (32,33). In contrast to stanols, bile acid synthesis may be disturbed by sterols (34). Bile acid absorption is not changed by plant stanol esters (17) or by plant sterols (35), even though no sterol balance studies have been performed during the current plant sterol ester investigations. Fecal bile acid concentration appears to be decreased (42), however, as shown above, short-term plant stanol and sterol ester studies showed roughly similar cholesterol-lowering effects, even though a slightly higher sterol intake tended to be more effective than stanols (36). Chronic plant sterol or stanol ester spread consumption is considered to reduce clinical signs of coronary heart disease by about 20% (37). However, as shown by clinical statin trials (1,2), long-term reduction of serum cholesterol (up to 1 year) is needed to reduce clinical signs of atheromatous arterial disease, and lipid lowering therapy has been recommended for the rest of patient's lifetime. Virtually all of the plant sterol studies deal with short-term consumption of different phytosterol preparations. For instance, 1-year and even shorter plant sterol ester studies showed inconsistent results in terms of cholesterol lowering (8,9,34). On the other hand, plant stanol ester consumption reduced cholesterol levels consistently in both men and women for at least 1 year (6). Thus, studies with added plant sterol preparations should be extended to continue for 1 year at least. It will be expensive to consume functional foods of long duration with no consistent cholesterol-lowering effect.

As noted above, plant sterol ester consumption increases serum plant sterol concentrations, the values of which are reduced by plant stanols. However, the values are not increased to the levels seen in patients with hereditary sitosterolemia (26), a condition associated with coronary heart disease developing in infancy in these families. It is owing to increased serum plant sterols that hemolysis can occur in these patients. An additional difference between the two phytosterols is that plant sterols, in contrast to plant stanols, are easily oxidized at increasing temperature (38), like food cooking or grilling.

Accordingly, in patients with sitosterolemia the large serum plant sterol pool with prolonged residence time seems also to be partly oxidized to 7α- and 7β-hydroxyphytosterols, corresponding 7β-hydroxycholesterol, which is known to be atherogenic (39). The increase of serum plant sterol pool during plant sterol ester consumption may also face the risks of oxidation. It is interesting to note that patients with a positive family history for coronary heart disease have increased serum levels of campesterol and sitosterol, suggesting that plant sterols might be an additional risk factor for coronary heart disease (40).

B. Aspects of Increasing Phytosterols in Plants

The clinical findings presented above favor the development of methods to increase stanols rather than sterols in plants during breeding for production of cholesterol-lowering functional foods. Some aspects of increasing sterols and stanols in plants have recently been presented suggesting a role of methyltransferases, plant sterol and stanol synthesis and esterification, and solubilization of phytosterols in general without esterification with fatty acids (13,24,25). To reduce the production costs of the plant stanol or sterol esters, nature should initially produce the unesterified forms in large enough concentrations to eliminate artificial saturation of phytosterol mixtures. Certainly the synthesis of esterified forms with unsaturated fatty acids in a plant would be ideal requiring only separation of the esters in pure concentrated form for human consumption. Consumption of phytosterols esterified with unsaturated fatty acids guarantees the daily intake of these healthy fatty acids known to have a clear antiatherogenic action (41). Wood sterols can contain up to one-third of stanols, but wood breeding could be time consuming and stanol isolation may not be easy. Based on selection of rapidly growing plants known to normally produce stanols (e.g., rice, rye, wheat, corn, even soya), it could be possible to develop by plant breeding, including gene modification, special varieties producing enhanced amounts of phytostanols or, even more desirable, their esters (rich in unsaturated fatty acids) mixed with minimal amounts of phytosterol esters. The plants in question should be manipulated to produce first large amounts of post-squalene side chain alkylated sterols, most likely ethyl alkylated ones. It is possible that up-regulation of HMG-CoA-R activity and subsequently increased squalene production could enhance the synthesis of cycloartenol only, but hypocholesterolemic activity of this sterol is not known. Further conversion of cycloartenol to side chain alkylated sterols could be obtained by up-regulating methyltransferase enzymes (SMT) to produce first 24-methylene intermediate by SMT1, and further 24-ethylidene derivatives by SMT2, and finally sitosterol with its very high ratio to campesterol. The high sitosterol/campesterol ratio could be obtained provided that the activity of SMT2 is relatively high in comparison to that of SMT1. Subsequently, conversion of $\Delta 5$ 24-alkylated plant sterols to corresponding 5α-saturated stanols seems to occur roughly similarly to conversion of cholesterol to cholestanol in mammalian liver (26). For instance, campesterol has been shown to be converted to 4-en-3β-ol, further to 4-en-3-one, then to 3-one, and finally to campestanol (27). Modification of the enzyme activities of this reaction scheme would apparently cause conversion of newly synthesized plant sterol mixture with principal sitosterol content to corresponding plant stanols.

REFERENCES

1. The Scandinavian Simvastatin Survival Study Group. Randomized trial of cholesterol lowering in 4444 patients with coronary heart disease: the Scandinavian Simvastatin Survival Study (4S). Lancet 344:1383–1389, 1994.
2. R Collins, J Armitage, S Parish, P Sleight, P Peto. MRC/BHF Heart Protection Study. Lancet 360:1783–1784, 2002.
3. OJ Pollak, D Kritchevsky. Sitosterol. In: TB Clarkson, D Kritchevsky, OJ Pollak, eds. Monographs on Atherosclerosis. Basel: Karger, 1981,Vol. 10, pp. 1–219.
4. JA Weststrate, GW Meijer. Plant sterol-enriched margarines and reduction of plasma total and LDL-cholesterol concentrations in normocholesterolemic and mildly hypercholesterolemic subjects. Eur J Clin Nutr 52:334–343, 1998.
5. MA Hallikainen, ES Sarkkinen, H Gylling, AT Erkkilä, MI Uusitupa. Comparison of the effects of plant sterol ester and plant stanol ester-enriched margarines in lowering serum cholesterol concentrations in hypercholesterolemic subjects on a low-fat diet. Eur J Clin Nutr 54:715–725, 2000.
6. TA Miettinen, P Puska, H Gylling, H Vanhanen, E Vartiainen. Reduction of serum cholesterol with sitostanol-ester margarine in a mildly hypercholesterolemic population. N Engl J Med 333:1308–1312, 1995.
7. HFJ Hendriks, FY Ntanios, EJ Brink, HMG Princen, R Buytenhek, GW Meijer. One year follow-up study on the use of a low fat spread enriched with plant sterols. Ann Nutr Metab 45(suppl 1):1–604, 2001.
8. EJ Brink, HFJ Hendriks. Long-term follow-up study on the use of a spread enriched with plant sterols. TNO report, V99.869. TNO Nutrition and Food Research Institute report, March 2000.
9. MH Davidson, KC Maki, DM Umporowicz, KA Ingram, MR Dicklin, E Schaefer, RW Lane, JR McNamara, JD Ribaya-Mercado, G Perrone, SJ Robins, WC France. Safety and tolerability of esterified phytosterols administered in reduced-fat spread and salad dressing to healthy adult men and women. J Am Coll Nutr 20:307–319, 2001.
10. H Gylling, P Puska, E Vartiainen, TA Miettinen. Serum sterols during stanol

ester feeding in a mildly hypercholesterolemic population. J Lipid Res 44:593–600, 1999.

11. HT Vanhanen, TA Miettinen. Effects of unsaturated and saturated dietary plant sterols on their serum concentrations. Clin Chim Acta 205:97–107, 1992.

12. JA Weststrate, GW Meijer. Plant sterol-enriched margarines and reduction of plasma total and LDL-cholesterol concentrations in normocholesterolemic and mildly hypercholesterolemic subjects. Eur J Clin Nutr 52:334–343, 1998.

13. TA Miettinen. Phytosterols—what plant breeders should focus on. J Sci Food Agric 81:895–903, 2001.

14. SA Spilburg. Food labeling: Health claims; Plant sterol/stanol esters and coronary heart disease. Docket Nos.00P-1275 and 00P-1276, 2001.

15. M Vuoristo, TA Miettinen. Absorption, metabolism, and serum concentrations of cholesterol in vegetarians: effects of cholesterol feeding. Am J Clin Nutr 59:1325–1331, 1994.

16. KM Phillips, MT Tarrago-Trani, KK Stewart. Phytosterol content of experimental diets differing in fatty acid composition. Food Chem 64:415–422, 1999.

17. H Gylling, TA Miettinen. Phytosterols, analytical and nutritional aspects. In: R Lasztity, W Pfannhauser, L Simon-Sarkadi, S Tömösközi, eds. Functional Foods. A New Challenge for the Food Chemist. Proceedings of the Euro Food Chem X. Budapest: Publishing Company of TUB, Vol. 1:109, 1999.

18. Gylling, TA Miettinen. New biologically active lipids in food, health food and pharmaceuticals. In: G Lambertsen, ed. Proceedings of the 19th Nordic Lipid Forum. Bergen: Lipid Forum, 1997, pp. 81–86.

19. RE Ostlund. Phytosterols in human nutrition. Annu Rev Nutr 22:533–549, 2002.

20. RE Ostlund, SB Recette, WF Stenson. Effects of trace components of dietary fat on cholesterol metabolism: phytosterols, oxysterols, and squalene. Nutr Rev 60:349–359, 2002.

21. F Czubayko, B Beumers, BS Lammsfus, D Lutjohann, K von Bergmann. A simplified micro-method for quantification of fecal excretion of neutral and acidic sterols for outpatient studies in humans. J Lipid Res 32:1861–1867, 1991.

22. PC Dutta, L-Å Appelqvist. Saturate sterols (stanols) in unhydrogenated and hydrogenated edible vegetable oils and cereal lipids. J Sci Food Agric 71:383–391, 1996.

23. TA Miettinen, H Gylling. Regulation of cholesterol metabolism by dietary plant sterols. Curr Opin Lipidol 10:9–14, 1998.

24. V Piironen, DG Lindsay, TA Miettinen, J Toivo, A-M Lampi, Plant sterols: biosynthesis, biological function and their importance to human nutrition. 80:939–966, 2000.

25. RA Moreau, BD Whitaker, KB Hicks. Phytosterols, phytostanols and their conjugates in foods: structural diversity, quantitative analysis, and health-promoting uses. Progr Lipid Res 41:457–500, 2002.

26. I Björkhem, KM Boberg, E Leitersdorf. Inborn errors in bile acid biosynthesis and storage of sterols other than cholesterol. In: CR Scriver, AL Beaudet, WS Sly, D Valle, eds. The Metabolic Bases of Inherited Disease. New York: Mc-Graw-Hill, 2001, pp. 2961–2988.

27. T Noguchi, S Fujioka, S Takatsuto, A Sakurai, S Yoshida, J Li, J Chory. Arabidopsis *det2* is defective in the conversion of (24R)-24-methylcholest-4-en-3-one to (24R)-24-methyl-5alfa-cholestan-3-one in brassinosteroid biosynthesis. Plant Physiol 120:833–839, 1999.

28. N Nakajima, S Fujioika, T Tanaka, S Takatsuto, S Yoshida. Biosynthesis of cholestanol in higher plants. Phytochemistry 60:275–279, 2002.

29. M Sugano, F Kamo, I Ikeda, H Marioka. Lipid-lowering activity of phytostanols in rats. Atherosclerosis 24:301–309, 1976.

30. M Sugano, H Morioka, I Ikeda. A comparison of hypocholesterolemic activity of β-sitosterol and β-sitostanol in rats. J Nutr 107:2011–2019, 1976.

31. I Ikeda, A Kawasaki, K Samezima, M Sugano. Antihypercholesterolemic activity of β-sitostanol in rabbits. J Nutr Sci Vitaminol 27:243–251, 1981.

32. T Heinemann, GA Kullak-Ublinck, B Pietruk, K von Bergmann. Mechanism of action of plant sterols on inhibition of cholesterol absorption: comparison of sitosterol and sitostanol. Eur J Clin Pharmacol 40(suppl):S59.63, 1991.

33. T Heinemann, G Axtmann, K von Bergmann. Comparison of intestinal absorption of cholesterol with different plant sterols in man. Eur J Clin Invest 23:827–831, 1993.

34. FH O'Neill, E Brynes, R Mandeno, M Seed, GR Thompson. Head to head comparison of the cholesterol-lowering efficasy of plant sterol and plant stanol esters. Circulation 106(suppl II):II-676, 2002.

35. SM Grundy, EH Ahrens, J Davignon. The interaction of cholesterol absorption and synthesis in man. J Lipid Res 10:304–315, 1969.

36. PJ Jones, M Raeini-Sarjaz, FY Ntanios, CA Vanstone, JY Feng, PW Earsons. Modulation of plasma lipid levels and cholesterol kinetics by phytosterol versus phytostanol. J Lipid Res 41:697–705, 2000.

37. M Law. Plant sterol and stanol margarines and health. Br Med J 320:861–864, 2000.

38. A-M Lampi, R-L Hovi, J Toivo, V Piironen. Functional Foods: A New Challenge for the Food Chemist. In: R Lasztity, W Pfannhauser, L Simon-Sarkadi, S Tömösközi, eds. Proceedings of the Euro Food Chem X. Budapest: Publishing Company of TUB, Vol. 2, pp. 513–516, 2000.

39. J Plat, H Brzezinka, D Lütjohann, RP Mensik, K von Bergmann. Oxidized plant sterols in human serum and lipid transfusions as measured by combined gas-liquid chromatography-mass spectrometry. J Lipid Res 42:2030–2038, 2001.

40. T Sudhop, BM Gottwald, K von Bergmann. Serum plant sterols as a potential risk factor for coronary heart disease. Metabolism 51:1519–1521, 2002.

41. H Gylling, TA Miettinen. A review of clinical trials in dietary interventions to decrease the incidence of coronary heart disease. Curr Control Trials Cardiovasc Med 2:123–128, 2001.

42. JA Weststrate, R Ayesh, C Bauer-Plank, PN Drewitt. Safety evaluation of phytosterol esters. Part 4. Faecal concentrations of bile acids and neutral sterols in healthy normolipidaemic volunteers consuming a controlled diet either with or without a phytosterol ester-enriched margarine. Food Chem Toxicol 37: 1063–1071, 1999.

Index

Milton Keynes UK
Ingram Content Group UK Ltd.
UKHW020010071024
449327UK00031B/2723